WORLD HEALTH ORGANIZATION
INTERNATIONAL AGENCY FOR RESEARCH ON CANCER

IARC Monographs on the Evaluation of Carcinogenic Risks to Humans

VOLUME 84

Some Drinking-water Disinfectants and Contaminants, including Arsenic

This publication represents the views and expert opinions
of an IARC Working Group on the
Evaluation of Carcinogenic Risks to Humans,
which met in Lyon,

15–22 October 2002

2004

IARC MONOGRAPHS

In 1969, the International Agency for Research on Cancer (IARC) initiated a programme on the evaluation of the carcinogenic risk of chemicals to humans involving the production of critically evaluated monographs on individual chemicals. The programme was subsequently expanded to include evaluations of carcinogenic risks associated with exposures to complex mixtures, life-style factors and biological and physical agents, as well as those in specific occupations.

The objective of the programme is to elaborate and publish in the form of monographs critical reviews of data on carcinogenicity for agents to which humans are known to be exposed and on specific exposure situations; to evaluate these data in terms of human risk with the help of international working groups of experts in chemical carcinogenesis and related fields; and to indicate where additional research efforts are needed.

The lists of IARC evaluations are regularly updated and are available on Internet: http://monographs.iarc.fr/

This project was supported by Cooperative Agreement 5 UO1 CA33193 awarded by the United States National Cancer Institute, Department of Health and Human Services, and was funded in part by the European Commission, Directorate-General EMPL (Employment, and Social Affairs), Health, Safety and Hygiene at Work Unit. Additional support has been provided since 1993 by the United States National Institute of Environmental Health Sciences.

This publication was made possible in part by a Cooperative Agreement between the United States Environmental Protection Agency, Office of Research and Development (US EPA, ORD) and IARC and does not necessarily express the views of the US EPA, ORD.

©International Agency for Research on Cancer, 2004

Distributed by IARC*Press* (Fax: +33 4 72 73 83 02; E-mail: press@iarc.fr) and by the World Health Organization Marketing and Dissemination (MDI), 1211 Geneva 27 (Fax: +41 22 791 4857; E-mail: publications@who.int)

IARC Library Cataloguing in Publication Data

Some drinking-water disinfectants and contaminants, including arsenic /
  IARC Working Group on the Evaluation of Carcinogenic Risks to Humans
  (2004 : Lyon, France)

  (IARC monographs on the evaluation of carcinogenic risks to humans ; 84)

  1. Carcinogens – congresses   2. Drinking-water disinfectants – congresses
  3. Drinking-water contaminants – congresses   4. Arsenic – congresses
  I. IARC Working Group on the Evaluation of Carcinogenic Risks to Humans   II. Series

  ISBN 92 832 1284 3   (NLM Classification: W1)
  ISSN 1017-1606

PRINTED IN FRANCE

1 Woman at a well in India
2 Sewage works in South-West England

Cover design by Georges Mollon, IARC

CONTENTS

NOTE TO THE READER

The term 'carcinogenic risk' in the *IARC Monographs* series is taken to mean the probability that exposure to an agent will lead to cancer in humans.

Inclusion of an agent in the *Monographs* does not imply that it is a carcinogen, only that the published data have been examined. Equally, the fact that an agent has not yet been evaluated in a monograph does not mean that it is not carcinogenic.

The evaluations of carcinogenic risk are made by international working groups of independent scientists and are qualitative in nature. No recommendation is given for regulation or legislation.

Anyone who is aware of published data that may alter the evaluation of the carcinogenic risk of an agent to humans is encouraged to make this information available to the Unit of Carcinogen Identification and Evaluation, International Agency for Research on Cancer, 150 cours Albert Thomas, 69372 Lyon Cedex 08, France, in order that the agent may be considered for re-evaluation by a future Working Group.

Although every effort is made to prepare the monographs as accurately as possible, mistakes may occur. Readers are requested to communicate any errors to the Unit of Carcinogen Identification and Evaluation, so that corrections can be reported in future volumes.

IARC WORKING GROUP ON THE EVALUATION OF CARCINOGENIC RISKS TO HUMANS: SOME DRINKING-WATER DISINFECTANTS AND CONTAMINANTS, INCLUDING ARSENIC

Lyon, 15–22 October 2002

LIST OF PARTICIPANTS

Members

M.W. Anders, Department of Pharmacology and Physiology, University of Rochester Medical Center, 601 Elmwood Avenue, Box 711, Rochester, NY 14642, USA

Richard J. Bull, Department of Environmental Sciences, Washington State University – Tri cities, 2710 University Drive, Richland, WA 99352-6534, USA

Kenneth P. Cantor, Occupational Epidemiology Branch, Division of Cancer Epidemiology and Genetics, National Cancer Institute, 6120 Executive Boulevard, EPS-8106, Bethesda, MD 20892-7240, USA (*Subgroup Chair: Cancer in Humans*)

Dipankar Chakraborti, School of Environmental Studies, Jadavpur University, Calcutta – 700 032, India

Chien-Jen Chen, Graduate Institute of Epidemiology, College of Public Health, National Taiwan University, 1 Jen-Ai Road Section 1, Room 1547, Taipei 10018, Taiwan, China

Anthony B. DeAngelo, Environmental Carcinogenesis Division, US Environmental Protection Agency, National Health and Environmental Effects Research Laboratory, MD-68 ERC, 86 TW Alexander Drive, Research Triangle Park, NC 27711, USA (*Subgroup Chair: Cancer in Experimental Animals*)

David M. DeMarini, Environmental Carcinogenesis Division (MD-68), US Environmental Protection Agency, 86 Alexander Drive, Research Triangle Park, NC 27711, USA

Catterina Ferreccio, Department of Pontifical Public Health, Catholic University of Chile, Marcoleta 352, Santiago, Chile (*Subgroup Chair: Other Relevant Data*)

Shoji Fukushima, Department of Pathology, Osaka City University Medical School, 1-4-3 Asahi-machi, Abeno-ku, Osaka 545-8585, Japan

Thomas W. Gebel, Federal Institute of Occupational Safety and Health, Division 4, Safety and Health with Chemical and Biological Agents, Friedrich-Henkel-Weg 1-25, Postfach 17 02 02, Haus IV, FB 4, 44149 Dortmund, Germany

Debendra Nath Guha Mazumder[1], Department of Gastroenterology, Ronald Ross Building, 4th Floor, Institute of Post-graduate Medical Education and Research, 244 Acharya Jagadish Chandra Bose Road, Kolkata 700 020, India

Margaret R. Karagas, Norris Cotton Cancer Center, Dartmouth Medical School, 7927 Rubin Building 462M3, One Medical Center Drive, Lebanon, NH 03756-0001, USA

Manolis Kogevinas, Respiratory and Environmental Health Research Unit, Municipal Institute of Medical Research, 80 Dr Aiguader Road, Barcelona 08003, Spain (*Chairman*)

Hannu Komulainen, Laboratory of Toxicology, National Public Health Institute, Neulanienmentie 4, P.O.B. 95, 70701 Kuopio, Finland

Frank Le Curieux[2], Laboratory of Toxicology, Pasteur Institute of Lille, 1, rue du Professeur Calmette, BP 245, 59019 Lille Cedex, France

Andy Meharg, Department of Plant and Soil Science, University of Aberdeen, Cruickshank Building, St Machar Drive, Aberdeen AB24 3UU, Scotland, United Kingdom (*Unable to attend*)

Jack C. Ng, National Research Centre for Environmental Toxicology (ENTOX), The University of Queensland, 39 Kessels Road, Coopers Plains, Brisbane, Queensland 4108, Australia

Mark J. Nieuwenhuijsen, Department of Environmental Science and Technology, Faculty of Life Sciences, Imperial College of Science, Technology and Medicine, Royal School of Mines, Prince Consort Road, London SW7 2BP, United Kingdom

Steve Olin, ILSI Risk Science Institute, One Thomas Circle, NW, 9th Floor, Washington DC 20005-5802, USA (*Subgroup Chair: Exposure Data*)

Michael Pereira, Department of Pathology, Medical College of Ohio, 3055 Arlington Avenue, Toledo, OH 43614-5806, USA

Mahfuzar Rahman, Public Health Sciences Division, Centre for Health and Population Research, International Centre for Diarrhoeal Disease Research, GPO Box 128, Mohakhali CA, Dhaka 1000, Bangladesh

J. Alan Roberson, Director of Regulatory Affairs, American Water Works Association, 1401 New York Avenue, NW, #640, Washington DC 20005, USA

Alan H. Smith, School of Public Health, University of California, Berkeley, 140 Warren Hall, Berkeley, CA 94720-7360, USA

Marie Vahter, Institute of Environmental Medicine, Karolinska Institute, Box 210, Nobels väg, 13, 5th Floor, 171 77 Stockholm, Sweden

---

[1] Contact address: 371C, Block 'B', New Alipore, Kolkata 700053, India

[2] Also at: Faculty of Pharmaceutical and Biological Sciences, Department of Toxicological Study and Research, Public Health, Environment, University of Lille 2, Law and Health, 3, rue du Professeur Laguesse, BP 83, 59006 Lille Cedex, France

Representatives/Observers

Representative of the US National Soft Drink Association
Richard H. Adamson, Scientific and Technical Affairs, National Soft Drink Association, 1101 Sixteenth Street NW, Washington DC 20036, USA

Observer, representing the Technical Resources International Incorporated
Ted Junghans, Technical Resources International Inc., 6500 Rock Spring Drive, Suite 650, Bethesda, MD 20817-1197, USA

Representative of the US National Cancer Institute
David Longfellow, Chemical and Physical Carcinogenesis Branch, Division of Cancer Biology, National Cancer Institute, 6130 Executive Boulevard, Suite 5000 MSC7368, Rockville, MD 20892, USA

Observer, representing the Natural Resources Defense Council
Jennifer Sass, Natural Resources Defense Council, 1200 New York Avenue, NW, Suite 400, Washington DC 20005, USA

Representative of the International Programme on Chemical Safety, WHO
Carolyn Vickers, International Programme on Chemical Safety, World Health Organization, 20 Avenue Appia, 1211 Geneva 27, Switzerland

IARC Secretariat
Robert Baan, Unit of Carcinogen Identification and Evaluation (*Co-rapporteur; Subgroup on Other Relevant Data*)
Raika Durusoy, Unit of Environmental Cancer Epidemiology
Marlin Friesen, Unit of Nutrition and Cancer
Yann Grosse, Unit of Carcinogen Identification and Evaluation (*Responsible Officer; Rapporteur, Subgroup on Cancer in Experimental Animals*)
Mia Hashibe, Unit of Environmental Cancer Epidemiology
Nikolai Napalkov[1]
Linda Northrup (*Co-editor*)
Christiane Partensky, Unit of Carcinogen Identification and Evaluation (*Rapporteur, Subgroup on Exposure Data*)
Jerry Rice[2], Unit of Carcinogen Identification and Evaluation (*Head of Programme*)

---

[1] Present address: Director Emeritus, Petrov Institute of Oncology, Pesochny-2, 197758 St Petersburg, Russia

[2] Present address: Georgetown University Medical Center, Department of Oncology, Box 571465, Lombardi Comprehensive Cancer Center, Level LL, Room S150, 3800 Reservoir Road, NW, Washington DC 20057-1465, USA

Béatrice Secretan, Unit of Carcinogen Identification and Evaluation (*Co-rapporteur,*
    Subgroup on Other Relevant Data)
Min Shen, Unit of Environmental Cancer Epidemiology
Kurt Straif, Unit of Carcinogen Identification and Evaluation (*Rapporteur; Subgroup*
    on Cancer in Humans)

Post-meeting scientific assistance
Véronique Bouvard
Catherine Cohet

Technical assistance
Sandrine Egraz
Martine Lézère
Jane Mitchell (*Co-editor*)
Elspeth Perez
Annick Rivoire

PREAMBLE

IARC MONOGRAPHS PROGRAMME ON THE EVALUATION OF CARCINOGENIC RISKS TO HUMANS

PREAMBLE

1. BACKGROUND

In 1969, the International Agency for Research on Cancer (IARC) initiated a programme to evaluate the carcinogenic risk of chemicals to humans and to produce monographs on individual chemicals. The *Monographs* programme has since been expanded to include consideration of exposures to complex mixtures of chemicals (which occur, for example, in some occupations and as a result of human habits) and of exposures to other agents, such as radiation and viruses. With Supplement 6 (IARC, 1987a), the title of the series was modified from *IARC Monographs on the Evaluation of the Carcinogenic Risk of Chemicals to Humans* to *IARC Monographs on the Evaluation of Carcinogenic Risks to Humans*, in order to reflect the widened scope of the programme.

The criteria established in 1971 to evaluate carcinogenic risk to humans were adopted by the working groups whose deliberations resulted in the first 16 volumes of the *IARC Monographs series*. Those criteria were subsequently updated by further ad-hoc working groups (IARC, 1977, 1978, 1979, 1982, 1983, 1987b, 1988, 1991a; Vainio *et al.*, 1992).

2. OBJECTIVE AND SCOPE

The objective of the programme is to prepare, with the help of international working groups of experts, and to publish in the form of monographs, critical reviews and evaluations of evidence on the carcinogenicity of a wide range of human exposures. The *Monographs* may also indicate where additional research efforts are needed.

The *Monographs* represent the first step in carcinogenic risk assessment, which involves examination of all relevant information in order to assess the strength of the available evidence that certain exposures could alter the incidence of cancer in humans. The second step is quantitative risk estimation. Detailed, quantitative evaluations of epidemiological data may be made in the *Monographs*, but without extrapolation beyond the range of the data available. Quantitative extrapolation from experimental data to the human situation is not undertaken.

The term 'carcinogen' is used in these monographs to denote an exposure that is capable of increasing the incidence of malignant neoplasms; the induction of benign neoplasms may in some circumstances (see p. 19) contribute to the judgement that the exposure is carcinogenic. The terms 'neoplasm' and 'tumour' are used interchangeably.

Some epidemiological and experimental studies indicate that different agents may act at different stages in the carcinogenic process, and several mechanisms may be involved. The aim of the *Monographs* has been, from their inception, to evaluate evidence of carcinogenicity at any stage in the carcinogenesis process, independently of the underlying mechanisms. Information on mechanisms may, however, be used in making the overall evaluation (IARC, 1991a; Vainio *et al.*, 1992; see also pp. 25–27).

The *Monographs* may assist national and international authorities in making risk assessments and in formulating decisions concerning any necessary preventive measures. The evaluations of IARC working groups are scientific, qualitative judgements about the evidence for or against carcinogenicity provided by the available data. These evaluations represent only one part of the body of information on which regulatory measures may be based. Other components of regulatory decisions vary from one situation to another and from country to country, responding to different socioeconomic and national priorities. **Therefore, no recommendation is given with regard to regulation or legislation, which are the responsibility of individual governments and/or other international organizations.**

The *IARC Monographs* are recognized as an authoritative source of information on the carcinogenicity of a wide range of human exposures. A survey of users in 1988 indicated that the *Monographs* are consulted by various agencies in 57 countries. About 2500 copies of each volume are printed, for distribution to governments, regulatory bodies and interested scientists. The Monographs are also available from IARC*Press* in Lyon and via the Marketing and Dissemination (MDI) of the World Health Organization in Geneva.

3.   SELECTION OF TOPICS FOR MONOGRAPHS

Topics are selected on the basis of two main criteria: (a) there is evidence of human exposure, and (b) there is some evidence or suspicion of carcinogenicity. The term 'agent' is used to include individual chemical compounds, groups of related chemical compounds, physical agents (such as radiation) and biological factors (such as viruses). Exposures to mixtures of agents may occur in occupational exposures and as a result of personal and cultural habits (such as smoking and dietary practices). Chemical analogues and compounds with biological or physical characteristics similar to those of suspected carcinogens may also be considered, even in the absence of data on a possible carcinogenic effect in humans or experimental animals.

The scientific literature is surveyed for published data relevant to an assessment of carcinogenicity. The IARC information bulletins on agents being tested for carcinogenicity (IARC, 1973–1996) and directories of on-going research in cancer epidemiology (IARC, 1976–1996) often indicate exposures that may be scheduled for future meetings. Ad-hoc working groups convened by IARC in 1984, 1989, 1991, 1993 and

1998 gave recommendations as to which agents should be evaluated in the IARC Monographs series (IARC, 1984, 1989, 1991b, 1993, 1998a,b).

As significant new data on subjects on which monographs have already been prepared become available, re-evaluations are made at subsequent meetings, and revised monographs are published.

4. DATA FOR MONOGRAPHS

The *Monographs* do not necessarily cite all the literature concerning the subject of an evaluation. Only those data considered by the Working Group to be relevant to making the evaluation are included.

With regard to biological and epidemiological data, only reports that have been published or accepted for publication in the openly available scientific literature are reviewed by the working groups. In certain instances, government agency reports that have undergone peer review and are widely available are considered. Exceptions may be made on an ad-hoc basis to include unpublished reports that are in their final form and publicly available, if their inclusion is considered pertinent to making a final evaluation (see pp. 25–27). In the sections on chemical and physical properties, on analysis, on production and use and on occurrence, unpublished sources of information may be used.

5. THE WORKING GROUP

Reviews and evaluations are formulated by a working group of experts. The tasks of the group are: (i) to ascertain that all appropriate data have been collected; (ii) to select the data relevant for the evaluation on the basis of scientific merit; (iii) to prepare accurate summaries of the data to enable the reader to follow the reasoning of the Working Group; (iv) to evaluate the results of epidemiological and experimental studies on cancer; (v) to evaluate data relevant to the understanding of mechanism of action; and (vi) to make an overall evaluation of the carcinogenicity of the exposure to humans.

Working Group participants who contributed to the considerations and evaluations within a particular volume are listed, with their addresses, at the beginning of each publication. Each participant who is a member of a working group serves as an individual scientist and not as a representative of any organization, government or industry. In addition, nominees of national and international agencies and industrial associations may be invited as observers.

6. WORKING PROCEDURES

Approximately one year in advance of a meeting of a working group, the topics of the monographs are announced and participants are selected by IARC staff in consultation with other experts. Subsequently, relevant biological and epidemiological data are collected by the Carcinogen Identification and Evaluation Unit of IARC from recognized sources of information on carcinogenesis, including data storage and retrieval systems such as MEDLINE and TOXLINE.

For chemicals and some complex mixtures, the major collection of data and the preparation of first drafts of the sections on chemical and physical properties, on analysis, on production and use and on occurrence are carried out under a separate contract funded by the United States National Cancer Institute. Representatives from industrial associations may assist in the preparation of sections on production and use. Information on production and trade is obtained from governmental and trade publications and, in some cases, by direct contact with industries. Separate production data on some agents may not be available because their publication could disclose confidential information. Information on uses may be obtained from published sources but is often complemented by direct contact with manufacturers. Efforts are made to supplement this information with data from other national and international sources.

Six months before the meeting, the material obtained is sent to meeting participants, or is used by IARC staff, to prepare sections for the first drafts of monographs. The first drafts are compiled by IARC staff and sent before the meeting to all participants of the Working Group for review.

The Working Group meets in Lyon for seven to eight days to discuss and finalize the texts of the monographs and to formulate the evaluations. After the meeting, the master copy of each monograph is verified by consulting the original literature, edited and prepared for publication. The aim is to publish monographs within six months of the Working Group meeting.

The available studies are summarized by the Working Group, with particular regard to the qualitative aspects discussed below. In general, numerical findings are indicated as they appear in the original report; units are converted when necessary for easier comparison. The Working Group may conduct additional analyses of the published data and use them in their assessment of the evidence; the results of such supplementary analyses are given in square brackets. When an important aspect of a study, directly impinging on its interpretation, should be brought to the attention of the reader, a comment is given in square brackets.

7.   EXPOSURE DATA

Sections that indicate the extent of past and present human exposure, the sources of exposure, the people most likely to be exposed and the factors that contribute to the exposure are included at the beginning of each monograph.

Most monographs on individual chemicals, groups of chemicals or complex mixtures include sections on chemical and physical data, on analysis, on production and use and on occurrence. In monographs on, for example, physical agents, occupational exposures and cultural habits, other sections may be included, such as: historical perspectives, description of an industry or habit, chemistry of the complex mixture or taxonomy.

Monographs on biological agents have sections on structure and biology, methods of detection, epidemiology of infection and clinical disease other than cancer.

For chemical exposures, the Chemical Abstracts Services Registry Number, the latest Chemical Abstracts primary name and the IUPAC systematic name are recorded; other synonyms are given, but the list is not necessarily comprehensive. For biological agents, taxonomy and structure are described, and the degree of variability is given, when applicable.

Information on chemical and physical properties and, in particular, data relevant to identification, occurrence and biological activity are included. For biological agents, mode of replication, life cycle, target cells, persistence and latency and host response are given. A description of technical products of chemicals includes trade names, relevant specifications and available information on composition and impurities. Some of the trade names given may be those of mixtures in which the agent being evaluated is only one of the ingredients.

The purpose of the section on analysis or detection is to give the reader an overview of current methods, with emphasis on those widely used for regulatory purposes. Methods for monitoring human exposure are also given, when available. No critical evaluation or recommendation of any of the methods is meant or implied. The IARC published a series of volumes, *Environmental Carcinogens: Methods of Analysis and Exposure Measurement* (IARC, 1978–93), that describe validated methods for analysing a wide variety of chemicals and mixtures. For biological agents, methods of detection and exposure assessment are described, including their sensitivity, specificity and reproducibility.

The dates of first synthesis and of first commercial production of a chemical or mixture are provided; for agents which do not occur naturally, this information may allow a reasonable estimate to be made of the date before which no human exposure to the agent could have occurred. The dates of first reported occurrence of an exposure are also provided. In addition, methods of synthesis used in past and present commercial production and different methods of production which may give rise to different impurities are described.

Data on production, international trade and uses are obtained for representative regions, which usually include Europe, Japan and the United States of America. It should not, however, be inferred that those areas or nations are necessarily the sole or major sources or users of the agent. Some identified uses may not be current or major applications, and the coverage is not necessarily comprehensive. In the case of drugs, mention of their therapeutic uses does not necessarily represent current practice, nor does it imply judgement as to their therapeutic efficacy.

Information on the occurrence of an agent or mixture in the environment is obtained from data derived from the monitoring and surveillance of levels in occupational environments, air, water, soil, foods and animal and human tissues. When available, data on the generation, persistence and bioaccumulation of the agent are also included. In the case of mixtures, industries, occupations or processes, information is given about all agents present. For processes, industries and occupations, a historical description is also given, noting variations in chemical composition, physical properties and levels of occupational exposure with time and place. For biological agents, the epidemiology of infection is described.

Statements concerning regulations and guidelines (e.g. pesticide registrations, maximal levels permitted in foods, occupational exposure limits) are included for some countries as indications of potential exposures, but they may not reflect the most recent situation, since such limits are continuously reviewed and modified. The absence of information on regulatory status for a country should not be taken to imply that that country does not have regulations with regard to the exposure. For biological agents, legislation and control, including vaccines and therapy, are described.

8.   STUDIES OF CANCER IN HUMANS

(a)   Types of studies considered

Three types of epidemiological studies of cancer contribute to the assessment of carcinogenicity in humans — cohort studies, case–control studies and correlation (or ecological) studies. Rarely, results from randomized trials may be available. Case series and case reports of cancer in humans may also be reviewed.

Cohort and case–control studies relate the exposures under study to the occurrence of cancer in individuals and provide an estimate of relative risk (ratio of incidence or mortality in those exposed to incidence or mortality in those not exposed) as the main measure of association.

In correlation studies, the units of investigation are usually whole populations (e.g. in particular geographical areas or at particular times), and cancer frequency is related to a summary measure of the exposure of the population to the agent, mixture or exposure circumstance under study. Because individual exposure is not documented, however, a causal relationship is less easy to infer from correlation studies than from cohort and case–control studies. Case reports generally arise from a suspicion, based on clinical experience, that the concurrence of two events — that is, a particular exposure and occurrence of a cancer — has happened rather more frequently than would be expected by chance. Case reports usually lack complete ascertainment of cases in any population, definition or enumeration of the population at risk and estimation of the expected number of cases in the absence of exposure. The uncertainties surrounding interpretation of case reports and correlation studies make them inadequate, except in rare instances, to form the sole basis for inferring a causal relationship. When taken together with case–control and cohort studies, however, relevant case reports or correlation studies may add materially to the judgement that a causal relationship is present.

Epidemiological studies of benign neoplasms, presumed preneoplastic lesions and other end-points thought to be relevant to cancer are also reviewed by working groups. They may, in some instances, strengthen inferences drawn from studies of cancer itself.

(b)   Quality of studies considered

The *Monographs* are not intended to summarize all published studies. Those that are judged to be inadequate or irrelevant to the evaluation are generally omitted. They may be mentioned briefly, particularly when the information is considered to be a useful supplement to that in other reports or when they provide the only data available. Their inclusion does not imply acceptance of the adequacy of the study design or of the analysis and interpretation of the results, and limitations are clearly outlined in square brackets at the end of the study description.

It is necessary to take into account the possible roles of bias, confounding and chance in the interpretation of epidemiological studies. By 'bias' is meant the operation of factors in study design or execution that lead erroneously to a stronger or weaker association than in fact exists between disease and an agent, mixture or exposure circumstance. By 'confounding' is meant a situation in which the relationship with disease is made to appear stronger or weaker than it truly is as a result of an association between the apparent causal factor and another factor that is associated with either an increase or decrease in the incidence of the disease. In evaluating the extent to which these factors have been minimized in an individual study, working groups consider a number of aspects of design and analysis as described in the report of the study. Most of these considerations apply equally to case–control, cohort and correlation studies. Lack of clarity of any of these aspects in the reporting of a study can decrease its credibility and the weight given to it in the final evaluation of the exposure.

Firstly, the study population, disease (or diseases) and exposure should have been well defined by the authors. Cases of disease in the study population should have been identified in a way that was independent of the exposure of interest, and exposure should have been assessed in a way that was not related to disease status.

Secondly, the authors should have taken account in the study design and analysis of other variables that can influence the risk of disease and may have been related to the exposure of interest. Potential confounding by such variables should have been dealt with either in the design of the study, such as by matching, or in the analysis, by statistical adjustment. In cohort studies, comparisons with local rates of disease may be more appropriate than those with national rates. Internal comparisons of disease frequency among individuals at different levels of exposure should also have been made in the study.

Thirdly, the authors should have reported the basic data on which the conclusions are founded, even if sophisticated statistical analyses were employed. At the very least, they should have given the numbers of exposed and unexposed cases and controls in a case–control study and the numbers of cases observed and expected in a cohort study. Further tabulations by time since exposure began and other temporal factors are also important. In a cohort study, data on all cancer sites and all causes of death should have been given, to reveal the possibility of reporting bias. In a case–control study, the effects of investigated factors other than the exposure of interest should have been reported.

Finally, the statistical methods used to obtain estimates of relative risk, absolute rates of cancer, confidence intervals and significance tests, and to adjust for confounding should have been clearly stated by the authors. The methods used should preferably have been the generally accepted techniques that have been refined since the mid-1970s. These methods have been reviewed for case–control studies (Breslow & Day, 1980) and for cohort studies (Breslow & Day, 1987).

(c)   Inferences about mechanism of action

Detailed analyses of both relative and absolute risks in relation to temporal variables, such as age at first exposure, time since first exposure, duration of exposure, cumulative exposure and time since exposure ceased, are reviewed and summarized when available. The analysis of temporal relationships can be useful in formulating models of carcino-genesis. In particular, such analyses may suggest whether a carcinogen acts early or late in the process of carcinogenesis, although at best they allow only indirect inferences about the mechanism of action. Special attention is given to measurements of biological markers of carcinogen exposure or action, such as DNA or protein adducts, as well as markers of early steps in the carcinogenic process, such as proto-oncogene mutation, when these are incorporated into epidemiological studies focused on cancer incidence or mortality. Such measurements may allow inferences to be made about putative mecha-nisms of action (IARC, 1991a; Vainio et al., 1992).

(d)   Criteria for causality

After the individual epidemiological studies of cancer have been summarized and the quality assessed, a judgement is made concerning the strength of evidence that the agent, mixture or exposure circumstance in question is carcinogenic for humans. In making its judgement, the Working Group considers several criteria for causality. A strong asso-ciation (a large relative risk) is more likely to indicate causality than a weak association, although it is recognized that relative risks of small magnitude do not imply lack of causality and may be important if the disease is common. Associations that are replicated in several studies of the same design or using different epidemiological approaches or under different circumstances of exposure are more likely to represent a causal relation-ship than isolated observations from single studies. If there are inconsistent results among investigations, possible reasons are sought (such as differences in amount of exposure), and results of studies judged to be of high quality are given more weight than those of studies judged to be methodologically less sound. When suspicion of carcino-genicity arises largely from a single study, these data are not combined with those from later studies in any subsequent reassessment of the strength of the evidence.

If the risk of the disease in question increases with the amount of exposure, this is considered to be a strong indication of causality, although absence of a graded response is not necessarily evidence against a causal relationship. Demonstration of a decline in risk after cessation of or reduction in exposure in individuals or in whole populations also supports a causal interpretation of the findings.

Although a carcinogen may act upon more than one target, the specificity of an association (an increased occurrence of cancer at one anatomical site or of one morphological type) adds plausibility to a causal relationship, particularly when excess cancer occurrence is limited to one morphological type within the same organ.

Although rarely available, results from randomized trials showing different rates among exposed and unexposed individuals provide particularly strong evidence for causality.

When several epidemiological studies show little or no indication of an association between an exposure and cancer, the judgement may be made that, in the aggregate, they show evidence of lack of carcinogenicity. Such a judgement requires first of all that the studies giving rise to it meet, to a sufficient degree, the standards of design and analysis described above. Specifically, the possibility that bias, confounding or misclassification of exposure or outcome could explain the observed results should be considered and excluded with reasonable certainty. In addition, all studies that are judged to be methodologically sound should be consistent with a relative risk of unity for any observed level of exposure and, when considered together, should provide a pooled estimate of relative risk which is at or near unity and has a narrow confidence interval, due to sufficient population size. Moreover, no individual study nor the pooled results of all the studies should show any consistent tendency for the relative risk of cancer to increase with increasing level of exposure. It is important to note that evidence of lack of carcinogenicity obtained in this way from several epidemiological studies can apply only to the type(s) of cancer studied and to dose levels and intervals between first exposure and observation of disease that are the same as or less than those observed in all the studies. Experience with human cancer indicates that, in some cases, the period from first exposure to the development of clinical cancer is seldom less than 20 years; studies with latent periods substantially shorter than 30 years cannot provide evidence for lack of carcinogenicity.

9.   STUDIES OF CANCER IN EXPERIMENTAL ANIMALS

All known human carcinogens that have been studied adequately in experimental animals have produced positive results in one or more animal species (Wilbourn *et al.*, 1986; Tomatis *et al.*, 1989). For several agents (aflatoxins, 4-aminobiphenyl, azathioprine, betel quid with tobacco, bischloromethyl ether and chloromethyl methyl ether (technical grade), chlorambucil, chlornaphazine, ciclosporin, coal-tar pitches, coal-tars, combined oral contraceptives, cyclophosphamide, diethylstilboestrol, melphalan, 8-methoxypsoralen plus ultraviolet A radiation, mustard gas, myleran, 2-naphthylamine, nonsteroidal estrogens, estrogen replacement therapy/steroidal estrogens, solar radiation, thiotepa and vinyl chloride), carcinogenicity in experimental animals was established or highly suspected before epidemiological studies confirmed their carcinogenicity in humans (Vainio *et al.*, 1995). Although this association cannot establish that all agents and mixtures that cause cancer in experimental animals also cause cancer in humans, nevertheless, **in the absence of adequate data on humans, it is biologically plausible and prudent to regard agents and mixtures for which there is *sufficient evidence* (see p. 24) of carcinogenicity in experimental animals as if they presented a carcinogenic risk to humans**. The possibility that a given agent may cause cancer through a species-specific mechanism which does not operate in humans (see p. 27) should also be taken into consideration.

The nature and extent of impurities or contaminants present in the chemical or mixture being evaluated are given when available. Animal strain, sex, numbers per group, age at start of treatment and survival are reported.

Other types of studies summarized include: experiments in which the agent or mixture was administered in conjunction with known carcinogens or factors that modify carcinogenic effects; studies in which the end-point was not cancer but a defined precancerous lesion; and experiments on the carcinogenicity of known metabolites and derivatives.

For experimental studies of mixtures, consideration is given to the possibility of changes in the physicochemical properties of the test substance during collection, storage, extraction, concentration and delivery. Chemical and toxicological interactions of the components of mixtures may result in nonlinear dose–response relationships.

An assessment is made as to the relevance to human exposure of samples tested in experimental animals, which may involve consideration of: (i) physical and chemical characteristics, (ii) constituent substances that indicate the presence of a class of substances, (iii) the results of tests for genetic and related effects, including studies on DNA adduct formation, proto-oncogene mutation and expression and suppressor gene inactivation. The relevance of results obtained, for example, with animal viruses analogous to the virus being evaluated in the monograph must also be considered. They may provide biological and mechanistic information relevant to the understanding of the process of carcinogenesis in humans and may strengthen the plausibility of a conclusion that the biological agent under evaluation is carcinogenic in humans.

(a)   Qualitative aspects

An assessment of carcinogenicity involves several considerations of qualitative importance, including (i) the experimental conditions under which the test was per-formed, including route and schedule of exposure, species, strain, sex, age, duration of follow-up; (ii) the consistency of the results, for example, across species and target organ(s); (iii) the spectrum of neoplastic response, from preneoplastic lesions and benign tumours to malignant neoplasms; and (iv) the possible role of modifying factors.

As mentioned earlier (p. 11), the *Monographs* are not intended to summarize all published studies. Those studies in experimental animals that are inadequate (e.g. too short a duration, too few animals, poor survival; see below) or are judged irrelevant to the evaluation are generally omitted. Guidelines for conducting adequate long-term carcinogenicity experiments have been outlined (e.g. Montesano *et al.*, 1986).

Considerations of importance to the Working Group in the interpretation and evaluation of a particular study include: (i) how clearly the agent was defined and, in the case of mixtures, how adequately the sample characterization was reported; (ii) whether the dose was adequately monitored, particularly in inhalation experiments; (iii) whether the doses and duration of treatment were appropriate and whether the survival of treated animals was similar to that of controls; (iv) whether there were adequate numbers of animals per group; (v) whether animals of each sex were used; (vi) whether animals were allocated randomly to groups; (vii) whether the duration of observation was adequate; and (viii) whether the data were adequately reported. If available, recent data on the incidence of specific tumours in historical controls, as well as in concurrent controls, should be taken into account in the evaluation of tumour response.

When benign tumours occur together with and originate from the same cell type in an organ or tissue as malignant tumours in a particular study and appear to represent a stage in the progression to malignancy, it may be valid to combine them in assessing tumour incidence (Huff *et al.*, 1989). The occurrence of lesions presumed to be preneoplastic may in certain instances aid in assessing the biological plausibility of any neoplastic response observed. If an agent or mixture induces only benign neoplasms that appear to be end-points that do not readily progress to malignancy, it should nevertheless be suspected of being a carcinogen and requires further investigation.

(b)   Quantitative aspects

The probability that tumours will occur may depend on the species, sex, strain and age of the animal, the dose of the carcinogen and the route and length of exposure. Evidence of an increased incidence of neoplasms with increased level of exposure strengthens the inference of a causal association between the exposure and the development of neoplasms.

The form of the dose–response relationship can vary widely, depending on the particular agent under study and the target organ. Both DNA damage and increased cell division are important aspects of carcinogenesis, and cell proliferation is a strong determinant of dose–response relationships for some carcinogens (Cohen & Ellwein, 1990). Since many chemicals require metabolic activation before being converted into their reactive intermediates, both metabolic and pharmacokinetic aspects are important in determining the dose–response pattern. Saturation of steps such as absorption, activation, inactivation and elimination may produce nonlinearity in the dose–response relationship, as could saturation of processes such as DNA repair (Hoel *et al.*, 1983; Gart *et al.*, 1986).

(c)   Statistical analysis of long-term experiments in animals

Factors considered by the Working Group include the adequacy of the information given for each treatment group: (i) the number of animals studied and the number examined histologically, (ii) the number of animals with a given tumour type and (iii) length of survival. The statistical methods used should be clearly stated and should be the generally accepted techniques refined for this purpose (Peto *et al.*, 1980; Gart *et al.*, 1986). When there is no difference in survival between control and treatment groups, the Working Group usually compares the proportions of animals developing each tumour type in each of the groups. Otherwise, consideration is given as to whether or not appropriate adjustments have been made for differences in survival. These adjustments can include: comparisons of the proportions of tumour-bearing animals among the effective number of animals (alive at the time the first tumour is discovered), in the case where most differences in survival occur before tumours appear; life-table methods, when tumours are visible or when they may be considered 'fatal' because mortality rapidly follows tumour development; and the Mantel-Haenszel test or logistic regression, when occult tumours do not affect the animals' risk of dying but are 'incidental' findings at autopsy.

In practice, classifying tumours as fatal or incidental may be difficult. Several survival-adjusted methods have been developed that do not require this distinction (Gart *et al.*, 1986), although they have not been fully evaluated.

10.   OTHER DATA RELEVANT TO AN EVALUATION OF CARCINOGENICITY AND ITS MECHANISMS

In coming to an overall evaluation of carcinogenicity in humans (see pp. 25–27), the Working Group also considers related data. The nature of the information selected for the summary depends on the agent being considered.

For chemicals and complex mixtures of chemicals such as those in some occupational situations or involving cultural habits (e.g. tobacco smoking), the other data considered to be relevant are divided into those on absorption, distribution, metabolism and excretion; toxic effects; reproductive and developmental effects; and genetic and related effects.

Concise information is given on absorption, distribution (including placental transfer) and excretion in both humans and experimental animals. Kinetic factors that may affect the dose–response relationship, such as saturation of uptake, protein binding, metabolic activation, detoxification and DNA repair processes, are mentioned. Studies that indicate the metabolic fate of the agent in humans and in experimental animals are summarized briefly, and comparisons of data on humans and on animals are made when possible. Comparative information on the relationship between exposure and the dose that reaches the target site may be of particular importance for extrapolation between species. Data are given on acute and chronic toxic effects (other than cancer), such as organ toxicity, increased cell proliferation, immunotoxicity and endocrine effects. The presence and toxicological significance of cellular receptors is described. Effects on reproduction, teratogenicity, fetotoxicity and embryotoxicity are also summarized briefly.

Tests of genetic and related effects are described in view of the relevance of gene mutation and chromosomal damage to carcinogenesis (Vainio *et al.*, 1992; McGregor *et al.*, 1999). The adequacy of the reporting of sample characterization is considered and, where necessary, commented upon; with regard to complex mixtures, such comments are similar to those described for animal carcinogenicity tests on p. 18. The available data are interpreted critically by phylogenetic group according to the end-points detected, which may include DNA damage, gene mutation, sister chromatid exchange, micro-nucleus formation, chromosomal aberrations, aneuploidy and cell transformation. The concentrations employed are given, and mention is made of whether use of an exogenous metabolic system *in vitro* affected the test result. These data are given as listings of test systems, data and references. The data on genetic and related effects presented in the *Monographs* are also available in the form of genetic activity profiles (GAP) prepared in collaboration with the United States Environmental Protection Agency (EPA) (see also Waters *et al.*, 1987) using software for personal computers that are Microsoft Windows® compatible. The EPA/IARC GAP software and database may be downloaded free of charge from *www.epa.gov/gapdb*.

Positive results in tests using prokaryotes, lower eukaryotes, plants, insects and cultured mammalian cells suggest that genetic and related effects could occur in mammals. Results from such tests may also give information about the types of genetic effect produced and about the involvement of metabolic activation. Some end-points described are clearly genetic in nature (e.g. gene mutations and chromosomal aberrations), while others are to a greater or lesser degree associated with genetic effects (e.g. unscheduled DNA synthesis). In-vitro tests for tumour-promoting activity and for cell transformation may be sensitive to changes that are not necessarily the result of genetic alterations but that may have specific relevance to the process of carcinogenesis. A critical appraisal of these tests has been published (Montesano *et al.*, 1986).

Genetic or other activity detected in experimental mammals and humans is regarded as being of greater relevance than that in other organisms. The demonstration that an agent or mixture can induce gene and chromosomal mutations in whole mammals indicates that it may have carcinogenic activity, although this activity may not be detectably expressed in any or all species. Relative potency in tests for mutagenicity and related effects is not a reliable indicator of carcinogenic potency. Negative results in tests for mutagenicity in selected tissues from animals treated *in vivo* provide less weight, partly because they do not exclude the possibility of an effect in tissues other than those examined. Moreover, negative results in short-term tests with genetic end-points cannot be considered to provide evidence to rule out carcinogenicity of agents or mixtures that act through other mechanisms (e.g. receptor-mediated effects, cellular toxicity with regenerative proliferation, peroxisome proliferation) (Vainio *et al.*, 1992). Factors that may lead to misleading results in short-term tests have been discussed in detail elsewhere (Montesano *et al.*, 1986).

When available, data relevant to mechanisms of carcinogenesis that do not involve structural changes at the level of the gene are also described.

The adequacy of epidemiological studies of reproductive outcome and genetic and related effects in humans is evaluated by the same criteria as are applied to epidemiological studies of cancer.

Structure–activity relationships that may be relevant to an evaluation of the carcinogenicity of an agent are also described.

For biological agents — viruses, bacteria and parasites — other data relevant to carcinogenicity include descriptions of the pathology of infection, molecular biology (integration and expression of viruses, and any genetic alterations seen in human tumours) and other observations, which might include cellular and tissue responses to infection, immune response and the presence of tumour markers.

11.   SUMMARY OF DATA REPORTED

In this section, the relevant epidemiological and experimental data are summarized. Only reports, other than in abstract form, that meet the criteria outlined on p. 11 are considered for evaluating carcinogenicity. Inadequate studies are generally not summarized: such studies are usually identified by a square-bracketed comment in the preceding text.

(*a*)   *Exposure*

Human exposure to chemicals and complex mixtures is summarized on the basis of elements such as production, use, occurrence in the environment and determinations in human tissues and body fluids. Quantitative data are given when available. Exposure to biological agents is described in terms of transmission and prevalence of infection.

(*b*)   *Carcinogenicity in humans*

Results of epidemiological studies that are considered to be pertinent to an assessment of human carcinogenicity are summarized. When relevant, case reports and correlation studies are also summarized.

(*c*)   *Carcinogenicity in experimental animals*

Data relevant to an evaluation of carcinogenicity in animals are summarized. For each animal species and route of administration, it is stated whether an increased incidence of neoplasms or preneoplastic lesions was observed, and the tumour sites are indicated. If the agent or mixture produced tumours after prenatal exposure or in single-dose experiments, this is also indicated. Negative findings are also summarized. Dose–response and other quantitative data may be given when available.

(*d*)   *Other data relevant to an evaluation of carcinogenicity and its mechanisms*

Data on biological effects in humans that are of particular relevance are summarized. These may include toxicological, kinetic and metabolic considerations and evidence of DNA binding, persistence of DNA lesions or genetic damage in exposed humans. Toxicological information, such as that on cytotoxicity and regeneration, receptor binding and hormonal and immunological effects, and data on kinetics and metabolism in experimental animals are given when considered relevant to the possible mechanism of the carcinogenic action of the agent. The results of tests for genetic and related effects are summarized for whole mammals, cultured mammalian cells and nonmammalian systems.

When available, comparisons of such data for humans and for animals, and particularly animals that have developed cancer, are described.

Structure–activity relationships are mentioned when relevant.

For the agent, mixture or exposure circumstance being evaluated, the available data on end-points or other phenomena relevant to mechanisms of carcinogenesis from studies in humans, experimental animals and tissue and cell test systems are summarized within one or more of the following descriptive dimensions:

(i) Evidence of genotoxicity (structural changes at the level of the gene): for example, structure–activity considerations, adduct formation, mutagenicity (effect on specific genes), chromosomal mutation/aneuploidy

(ii) Evidence of effects on the expression of relevant genes (functional changes at the intracellular level): for example, alterations to the structure or quantity of the product of a proto-oncogene or tumour-suppressor gene, alterations to metabolic activation/inactivation/DNA repair

(iii) Evidence of relevant effects on cell behaviour (morphological or behavioural changes at the cellular or tissue level): for example, induction of mitogenesis, compensatory cell proliferation, preneoplasia and hyperplasia, survival of premalignant or malignant cells (immortalization, immunosuppression), effects on metastatic potential

(iv) Evidence from dose and time relationships of carcinogenic effects and interactions between agents: for example, early/late stage, as inferred from epidemiological studies; initiation/promotion/progression/malignant conversion, as defined in animal carcinogenicity experiments; toxicokinetics

These dimensions are not mutually exclusive, and an agent may fall within more than one of them. Thus, for example, the action of an agent on the expression of relevant genes could be summarized under both the first and second dimensions, even if it were known with reasonable certainty that those effects resulted from genotoxicity.

12.   EVALUATION

Evaluations of the strength of the evidence for carcinogenicity arising from human and experimental animal data are made, using standard terms.

It is recognized that the criteria for these evaluations, described below, cannot encompass all of the factors that may be relevant to an evaluation of carcinogenicity. In considering all of the relevant scientific data, the Working Group may assign the agent, mixture or exposure circumstance to a higher or lower category than a strict interpretation of these criteria would indicate.

(a)   Degrees of evidence for carcinogenicity in humans and in experimental animals and supporting evidence

These categories refer only to the strength of the evidence that an exposure is carcinogenic and not to the extent of its carcinogenic activity (potency) nor to the mechanisms involved. A classification may change as new information becomes available.

An evaluation of degree of evidence, whether for a single agent or a mixture, is limited to the materials tested, as defined physically, chemically or biologically. When the agents evaluated are considered by the Working Group to be sufficiently closely related, they may be grouped together for the purpose of a single evaluation of degree of evidence.

(i)   Carcinogenicity in humans

The applicability of an evaluation of the carcinogenicity of a mixture, process, occupation or industry on the basis of evidence from epidemiological studies depends on the variability over time and place of the mixtures, processes, occupations and industries. The Working Group seeks to identify the specific exposure, process or activity which is considered most likely to be responsible for any excess risk. The evaluation is focused as narrowly as the available data on exposure and other aspects permit.

The evidence relevant to carcinogenicity from studies in humans is classified into one of the following categories:

Sufficient evidence of carcinogenicity: The Working Group considers that a causal relationship has been established between exposure to the agent, mixture or exposure circumstance and human cancer. That is, a positive relationship has been observed between the exposure and cancer in studies in which chance, bias and confounding could be ruled out with reasonable confidence.

Limited evidence of carcinogenicity: A positive association has been observed between exposure to the agent, mixture or exposure circumstance and cancer for which a causal interpretation is considered by the Working Group to be credible, but chance, bias or confounding could not be ruled out with reasonable confidence.

Inadequate evidence of carcinogenicity: The available studies are of insufficient quality, consistency or statistical power to permit a conclusion regarding the presence or absence of a causal association between exposure and cancer, or no data on cancer in humans are available.

Evidence suggesting lack of carcinogenicity: There are several adequate studies covering the full range of levels of exposure that human beings are known to encounter, which are mutually consistent in not showing a positive association between exposure to the agent, mixture or exposure circumstance and any studied cancer at any observed level of exposure. A conclusion of 'evidence suggesting lack of carcinogenicity' is inevitably limited to the cancer sites, conditions and levels of exposure and length of observation covered by the available studies. In addition, the possibility of a very small risk at the levels of exposure studied can never be excluded.

In some instances, the above categories may be used to classify the degree of evidence related to carcinogenicity in specific organs or tissues.

(ii)   *Carcinogenicity in experimental animals*

The evidence relevant to carcinogenicity in experimental animals is classified into one of the following categories:

Sufficient evidence of carcinogenicity: The Working Group considers that a causal relationship has been established between the agent or mixture and an increased incidence of malignant neoplasms or of an appropriate combination of benign and malignant neoplasms in (a) two or more species of animals or (b) in two or more independent studies in one species carried out at different times or in different laboratories or under different protocols.

Exceptionally, a single study in one species might be considered to provide sufficient evidence of carcinogenicity when malignant neoplasms occur to an unusual degree with regard to incidence, site, type of tumour or age at onset.

Limited evidence of carcinogenicity: The data suggest a carcinogenic effect but are limited for making a definitive evaluation because, e.g. (a) the evidence of carcinogenicity is restricted to a single experiment; or (b) there are unresolved questions regarding the adequacy of the design, conduct or interpretation of the study; or (c) the agent or mixture increases the incidence only of benign neoplasms or lesions of uncertain neoplastic potential, or of certain neoplasms which may occur spontaneously in high incidences in certain strains.

Inadequate evidence of carcinogenicity: The studies cannot be interpreted as showing either the presence or absence of a carcinogenic effect because of major qualitative or quantitative limitations, or no data on cancer in experimental animals are available.

Evidence suggesting lack of carcinogenicity: Adequate studies involving at least two species are available which show that, within the limits of the tests used, the agent or mixture is not carcinogenic. A conclusion of evidence suggesting lack of carcinogenicity is inevitably limited to the species, tumour sites and levels of exposure studied.

(b)   *Other data relevant to the evaluation of carcinogenicity and its mechanisms*

Other evidence judged to be relevant to an evaluation of carcinogenicity and of sufficient importance to affect the overall evaluation is then described. This may include data on preneoplastic lesions, tumour pathology, genetic and related effects, structure–activity relationships, metabolism and pharmacokinetics, physicochemical parameters and analogous biological agents.

Data relevant to mechanisms of the carcinogenic action are also evaluated. The strength of the evidence that any carcinogenic effect observed is due to a particular mechanism is assessed, using terms such as weak, moderate or strong. Then, the Working Group assesses if that particular mechanism is likely to be operative in humans. The strongest indications that a particular mechanism operates in humans come from data on humans or biological specimens obtained from exposed humans. The data may be considered to be especially relevant if they show that the agent in question has caused changes in exposed humans that are on the causal pathway to carcinogenesis. Such data may, however, never become available, because it is at least conceivable that certain compounds may be kept from human use solely on the basis of evidence of their toxicity and/or carcinogenicity in experimental systems.

For complex exposures, including occupational and industrial exposures, the chemical composition and the potential contribution of carcinogens known to be present are considered by the Working Group in its overall evaluation of human carcinogenicity. The Working Group also determines the extent to which the materials tested in experimental systems are related to those to which humans are exposed.

(c)   Overall evaluation

Finally, the body of evidence is considered as a whole, in order to reach an overall evaluation of the carcinogenicity to humans of an agent, mixture or circumstance of exposure.

An evaluation may be made for a group of chemical compounds that have been evaluated by the Working Group. In addition, when supporting data indicate that other, related compounds for which there is no direct evidence of capacity to induce cancer in humans or in animals may also be carcinogenic, a statement describing the rationale for this conclusion is added to the evaluation narrative; an additional evaluation may be made for this broader group of compounds if the strength of the evidence warrants it.

The agent, mixture or exposure circumstance is described according to the wording of one of the following categories, and the designated group is given. The categorization of an agent, mixture or exposure circumstance is a matter of scientific judgement, reflecting the strength of the evidence derived from studies in humans and in experimental animals and from other relevant data.

Group 1 — The agent (mixture) is carcinogenic to humans.
The exposure circumstance entails exposures that are carcinogenic to humans.

This category is used when there is *sufficient evidence* of carcinogenicity in humans. Exceptionally, an agent (mixture) may be placed in this category when evidence of carcinogenicity in humans is less than sufficient but there is *sufficient evidence* of carcinogenicity in experimental animals and strong evidence in exposed humans that the agent (mixture) acts through a relevant mechanism of carcinogenicity.

Group 2

This category includes agents, mixtures and exposure circumstances for which, at one extreme, the degree of evidence of carcinogenicity in humans is almost sufficient, as well as those for which, at the other extreme, there are no human data but for which there is evidence of carcinogenicity in experimental animals. Agents, mixtures and exposure circumstances are assigned to either group 2A (probably carcinogenic to humans) or group 2B (possibly carcinogenic to humans) on the basis of epidemiological and experimental evidence of carcinogenicity and other relevant data.

Group 2A — The agent (mixture) is probably carcinogenic to humans.
The exposure circumstance entails exposures that are probably carcinogenic to humans.

This category is used when there is *limited evidence* of carcinogenicity in humans and *sufficient evidence* of carcinogenicity in experimental animals. In some cases, an agent (mixture) may be classified in this category when there is *inadequate evidence* of carcinogenicity in humans, *sufficient evidence* of carcinogenicity in experimental animals and strong evidence that the carcinogenesis is mediated by a mechanism that also operates in humans. Exceptionally, an agent, mixture or exposure circumstance may be classified in this category solely on the basis of *limited evidence* of carcinogenicity in humans.

Group 2B — The agent (mixture) is possibly carcinogenic to humans.
The exposure circumstance entails exposures that are possibly carcinogenic to humans.

This category is used for agents, mixtures and exposure circumstances for which there is *limited evidence* of carcinogenicity in humans and less than *sufficient evidence* of carcinogenicity in experimental animals. It may also be used when there is *inadequate evidence* of carcinogenicity in humans but there is *sufficient evidence* of carcinogenicity in experimental animals. In some instances, an agent, mixture or exposure circumstance for which there is *inadequate evidence* of carcinogenicity in humans but *limited evidence* of carcinogenicity in experimental animals together with supporting evidence from other relevant data may be placed in this group.

Group 3 — The agent (mixture or exposure circumstance) is not classifiable as to its carcinogenicity to humans.

This category is used most commonly for agents, mixtures and exposure circumstances for which the *evidence of carcinogenicity* is *inadequate* in humans and *inadequate* or *limited* in experimental animals.

Exceptionally, agents (mixtures) for which the *evidence of carcinogenicity* is *inadequate* in humans but *sufficient* in experimental animals may be placed in this category when there is strong evidence that the mechanism of carcinogenicity in experimental animals does not operate in humans.

Agents, mixtures and exposure circumstances that do not fall into any other group are also placed in this category.

Group 4 — The agent (mixture) is probably not carcinogenic to humans.

This category is used for agents or mixtures for which there is *evidence suggesting lack of carcinogenicity* in humans and in experimental animals. In some instances, agents or mixtures for which there is *inadequate evidence* of carcinogenicity in humans but *evidence suggesting lack of carcinogenicity* in experimental animals, consistently and strongly supported by a broad range of other relevant data, may be classified in this group.

13.   REFERENCES

Breslow, N.E. & Day, N.E. (1980) *Statistical Methods in Cancer Research, Vol 1, The Analysis of Case–Control Studies* (IARC Scientific Publications No. 32), Lyon, IARC*Press*

Breslow, N.E. & Day, N.E. (1987) *Statistical Methods in Cancer Research*, Vol. 2, *The Design and Analysis of Cohort Studies* (IARC Scientific Publications No. 82), Lyon, IARC*Press*

Cohen, S.M. & Ellwein, L.B. (1990) Cell proliferation in carcinogenesis. *Science*, **249**, 1007–1011

Gart, J.J., Krewski, D., Lee, P.N., Tarone, R.E. & Wahrendorf, J. (1986) *Statistical Methods in Cancer Research*, Vol. 3, *The Design and Analysis of Long-term Animal Experiments* (IARC Scientific Publications No. 79), Lyon, IARC*Press*

Hoel, D.G., Kaplan, N.L. & Anderson, M.W. (1983) Implication of nonlinear kinetics on risk estimation in carcinogenesis. *Science*, **219**, 1032–1037

Huff, J.E., Eustis, S.L. & Haseman, J.K. (1989) Occurrence and relevance of chemically induced benign neoplasms in long-term carcinogenicity studies. *Cancer Metastasis Rev.*, **8**, 1–21

IARC (1973–1996) *Information Bulletin on the Survey of Chemicals Being Tested for Carcinogenicity/Directory of Agents Being Tested for Carcinogenicity*, Numbers 1–17, Lyon, IARC*Press*

IARC (1976–1996), Lyon, IARC*Press*

   Directory of On-going Research in Cancer Epidemiology 1976. Edited by C.S. Muir & G. Wagner

   Directory of On-going Research in Cancer Epidemiology 1977 (IARC Scientific Publications No. 17). Edited by C.S. Muir & G. Wagner

   Directory of On-going Research in Cancer Epidemiology 1978 (IARC Scientific Publications No. 26). Edited by C.S. Muir & G. Wagner

   Directory of On-going Research in Cancer Epidemiology 1979 (IARC Scientific Publications No. 28). Edited by C.S. Muir & G. Wagner

   Directory of On-going Research in Cancer Epidemiology 1980 (IARC Scientific Publications No. 35). Edited by C.S. Muir & G. Wagner

   Directory of On-going Research in Cancer Epidemiology 1981 (IARC Scientific Publications No. 38). Edited by C.S. Muir & G. Wagner

Directory of On-going Research in Cancer Epidemiology 1982 (IARC Scientific Publications No. 46). Edited by C.S. Muir & G. Wagner

Directory of On-going Research in Cancer Epidemiology 1983 (IARC Scientific Publications No. 50). Edited by C.S. Muir & G. Wagner

Directory of On-going Research in Cancer Epidemiology 1984 (IARC Scientific Publications No. 62). Edited by C.S. Muir & G. Wagner

Directory of On-going Research in Cancer Epidemiology 1985 (IARC Scientific Publications No. 69). Edited by C.S. Muir & G. Wagner

Directory of On-going Research in Cancer Epidemiology 1986 (IARC Scientific Publications No. 80). Edited by C.S. Muir & G. Wagner

Directory of On-going Research in Cancer Epidemiology 1987 (IARC Scientific Publications No. 86). Edited by D.M. Parkin & J. Wahrendorf

Directory of On-going Research in Cancer Epidemiology 1988 (IARC Scientific Publications No. 93). Edited by M. Coleman & J. Wahrendorf

Directory of On-going Research in Cancer Epidemiology 1989/90 (IARC Scientific Publications No. 101). Edited by M. Coleman & J. Wahrendorf

Directory of On-going Research in Cancer Epidemiology 1991 (IARC Scientific Publications No.110). Edited by M. Coleman & J. Wahrendorf

Directory of On-going Research in Cancer Epidemiology 1992 (IARC Scientific Publications No. 117). Edited by M. Coleman, J. Wahrendorf & E. Démaret

Directory of On-going Research in Cancer Epidemiology 1994 (IARC Scientific Publications No. 130). Edited by R. Sankaranarayanan, J. Wahrendorf & E. Démaret

Directory of On-going Research in Cancer Epidemiology 1996 (IARC Scientific Publications No. 137). Edited by R. Sankaranarayanan, J. Wahrendorf & E. Démaret

IARC (1977) *IARC Monographs Programme on the Evaluation of the Carcinogenic Risk of Chemicals to Humans.* Preamble (IARC intern. tech. Rep. No. 77/002)

IARC (1978) *Chemicals with* Sufficient Evidence *of Carcinogenicity in Experimental Animals —* IARC Monographs *Volumes 1–17* (IARC intern. tech. Rep. No. 78/003)

IARC (1978–1993) *Environmental Carcinogens. Methods of Analysis and Exposure Measurement,* Lyon, IARC*Press*

Vol. 1. Analysis of Volatile Nitrosamines in Food (IARC Scientific Publications No. 18). Edited by R. Preussmann, M. Castegnaro, E.A. Walker & A.E. Wasserman (1978)

Vol. 2. Methods for the Measurement of Vinyl Chloride in Poly(vinyl chloride), Air, Water and Foodstuffs (IARC Scientific Publications No. 22). Edited by D.C.M. Squirrell & W. Thain (1978)

Vol. 3. Analysis of Polycyclic Aromatic Hydrocarbons in Environmental Samples (IARC Scientific Publications No. 29). Edited by M. Castegnaro, P. Bogovski, H. Kunte & E.A. Walker (1979)

Vol. 4. Some Aromatic Amines and Azo Dyes in the General and Industrial Environment (IARC Scientific Publications No. 40). Edited by L. Fishbein, M. Castegnaro, I.K. O'Neill & H. Bartsch (1981)

Vol. 5. Some Mycotoxins (IARC Scientific Publications No. 44). Edited by L. Stoloff, M. Castegnaro, P. Scott, I.K. O'Neill & H. Bartsch (1983)

Vol. 6. N-Nitroso Compounds (IARC Scientific Publications No. 45). Edited by R. Preussmann, I.K. O'Neill, G. Eisenbrand, B. Spiegelhalder & H. Bartsch (1983)

Vol. 7. Some Volatile Halogenated Hydrocarbons (IARC Scientific Publications No. 68). Edited by L. Fishbein & I.K. O'Neill (1985)

Vol. 8. Some Metals: As, Be, Cd, Cr, Ni, Pb, Se, Zn (IARC Scientific Publications No. 71). Edited by I.K. O'Neill, P. Schuller & L. Fishbein (1986)

Vol. 9. Passive Smoking (IARC Scientific Publications No. 81). Edited by I.K. O'Neill, K.D. Brunnemann, B. Dodet & D. Hoffmann (1987)

*Vol. 10. Benzene and Alkylated Benzenes (*IARC Scientific Publications No. 85). Edited by L. Fishbein & I.K. O'Neill (1988)

Vol. 11. Polychlorinated Dioxins and Dibenzofurans (IARC Scientific Publications No. 108). Edited by C. Rappe, H.R. Buser, B. Dodet & I.K. O'Neill (1991)

Vol. 12. Indoor Air (IARC Scientific Publications No. 109). Edited by B. Seifert, H. van de Wiel, B. Dodet & I.K. O'Neill (1993)

IARC (1979) *Criteria to Select Chemicals for* IARC Monographs (IARC intern. tech. Rep. No. 79/003)

IARC (1982) *IARC Monographs on the Evaluation of the Carcinogenic Risk of Chemicals to Humans,* Supplement 4, *Chemicals, Industrial Processes and Industries Associated with Cancer in Humans* (IARC Monographs, Volumes 1 to 29), Lyon, IARC*Press*

IARC (1983) *Approaches to Classifying Chemical Carcinogens According to Mechanism of Action* (IARC intern. tech. Rep. No. 83/001)

IARC (1984) *Chemicals and Exposures to Complex Mixtures Recommended for Evaluation in IARC Monographs and Chemicals and Complex Mixtures Recommended for Long-term Carcinogenicity Testing* (IARC intern. tech. Rep. No. 84/002)

IARC (1987a) *IARC Monographs on the Evaluation of Carcinogenic Risks to Humans,* Supplement 6, *Genetic and Related Effects: An Updating of Selected* IARC Monographs *from Volumes 1 to 42,* Lyon, IARC*Press*

IARC (1987b) *IARC Monographs on the Evaluation of Carcinogenic Risks to Humans,* Supplement 7, *Overall Evaluations of Carcinogenicity: An Updating of* IARC Monographs *Volumes 1 to 42,* Lyon, IARC*Press*

IARC (1988) *Report of an IARC Working Group to Review the Approaches and Processes Used to Evaluate the Carcinogenicity of Mixtures and Groups of Chemicals* (IARC intern. tech. Rep. No. 88/002)

IARC (1989) *Chemicals, Groups of Chemicals, Mixtures and Exposure Circumstances to be Evaluated in Future IARC Monographs, Report of an ad hoc Working Group* (IARC intern. tech. Rep. No. 89/004)

IARC (1991a) *A Consensus Report of an IARC Monographs Working Group on the Use of Mechanisms of Carcinogenesis in Risk Identification* (IARC intern. tech. Rep. No. 91/002)

IARC (1991b) *Report of an ad-hoc* IARC Monographs *Advisory Group on Viruses and Other Biological Agents Such as Parasites* (IARC intern. tech. Rep. No. 91/001)

IARC (1993) *Chemicals, Groups of Chemicals, Complex Mixtures, Physical and Biological Agents and Exposure Circumstances to be Evaluated in Future* IARC Monographs, *Report of an ad-hoc Working Group* (IARC intern. Rep. No. 93/005)

IARC (1998a) *Report of an ad-hoc* IARC Monographs *Advisory Group on Physical Agents* (IARC Internal Report No. 98/002)

IARC (1998b) *Report of an ad-hoc* IARC Monographs *Advisory Group on Priorities for Future Evaluations* (IARC Internal Report No. 98/004)

McGregor, D.B., Rice, J.M. & Venitt, S., eds (1999) *The Use of Short and Medium-term Tests for Carcinogens and Data on Genetic Effects in Carcinogenic Hazard Evaluation* (IARC Scientific Publications No. 146), Lyon, IARC*Press*

Montesano, R., Bartsch, H., Vainio, H., Wilbourn, J. & Yamasaki, H., eds (1986) *Long-term and Short-term Assays for Carcinogenesis — A Critical Appraisal* (IARC Scientific Publications No. 83), Lyon, IARC*Press*

Peto, R., Pike, M.C., Day, N.E., Gray, R.G., Lee, P.N., Parish, S., Peto, J., Richards, S. & Wahrendorf, J. (1980) Guidelines for simple, sensitive significance tests for carcinogenic effects in long-term animal experiments. In: *IARC Monographs on the Evaluation of the Carcinogenic Risk of Chemicals to Humans*, Supplement 2, *Long-term and Short-term Screening Assays for Carcinogens: A Critical Appraisal*, Lyon, IARC*Press*, pp. 311–426

Tomatis, L., Aitio, A., Wilbourn, J. & Shuker, L. (1989) Human carcinogens so far identified. *Jpn. J. Cancer Res.*, **80**, 795–807

Vainio, H., Magee, P.N., McGregor, D.B. & McMichael, A.J., eds (1992) *Mechanisms of Carcinogenesis in Risk Identification* (IARC Scientific Publications No. 116), Lyon, IARC*Press*

Vainio, H., Wilbourn, J.D., Sasco, A.J., Partensky, C., Gaudin, N., Heseltine, E. & Eragne, I. (1995) Identification of human carcinogenic risk in IARC Monographs. *Bull. Cancer,* **82**, 339–348 (in French)

Waters, M.D., Stack, H.F., Brady, A.L., Lohman, P.H.M., Haroun, L. & Vainio, H. (1987) Appendix 1. Activity profiles for genetic and related tests. In: *IARC Monographs on the Evaluation of Carcinogenic Risks to Humans*, Suppl. 6, *Genetic and Related Effects: An Updating of Selected IARC Monographs from Volumes 1 to 42*, Lyon, IARC*Press*, pp. 687–696

Wilbourn, J., Haroun, L., Heseltine, E., Kaldor, J., Partensky, C. & Vainio, H. (1986) Response of experimental animals to human carcinogens: an analysis based upon the IARC Monographs Programme. *Carcinogenesis*, **7**, 1853–1863

GENERAL REMARKS

In this eighty-fourth volume of the *IARC Monographs*, evaluations are made of evidence in relation to the carcinogenicity of arsenic (mostly naturally occurring) as a contaminant of drinking-water, as well as that of the water disinfectant chloramine and some chlorination by-products found in drinking-water. Some of the chlorination by-products evaluated here (chloral hydrate and di- and trichloroacetic acids) were evaluated previously (IARC, 1995). Chlorinated drinking-water and some chlorination by-products which are not re-evaluated in the present volume of the *Monographs* were previously evaluated in *IARC Monographs* Volume 52 (IARC, 1991). Those evaluations are listed in Table 1.

Table 1. Evaluations of chlorinated drinking-water, some chemicals used in the chlorination of drinking-water and some chlorination by-products, from *IARC Monographs* Volume 52 (1991)

Agent	Degree of evidence[a,b] of carcinogenicity		Overall evaluation of carcinogenicity to humans
	Human	Animal	
Chlorinated drinking-water	I	I	3
Some chemicals used in the chlorination of drinking-water			
Sodium chlorite	ND	I	3
Hypochlorite salts	ND	I	3
Chlorination by-products			
Bromodichloromethane	I	S	2B
Bromoform	I	L	3
Chlorodibromomethane	I	L	3
Halogenated acetonitriles			
Bromochloroacetonitrile	ND	I	3
Chloroacetonitrile	ND	I	3
Dibromoacetonitrile	ND	I	3
Dichloroacetonitrile	ND	I	3
Trichloroacetonitrile	ND	I	3

[a] I, inadequate evidence; S, sufficient evidence; L, limited evidence; ND, no data.
[b] For definitions of degrees of evidence and groupings of evaluations, see Preamble.
[Modified from the General Remarks to Vol. 52]

Water is essential to life, and the maintenance of an adequate supply of unpolluted water is a requirement for both human health and good environmental quality. Human demands upon the earth's water are great, and in some regions dangerously so. Water is taken for human use, irrigation and industry and is returned as industrial discharge, agricultural run-off and microbiologically contaminated, treated or untreated sewage. Water quality varies according to these discharges, the season and the geology of an area. The most critical characteristic of water for human health is its microbiology.

The microbiological quality of water can be controlled effectively by disinfection methods, which normally involve the introduction of chemical oxidants into the water supply. Chemicals used on a substantial scale as disinfectants are chlorine, hypochlorite, chloramine, chlorine dioxide and ozone. Chlorination is almost universally accepted as the method of choice for purifying water supplies. It was first used on a continuous basis for this purpose at the beginning of the twentieth century, and it is also used for sewage treatment in a few countries. Since some water suppliers have difficulty in maintaining acceptable water quality, particularly with regard to taste and odour, chlorine may be used in combination with ozone, chlorine dioxide, ammonia and activated charcoal. These treatments are sometimes followed by dechlorination, for example with sulfur dioxide.

There are substantial and irrefutable benefits of disinfection of water supplies by chemical methods, including chlorination. Any major change to these programmes would need to be evaluated fully as to its costs and benefits with regard not only to the need to maintain microbiological safety but also to the possible long-term adverse effects of alternatives to chlorination.

The investigation of possible risks for cancer from consuming chlorinated drinking-water in human populations is difficult, and hindered by a number of methodological obstacles. Chlorination may produce quite different profiles of chemical by-products in different areas. Characterization of a person's water consumption is complicated by the fact that, in many parts of the world, people change residence from time to time, and the nature of their source of domestic water changes as a consequence. Furthermore, people may consume water not only at home but also at work and elsewhere, and may drink not only chlorinated water but also unchlorinated water, bottled water, boiled water and other liquids, which will greatly influence their exposure to chlorination by-products. Exposure to constituents of water other than by ingestion — by inhalation or skin absorption — may also occur. Even if associations between human cancer risk and exposure to residential chlorinated water supplies can be demonstrated, they may be due to other constituents of the water that is chlorinated or to particular characteristics of the populations who live in areas served by such water supplies.

Evaluation of chlorinated water for carcinogenicity in experimental animals is similarly challenging. Few studies have been conducted in which the effects, if any, of constituents of chlorinated water have been compared with those of constituents of water from the same supply collected before chlorination.

Chlorination/disinfection by-products in water are formed when chlorine-based disinfectants react with the available organic matter in water. As many as 500 chlorination by-products have been identified, including trihalomethanes (THMs), and halogenated acids and aldehydes (HAAs), some of which are carcinogenic to experimental animals. Water in which these compounds and other trace contaminants are present is a very complex mixture. Formation and occurrence of the by-products is dependent on a number of factors, including the method of disinfection, the level and content of organic matter, pH, temperature and duration of treatment. THMs and HAAs are generally the most prevalent by-products; others occur at lower levels, but current knowledge on them is limited due to the small number of studies. Little is known about how various by-products are correlated; they may differ by region due to differences in determinants of their formation. THMs have regularly been used as a marker for the mixture of by-products in water, because they are measured routinely, but little is known as to whether they are a good marker for individual by-products such as chloral hydrate, trichloroacetic acid, dichloroacetic acid and MX ('mutagen X', 3-chloro-4-(dichloromethyl)-5-hydroxy-2(5*H*)-furanone). Only a small fraction of the chlorination by-products that have been identified in drinking-water have been tested for carcinogenicity.

Chlorinated drinking-water has been reported to be associated with the development of some human cancers, most particularly cancer of the urinary bladder. The epidemiological studies have examined exposure to chlorinated drinking-water rather than to specific disinfection by-products, and they are described in the Introduction section of the Monographs on chloramine, chloral hydrate, dichloroacetic acid, trichloroacetic acid and MX. However, they were not formally reviewed by this Working Group. It is important to recognize that none of the disinfectant by-products reviewed by IARC in this or previous volumes of the *Monographs* can be considered individually as the plausible causes for the cancers observed in human studies. When the potency of the by-products that have been evaluated is viewed in the context of their concentrations in drinking-water, it appears that these are at least three orders of magnitude too low to account for the risks implied. In most circumstances, however, the toxicity of a mixture cannot be assessed solely by evaluating individual compounds identified in this mixture. This is particularly the case for chlorinated drinking-water that contains hundreds of disinfection by-products, none of which dominates the toxicity of the mixture.

Arsenic and arsenic compounds were evaluated previously (IARC, 1980, 1987) as being *carcinogenic to humans* (Group 1) on the basis of *sufficient evidence* of an increased risk for skin cancer among patients exposed to inorganic arsenic through medical treatment, and an increased risk for lung cancer among workers involved in mining and smelting, who inhaled inorganic arsenic.

The natural and anthropogenic occurrence of arsenic in drinking-water has been increasingly recognized as a major public health concern in several regions of the world over the past several decades. Significant exposures have been documented in Argentina, Bangladesh, Chile, China, India (West Bengal), Mexico, Taiwan, China, and parts of the south-western USA. As an example, tube-wells began to be used for drinking-water in the 1970s in Bangladesh to control cholera and other gastrointestinal diseases that can be spread by drinking contaminated surface water. In the 1990s, it was discovered that the water from many of these wells was contaminated with arsenic. Worldwide, an estimated 160 million people live in regions with naturally elevated levels of arsenic in the drinking-water, due to the presence of arsenic-rich geological formations.

In this Volume, informative epidemiological studies of the prevalence of human cancer are reviewed — mainly ecological studies in Taiwan, China, and Chile, and several case–control and cohort studies — in relation to arsenic in drinking-water, which occurs primarily as inorganic arsenic (arsenate and, to a lesser extent, arsenite). Also in this Volume, studies of the carcinogenicity in experimental animals of dimethylarsinic acid, a methylated metabolite of inorganic arsenic, are reviewed for the first time.

Substances evaluated in this Volume of *Monographs* are those for which either there is published evidence of a causal association with human cancer when present in drinking-water (arsenic) or bioassays for carcinogenicity in experimental animals have been conducted (dimethylarsenic acid, chloral hydrate, di- and trichloroacetic acid, MX and chloramine).

References

IARC (1980) *IARC Monographs on the Evaluation of the Carcinogenic Risk of Chemicals to Humans*, Volume 23, *Some Metals and Metallic Compounds*, Lyon, IARCPress

IARC (1987) *IARC Monographs on the Evaluation of the Carcinogenic Risks to Humans*, Suppl. 7, *Overall Evaluations of Carcinogenicity: An Updating of IARC Monographs Volumes 1 to 42*, Lyon, IARCPress

IARC (1991) *IARC Monographs on the Evaluation of Carcinogenic Risks to Humans*, Vol. 52, *Chlorinated Drinking-water; Chlorination By-products; Some Other Halogenated Compounds; Cobalt and Cobalt Compounds*, Lyon, IARCPress

IARC (1995) *IARC Monographs on the Evaluation of Carcinogenic Risks to Humans*, Vol. 63, *Dry Cleaning, Some Chlorinated Solvents and Other Industrial Chemicals*, Lyon, IARCPress

THE MONOGRAPHS

ARSENIC IN DRINKING-WATER

ARSENIC IN DRINKING-WATER

1. Exposure Data

1.1 Chemical and physical data

Arsenic is the 20th most common element in the earth's crust, and is associated with igneous and sedimentary rocks, particularly sulfidic ores. Arsenic compounds are found in rock, soil, water and air as well as in plant and animal tissues. Although elemental arsenic is not soluble in water, arsenic salts exhibit a wide range of solubilities depending on pH and the ionic environment. Arsenic can exist in four valency states: –3, 0, +3 and +5. Under reducing conditions, the +3 valency state as arsenite (As^{III}) is the dominant form; the +5 valency state as arsenate (As^{V}) is generally the more stable form in oxygenized environments (Boyle & Jonasson, 1973; National Research Council, 1999; O'Neil, 2001; WHO, 2001).

Arsenic species identified in water are listed in Table 1. Inorganic As^{III} and As^{V} are the major arsenic species in natural water, whereas minor amounts of monomethylarsonic acid (MMA) and dimethylarsinic acid (DMA) can also be present. The trivalent mono-methylated (MMA^{III}) and dimethylated (DMA^{III}) arsenic species have been detected in lake water (Hasegawa et al., 1994, 1999). The presence of these trivalent methylated arsenical species is possibly underestimated since only few water analyses include a solvent sepa-ration step required to identify these trivalent species independently from their respective

Table 1. Some arsenic species identified in water[a]

Name	Abbreviation	Chemical formula	CAS No.	pKa
Arsenous acid (arsenite)	As^{III}	$As(OH)_3$	13464-58-9	9.23, 12.13, 13.4
Arsenic acid (arsenate)	As^{V}	$AsO(OH)_3$	7778-39-4	2.22, 6.98, 11.53
Monomethylarsonic acid	MMA^{V}	$CH_3AsO(OH)_2$	124-58-3	4.1, 8.7
Monomethylarsonous acid	MMA^{III}	$CH_3As(OH)_2$	25400-23-1	
Dimethylarsinic acid	DMA^{V}	$(CH_3)_2AsO(OH)$	75-60-5	6.2
Dimethylarsinous acid	DMA^{III}	$(CH_3)_2AsOH$	55094-22-9	
Trimethylarsine oxide	TMAO	$(CH_3)_3AsO$	4964-14-1	

[a] From National Research Council (1999); Francesconi & Kuehnelt (2002); Le (2002)

pentavalent analogues. Other unidentified arsenic species have also been reported in seawater and fresh water, and could represent up to 20% of the total arsenic (Francesconi & Kuehnelt, 2002; Le, 2002).

1.2   Analysis

Studies of human exposure to arsenic and its consequences for human health require two different kinds of arsenic analyses depending on whether quantitative or qualitative results are required. Several methods have been developed and improved for the measurement of total arsenic, and have been widely used for the evaluation of drinking-water contamination and the resulting concentrations of arsenic in humans. On the other hand, analytical methods allowing arsenic speciation have gained increasing interest. The environmental fate and behaviour, bioavailability and toxicity of arsenic vary dramatically with the chemical form (species) in which it exists, the inorganic As^{III} and As^V being, for example, far more toxic than MMA and DMA. Thus selective methods that determine the relative concentration of the different arsenic species in drinking-water are required when more precise assessments of their impact on human health are needed.

Analytical methods for arsenic have been reviewed (National Research Council, 1999; WHO, 2001; Goessler & Kuehnelt, 2002).

The most commonly used methods for the analysis of arsenic and arsenic compounds in water and biological samples are described below, and their characteristics are summarized in Table 2.

1.2.1   *Preservation of samples*

Assessment of human exposure to arsenic through drinking-water relies on the analysis of arsenic in water and in biological samples. Biological markers may more accurately reflect total dose of exposure in populations exposed to low, but potentially carcinogenic levels of arsenic in drinking-water. Many tissues contain arsenic following exposure to the element, but not all represent useful biomarkers. For example, arsenic is removed from blood within a few hours and excreted through the kidneys and urine within a few days. Determination of arsenic in urine is commonly used as a measure of recent exposure. Hair and nails have been shown to provide reliable biomarkers for long-term chronic exposure to arsenic in humans (Karagas et al., 1996, 2000). However, nails are preferred to hair since their contamination with arsenic from the air is negligible, whereas hair can adsorb 9–16% exogenous inorganic arsenic (Mandal et al., 2003). Karagas et al. (2001a) found that measurements of arsenic in both toenails and water were reproducible over a 3–5-year period.

Depending on the sample studied and the type of analysis to be performed, particular caution must be taken to overcome problems related to sample contamination and stability of the arsenic species. For determining total element concentrations, the main considerations for sample collection and storage are to prevent contamination and to minimize

Table 2. Most commonly used analytical methods for arsenic and arsenic compounds in water and biological samples

Methodology	Sample analysed	Detection	Detection limit	Advantages	Disadvantages	References
Colorimetric/spectro-photometric methods	Water, Urine, serum, Hair, nails	Total arsenic	~ 40 µg/L	Low cost, very simple, uses a simple spectrophotometer		Kingsley & Schaffert (1951); Vogel et al. (1954); Dahr et al. (1997); Pillai et al. (2000); Goessler & Kuehnelt (2002)
Inductively coupled plasma–atomic emission spectrometry (ICP–AES)	Water	Total arsenic	~ 30 µg/L			SM 3120 (1999); Environmental Protection Agency (1994a); Goessler & Kuehnelt (2002)
Inductively coupled plasma–mass spectrometry (ICP–MS)	Water, Nails	Total arsenic	0.1 µg/L	Analytical method approved by US EPA	Spectral and matrix interference	Environmental Protection Agency (1994b); Chen et al., 1999; Goessler & Kuehnelt (2002)
High resolution (HR)–ICP-MS	Water, Urine, Nails	Total arsenic	0.01 µg/L	Solves spectral interferences in samples with complex matrices		Gallagher et al. (2001); Karagas et al. (2001, 2002)
Instrumental neutron activation analysis (INAA)	Hair, nails, Tissues	Total arsenic	~ 0.001 µg/g	Reference method for detection of arsenic		Garland et al. (1993); Nichols et al. (1993); Pan et al. (1993); Pazirandeh et al. (1998); Karagas et al. (2001)
Electrothermal atomization laser–excited atomic fluorescence spectrometry (ETA–LEAFS)	Serum	Total arsenic	0.065 µg/L	Requires only minimal sample volume, sample pretreatment and measurement time		Swart & Simeonsson (1999)
Graphite furnace–atomic absorption spectrometry (GF–AAS)	Water, urine, Hair, nails, tissues	Total arsenic	~ 0.025 µg/g	Analytical method approved by US EPA	Pre-atomization losses, requires the use of matrix modifyers	Agahian et al. (1990); SM 3113 (1999); WHO (2001)

Table 2 (contd)

Methodology	Sample analysed	Detection	Detection limit	Advantages	Disadvantages	References
Hydride generation–atomic absorption spectrometry (HG–AAS)	Water Urine Hair, nails	Total arsenic and arsenic speciation	0.6–6 µg/L	Analytical method approved by US EPA		Braman & Foreback (1973); Crecelius (1978); Le et al. (1994a,b); Chatterjee et al. (1995); Lin et al. (1998); Ng et al. (1998); Wyatt et al. (1998a,b); Shraim et al. (1999, 2000); SM 3114 (1999)
Hydride generation–quartz furnace–atomic absorption spectrometry (HG–QF–AAS)	Water Tissues	Total arsenic and arsenic speciation	0.003–0.015 µg/L	Inexpensive		Environmental Protection Agency (1996c)
High-performance liquid chromatography (HPLC)–HG–AAS	Urine	Total arsenic and arsenic speciation	1–47 µg/L			Lamble & Hill (1996); Kurttio et al. (1998)
HPLC or solid-phase cartridge separation combined with hydride generation–atomic fluorescence spectrometry (HPLC–HG–AFS)	Water, urine	Arsenic speciation	0.05–0.8 µg/L	Rapid, inexpensive No need for sample pretreatment		Le & Ma (1997); Aposhian et al. (2000); Le et al. (2000a,b); Gong et al. (2001); Yalcin & Le (2001)
HPLC–ICP-MS	Water Water, urine Hair, nails	Total arsenic	0.01 µg/L 0.14–0.33 µg/L	No need for sample pretreatment	Expensive and often time-consuming Spectral and matrix interference	Shibata & Morita (1989); Londesborough et al. (1999); Chatterjee et al. (2000); Mandal et al. (2001); Shraim et al. (2001); Karagas et al. (2002); Mandal et al. (2003)

loss of trace amounts of analytes. High-density polyethylene containers are usually preferred to glass containers because they are less adsorptive for arsenic. These are pre-cleaned with nitric acid and then rinsed with distilled water.

Groundwater sampling is carried out by allowing the well-water to flow through the pumping pipe for approximately 10 min before collection.

Traditionally, water and urine samples are acidified with sulfuric or nitric acid to reduce potential adsorption of trace elements onto the surface of the sample container and to prevent bacterial proliferation. Samples can then be kept at +4 °C or at room temperature and preferably measured within 7 days (Lin et al., 1998; Rahman et al., 2002). Pande et al. (2001) reported, however, that all the field kits they evaluated were subject to negative inter-ference if samples were acidified with nitric acid for preservation; they showed that acidifi-cation using 5% ascorbic acid instead of nitric acid eliminates interference.

In iron-rich waters, the stability of As^{III} and As^V can be affected by the formation of iron precipitates (iron oxides and/or hydroxides designated by 'FeOOH'). These precipi-tates can form during transport to the laboratory for analysis of arsenic. Studies of labo-ratory reagent water containing both As^{III} and Fe^{III} indicated that, within 18 h at room tem-perature, the resulting FeOOH precipitates contained a mixture of As^{III} and As^V with near quantitative removal of aqueous arsenic. Addition of a chelating agent such as ethylene-diamine tetraacetic acid (EDTA), by sequestering Fe^{III}, inhibits the formation of FeAsOH precipitates and preserves the stability of arsenic species in iron-rich waters for more than 10 days (Gallagher et al., 2001).

Reliable information from speciation analysis requires that the concentration of indi-vidual species of the element be unchanged by handling and treatment of the sample. Although traditionally used for their preservation, acidification of samples is not suitable since it leads to changes in arsenic speciation.

For urine specimens, low temperature (4 °C and –20 °C) conditions are required if they are to be stored up to 2 months without substantial changes in arsenic speciation (except for MMA^{III} and DMA^{III} species). For longer storage times, the stability of arsenic species varies with the complex matrix and pH of the urine, and accurate measurement of inorganic As^{III} and As^V separately is more difficult since As^V is rapidly reduced to As^{III}. MMA^V and DMA^V are more stable (for up to 4.5 months). The trivalent arsenic species, monomethylarsonous acid (MMA^{III}) and dimethylarsinous acid (DMA^{III}), suspected to be key metabolic inter-mediates in human urine, are extremely unstable. It was shown that over 90% of MMA^{III} was rapidly oxidized to MMA^V in urine samples when stored at +4 °C or –20 °C over a 5-month period, while DMA^{III} was completely oxidized to DMA^V within 1 day (Gong et al., 2001). In a recent review, these authors found that the use of a complexing agent, diethyl-ammonium diethyldithiocarbamate (DDDC), improved the stability of MMA^{III} and DMA^{III} in urine samples. In the presence of DDDC (1–10 mM), MMA^{III} was found to be stable for 4 months at –20 °C (with a recovery of 85–95%) and DMA^{III} was partially preserved. Approximately 80% of DMA^{III} remained after 3 weeks of storage and 10–24% remained after 4 months (Jiang et al., 2003). The use of other additives (such as hydrochloric acid, sodium azide, benzoic acid, benzyltrimethylammonium chloride and cetylpyridinium chloride) has no particular benefit (Feldman *et al.*, 1999; Chen *et al.*, 2002).

For arsenic speciation, well-water is usually filtered at the sampling site using a 0.45 μm filter (Lin *et al.*, 1998).

Methods for on-site separation of As^{III} and As^{V} species immediately after water-sample collection using solid disposable cartridges can be efficiently used for speciation of particulate and soluble arsenic. A measured volume of the sample is passed through the 0.45-μm membrane filter, then serially through a connected silica-based strong anion-exchange cartridge. The filter captures particulate arsenic, while the anion-exchange cartridge retains As^{V}. Arsenite is not retained and is detected in the effluent. Arsenate is subsequently eluted with 1 M hydrochloric acid (HCl) from the anion-exchange cartridge and then analysed for concentration (Le *et al.*, 2000a).

In hair and nail samples, the arsenic species are less prone to change. For analysis of total arsenic, as for speciation methods, these specimens are usually prepared according to the International Atomic Energy Agency (IAEA) procedure (Ryabukhin, 1978).

Following extensive washing to eliminate exogenous arsenic resulting from air contamination, approximately 100 mg of each hair sample are usually placed in a Teflon beaker, mixed with acetone and then washed with distilled water. Nails are treated similarly to hair following brushing. Samples are weighed prior to analysis (Lin et *al.*, 1998; Mandal *et al.*, 2003). More stringent washing procedures have also been described for complete removal of surface contamination, by incubating nails for 20 min in 1% Triton X100 before analysis (Chen *et al.*, 1999).

1.2.2   *Analytical methods for measurement of total arsenic*

Determination of total arsenic in biological samples in most cases requires the complete destruction of the organic matrix. During this process, all the organic arsenic compounds should be converted into inorganic arsenic by oxidative digestion. Acid digestion (or wet ashing) (Kingsley & Schaffert, 1951) and dry ashing (George *et al.*, 1973) are the two basic methods that have been widely employed for oxidative digestion of samples prior to analysis. A microwave-assisted digestion technique has been developed recently and is currently used as a rapid preparation for sample analysis (Le *et al.*, 1994c; Goessler & Kuehnelt, 2002). For analysis of soft biological tissues using inductively coupled plasma (ICP) techniques, a simple partial digestion in a closed vessel at low temperature and pressure is often sufficient for the sample preparation and pretreatment step (WHO, 2001).

Historically, colorimetric/spectrophotometric methods have been used to determine total arsenic concentration. Several commercial field kits have been based on these methods. At present, laboratories often prefer more sensitive methods such as atomic absorption spectrometry (AAS), neutron activation analysis (NAA), atomic emission spectrometry (AES), mass spectrometry (MS) or atomic fluorescence spectrometry (AFS).

(a) Colorimetric/spectrometric methods

These methods take advantage of the formation of volatile arsine (AsH_3) gas to separate arsenic from other possible interference with the sample matrix. The colorimetric methods are easy to use and inexpensive in terms of equipment and operator cost. They are useful for the semi-quantitative determination of high concentrations of arsenic in water.

The silver diethylthiocarbamate (AgDDTC) method is the most popular spectrophotometric method for the determination of arsenic in water. The method is based on the generation of arsine either with zinc and hydrochloric acid or sodium borohydride in acidic solutions. The arsine gas is then flushed through a solution of diethylthiocarbamate in pyridine or pyridine/chloroform. The red-coloured complex can be measured at 520 nm. Using a modification of this method, Dhar et al. (1997) reported a detection limit of 40 µg/L for arsenic in water samples, with a 95% confidence.

Pillai et al. (2000) reported a new simple and reliable spectrophotometric method to determine total arsenic in environmental and biological samples. It involves bleaching the pinkish-red dye Rhodamine-B (measured at 553 nm) by the action of iodine released from the reaction between potassium iodate and arsenic in a slightly acidic medium.

The classic Gutzeit test (Vogel, 1954) is derived from the historical Marsh test. It is based on the generation of arsine (AsH_3) from arsenic compounds by the addition of zinc granules to concentrated sulfuric acid. The arsine can be detected by its reaction on a strip of filter moistened with silver nitrate or mercuric chloride, which produces a grey or a yellow to reddish-brown spot, respectively.

Field test kits

The high concentrations of arsenic currently found in groundwater in many parts of the world pose an important challenge because of the large number of wells that must be tested. This is particularly true in Bangladesh and other Asian hot spots such as Myanmar, Nepal, Cambodia, Laos, Viet Nam and India. Although less accurate than laboratory-based methods, field kits that allow on-site semi-quantitative determination of arsenic concentrations in well-water are of vital importance, since in these countries, the current laboratory capacity cannot cover the high level of analytical needs. Field testing has several advantages. In Bangladesh and other hot climates, attempts to keep samples cool over a long period of transport to a laboratory can be difficult. With field kits, there is no need for transport, no storage and therefore no need for preservation, which in addition reduces the cost of analysis and the time required for the well owner to be informed. Field kits are also simple to use after reasonable training of technicians.

These tests, however, must be accurate and sensitive enough to assess the level of arsenic contamination.

Much concern about the reliability of field kits recently led to careful evaluations of commercially available kits (Pande et al., 2001; Rahman et al., 2002a,b; Environmental Protection Agency-Battelle, 2002a,b; Erickson, 2003). The original field kit widely used in Bangladesh had a stated minimal detectable concentration of 100 µg/L, which largely exceeded the maximum permissible arsenic concentration defined by WHO (10 µg/L) and even the maximum stated by most developing countries (50 µg/L). Fortunately, the newer field test kits are more sensitive. Evaluations of these kits are summarized in Table 3.

A modification of the Gutzeit method using mercuric bromide is the basis of most commercial field kits. A test strip moistened with mercuric bromide is exposed to arsine gas derived from the sample solution, to form complex salts of arsenic and mercury. These reactions give a yellow [H(HgBr$_2$)As] to brown [(HgBr)$_3$As] to black [Hg$_3$As$_2$] stain. The intensity of the yellowish-brown colour developed on the test strip is proportional to the arsenic concentration in the sample. When the reaction is completed, the test strip is compared with a colour chart provided with the kit and allows semi-quantitative determination of total arsenic concentration.

More recent field kits include digital measurement of arsenic levels without depending on the judgement of the technician's eyes to detect the difference between colour shades of the coloured strip (Arsenator, PeCo test). The improvement in reading results in higher sensitivity and reliability (Environmental Protection Agency-Battelle, 2002a,b; Durham & Kosmus, 2003).

In addition, promising biological tools (bacterial biosensors) may lead to new kits for quantitative and qualitative measurement of arsenite and arsenate in aqueous solution (Flynn *et al.*, 2002; Stocker *et al.*, 2003).

(b) Inductively coupled plasma–atomic emission spectrometry (ICP–AES)

ICP–AES involves the use of plasma, usually argon, at temperatures between 6000 and 8000 °K as the excitation source. The analyte is introduced into the plasma as an aerosol. A typical detection limit achievable for arsenic with this technique is 30 µg/L. Because of the rather high detection limits, ICP–AES is not frequently used for the determination of arsenic in biological samples (Goessler & Kuehnelt, 2002).

In August 2002, ICP–AES was withdrawn from the US Environmental Protection Agency (US EPA)-approved analytical methods for arsenic since this technique is inadequate to meet the requirements of the new EPA standard for arsenic in drinking-water of 10 µg/L (10 ppb), effective since February 2002 (Environmental Protection Agency, 2002).

(c) Inductively coupled plasma–mass spectrometry (ICP–MS)

ICP–MS is superior to ICP–AES with respect to detection limits, multi-element capabilities and wide linear dynamic range. This technique combines the ICP as the ion source with a mass analyser. Quadrupole mass filters are the most common mass analyser; double-focusing magnetic/electrostatic sector instruments and time-of-flight mass analysers are also used (Goessler & Kuehnelt, 2002).

ICP–MS is classified among the US EPA-approved analytical methods for arsenic (Environmental Protection Agency, 2002), with a detection limit of 0.1 µg/L.

The sensitivity can be further improved by the use of hydride generation (HG) techniques leading to a more efficient sample introduction and to matrix removal. The use of a high-resolution mode with HG–ICP–MS allows a 10-fold decrease in the detection

Table 3. Evaluation of some field test kits for analysing arsenic in water

Field test kits	Kit capability	Minimum detection limit of arsenic	Detection range	Rate of false positive/false negative	Effects of interferences (sodium chloride, iron, sulfate, acidity)	Occupational hazard potential (OH)	Time required per test	Evaluation reference
Quick™ (industrial test kit, Rock Hill, USA)	Semi-quantitative	~ 5–20 µg/L	5, 10, 20, 40, 60, 100, 200, … 500 µg/L	0–4%/5–16%	ND	Safe	< 15 min	Environmental Protection Agency-Battelle (2002a)
AS75 (PeCo test kit) (Peters Engineering, Graz, Austria)	Semi-quantitative	~ 15–50 µg/L	10, 20, 30, … 100 µg/L 2.5, 5, 10, 20, … 60 µg/L	0–3%/0%	None	Safe	ND	Environmental Protection Agency-Battelle (2002b)
AAN (Asia Arsenic Network, Japan)	Semi-quantitative	~ 20 µg/L	20, 50, 100, 200,… 700 µg/L	19%/71%	Some with sulfide	Accidental escape of arsine gas may cause OH.	15 min	Pande et al. (2001); Rahman et al. (2002)
E. Merck (Germany)	Qualitative for arsenic concentration > 50 µg/L	~ 50–100 µg/L	100, 500, 1000, 1700, 3000 µg/L	21%/60%	Some with sulfide	Accidental spillage of acid and escape of arsine gas may cause OH.	30 min	Pande et al. (2001); Rahman et al. (2002)
NIPSOM (National Institute of Preventive and Social Medicine, Bangladesh)	Qualitative for arsenic concentration > 50 µg/L	~ 10–20 µg/L	10, 20, 50, 100, 200, 300 … 700 µg/L	21%/33%	Some with sulfide	Accidental spillage of acid and escape of arsine gas may cause OH.	5 min	Pande et al. (2001); Rahman et al. (2002)
AIIH-PH (All India Institute of Hygiene and Public Health, India)	Semi-quantitative	~ 50 µg/L	> 50 µg/L	25%/1%	Sulfide interference eliminated	Accidental spillage of acid and escape of arsine gas may cause OH.	30 min	Pande et al. (2001); Rahman et al. (2002)
GPL (General Pharmaceuticals Ltd, USA)	Semi-quantitative	~ 10 µg/L	10, 50, 100, 200, 400, 500 … 1500 µg/L	10%/32%	ND	Accidental spillage of acid and escape of arsine gas may cause OH.	20 min	Rahman et al. (2002)
Aqua (Aqua Consortium, Calcutta, India)	Semi-quantitative	~ 100 µg/L	> 50 µg/L	ND	Sulfide interference eliminated	Accidental spillage of acid and escape of arsine gas may cause OH. Contact with HgBr2 paper affects fingers of the user.	15 min	Pande et al. (2001)

limit (0.01 µg/L) for arsenic in water samples. HG–ICP–MS can be used for biological samples such as urine and nails (Chen *et al.*, 1999; Gallagher *et al.*, 2001; Karagas *et al.*, 2001a, 2002).

(*d*)   *Neutron activation analysis (NAA)*

Instrumental NAA is an accurate and sensitive means to measure arsenic. The method can analyse relatively small biological samples, and has been used efficiently to measure total arsenic in hair, nails and other tissues, with a detection limit of approximately 0.001 µg/g (Pan *et al.*, 1993; Garland *et al.*, 1996; Nichols *et al.*, 1998; Pazirandeh *et al.*, 1998; Karagas *et al.*, 2001a).

(*e*)   *Electro-thermal atomization laser–excited atomic fluorescence spectrometry (ETA–LEAFS)*

ETA–LEAFS is a highly sensitive and selective method that has been developed by the combination of laser-excited atomic fluorescence spectrometry with electro-thermal atomization in graphite cup or tube furnaces. The technique provides excellent analytical performance at ultra-trace levels, with a detection limit of 0.065 µg/L for arsenic in undiluted serum. This approach allows measurements to be taken directly on the serum samples after a simple dilution step. It also minimizes the amounts of sample required and can provide multiple measurements when only limited amounts of sample are available (Swart & Simeonsson, 1999).

(*f*)   *Atomic absorption spectrometry (AAS)*

AAS is one of the most common analytical procedures for measuring arsenic in both environmental and biological materials. The main methods are flame AAS (FAAS), electro-thermal AAS (ET–AAS), also referred to as graphite furnace AAS (GF–AAS), and HG–AAS.

FAAS, with a relatively high detection limit (~1 mg/L), was never seriously considered for determining arsenic in environmental and biological samples.

The principal difference among the various AAS techniques is the means and form of presentation and atomization of the sample.

In GF–AAS, a small aliquot, rather than a continuous flow of sample, is deposited in a graphite furnace in which it is completely dissolved and mineralized *in situ*. The analyte is vaporized to form volatile hybrids. Matrix modifiers, such as a mixture of palladium and magnesium, must be used to protect the analyte from premature volatilization before vaporization, and therefore loss of arsenic. GF–AAS is classified among the approved US EPA analytical methods for arsenic in water (Environmental Protection Agency, 2002). It has been used for the determination of total arsenic in water and many biological samples (Agahian *et al.*, 1990).

HG–AAS uses the hydride generation technique, which can easily be connected to various detection systems and greatly improves the detection limit of all methods. The HG technique is based on the production of volatile arsines (by the addition of either zinc/hydrochloric acid or a sodium borohydrate/acid mixture) which are transported by an inert gas to the detection system. HG–AAS is probably the most widely used method to determine total arsenic in water (Rahman *et al.*, 2001; Chakraborti *et al.*, 2002) and various matrices (Wyatt *et al.*, 1998a; Das *et al.*, 1995). HG–AAS is also classified among the US EPA-approved analytical methods for arsenic in water (Environmental Protection Agency, 2002). Detection limits for total arsenic in water achievable by this technique are around 0.6 μg/L.

1.2.3   *Analytical methods for arsenic speciation*

The combination of high-performance separation methods with highly sensitive instrumental detection systems is necessary to determine arsenic species (arsenic speciation) at trace levels. These combinations, referred to as hyphenated techniques, have been extensively described by Goessler and Kuehnelt (2002).

Three steps are required for arsenic speciation: the extraction of arsenic from the sample, the separation of the different arsenic species and their detection/quantification. The extraction procedure should be as mild and complete as possible. A combination of various extractants is often necessary to remove all the arsenic; polar and organic solvents or water are commonly used for this purpose. In many cases (water or urine samples), extraction may not be necessary. In the next step, a combination of separation procedures is usually required because of the different chemical properties of the arsenic compounds (anionic, neutral, cationic). Selective HG and high-performance liquid chromatography (HPLC) are the most commonly used. After the different arsenic compounds have been separated, they must be detected with a suitable detector. All the methods cited in Section 1.2.2 have been used more or less successfully to identify and determine arsenic compounds. Some efficient and sensitive hyphenated methods, commonly used or recently developed, are described below and presented in Table 2.

(*a*)   *AAS–derived hyphenated methods*

Hydride generation quartz furnace atomic absorption spectrometry (HG–QF–AAS) is an improved modification of GF–AAS, described by the US Environmental Protection Agency (Environmental Protection Agency, 1996c), in which the graphite furnace is replaced by a quartz furnace. The method is designed to measure both total arsenic and arsenic species in water (range, 0.01–50 μg/L) and in tissue (range, 0.01–500 μg/g dry weight for arsenic and arsenic species). The detection limits for total inorganic arsenic, As[III] and As[V] have been determined to be 3 ng/L and 15 ng/L for DMA and MMA, respectively, when no background element or interference is present.

Modifications of the HG–AAS method have also been described that allow the determination of arsenic species (As[III], As[V], MMA, DMA) in water and biological samples (Braman & Foreback, 1973; Crecelius, 1978; Le *et al.*, 1994a,b,c; Hasegawa *et al.*, 1994; Lin *et al.*, 1998; Ng *et al.*, 1998). These modifications, which involve trapping the arsine species at liquid nitrogen temperature (–196 °C), allow the elution by chromatography of each compound at room temperature. Ng *et al.* (1998) described, for example, an optimized HG–cold trap–AAS procedure for the speciation of arsenic in urine, with detection limits of 0.25 μg/L, 0.325 μg/L and 0.75 μg/L for inorganic arsenic species, MMA and DMA, respectively. On the other hand, using the HG–AAS method after cold trapping and chromatographic separation, Hasegawa *et al.* (1994) were able, for the first time, to separate the trivalent MMA^{III} and DMA^{III} species from the pentavalent DMA and MMA species in natural water following solvent extraction using DDDC.

A system that can separate arsenic species using on-line HPLC prior to their on-line decomposition by microwave digestion, prereduction with L-cysteine and analysis by HG–AAS (HPLC–HG–AAS) has been developed (Lamble & Hill, 1996), and enables the full speciation of arsenobetaine, MMA, DMA, As^{III} and As^{V} in biological samples. A simple modification of the system can determine total arsenic in the sample. A comparable system was used to determine total arsenic and arsenic species in urine specimens, with detection limits of 1.0, 1.6, 1.2 and 4.7 μg/L for As^{III}, As^{V}, MMA and DMA, respectively (Kurttio *et al.*, 1998).

(b)   *Atomic fluorescence spectrometry (AFS)-derived hyphenated techniques*

AFS is an excellent detector of arsenic compounds; it is, in addition, rather simple and inexpensive. AFS has been used to detect arsenic hybrids in the ultraviolet spectral region because of the small background emission produced by the relatively cool hydrogen diffusion flame. The use of cold vapour or HG, together with an intense light source, enables very low detection limits to be reached.

A rapid method for speciation of As^{III}, As^{V}, MMA and DMA (and also arsenobetaine) has been developed based on the rapid separation of the target arsenic species on one or two 3-cm HPLC guard columns, followed by HG–AFS (Le & Ma, 1997). This simple method provides the complete speciation of arsenic present in water and urine samples within 1.5 min with no need for treatment of the sample. Detection limits for the four arsenic species in urine samples are 0.4–0.8 μg/L.

More recently, a solid-phase extraction cartridge linked to HG–AFS was described for speciation of arsenic in water and urine, with detection limits of 0.05 μg/L in water. The disposable cartridges are inexpensive and specific for selective retention of arsenic species, and the method is suitable for routine determination of trace levels of arsenic species in drinking-water to comply with the more stringent environmental regulations (Yalcin & Le, 2001).

HPLC–HG–AFS has led to the speciation in urine of trace levels of trivalent MMA^{III} and DMA^{III} together with the other arsenic species (Gong *et al.*, 2001).

(c)   *Inductively coupled plasma–mass spectrometry (ICP–MS)-derived*
       hyphenated methods

Among the detector methods, ICP–MS is certainly not the cheapest. The advantage of ICP–MS lies in its multi-element capabilities, excellent detection limits and wide linear range. Moreover, low detection limits are not restricted to the hybrid-forming arsenic compounds (Goessler & Kuehnelt, 2002).

Numerous methods have been developed for the speciation of arsenic using the separation power of HPLC combined with the sensitivity of ICP–MS detection (Shibata & Morita, 1989; Le *et al.*, 1998; Londesborough *et al.*, 1999; Chen *et al.*, 1999; Chatterjee *et al.*, 2000; Mandal *et al.*, 2001, 2003).

High-temperature (column temperature at 70 °C) HPLC–ICP–MS was used to determine 13 arsenic and selenium species in urine (Le *et al.*, 1998). The high temperature achieved an improved resolution and faster separation. The speciation of six arsenosugar metabolites in urine can be completed in 19 min at 70 °C compared with 37 min at room temperature.

Londesborough *et al.* (1999) reported an improved HPLC–ICP–MS method for the speciation of eight anionic, cationic or neutral arsenic species (As^{III}, As^V, MMA, DMA, arsenobetaine, arsenocholine, trimethylarsine oxide (TMAO) and tetramethylarsonium ion (TMA)) using a single ion-exchange column, with detection limits of 0.19, 0.52, 0.29, 0.16, 0.16, 0.58, 0.6 and 0.38 µg/L, respectively. In this method, the matrix of biological samples noticeably affects the column efficiency.

High sensitivity was also obtained with the development of the HPLC–ultrasonic nebulizer high-power nitrogen-microwave–ICP–MS method, which could be particularly useful for arsenic speciation in samples with high chloride concentrations since no chloride interference (as $^{40}Ar^{35}Cl$) was observed in urine with a chloride matrix of up to 10 000 mg/L (Chatterjee *et al.*, 2000).

Using optimized HPLC–ICP–MS, Mandal *et al.* (2001) detected the trivalent MMA^{III} and DMA^{III} species for the first time in urine samples, with no prechemical treatment, with detection limits in the range of 0.14–0.33 µg/L.

In conclusion, depending on the specific need, reliable results should be obtainable provided that special care is taken in the preservation and preparation of samples and the method of analysis is chosen carefully.

1.3   Natural occurrence

Arsenic is a metalloid that occurs naturally; it is the component of more than 245 minerals. Examples of arsenic levels in some geological materials are given in Table 4. Arsenic is commonly concentrated in sulfide-bearing mineral deposits, especially those associated with gold mineralization, and it has a strong affinity for pyrite, one of the more ubiquitous minerals in the earth's crust. It is also concentrated in hydrous iron oxides. Arsenic and its compounds are mobile in the environment. Weathering of rocks converts

Table 4. Levels of arsenic in geological materials

Materials	Concentration (mg/kg)	Source
Earth crust total	1–1.8	Matschullat (2000)
Upper crust	1.5–2	Matschullat (2000)
Igneous rocks		
Basic basalt	02–113	Mandal & Suzuki (2002); Smedley & Kinniburgh (2002)
Gabbro, dolorite	0.06–28	Mandal & Suzuki (2002); Smedley & Kinniburgh (2002)
Acidic granite	0.2–13.8	Mandal & Suzuki (2002); Smedley & Kinniburgh (2002)
Sedimentary rocks		
Phosphorites	0.4–188	Smedley & Kinniburgh (2002)
Sandstones	0.6–120	WHO (1981); Mandal & Suzuki (2002)
Shale and argillite	0.3–500	Hale (1981)
Schist and phyllite	0.5–143	Hale (1981)
Carbonates	0.1–20	Matschullat (2000); Mandal & Suzuki (2002)
Coals	0.3–35 000	Smedley & Kinniburgh (2002)
Sulfide minerals		
Pyrite	100–77 000	Smedley & Kinniburgh (2002)
Pyrrhotite	5–100	Boyle & Jonasson (1973)
Chalcopyrite	10–5000	Smedley & Kinniburgh (2002)
Galena	5–10 000	Smedley & Kinniburgh (2002)
Sphalerite	5–17 000	Smedley & Kinniburgh (2002)
Marcasite	20–126 000	Smedley & Kinniburgh (2002)
Oxide minerals		
Haematite	up to 160	Smedley & Kinniburgh (2002)
Iron oxide	up to 2000	Smedley & Kinniburgh (2002)
Iron(III) oxyhydroxide	up to 76 000	Smedley & Kinniburgh (2002)
Sulfate minerals		
Jarosite	34–1000	Smedley & Kinniburgh (2002)

arsenic sulfides to arsenic trioxide, which enters the arsenic cycle as dust or by dissolution in rain, rivers or groundwater. Arsenic can also enter the food chain, causing widespread distribution throughout the plant and animal kingdoms. The occurrence and behaviour of arsenic in the environment have been extensively reviewed (Cullen & Reimer, 1989; Tamaki & Frankenberger, 1992; Matschullat, 2000; Mandal & Suzuki, 2002; Nordstrom, 2002; Smedley & Kinniburgh, 2002).

A limited range of geological environments can result in significant natural elevation of arsenic in water supplies (Nordstrom, 2002). These include: organic rich (black) shales, Holocene alluvial sediments with slow flushing rates, mineralized and mined zones (most often gold deposits), volcanogenic sources, thermal springs, closed basins in arid-to-semi-arid climates, particularly in volcanic regions, and strongly reducing aquifers with low sulfate concentrations.

Depending on prevailing climatic and hydrological conditions, soils and sediments, surface waters, groundwaters and air can become enriched in arsenic where these geological conditions prevail.

1.3.1 *Arsenic speciation in natural materials*

Mineral forms in which arsenic is present in soils are approximately 60% arsenates and 20% sulfides and sulfosalts; the remaining 20% includes arsenides, arsenites, oxides, silicates and elemental arsenic.

These mineral forms are generally weathered to the inorganic water-soluble species, arsenate (As^V) and arsenite (As^{III}), with arsenate dominating under oxidized conditions and arsenite under reduced conditions (Cullen & Reimer, 1989). Under both aerobic and anaerobic conditions, micro-organisms can transform inorganic arsenic into organic forms such as MMA, DMA and volatile TMA. TMA in the air is then rapidly converted into water-soluble species, As^V and TMAO (Pongratz, 1998; Turpeinen *et al.*, 1999, 2002). These compounds can also be degraded by microflora. In certain materials, organic arsenic compounds naturally build up to high concentrations (Mandel & Suzuki, 2002; Smedley & Kinniburgh, 2002).

1.3.2 *Abundance and distribution of arsenic*

(*a*) *Soils and sediments*

Measurements of background arsenic levels in surface soil are all compromised by atmospheric deposition of anthropogenically derived arsenic. Anthropogenic sources to soil include use and resuspension of arsenic-based pesticides, mining, smelting, manufacturing and waste-disposal activities. Shotyk *et al.* (1996) showed that arsenic levels were 20-fold higher in surface horizons of ombrotrophic (rain-fed) peat bogs than in lower horizons. This high level was due to industrially derived inputs of arsenic. Centuries of mining activities can result in an extremely high concentration of arsenic in soils. This is the case in South-West England where arsenic concentrations in some old smelter and/or mine areas range from 24 to 161 000 mg/kg (Farago *et al.*, 1997).

Koljonen (1992) estimated a global average level of arsenic in soils of 5 mg/kg, but concentrations vary considerably among geographical regions. Arsenic concentrations in sediments in lakes, rivers and streams in the USA ranged from 0.1 to 4000 mg/kg. Levels of arsenic in a detailed survey of Finland, which has a low population density and is remote from major centres of pollution, ranged up to 60 mg/kg for the 1164 samples tested (Lahermo *et al.*, 1998). Soils formed from arsenic-enriched geological substrates can have naturally higher levels than the ranges quoted. These ranges must therefore be considered as typical background levels rather than absolute ranges.

Soils formed on top of arsenic-rich bedrocks have elevated levels of this element. Colbourn *et al.* (1975) reported mean arsenic levels of 88 mg/kg (range, 24–250 mg/kg; *n* = 18) in soils formed naturally from parent material consisting of metamorphic aureole around a granitic intrusion. The Strassegg area in Gasen (Styria, Austria) has extensive arsenopyrite (FeAsS) mineralization, with the ore body running close to the surface (Geiszinger *et al.*, 2002). The soils formed on top of this ore vein are enriched in arsenic, with levels ranging from 700 to 4000 mg/kg, and are used for agronomic cultivation.

Soils formed in and around ancient and modern hot springs with elevated arsenic in geothermal fluids have naturally elevated levels of arsenic due to enrichment of the parent material of the soil (Ballantyne & Moore, 1988). The ancient hot-spring system at Rhynie, north-eastern Scotland, has cherts with arsenic levels ranging from 15 to 300 mg/kg (Rice *et al.*, 1995). Sinter from active hot springs in the Taupo Volcanic Zone, New Zealand, have arsenic levels ranging from below detection limits to 1646 mg/kg (McKenzie *et al.*, 2001). An area of at least 10 km^2 in St Elizabeth, Jamaica, has a geochemical anomaly, whereby arsenic concentrations in soil reach 400 mg/kg (Lalor *et al.*, 1999). The anomalous values may result from an ancient hot-spring environment responsible for the introduction and deposition of pyrite and arsenopyrite in the limestone bedrock, which were subsequently oxidized and weathered, leading to arsenic-rich soils.

Sediment levels of arsenic in the Waikato River, New Zealand, ranged from 7.9 to 1520 mg/kg dry wt, resulting in high levels of arsenic in sediment living biota, such as the freshwater mussel, *Hyridella menziesi* (Hickey *et al.*, 1995).

In a number of delta environments in South-East Asia, deep fluvial and deltaic Pleistocene-Holocene sediments have accumulated (up to 10 km thick in Bangladesh) (Nickson *et al.*, 2000). During glaciation, river levels were 100 m lower than in inter-glacial times, and at this time of low sea level, the sediments were flushed and oxidized, leading to iron (Fe^{III}) oxyhydroxide precipitation on sediment surfaces. These sedimentary iron oxyhydroxides scavenge arsenic, with arsenic levels reaching up to 517 mg/kg in FeOOH phases (Nickson *et al.*, 2000). Under reducing conditions caused by microbial metabolism of sedimentary organic matter (present at up to 6% as C), in which sulfate levels are low, insoluble Fe^{III} is converted to soluble Fe^{II}, leading to the mobilization of arsenic from the dissolved FeOOH phase. Although traces of arsenic-rich pyrites are found in the sediments, they are present in quantities that are too small for pyrite oxidation to contribute significantly to arsenic in groundwaters.

Water percolating from hot-spring systems into the surrounding soil or sediment also causes a rise in arsenic concentrations (Langner *et al.*, 2001; Koch *et al.*, 1999).

The Antofagasta Region, northern Chile, is characterized by volcanism (Queirolo *et al.*, 2000a). High levels of arsenic are found in soils and river sediments in this region (Caceres *et al.*, 1992), and crops (maize and potato) grown on these soils have high levels of arsenic, reaching 2 mg/kg in maize (Queirolo *et al.*, 2000b).

Arsenic concentrations in mineralized zones rich in arsenic are further elevated, often severely, by mineral extraction and processing (Smedley & Kinniburgh, 2002).

(*b*)   *Groundwaters*

Under natural conditions, the greatest range and the highest concentrations of arsenic are found in groundwater as a result of the strong influence of the water–rock interactions and the favourable physical and geochemical conditions in aquifers for the mobilization and accumulation of arsenic. Arsenic is particularly mobile at pH values typically found in groundwater (pH, 6.5–8.5) under both oxidizing and reducing conditions.

Background concentrations of arsenic in groundwater in most countries are less than 10 µg/L and sometimes substantially lower. However, values quoted in the literature show a very wide range, from < 0.5 to 5000 µg/L. Most high levels of arsenic in groundwater are the result of natural occurrences of arsenic. Cases of arsenic pollution caused by mining are numerous but tend to be localized.

Arsenic can occur in the environment in several oxidation states (–3, 0, +3 and +5) but, in natural waters, is mostly found in inorganic forms as oxyanions of trivalent arsenite (As^{III}) or pentavalent arsenate (As^V). Redox potential (Eh) and pH are the most important factors controlling arsenic speciation. Under oxidizing conditions, arsenate is dominant, as the $H_2AsO_4^-$ form at low pH (less than approximately 6.9), or as the $HAsO_4^{2-}$ form at higher pH. Under reducing conditions at pH less than approximately 9.2, the uncharged arsenite species H_3AsO_3 predominates (Smedley *et al.*, 2002).

In two recent reviews, Smedley and Kinniburgh (2002) and Smedley *et al.* (2002) focused extensively on the factors that control arsenic concentration in groundwater.

In relatively pristine habitats where anthropogenic activity can be excluded as a contributor to arsenic levels in aquifers, Lahermo *et al.* (1998) found that arsenic levels in groundwaters in Finland reached up to 1040 µg/L, with a median of 0.65 µg/L (*n* = 472). The highest levels of arsenic were found in groundwaters from wells drilled in Precambrian bedrock.

In an extensive groundwater survey in the USA, Welch *et al.* (2000) reported that approximately half of the 30 000 samples analysed had naturally occurring arsenic levels ≤ 1 µg/L, with about 10% exceeding 10 µg/L. Geothermal water and high evaporation rates are associated with arsenic concentrations ≥ 10 µg/L in ground- and surface waters.

There are three major types of natural geological condition giving rise to high levels of arsenic in groundwaters:

(i)   aquifers composed of rocks or sediments enriched with arsenic-containing minerals of geogenic origin, such as sulfide mineralization;

(ii)   aquifers containing sediments coated with iron oxyhydroxide-(FeOOH) phases enriched in arsenic through hydrological action, where arsenic is mobilized into porewater by reducing conditions;

(iii)   aquifers enriched in arsenic through high rates of evaporation in arid areas, leading to increased mineral concentration in groundwaters; the arsenic is mobile in such aquifers because of the high pH (> 8.5) caused by concentration of alkali and alkali earth metals in solution.

Geochemical conditions similar to the alluvial sediments in Bangladesh exist in the Red River alluvial tract in the city of Hanoi, Viet Nam, where FeOOH reduction is thought to have led to the high arsenic levels recorded in groundwaters (Berg *et al.*, 2001). Smedley and Kinniburgh (2002) outline that the reducing conditions observed in Bangladesh/West Bengal and Viet Nam aquifers are similar to those in the regions of Taiwan, China, northern China and Hungary that suffer from high levels of arsenic in groundwaters.

Smedley *et al.* (2002) studied the geochemistry of arsenic in groundwaters from Quaternary loess aquifers, which were high in arsenic, in an area thought to spread over 10^6 km^2 in La Pampa province, central Argentina. Dissolved arsenic ranged from 4 to 5300 µg/L, with 73% of samples exceeding 50 µg/L. The conclusions drawn for La Pampa province may be applicable elsewhere in determining which regions are vulnerable to arsenic and related water-quality problems: "Under oxidising conditions, vulnerable aquifers potentially occur where several important criteria coincide: semi-arid climatic conditions with limited recharge where high-pH groundwater can be generated; young (Quaternary) sediments or volcanic sediments; and slow groundwater-flow conditions. Such aquifers are likely to have been poorly flushed over the geologically-short timescale since deposition and hence will have had little opportunity for removal of trace elements such as arsenic from the aquifer." Similar conditions exist in the Lagunera and Sonora regions of Mexico and in the Atacama Desert, Chile (Smedley & Kinniburgh, 2002).

(c)   Surface waters

Matschullat (2000) collated measurements of arsenic in surface waters. Levels of arsenic dissolved in uncontaminated stream waters ranged from 0.1 to 1.7 µg/L, and those in seawaters were 1.5–1.7 µg/L. Concentrations in open seawater show little variation from the value of 1.5 µg/L (Smedley & Kinniburgh, 2002).

Arsenic in surface stream waters in Finland, which could be considered a pristine environment because of its low population density and remote geographical location, ranged from 0.06 to 1.6 µg/L (median, 0.36 µg/L; n = 1157) (Lahermo *et al.* 1998). These levels correlated well with arsenic levels in glacial till, with the highest stream water levels occurring in catchments with metamorphic, volcanic and sedimentary geologies. Levels in the more geographically remote part of Finland were lower than those in the south, which is nearer to continental Europe. Arsenic levels in Finnish water were lower than those for continental Europe, again emphasizing the pristine nature of the Finnish environment.

The Ciwidey River, West Java, drains a catchment dominated by the Quaternary volcano Patuha, which contains an acid crater lake (pH < 1) (Sriwana *et al.*, 1998). Arsenic in the crater lake was recorded to be 279 µg/L, with the stream draining this lake having levels of 57 µg/L. In the tributary river of the stream, levels dropped to below 1 µg/L. In a crater lake with naturally elevated levels of arsenic, such as Lake Xolotlan in Nicaragua, mean arsenic concentrations ranged from 10.23 to 30.13 µg/L (Lacayo *et al.* 1992).

Takatsu and Uchiumi (1998) studied water from Lake Usoriko, Japan, which is acidified by hot springs. The sediments of this lake contained 1.6% by mass of arsenic, with arsenic levels in the open lake waters ranging from 10 to 450 µg/L.

Levels of arsenic in drinking-water extracted from the Waikato River, New Zealand, for the city of Hamilton averaged 32 µg/L. Arsenic concentrations appear to follow a regular seasonal variation, being approximately 10–25 µg/L higher in the summer months, and fall to 6 µg/L after water treatment (McLaren & Kim, 1995). The elevated levels of arsenic in the Waikato river are of natural origin, as its catchment is the volcanic region of the Central Plains (Hickey *et al.*, 1995).

Natural surface waters in the Antofagasta region of Chile, originating from springs, have very high levels of arsenic because of zones mineralization associated with volcanic activity (eruptions, vents, geysers and thermal springs). Surface water is used as drinking-water and to irrigate crops (Queirolo *et al.*, 2000a,b). Arsenic levels reached 3000 µg/L in rivers and canals in this region, with many rivers routinely having levels over 100 µg/L.

In an area with similar volcanic activity in the Salta Province, Argentina, high levels of arsenic have been recorded in thermal springs, tap-water and river water (Vahter *et al.*, 1995).

High levels of arsenic have been recorded in rivers in arid areas of Chile and Argentina where surface water is dominated by base-flow (whereby groundwater flows into the river from surrounding rock) (Caceres *et al.*, 1992; Lerda & Prosperi, 1996). Caceres *et al.* (1992) found concentrations in surface water up to 22 mg/L. The high degree of evaporation that occurs in these regions concentrates the arsenic leached from weathered rocks. Such surface waters have high pH, due again to high rates of evaporation that lead to concentration of alkaline and alkaline earth cations leached from the rocks.

(d)   Air

Concentrations of arsenic in ambient air in remote locations range from < 1 to 3 ng/m^3, but concentrations in cities may range up to 100 ng/m^3. Arsenic in ambient air is usually a mixture of arsenite and arsenate, with organic species being of negligible importance except in areas of arsenical pesticide application or other industrial activity (WHO, 2001). Sources of arsenic to air include use and resuspension of arsenic-based pesticides, mining, smelting, manufacturing and waste-disposal activities. Arsenic may be introduced into the atmosphere directly from these processes, or it may be derived from sediment and soil particles being entrained into the atmosphere or the production of volatile arsenic metabolites, such as arsines, from soils (Woolson, 1977; Turpeinen *et al.*, 2002). Defining what constitutes natural levels is, therefore, difficult.

(e)   Other

Arsenic has been detected in rainwater at concentrations ranging from < 0.005 to 45 µg/L, with higher levels occurring in contaminated areas (WHO, 2001).

Arsenic compounds are abundant in certain seafoods at concentrations as high as several hundred milligrams per kilogram. Although marine animals contain many arsenic compounds, most species contain arsenobetaine as the major arsenical. Arsenobetaine is not metabolized by humans and is believed to have low or negligible toxicity. Inorganic arsenic and arsenosugars can, however, be present in some marine algae, seaweeds, oysters, mussels and clams (reviewed by Francesconi & Kuehnelt, 2002).

Dimethylarsinate is often the major arsenical constituent of species of fungi. Arsenite and arsenate are also commonly found in fungi (Francesconi & Kuehnelt, 2002).

Inorganic arsenic species are dominant in the chemistry of arsenic in terrestrial plants (Francesconi & Kuehnelt, 2002) and, although less studied, the concentration of arsenic in wheat and vegetables grown in countries highly contaminated with arsenic could be relevant to human health. Most of the vegetables cultivated in the Antofagasta Region (northern Chile), which is characterized by volcanic events (eruptions, thermal springs), are found at local markets of a population of approximately 4000 people. In this region, very high arsenic contents have been reported in Socaire and Talabre (1850 µg/kg in corn and 860 µg/kg in potatoes, including potato skins, respectively), two towns situated close to the Lascar volcano (Queirolo et al., 2000b). These values exceed the national standard for arsenic (500 µg/kg) by approximately 400% and 180%, respectively.

In Bangladesh, contamination of agricultural soils from long-term irrigation with arsenic-contaminated groundwater led to phyto-accumulation in food crops. Various vegetables harvested in Samta village in the Jessore district have been reported to contain high concentrations of arsenic (range, 306–489 µg/kg) (Alam et al., 2003). In West Bengal (India), high arsenic contents have also been reported in many vegetables and spices, especially in the skin of most vegetables, as a result of the dependence of the agricultural system on groundwater (Roychowdhury et al., 2002, 2003).

Moreover, high concentrations of arsenic have been reported in fruit, vegetables, grain and meat in regions contaminated by anthropogenic pollution; this is the case in the Moscow region (Russia), which has been shown to be contaminated by fertilizer industry plants (Zakharova et al., 2002). High levels of arsenic have also been reported in plants, vegetables and cow's milk, as a consequence of heavy contamination of soils, surface and groundwaters by arsenic attributed to industrial sources (veterinary chemicals, pharmaceuticals, pesticide industries) in the area of Patancheru, Andhra Pradesh (India) (Sekhar et al., 2003).

Interestingly, rare plants are able to accumulate exceedingly high concentrations of arsenic (in the order of 1% dry mass). Brake fern (*Pteris vittata*) in particular is extremely efficient at extracting arsenic from soils and translocating it into its fronds. Arsenic concentrations in fern fronds, growing in soil spiked with 1500 mg/kg arsenic, increased from 29.4 to 150 861 mg/kg within 2 weeks. Since it acts as an arsenic hyperaccumulator, brake fern could be used in the remediation of arsenic-contaminated soils (Ma et al., 2001).

1.4   Human exposure

The natural and anthropogenic occurrence of arsenic in drinking-water has been recognized as a major public health issue in several regions of the world over the past two or three decades. Areas affected by arsenic span the globe, and significant exposures have been identified in Bangladesh, India, Taiwan, China, Mexico, Argentina, Chile and the

USA. Table 5 summarizes the geological characteristics of the regions of the world with naturally elevated levels of arsenic in the drinking-water.

Recent reviews have outlined the worldwide problem of arsenic in drinking-water (WHO, 2001; Mandal & Suzuki, 2002; Nordstrom, 2002; Smedley & Kinniburgh, 2002; Chakraborti *et al.*, 2003b).

1.4.1   *Exposure in Bangladesh*

In terms of the population exposed, the problem of arsenic contamination in much of southern and eastern Bangladesh is the most serious in the world, and occurs in ground-water from the alluvial and deltaic sediments that make up much of the area. In addition, it is complicated by large variability in arsenic levels at both local and regional scales.

In Bangladesh, tubewells began to be used for drinking-water in the 1970s to control the problem of gastrointestinal disease linked to contamination of shallow wells and surface waters. In the 1990s, it was discovered that the water from many of these wells was contaminated with arsenic. Since then, extensive research has been carried out to characterize the extent of the problem. Figure 1 shows the districts in Bangladesh affected by arsenic and Table 6 gives an overall picture of the database. Table 7 shows the distribution of concentrations of arsenic in hand tubewells, and Table 8 summarizes the levels of arsenic measured in biological samples.

The level of contamination with arsenic of tubewells in Bangladesh exceeded both the World Health Organization guideline of 10 µg/L and the Bangladesh permissible limit of 50 µg/L (Dhar *et al.*, 1997; Smith *et al.*, 2000a; Kinniburgh & Smedley, 2001: Alam *et al.*, 2002).

A survey of 27 districts in Bangladesh up to January 1997 analysed over 3000 water samples and revealed that 38% of them contained more than 50 µg/L arsenic (Dhar *et al.*, 1997). In another survey examining 294 tubewells, 85 samples (29%) were contaminated by arsenic at levels above 50 µg/L (Ahmad *et al.*, 1997). Between September 1996 and June 1997, all functioning wells ($n = 265$) in the village of Samta in the Jessore District were tested for arsenic (Biswas *et al.*, 1998). Approximately 91% of the wells contained arsenic at levels higher than 50 µg/L. Furthermore, 600 people were examined clinically, and a few hundred hair, nail and urine samples were tested using flow injection HG–AAS. The data obtained showed that 99% of urine samples and 98% of nail samples of the population studied in Samta village contained levels of arsenic above normal and 78% of hair samples above toxic levels. The arsenic problem of Bangladesh became highlighted when an international conference was held in Dhaka, Bangladesh, in 1998 (Dhaka Community Hospital Trust and School of Environmental Studies, 1998).

By March 1998, it was reported that 4196 of 9024 wells in Bangladesh tested for arsenic contained levels higher than 50 µg/L and 884 wells had levels higher than 500 µg/L (Mandal *et al.*, 1999). A Rapid Action Programme (RAP) was performed by field kit in a sample of 500 villages with a total population of 469 424. Approximately 62% of the 32 651 tubewells sampled had levels of arsenic above 100 µg/L (Quamruzzaman *et al.*, 1999).

Table 5. Regions of the world with naturally elevated levels of arsenic in groundwater

Country/region	Affected area (km²)	Potentially exposed population	Arsenic concentration (µg/L)	Environmental conditions	Reference
Bangladesh	118 849	$\sim 3 \times 10^7$	<0.5–2500	Hollocene alluvial/deltaic sediments; abundance of organic matter; strongly reducing, neutral pH, high alkalinity, slow groundwater flow rates	Chakraborti et al. (2002); Smedley & Kinniburgh (2002)
India/West Bengal	38 865	6×10^6	<10–3200	Same as Bangladesh	Chakraborti et al. (2002); Smedley & Kinniburgh (2002)
Viet Nam				Pleistocene and Holocene sediments; strongly reducing conditions	Berg et al. (2001)
China/Taiwan	4 000	$\sim 10^5$	10–1820	Coastal zones, sediments, including black shales; strongly reducing, artesian conditions, some groundwaters contain humic acids	Smedley & Kinniburgh (2002)
China/Xinjiang, Shanxi	38 000	~ 500	40–750	Holocene alluvial plain; reducing	Smedley & Kinniburgh (2002); Cao (1996)
Thailand	100	1.5×10^4	1–<5000	Dredge quaternary alluvium; oxidation of disseminated arsenopyrite due to mining	Smedley & Kinniburgh (2002)
Mongolia/Inner Mongolia	4 300	$\sim 10^5$	<1–2400	Holocene alluvial and lacustrine sediments; strongly reducing, neutral pH, high alkalinity, some groundwaters contain humic acids	Cao (1996); Smedley & Kinniburgh (2002); Sun et al. (2001)

Table 5 (contd)

Country/region	Affected area (km^2)	Potentially exposed population	Arsenic concentration (µg/L)	Environmental conditions	Reference
Argentina/ Chaco- Pampean Plain	10^6	2 × 10^6	< 1–7550	Holocene and earlier loess with rhyolitic volcanic ash; oxidizing, neutral to high pH, high alkalinity; groundwaters often saline	Nordstrom (2002); Smedley & Kinniburgh (2002)
Northern Chile/ Antofagasta	35 000	5 × 10^5	100–1000	Quaternary volcanogenic sediments; generally oxidizing, arid conditions, high salinity	Queirolo et al. (2000a); Smedley & Kinniburgh (2002)
Bolivia		5 × 10^4		Same as Argentina and Northern Chile	Nordstrom (2002)
Mexico	32 000	4 × 10^5	8–620	Volcanic sediments; oxidizing, neutral to high pH	Smedley & Kinniburgh (2002)
Germany/ Bavaria	2 500		< 10–150	Mineralized sandstone	Nordstrom (2002)
Hungary, Romania/ Danube Basin	110 000	4 × 10^5		Quaternary alluvial plain; reducing conditions, some high in humic acid	Smedley & Kinniburgh (2002)
Spain		> 5 × 10^4	< 1–100	Mineralization; alluvial sediments	Nordstrom (2002)
Greece		1.5 × 10^5		Mineralization; thermal springs; mining	Nordstrom (2002)
Ghana		< 1 × 10^5	< 1–175	Sulfide mineralization, particularly arsenopyrite; gold mining	Nordstrom (2002)

Table 5 (contd)

Country/region	Affected area (km²)	Potentially exposed population	Arsenic concentration (µg/L)	Environmental conditions	Reference
Canada/Moira Lake, Ontario	100		50–3000	Mine tailing; ore mining	Smedley & Kinniburgh (2002)
Canada/British Columbia	50		0.5–580	Sulfide mineralization in volcanic rocks; neutral to high pH groundwater	Smedley & Kinniburgh (2002)
USA/Arizona	200 000		< 1300	Alluvial basins, some evaporites; oxidizing, high pH	Smedley & Kinniburgh (2002)
USA/California	5 000		< 1–2600	Holocene and older basin-fill sediments; internally drained basin, mixed redox conditions, high salinity	Smedley & Kinniburgh (2002)
USA/Nevada	1 300		< 2600	Holocene mixed aeolian, alluvial and lacustrine sediments; mainly reducing, some high pH, some with high salinity due to evaporation	Smedley & Kinniburgh (2002)

Figure 1. Degree of arsenic contamination in 64 districts in Bangladesh

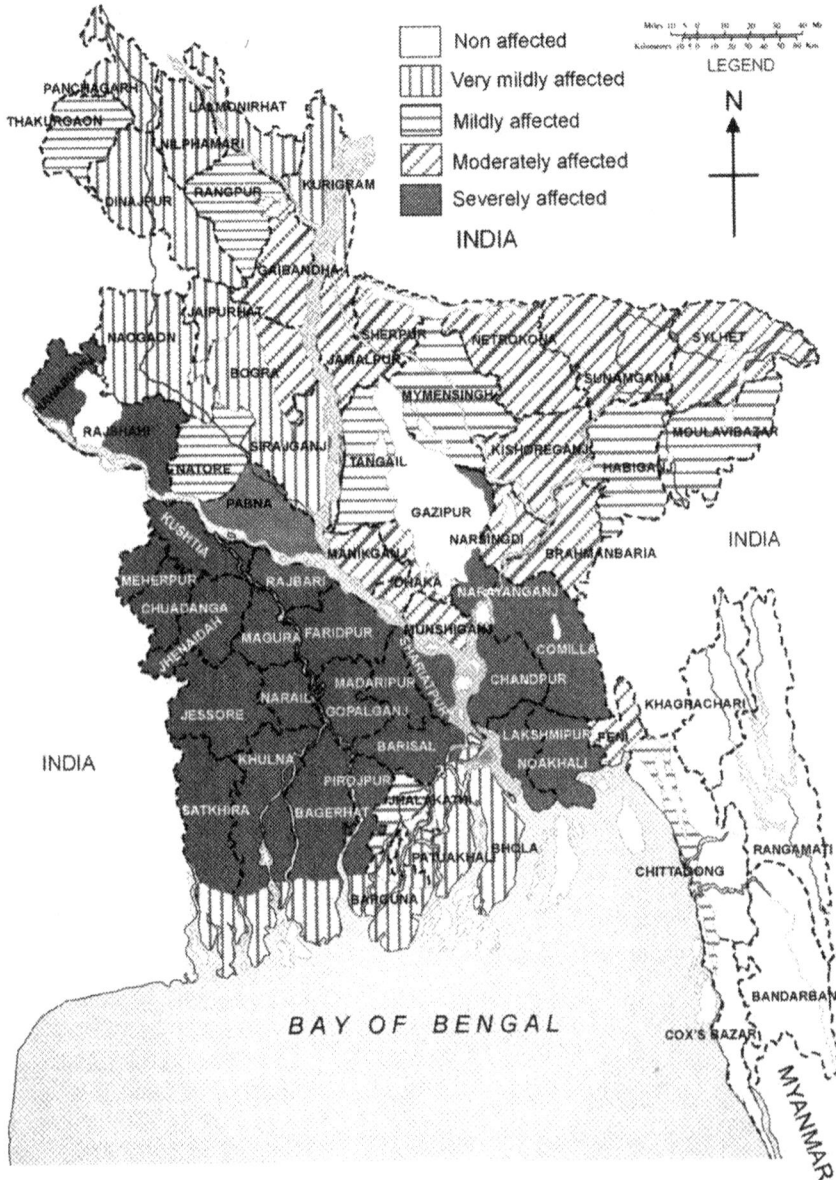

From Chakraborti *et al.* (2002)

Table 6. Status of contamination of groundwater by arsenic in Bangladesh

	Bangladesh
Total area (km^2)	148 393
Population (millions)	120
Total number of districts	64
Total number of water samples analysed	34000
Samples containing > 10 µg/L arsenic (%)	56.35
Samples containing > 50 µg/L arsenic (%)	37.38
Number of districts affected by arsenic (> 50 µg/L)	50
Population of districts affected by arsenic (millions)	104.9
Area of districts affected by arsenic (km^2)	118 849
Number of villages affected by arsenic (arsenic in drinking-water > 50 µg/L)	2000
Number of people drinking arsenic-contaminated water > 50 µg/L (millions)	25

From Chakraborti *et al.* (2002)

Table 7. Distribution of arsenic concentrations in water samples from hand tubewells

Total no. of water samples analysed	Arsenic concentration range (µg/L)							
	< 10	10–50	51–99	100–299	300–499	500–699	700–1000	> 1000
34 000	14 991	6429	2949	5812	2174	894	479	272
	44.1%	18.9%	8.7%	17.1%	6.4%	2.6%	1.4%	0.8%

From Rahman *et al.* (2001)

In continuing surveys of 42 districts affected by arsenic in Bangladesh, Chowdhury *et al.* (2000a,b) reported the analysis of 10 991 water samples of which 59% contained arsenic levels above 50 µg/L.

Of the 34 000 drinking-water samples collected in Bangladesh up to August 2001, 272 contained ≥ 1000 µg/L arsenic (Table 6; Chakraborti *et al.*, 2002). The highest concentration of arsenic measured in drinking-water in Bangladesh was 4700 µg/L. In the Chiladi village of Senbagh Thana in the Noakhali district, 100% of tubewell-water samples contained arsenic concentrations ≥ 50 µg/L, 94% contained ≥ 300 µg/L and 28% contained ≥ 1000 µg/L.

Table 8. Concentrations of arsenic in samples of hair, nails, urine (metabolites) and skin scale collected from the areas in Bangladesh affected by arsenic

Parameter	Arsenic in hair[a] (µg/kg)	Arsenic in nails[b] (µg/kg)	Arsenic in urine[c] (µg/L)	Arsenic in skin scale[d] (µg/kg)
No. of observations	4 386	4 321	1 084	705
Mean	3 390	8 570	280	5 730
Median	2 340	6 400	116	4 800
Minimum	280	260	24	600
Maximum	28 060	79 490	3 086	53 390
Standard deviation	3 330	7 630	410	9 790
% of samples having arsenic above normal	83.15	93.77	95.11	–

From Rahman *et al.* (2001)

[a] Normal levels of arsenic in hair range from 80 to 250 µg/kg; 1000 µg/kg indicates toxicity.

[b] Normal levels of arsenic in nails range from 430 to 1080 µg/kg

[c] Normal levels of arsenic in urine range from 5 to 50 µg/1.5 L (per day)

[d] Normal value for skin scale arsenic not defined

Thousands of hair, nail and urine samples from people living in villages affected by arsenic have been analysed (Table 8). Approximately 90% of children under 11 years of age living in the affected areas show levels of arsenic in hair and nails above the normal level (Rahman *et al.*, 2001).

A comparative study reported analyses of arsenic species in urine samples ($n = 42$) from one affected village of Madaripur district, where the average concentration of arsenic in drinking-water was 376 µg/L, and a non-affected village ($n = 27$), where the concentration of arsenic in drinking-water is known to be below 3 µg/L (Chowdhury *et al.*, 2003). The average urinary levels of arsenic of children were higher than those of adults. The ratios of MMA to inorganic arsenic and of DMA to MMA were 0.93 and 4.11 in adults and 0.74 and 8.15 in children, respectively.

Chakraborti *et al.* (1999a) reported arsenic concentrations in hand tubewells from 100 to 415 m in depth in all geographical regions in Bangladesh. The report indicated that 99% of the tubewells analysed that were deeper than 300 m had an arsenic concentration below 50 µg/L. Understanding the mechanism of arsenic release to groundwater in Bangladesh should help to provide guidance for the placement of safe new water wells (Nickson *et al.,* 1998, 2000).

1.4.2   *Exposure in India*

(*a*)   *Contamination by arsenic of groundwater in northern India*

A preliminary study was reported in 1976 on arsenic in dug wells, hand pumps and spring water from Chandigarh and different villages of the Punjab, Haryana and Himachal Pradesh in northern India (Datta & Kaul, 1976). A value as high as 545 µg/L arsenic was obtained in one water sample from a hand pump. Datta (1976) further reported high arsenic content in the liver of five of nine patients with non-cirrhotic portal hypertension who had been drinking arsenic-contaminated water. To date no further information on arsenic poisoning in northern India is available.

(*b*)   *Contamination by arsenic of groundwater in West Bengal*

Since 1984, extensive research in West Bengal has revealed that this region has one of the most serious problems with groundwater contamination by arsenic in wells used for drinking-water. Figure 2 shows the districts in West Bengal affected by arsenic and Table 9 gives an overall picture of the database and the extent of the problem. Table 10 shows the distribution of concentrations of arsenic in hand tubewells in areas of West Bengal, and Table 11 summarizes the levels of arsenic measured in biological samples.

Contamination of groundwater by arsenic was first detected in the state of West Bengal, India, in 1983 (Garai *et al.*, 1984). Sixteen people whose drinking-water came from two hand tubewells in one village in the district of 24-Parganas were identified as having arsenical skin lesions. Arsenic concentrations in these tubewells were 1250 and

Table 9. Status of contamination of groundwater by arsenic in West Bengal, India

	West Bengal
Total area (km^2)	89 193
Population (millions; according to 1991 Census)	68
Total number of districts	18
Total number of water samples analysed	105 000
Samples containing > 10 µg/L arsenic (%)	51
Samples containing > 50 µg/L arsenic (%)	25
Number of districts affected by arsenic (> 50 µg/L)	9
Population of districts affected by arsenic (millions)	42.7
Area of districts affected by arsenic (km^2)	38 865
Number of blocks/police stations affected by arsenic	74
Number of villages (approx.) affected by arsenic (arsenic in groundwater > 50 µg/L)	2700
Number of people drinking arsenic-contaminated water > 50 µg/L (millions)	6

From Chakraborti *et al* (2002)

Figure 2. Areas of West Bengal in which drinking-water is contaminated with arsenic

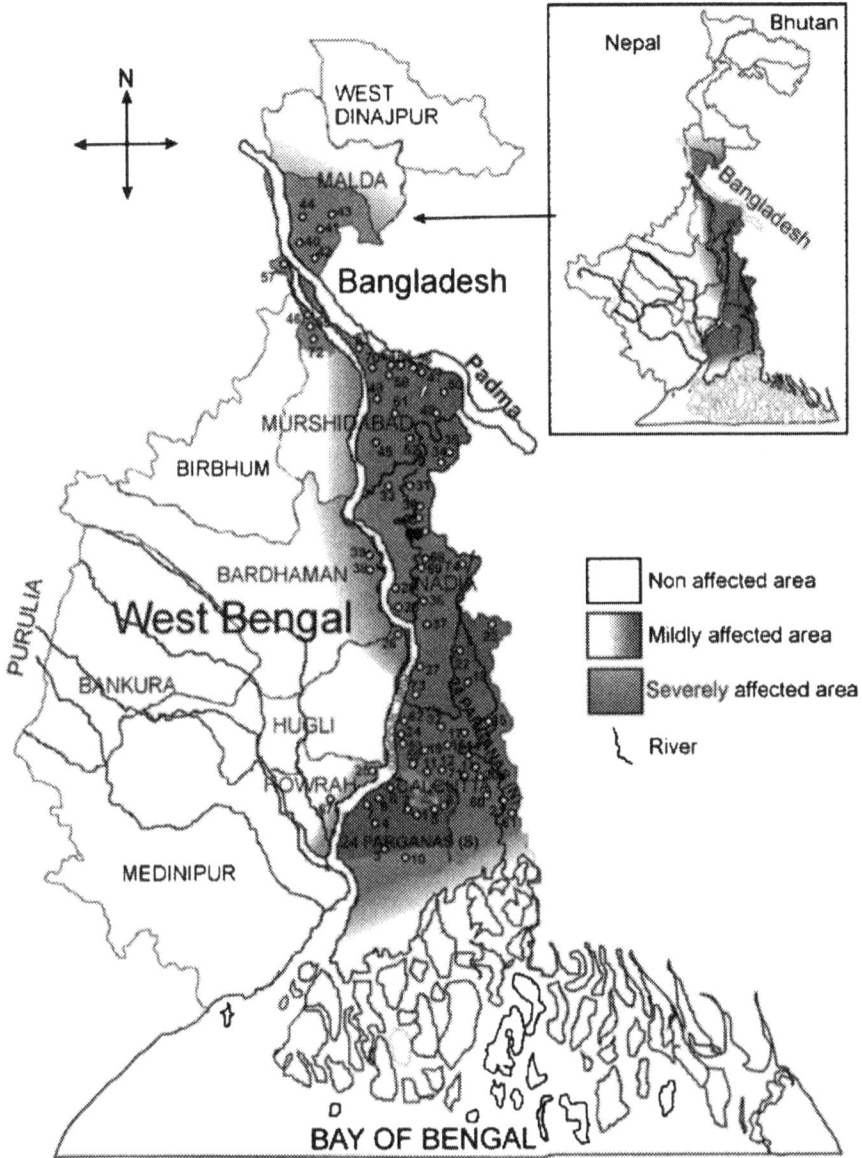

From Chakraborti *et al.* (2002)

Table 10. Concentrations of arsenic in water samples from hand tubewells in West Bengal, India

No. of water samples analysed	Arsenic concentration range (µg/L)							
	< 10	10–50	51–99	100–299	300–499	500–699	700–1000	> 1000
101 934	49 310	27 309	10 005	11 782	2354	724	334	116
	48.4%	26.8%	9.8%	11.6%	2.3%	0.7%	0.3%	0.1%

From Rahman *et al.* (2001)

Table 11. Concentrations of arsenic in samples of hair, nails, urine (metabolites) and skin scale collected from the areas in West Bengal (India) affected by arsenic

Parameters	Arsenic in hair[a] (µg/kg)	Arsenic in nails[b] (µg/kg)	Arsenic in urine[c] (µg/L)	Arsenic in skin scale[d] (µg/L)
No. of observations	7 135	7 381	9 795	165
Mean	1 480	4 560	180	6 820
Median	1 320	3 870	115	4 460
Minimum	180	380	10	1 280
Maximum	20 340	44 890	3 147	15 510
Standard deviation	1 550	3 980	268	4 750
% of samples having arsenic above normal	57	83	89	–

From Rahman *et al.* (2001)

[a] Normal levels of arsenic in hair range from 80 to 250 µg/kg; 1000 µg/kg indicates toxicity.

[b] Normal levels of arsenic in nails range from 430 to 1080 µg/kg

[c] Normal excretion of arsenic in urine ranges from 5 to 40 µg/1.5 L (per day)

[d] Normal value for skin scale arsenic not defined

700 µg/L. Saha and Poddar (1986) reported that 36 villages from 18 police stations/blocks of six districts were affected in 24-Parganas, Murshidabad, Nadia, Barddhaman, Midnapur and Maldah. Water samples from 207 hand tubewells were analysed and 105 (50.7%) showed arsenic concentrations above 50 µg/L; the highest concentration recorded was 568 µg/L. Analysis of arsenic in hair, nails and skin-scale from people in the affected villages confirmed exposure to arsenic.

In 1987, an epidemiological survey in six villages of three districts (24-Parganas, Barddhaman and Nadia) revealed 197 patients with arsenical dermatosis in 48 families

(Chakraborty & Saha, 1987). Of 71 water samples collected from tubewells of the affected villages, the concentration of arsenic in 55 (77.5%) was higher than the permissible limit (50 µg/L) for arsenic in drinking-water in India. The mean arsenic concentration in 31 water samples collected from tubewells of affected families was 640 µg/L and that in 40 water samples collected from tubewells of unaffected families was 210 µg/L. Another epidemiological investigation (Guha Mazumder et al., 1988) in a village in 24-Parganas also found evidence of effects of arsenic in 62 (92.5%) of 67 members of families who drank contaminated tubewell-water (level of arsenic, 200–2000 µg/L). In contrast, only six (6.25%) of 96 persons from the same area who drank water with a level of arsenic < 50 µg/L showed any effects.

In 1991, a report from the government of West Bengal (Steering Committee, Arsenic Investigation Project, 1991) concluded that water of the intermediate aquifer in areas of West Bengal was polluted with arsenic. Neither the shallow (first) nor the deep (third) aquifers had reported arsenic contamination. The sand grains in the arsenic-contaminated aquifer were generally coated with iron and material rich in arsenic.

In October 1994, a committee constituted by the government of West Bengal (Committee Constituted by Government of West Bengal, 1994) reported arsenic contamination in 41 blocks in six districts of West Bengal. The committee analysed about 1200 water samples from these six districts for arsenic and other common water-quality parameters, and the highest concentration of arsenic reported was 3200 µg/L.

The expanding database on the problem of arsenic contamination in West Bengal has been documented in a continuing series of publications. By December 1994, it was reported that 312 villages from 37 blocks/police stations in six districts in West Bengal were affected by contamination of groundwater with arsenic. From extrapolation of the data, it was predicted that more than 800 000 people were drinking arsenic-contaminated water from these districts, and based on the analysis of several thousand water samples, average arsenic concentrations in the wells sampled ranged from 193 to 737 µg/L (Das et al., 1994; Chatterjee et al., 1995). The highest arsenic concentration of 3700 µg/L was found in a hand tubewell from a village in South 24-Parganas district. Groundwater and urine samples from affected villages were also analysed for arsenite, arsenate, MMA and DMA. Groundwater contained arsenate and arsenite in a ratio of approximately 1:1. In urine, DMA and MMA were the predominant species, together with some arsenite and arsenate. Das et al. (1995) reported high arsenic levels in the hair, nails, urine, skin-scale and a few liver tissues (biopsy) of people from arsenic-affected villages who had arsenical skin lesions.

Based on the analysis of 20 000 water samples from areas of West Bengal, Mandal et al. (1996) reported that seven districts (North 24-Parganas, South 24-Parganas, Nadia, Barddhaman, Murshidabad, Maldah, Hugli) were affected by arsenic. Approximately 45% of these samples had arsenic concentrations above 50 µg/L, and the average concentration was approximately 200 µg/L.

Groundwater contamination was reported in 985 villages from 69 police stations/ blocks in nine districts of West Bengal on the basis of analyses of 58 166 water samples.

The nine districts were Maldah, Murshidabad, Barddhaman, Hugli, Howrah, Nadia, North 24-Parganas, South 24-Parganas and Calcutta. After extrapolation of data from the water analyses and screening villagers for arsenical skin lesions, it was estimated that about 5 million people were drinking-water contaminated with levels of arsenic above 50 µg/L. The total population in the nine districts of West Bengal affected by arsenic is about 43 million (Chowdhury et al., 2000a,b).

On the basis of an analysis of 101 394 hand tubewells and approximately 25 000 biological samples, and screening of 86 000 persons in affected villages of West Bengal, Rahman et al. (2001) reported that 2600 villages were affected by arsenic in groundwater at levels of > 50 µg/L and that approximately 6 million people drank water contaminated with arsenic at levels above 50 µg/L. Mandal et al. (2001) identified DMA[III] and MMA[III] for the first time in urine from the affected areas of West Bengal.

Roychowdhury et al. (2002) reported total arsenic in food composites collected from a few arsenic-affected villages in Murshidabad district, West Bengal, where arsenic-contaminated groundwater was used for agricultural irrigation. The report showed average daily dietary intake of arsenic from foodstuffs for adults and children of 180 and 96.5 µg, respectively.

Rahman et al. (2003) studied North 24-Parganas, one of the nine affected districts of West Bengal, for 7 years. On the basis of analyses of 48 030 water samples and 21 000 hair, nail and urine samples, and screening of 33 000 people in North 24-Parganas, it was estimated that about 2 million and 1 million people are drinking water contaminated with arsenic at levels above 10 and 50 µg/L, respectively.

(i)   *Source of contamination of groundwater by arsenic in West Bengal*

When the contamination of drinking-water by arsenic was first discovered in West Bengal, tubewell strainers, pesticides, insecticides and other anthropogenic sources were first considered as possible origins of the groundwater contamination (Chakraborty & Saha, 1987). However, Das et al. (1994) showed that a single deep tubewell supplying water to a few villages in Maldah, one of the nine arsenic-affected districts, was drawing nearly 150 kg arsenic per year, indicating that the source of arsenic was geological. Analyses of bore-hole sediments showed high concentrations of arsenic in only a few soil layers and the arsenic therein was found to be associated with iron pyrites. Das et al. (1995, 1996) also confirmed analytically the existence of arsenic-rich pyrites in bore-hole sediment. It was proposed that heavy drawing of groundwater and aeration of the aquifer leads to the decomposition of arsenic-rich pyrites and consequently contamination of groundwater with arsenic. Similar conclusions were reached by Mallick and Rajagopal (1995).

Bhattacharya et al. (1997, 1998) reported an association between arsenic and hydrated ferric oxide (HFO) and its mobilization to the aquifer due to changes in redox conditions during the development of groundwater. Ahmed et al. (1998) and Nickson et al. (1998, 2000) also suggested that reduction of HFO resulted in the mobilization of arsenic from absorbed HFO.

(ii)   *Contamination of groundwater by arsenic in the residential
area of Behala-Calcutta due to industrial pollution*

In Calcutta, chronic arsenicosis was first reported by Guha Mazumder *et al.* (1992). The study of Chatterjee *et al.* (1993) on the source of arsenic and the magnitude of the contamination revealed that a chemical factory producing several chemical compounds, including the insecticide Paris-Green (copper acetoarsenite), was responsible for the contamination. This factory had been producing about 20 tonnes of Paris-Green per year for approximately 20 years. Analysis of soil surrounding the production waste-dumping ground showed very high concentrations of arsenic (as high as 10 000 µg/g). Nineteen hand tubewells, used for drinking and cooking in the immediate area, showed very high concentrations of arsenic (up to 39 000 µg/L). The concentration of arsenic in hand tube-wells decreased the farther the wells were located from the dumping ground. A follow-up study in the affected areas (Chakraborti *et al.*, 1998) showed that the total average concentration of arsenic in the 19 hand tubewells sampled previously had decreased by only 10–15% from the levels observed 8 years before.

(c)   *Contamination of groundwater by arsenic in Chhattisgarh State*

Contamination of groundwater by arsenic was reported in a few villages of Rajnandgaon district of Chhattisgarh by Chakraborti *et al.* (1999b). The present State of Chhattisgarh had been within the State of Madhya Pradesh 2 years previously. The source of arsenic in groundwater is natural and geological both for the alluvial Bengal Basin and the rocky belt of Dongargarh-Kotri zone of Rajnandgaon district. The total population of the district is 1.5 million. Except for two towns — Rajnandgaon and Khairagarh — the entire district depends on tubewells and dugwells. Water samples ($n = 146$) were collected from 22 villages of Chowki block, Rajnandgaon district, and levels of arsenic in ground-water were found to be above 10 µg/L in eight villages and above 50 µg/L in five villages, with the highest concentration being 880 µg/L. From 150 hair samples examined, approximately 75% of people were found to have levels of arsenic in hair above toxic levels. Pandey *et al.* (1999) also reported contamination of groundwater by arsenic in the Rajnandgaon district of Chhattisgarh. Of 390 samples analysed, 26 sites were found to be contaminated with arsenic, with the highest concentration being 1010 µg/L. The number of people at risk was estimated at 10 000. Pandey *et al.* (2002) established that the extent of the arsenic contamination in this area is even greater; about 30 000 people residing in 30 villages and towns are directly exposed to high levels of arsenic in drinking-water (up to 3050 µg/L arsenic) and more than 200 000 people are at risk. The source and mobilization process of arsenic from affected areas of Rajnandgaon district Chhattisgarh was reported by Acharyya (2002).

(d)   *Contamination of groundwater by arsenic in Middle Ganga Plain, Bihar*

In the Middle Ganga Plain, Bihar, tubewells replaced dugwells about 20 years ago. Analyses of the arsenic content of 206 tubewells from Semria Ojha Patti (95% of the total in the village) showed that 56.8% exceeded concentrations of 50 µg/L, with 19.9% > 300 µg/L. The distribution indicated that, of the 5000 residents of Semria Ojha Patti, 18.8% used safe water (< 10 µg/L arsenic), 24.7% used water containing 10–50 µg/L arsenic, 56.8% used water containing > 50 µg/L, and 19.9% used water containing > 300 µg/L. The concentrations of arsenic in urine, hair and nail correlated significantly ($r = 0.72$–0.77) with concentrations in drinking-water. Of the 51 urine samples analysed, 98% had levels of arsenic above that of the normal secretion, with 47% > 500 mg/L, 33% > 1000 mg/L and 5.9% > 3000 mg/L; 57.6% of hair samples and 76.3% of nail samples were found to be above the normal range (Chakraborti *et al.*, 2003a).

(e)   *Contamination of groundwater and surface water by arsenic in the industrial region of Patancheru, Andhra Pradesh*

Patancheru, in the Medak District of Andhra Pradesh, is one of the major industrial estates, situated 30 km from Hyderabad. The main source of arsenic has been identified as Park Trade Centre, Gaddapotharam Bulk Drug Factory, which makes veterinary drugs based on arsonic acid, as well as other sources such as the pesticide and drug intermediate industries. The solid wastes of these industries are dumped indiscriminately near Kazipally Lake, and represent a source of contamination of nearby waters and soils. Arsenic contamination was evaluated in 14 villages in this area. Very high levels of arsenic were found in the range of 80–8960 µg/L and 140–7350 µg/L in surface water and groundwater, respectively. In both surface water and groundwater, the average arsenite (As^{III}) concentration was about 20% of total arsenic (Sekhar *et al.*, 2003).

Samples of blood, urine, hair and nails from 193 inhabitants of these 14 contaminated villages were analysed. Arsenic levels in the biological samples were very high, ranging from 400 to 1400 µg/kg in blood (control, 6–10 µg/kg), from 60 to 160 µg/L in urine (control, 6–10 µg/L), from 300 to 940 µg/kg in hair (control, 10–130 µg/kg) and from 500 to 1630 µg/kg in nails (control, 120–160 µg/kg). High concentrations of arsenic were also detected in vegetables, plants and cow's milk in this area and represent a second possible source of exposure for the population (Sekhar *et al.*, 2003).

1.4.3   *Exposure in Central and South America*

In South America, the main source of exposure to arsenic has been the natural contamination of drinking-water. In this area, arsenic originates from the geological formations associated with volcanoes, affecting Chile, Bolivia, Peru and Argentina in the Andean region (Queirolo *et al.*, 2000a). The largest populations affected are the Antofagasta Region in northern Chile, with approximately 400 000 exposed inhabitants, and the Córdoba Province in Argentina, with approximately 630 000 people exposed. Mexico also has naturally occurring arsenic in drinking-water, which is best characterized in the Lagunera region, in central northern Mexico, where approximately 400 000 people are exposed. The mean levels in drinking-water for these populations range from 50 to 500 µg/L; in isolated wells, levels reach as high as 6897 µg/L arsenic. Exposure has been recorded since the beginning of the last century. Currently, most areas in these regions receive water with levels of arsenic below 50 µg/L.

Exposures to arsenic due to contaminated air, soils and water as a result of copper, gold or silver mining have been described in Mexico (Díaz-Barriga *et al.*, 1993; Calderón, 1999; Mejía *et al.*, 1999), Chile (Romo-Kroger & Llona, 1993; Romo-Kroger *et al.*, 1994; Santolaya *et al.*, 1995; Sancha, 1997; Flynn *et al.*, 2002), Brazil (Romo-Kroeger & Llona, 1993) and Nicaragua (Cruz *et al.*, 1994). The area affected in central Mexico is San Luis de Potosí. In Chile, environmental and occupational exposures to arsenic in air have been reported in the Andes Mountains in Regions II, III, V and VI and the Metropolitan Region, where five major copper mining plants are located (Ministerio de Salud, 1986; Santolaya *et al.*, 1995; Ferreccio *et al.*, 1996), but no secondary contamination of drinking-water.

(a)   Mexico

In Mexico, most studies of arsenic in drinking-water have been conducted in the States of Durango and Coahuila, which constitute the Lagunera Region (Table 12). Del Razo *et al.* (1990) studied 128 wells from 11 counties and found arsenic contents of 8–624 µg/L; 50% of the wells had arsenic levels > 50 µg/L. They estimated that at least 400 000 people, mostly from the rural areas of the region, have been exposed to levels of arsenic > 50 µg/L. Since the 1960s, when arsenic contamination was first identified, the polluted wells have gradually been replaced and, by the end of 1989, most of the population was receiving water with arsenic levels below 20 µg/L (Cebrián *et al.*, 1994). Some contamination of drinking-water by arsenic has been reported in the State of Hidalgo in the Zimapan Valley, where the exposed population has been estimated at 35 000, and levels of arsenic in the drinking-water ranged from 21 to 1070 µg/L (Armienta *et al.*, 1997; Gomez-Arroyo *et al.*, 1997).

In San Luis de Potosí, in central Mexico, exposure to arsenic associated with mining activities arises from drinking-water, soil and dust, and the estimated exposed population is 600 000. Mean arsenic concentrations in air have been measured at 0.48 µg/m^3 (0.36–0.88 µg/m^3) (Díaz-Barriga *et al.*, 1993), and concentrations of arsenic in drinking-water vary from 9.9 to 20.9 µg/L (with some wells near the smelter having concentrations that range from 105 to 6897 µg/L). Studies of soil in San Luis de Potosí have demonstrated extremely high levels of arsenic in the vicinity of the mines (188–944 µg/g, Díaz-Barriga *et al.*, 1993; 2215–2675 µg/g, Mejía *et al.*, 1999), and also in the dust of the nearby households (800–1182 µg/g, Díaz-Barriga *et al.*, 1993; 1780–9950 µg/g, Mejía *et al.*, 1999). By 1991, the copper mining companies that caused the air, soil and water contamination of the area implemented dust control technologies and other measures to control soil pollution (Cebrián *et al.*, 1994).

Table 12. Exposure to arsenic in drinking-water in Mexico

Location	Source of water	No. of samples studied	Year	Total arsenic in water (µg/L; range)	Reference
Sonora, Hermosillo	Wells of 29 cities	173	NR	2–305	Wyatt *et al.* (1998a)
Lagunera Region	Wells, different towns	171	1970s–1980s	7–624	Cebrían *et al.* (1983); Del Razo *et al.* (1990); García-Vargas *et al.* (1994); Gonsebatt *et al.* (1997); Hernández-Zabala *et al.* (1999)
Zimapán, Hidalgo	Aquifer, 6 different towns		(Since 1970)	21–1070	Gomez- Arroyo *et al.* (1997)
San Luis de Potosí	Tap-water, Morales and Graciano	19	NR	9.9–20.9	Díaz-Barriga *et al.* (1993)
	Wells near smelter	NR	NR	106 6897	Meíja *et al.* (1999)

NR, not reported

Levels of arsenic in urine and hair are presented in Tables 13–14. Levels of arsenic in hair were high in samples from subjects exposed to arsenic in water in Zimapán, and were twice those in samples from subjects in Mexico City, which were also above the reference value, probably due to air pollution (Armienta *et al.*, 1997).

(*b*)   *Argentina*

In Argentina, the main source of arsenic in drinking-water has been from wells, with concentrations ranging from 40 to > 4500 µg/L (Table 15), and arsenic was first reported in well-water in 1917 (Arguello *et al.*, 1938). In 1970 and 1980, aqueducts from rivers with low levels of arsenic were built to replace the use of well-water, but some populations continued to be exposed (Hopenahyn-Rich *et al.*, 1996a,b,c). The provinces with high levels of arsenic in their well-water are: Córdoba, Salta, La Pampa, Santa Fé, Tucuman, Santiago del Estero, San Luis and part of Buenos Aires. The best characterized is Córdoba, a region in central Argentina, that extends over an area of 165 000 km² and has a population of 2 750 000, distributed in 26 counties. In some counties of Córdoba, high levels (between 100 and 2000 µg/L) of arsenic were recorded in drinking-water during the 1930s (Hopenhayn-Rich *et al.*, 1996a).

In Córdoba, Hopenhayn-Rich *et al.* (1996a) obtained data from various sources, including measurements of arsenic in drinking-water from official national health reports made in the 1930s, a survey in 1942, two studies reported in 1968 and 1985 and a water survey reported in 1973. Based on the available measurements, average exposure of the population of each town was estimated, assuming that all people drank the same concentration

Table 13. Total arsenic in human urine samples in Mexico, Argentina and Chile

Location	No. of exposed subjects studied	Year	Mean arsenic in urine	Range	Reference
Mexico					
Lagunera, Santa Ana	36 adults		489 µg/gc	109–1829 µg/gc	García-Vargas et al. (1994);
	35 adults		548 µg/gc	295–849 µg/gc	Del Razo et al. (1997);
	37 adults		848 µg/L	88–2058 µg/L	Hernández-Zavala et al. (1999)
San Luis de Potosí	80 children		51.6 µg/gc	18.2–186.2 µg/gc	Calderón et al. (2001)
	112 children		70.5 µg/gc	17.7–497.7 µg/gc	Mejía (1999)
	133 children		117.6 µg/gc	33–594 µg/gc	Díaz-Barriga et al. (1993)
Argentina					
Córdoba province, MJ	282		160 µg/L	60–410 µg/L	Lerda (1994)
Santa Fe, Tortugas	155		70 µg/L	10–600 µg/L	Lerda (1994)
San Antonio	11		274 µg/L	126–440 µg/L	Vahter et al. (1995)
Other towns	15		36 µg/L	13–89 µg/L	Vahter et al. (1995)
San Antonio	10 lactating women		400 µg/L	250–610 µg/L	Concha et al. (1998a)
	11 pregnant women		335 µg/L	116–439 µg/L	Concha et al. (1998c)
San Antonio and Taco	34 children	1994	382 µg/L	125–621 µg/L	Concha et al. (1998b)
San Antonio	23 women		344 µg/L	90–606 µg/L	
Chile					
Region I	93 general population	1984–95	45 µg/L	10–92 µg/L	Venturino (1987); Sancha (1997)
Rest of Chile	2472 general population	1984–2000	13 µg/L	5–49 µg/L	Venturino (1987); Sancha (1997); CONAMA (2000)
Antofagasta	164 general population	1968	NR	1–700 µg/L	Gonzalez (1970)
Antofagasta	262 general population	1994–2000	69 µg/L	18–99 µg/L	Sancha (1997); CONAMA (2000)
Calama	239 general population	1977–95	76 µg/L	21–124 µg/L	Borgoño et al. (1980); Sancha (1997)
San Pedro	265 general population	1997	611.7 µg/L	61–1893 µg/L	Hopenhayn-Rich et al. (1996b,c); Moore et al. (1997a,b)

gc, grams of creatinine; NR, not reported

Table 14. Arsenic in human hair samples in Mexico and Chile

Location (source of exposure)	Year of sample	No. of subjects	Total arsenic in sample (µg/g)		Reference
			Mean	SD (range)	
Mexico					
Zimapán	NR	120	8.5	3.56	Armienta *et al.* (1997)
Mexico City (water)		17	4.6	1.96	
San Luis de Potosí	NR				Díaz-Barriga *et al.* (1993)
(smelter (1.5 km))		75	9.9	(1.4–57.3)	
(smelter (25 km))		25	0.5	(0.2–1.2)	
Lagunera (wells)	NR	35	NR	(0–23.3)	Chávez *et al.* (1964)
Chile					
Iquique	1969	26	0.8	NR	Borgoño & Greiber (1971)
Antofagasta	1968–76	607	7.7	4.2–14.8	Gonzalez (1970); Borgoño & Greiber (1971); Sandoval & Venturino (1987)
Antofagasta	1986–92	293	0.42	0.01–3.68	Jamett *et al.* (1992); Peña *et al.* (1992)
Calama	1977	203	3.75	0–10	Borgoño *et al.* (1980)
Calama	1986–92	60	4.28	0.98–14.2	Jamett *et al.* (1992); Peña *et al.* (1992)
Chuquicamata	1986–92	60	17.19	3.03–54.77	Jamett *et al.* (1992); Peña *et al.* (1992)
Puchuncaví	1990	151	2.178	0.103–18.023	Chiang *et al.* (1990)
Valparaíso	1990	NR	0.434	0.015–1.525	Chiang *et al.* (1990)

SD, standard deviation

of arsenic. It was estimated that 273 014 people had been exposed to an average of 178 µg/L arsenic and another 406 000 people had been exposed to some arsenic (at least one measurement of 120 µg/L in water). A report available through CEPIS/PAHO (Penedo & Zigarán, 2002) described the arsenic content of 100 water samples from wells in Córdoba and confirmed Hopenhayn's estimations: they estimated that 625 861 people were exposed to arsenic, with regional averages ranging from 70 to 180 µg/L and individual well measurements from 10 to 1900 µg/L.

The Salta Province is the only area where high levels of arsenic have also been found in surface waters (Penedo & Zigarán, 2002). In the provinces of Salta and Jujui, in north-western Argentina, samples from five rivers had arsenic levels ranging from 52 to 1045 µg/L, and samples from three surging thermal springs had arsenic levels of 128–10 650 µg/L (de Sastre *et al.*, 1992). The population of San Antonio de los Cobres is the best studied in this province (Vahter *et al.*, 1995; Concha *et al.*, 1998a,b,c). San Antonio de los Cobres is a village in the Salta Province, 3800 m above sea level, with

Table 15. Exposure to arsenic in drinking-water in Argentina

Location	Source of drinking-water wells	No. of samples (year)	Total arsenic in water (µg/L)		Reference
			Average	Range	
Córdoba Province	Bell-Ville	NR (1917–20)	NR	1120–4500	Arguello *et al.* (1938)
Córdoba Province	Marcos Juárez	282 (NR)	130	10–660	Lerda (1994)
Córdoba Province	2 counties	118 (1942)	178	40–533	Hopenhayn-Rich *et al.* (1996a)
Córdoba Province	5 counties	67 (NR)	70–180	10–1900	Penedo & Zigarán (2002)
Santa Fe Province	Tortugas	155 (NR)	20	0–70	Lerda (1994)
Salta Province	School pipes	18 (NR)	592	4–1490	Astolfi (1971)
Salta Province	San Antonio	2 areas (NR)	NR	93–440	de Sastre *et al.* (1992)
	San Antonio	1 well	NR	8250–10 650	
	San Antonio	10 (1994)	167	117–219	Vahter *et al.* (1995); Concha *et al.* (1998a); Del Razo *et al.* (1999)

NR, not reported

approximately 5000 inhabitants (Vahter *et al.*, 1995). Until recently, this population had been drinking-water from wells with arsenic contents varying from < 1 to 440 µg/L, with one well reaching 9450 µg/L on average. Arsenic levels in urine are presented in Table 13 and other biomarkers in Table 16.

There are no studies of arsenic in air in Argentina. High levels of arsenic have been found in prepared food and soups in San Antonio de los Cobres (soup, 259–427 µg/g; prepared food, 131–418 µg/g; Concha *et al.*, 1998b).

(c)   Chile

Northern Chile (Regions I–III) is an expanse of 250 000 km², of which 35 000 are quaternary volcanic rocks rich in arsenic (Queirolo *et al.*, 2000a). Arsenic reaches the population through drinking-water and through contamination of air and soil, as a result of mining activities.

In Chile, the main sources of drinking-water are rivers that originate in Cordillera de los Andes and reach the Pacific Ocean. Rivers in northern Chile (Regions I and II) have high natural arsenic concentrations, particularly those from the Region of Antofagasta. Arsenic concentrations in rivers in Region II vary along its course, depending on the arsenic content of its tributaries, and range from 30 to 3310 µg/L but reach 14 250 µg/L in some hot springs (Table 17) (Alonso, 1992; Queirolo *et al.*, 2000a). Exposure of the

Table 16. Arsenic in other biological samples in Salta Province, Argentina

Location	No. of subjects	Type of sample (year)	Total arsenic in sample		Reference
			Median	Range	
San Antonio de los Cobres	9 women	Breast milk	2.3 μg/kg[a]	0.83–7.6 μg/kg	Concha et al. (1998a)
		Blood	9.8 μg/L	4.4–19 μg/L	
		Urine	390 μg/L	250–610 μg/L	
		Blood	μg/L	μg/L	Vahter et al. (1995)
San Antonio	15		7.6	2.7–18.3	
Santa Rosa	5		1.5	1.1–2.0	
Olacapato	5		1.3	1.2–2.4	
Tolar Grande	5		1.3	1.0–1.3	
	Children	Blood (1994)	μg/L	μg/L	Concha et al. (1998b)
San Antonio and T Pozo	36		9.1	5.5–17	
Rosario de Lerma	20		0.8	0.27–1.5	
	Women				
San Antonio and T Pozo	27		9.3	2.7–18	
Rosario de Lerma	11		0.95	0.69–1.8	
San Antonio de los Cobres	Pregnant women	Blood	11 μg/L	5.6–13 μg/L	Concha et al. (1998c)
	11	Cord blood	9.0 μg/L	6.0–12 μg/L	
	11	Placenta	34 μg/kg	17–54 μg/kg	
	10	Maternal milk	3.0 μg/kg[a]	2.3–4.8 μg/kg	

[a] μg/kg fresh weight

Table 17. Concentration of arsenic in surface waters in Chile, Region II, 1983–86

River studied and location	Mean total arsenic in water (µg/L)
Salado River	
Tatio Hot Springs	14 250
Codelco Mine Reservoir	7 500
Before Toconce River	3 310
Toconce River Before Salado	600
Before Curti	860
Ayquina	980
Before Loa River	760
Loa River	
Before Salado	270
Yalquincha	800
Finca	910
Before San Salvador River	1 380
San Salvador Before Loa	1 270
La Posada Bridge	1 500
Quillagua	1 440
Outlet of River	1 360
Upper Loa River Basin	210–330
Gorges south of Salado river	30–60
Spring north of Salado river	190–370

From Alonso (1992)

population in this region has ranged from 40 to 860 µg/L, depending on the rivers used for its water supply; 1958–70 was the highest exposure period for the largest population (approximately 300 000) (Table 18).

Sancha (1997) estimated the total number of people exposed in 1996 to specific levels of arsenic in the drinking-water in Chile: 7 million inhabitants (53.3%) were exposed to less than 10 µg/L; 5.5 million (41.9%) were exposed to 10–30 µg/L; 450 000 inhabitants (3.4%) were exposed to 30–50 µg/L; 170 000 inhabitants (1.3%) were exposed to 50–60 µg/L; and 1500 (0.01%) were exposed to 600–800 µg/L.

There are a few studies of arsenic in general environmental air in Chile (Romo-Kroger & Llona, 1993; Romo-Kroger *et al.*, 1994; Sancha, 1997; COSUDE, 2000). Sancha (1997) and COSUDE (2000) covered a large part of the country from 1994 to 1999. They found that cities not in the vicinity of copper smelting operations had arsenic levels in the air ranging from 0.001 to 0.057 µg/m^3, with a population of approximately 6 million people. The cities located 30–45 km from a copper smelter had arsenic levels in the air ranging from 0.01 to 0.14 µg/m^3 and had approximately 755 000 inhabitants. The cities in the vicinity (within 10 km) of smelters had arsenic levels in air ranging from 0.03

Table 18. Average concentration (μg/L) of arsenic in drinking-water in Regions II and I and the rest of Chile

Region	Town	Population (2002 census) living in exposed areas	1930–57	1958–70	1971–77	1978–80	1981–87	1988–94	1995–2002	Reference
II	Tocopilla and Elena	31 175	250	250	636	110	110	40	–	Ferreccio et al. (2000)
	Calama	136 739	150	150	287	110	110	40	–	Ferreccio et al. (2000)
									38	Sancha (1997)
	San Pedro	4 883	600	600	600	600	600	600	–	Ferreccio et al. (2000)
	Chiu Chiu	250	–	–	–	–	–	–	753	Smith et al. (2000b)
	Caspana	275	–	–	–	–	–	–	13	Smith et al. (2000b)
	Antofagasta and Mejillones	306 548	90	860	110	110	70	40	–	Ferreccio et al. (2000)
									32	Sancha (1997)
I	Arica-Iquique	426 351	–	–	–	–	–	–	32	Sancha (1997)
III–XIII	Rest of Chile	14 213 266	–	–	–	–	–	–	5	Sancha (1997)

Averages supplied by Empresa Servicios Sanitarios de Antofagasta for 1950–67 and Servicio de Salud Antofagasta for 1968–94
For 1995–2000, data and ranges published in studies

to 2.4 µg/m³, with an estimated 60 000 people exposed. The rest of the country that was not sampled has a population of approximately 6 million people whose exposure is estimated to be in the lowest range of < 0.010 µg/m³ (COSUDE, 2000).

Mean levels of arsenic in the air inside the Chuquicamata copper mine, the world's largest open copper mine, for the period 1952–91 ranged from 1.6 µg/m³ in the administrative areas to 201.72 µg/m³ in the smelting areas (Ferreccio et al., 1996). This exposure has decreased in the last decade in correlation with the implementation of new technologies in smelting to avoid arsenic contamination. Workers exposed to arsenic had urinary levels ranging from 40 to 490 µg/L in 1992; between 1987 and 1990, 32–58% of workers in exposed areas had levels of urinary arsenic above 300 µg/L.

In Region II, 1020 people were examined between 1987 and 2002; the mean total arsenic in urine was 225 µg/L, ranging from 1 to 1893 µg/L. In comparison, in Region I which has intermediate exposure to arsenic in water (40 µg/L), 91 people were examined and their urinary arsenic averaged 45.5 µg/L, ranging from 10 to 92 µg/L. In the rest of Chile (arsenic in drinking-water, < 10 µg/L), 2472 people were sampled and mean urinary arsenic was 13 µg/L, ranging from 5 to 49 µg/L (Table 13). Arsenic measured in hair from people in Chile is presented in Table 14.

(d) Other

In Nicaragua, there has been concern regarding contamination with heavy metal of Lake Asososca, which is a source of drinking-water for Managua. The level of arsenic in sediment was found to be 4.1 µg/g, and that in water ranged from 0 to 18.07 µg/L, with a mean concentration of 5.86 µg/L, well below current water standards. Higher values of 25 µg/L were found in Lake Monte Galán (Cruz et al., 1994). An earlier study conducted in Lake Xolotlán found arsenic levels in surface water ranging from 10.2 to 30.1 µg/L; wastewater from a thermal plant discharging into the river contained concentrations of 5295–16 700 µg/L (Lacayo et al., 1992).

In Brazil, concerns have been raised regarding arsenic contamination as a result of gold mining in the zone of Minas Gerais, in south-eastern Brazil. In 1998, urinary arsenic was measured in 126 schoolchildren, and a mean concentration of 25.7 µg/L (range, 2.2–106 µg/L) was found. Environmental studies in the surrounding areas found mean levels of arsenic in surface water of 30.5 µg/L (range, 0.4–350 µg/L); levels of arsenic in soils ranged from 200 to 860 mg/kg; and sediments had a mean concentration of 350 mg/kg, ranging from 22 to 3200 mg/kg (Matschullat et al., 2000).

1.4.4 Exposure in South-East Asia

There are many reports on the human exposure to arsenic in the drinking-water in South-East Asia. High concentrations of arsenic in drinking-water have been documented in China (Cao, 1996), Taiwan, China (Tseng et al., 1968; Chiou et al., 1997a), Thailand (Choprapawon & Rodcline, 1997), and Viet Nam (Berg et al., 2001). The use of artesian wells, which were later shown to have high levels of arsenic in the water, began in the early 1920s in southern Taiwan, China (Tseng *et al.*, 1968), in the early 1950s in Inner Mongolia (Ma *et al.*, 1996), in the late 1950s in north-eastern Taiwan, China (Chiou *et al.*, 1997a, 2001), in the early 1960s in Xinjiang, China (Wang, 1996), in the late 1980s in Ronbipool, Thailand (Choprapawon & Rodcline, 1997), in the early 1990s in Shanxi, China (Cao, 1996) and in the mid-1990s in Viet Nam (Berg *et al.*, 2001). There have been several reports on industry-related exposure to arsenic through drinking-water contamination from tin mining in Ronpibool, Thailand (Choprapawon & Rodcline, 1997).

Table 19 summarizes data on arsenic contamination of drinking-water in various regions of South-East Asia.

(a)   China

Several geographical areas in mainland China have a high content of arsenic in the drinking-water, including Xinjiang, Inner Mongolia and Shanxi (Cao, 1996). The villages with high concentrations of arsenic in the drinking-water in Inner Mongolia are clustered in Bamen and Huhehot. Ma *et al.* (1996) reported the arsenic concentration in the water of 9733 wells in Bamen: 2465 had levels of arsenic > 50 µg/L; in five counties of Bamen, the percentage of wells with an arsenic concentration > 50 µg/L varied, ranging from 11 to 59%; more than 500 villages had at least one well with an arsenic concentration > 50 µg/L; and the level of arsenic in drinking-water from all wells from the two areas ranged from < 50 to 890 µg/L. The water from a total of 497 wells in Huhehot were tested for arsenic: 111 had an arsenic level > 50 µg/L; 48 villages had at least one well with arsenic concentration > 50 µg/L; and the level of arsenic in drinking-water ranged from < 81 to 890 µg/L (Ma *et al.*, 1996). Sun *et al.* (2001) reported a survey on the concentration of arsenic in 303 wells in a village in Inner Mongolia: 77 wells (25.4%) had a level of arsenic < 10 µg/L, 85 (28.1%) had levels of 10–49 µg/L, 131 (43.2%) had levels of 50–499 µg/L and 10 (3.3%) had levels of ≥ 500 µg/L.

In the highly contaminated area of Xinjiang, located in Tunguei, arsenic concentrations in well-water in 15 villages of the area ranged from 50 to 850 µg/L, and were mostly between 100 and 500 µg/L (Wang, 1996).

Sun *et al.* (2001) reported a survey of 2373 wells in 129 villages in the Basin of Datong and Jinzhong, Shanxi, in 1994–95. Levels of arsenic in drinking-water ranged from < 50 to 4440 µg/L and 833 wells had an arsenic concentration > 50 µg/L. The percentage of wells with an arsenic concentration > 50 µg/L in seven counties of the area varied from 6.3 to 54.7%.

(b)   Taiwan, China

There are two endemic areas of arseniasis in Taiwan, China. One is located in the south-western coastal area where Blackfoot disease, a unique peripheral vascular disease associated with long-term ingestion of arsenic from artesian well-water, is endemic. There are four townships in this area: Peimen, Hsuehchia, Putai and Ichu. High levels of arsenic in artesian wells and patients with Blackfoot disease have also been documented in two neighbouring townships, Hsiayin and Yensui. Another endemic area of chronic arsenic

Table 19. Contamination of drinking-water by arsenic in various regions of South-East Asia

Country	Area/population	Sample	Level of arsenic (range; µg/L)	Source of arsenic	Reference
Taiwan (China)	South-western Blackfoot disease-endemic area (Peimen, Hsuehchia, Putai, Ichu)	13 artesian well-water 34 artesian well-water 11 artesian well-water 97 artesian well-water	240–960 350–1100 340–896 10–1100	Natural Natural Natural Natural	Blackwell et al. (1961) Chen et al. (1962) Yeh (1963) Kuo (1968)
	North-eastern endemic area of chronic arsenic poisoning (Chuangwei, Wuchieh, Chiaohsi, Tungshan)	3901 well-water	< 0.15–3590	Natural	Chiou et al. (2001)
	Taiwan (314 townships)	83 656 well-water	< 10–> 1000	Natural	Lo (1975)
Thailand	Thammarat Province	Surface water	< 0.5–583 < 0.5–28.4 As^{III}	Arsenopyrite wastes	Williams et al. (1996)
		Shallow water	1.25–5114 < 0.5–125 As^{III}		
		Surface water	4.8–583	Mining	Choprapawon & Porapakkham (2001)
		River	541–583		
China	Inner Mongolia	497 well-water (Huhhot)	< 10–1860	Natural	Ma et al. (1996); Luo et al. (1997)
		9733 well-water (Bamen)	< 50–890	Natural	Ma et al. (1996)
	Xinjiang	Well-water in 15 villages (Tunguei)	50–850	Natural	Wang (1996)
	Shanxi	2373 well-water in 129 villages (Datong, Jinzhong)	< 50–4440	Natural	Sun et al. (2001)

Table 19 (contd)

Country	Area/population	Sample	Level of arsenic (range; μg/L)	Source of arsenic	Reference
Japan	Fukuoka	67 well-water	1–293 11–220 As^V 15–70 As^{III}	Natural	Kondo et al. (1999)
	Sendai		1–35		
	Takatsuki		3–60		
	Kumamoto		5–66		
Viet Nam	Red River Basin	68 tubewells,	1–3050 (72% > 10 μg/L)	Natural	Berg et al. (2001)
		8 treatment plants	11–190		

toxicity is located in the Lanyang Basin of north-eastern Taiwan, in which there are four townships: Chiaohsi, Chuangwei, Tungshan and Wuchieh (Table 19).

In the area of south-western Taiwan where Blackfoot disease is endemic, Blackwell *et al.* (1961) reported levels of arsenic of 240–960 µg/L in 13 artesian wells, Chen *et al.* (1962) reported levels ranging from 350 to 1100 µg/L in 34 artesian wells and Yeh (1963) found levels ranging from 340 to 900 µg/L in water samples from 11 wells. Kuo (1968) carried out a larger survey of 97 artesian wells in 42 villages of the six townships in the endemic area and found concentrations in well-water ranging from 10 to 1100 µg/L, with a median of 500 µg/L. In this south-western area, the arsenic concentration was higher in water from deep artesian wells than in that from shallow wells, showing a correlation coefficient of $r = 0.627$ ($p < 0.01$). Arsenate was the dominant species of arsenic in the artesian well-water.

Lo (1975) reported a nationwide survey of arsenic content in drinking-water from 83 656 wells in 314 precincts and townships. In total, 15 649 (18.7%) wells had an arsenic concentration ≥ 50 µg/L and 2224 (2.7%) had an arsenic concentration ≥ 350 µg/L. Most townships with high arsenic concentration in well-water were found to cluster in south-western and north-eastern Taiwan.

Chiou *et al.* (2001) tested the water from 3901 tubewells in 18 villages of four townships in the north-eastern endemic area of chronic arsenic toxicity by the HG–FAAS method in 1991–94. The arsenic content ranged from undetectable (< 0.15 µg/L) to 3590 µg/L.

(c) Thailand

The Ronpibool district is situated approximately 70 km south of Nakorn Sri Thammarat Province, in the southern part of Thailand, and had a total population of approximately 23 000 in 1998. A geological survey found that the potential sources of arsenic contamination in the mining areas were from high-grade arsenopyrite waste piles in bedrock mining localities, sub-ore grade waste-rock piles, sulfide-rich wastes from ore-dressing plants, disseminated sulfide waste from small-scale prospecting and floatation activities and alluvial tin workings (Choprapawon & Porapakkham, 2001).

In 1994, a collaborative study was initiated to establish the distribution and geo-chemical form of arsenic in surface drainage and aquifer systems in the area. Surface waters were sampled at 26 stations and groundwater samples were collected from 23 shallow wells and 13 deep boreholes. Concentrations of arsenic in samples of surface water ranged from < 0.5 (limit of detection) to 583 µg/L and As^{III} levels ranged from < 0.5 to 28.4 µg/L. Concentrations in shallow groundwater samples ranged from 1.25 to 5114 µg/L and As^{III} levels ranged from < 0.5 to 125 µg/L. Concentrations in deep borehole samples ranged from 1.25 to 1032 µg/L and As^{III} levels ranged from < 0.5 to 53.6 µg/L (Williams *et al.*, 1996).

In another study, significant concentrations of arsenite (As^{III}) were detected in several of the water samples with the highest levels of arsenic (28.4, 25.6 and 24.9 µg/L), although arsenate (As^V) remained the dominant species (more than 92% of the total) (Choprapawon & Porapakkham, 2001).

(d)   Viet Nam

Berg et al. (2001) reported arsenic contamination of the Red River alluvial tract in the city of Hanoi, Viet Nam, and in the surrounding rural districts. Because of naturally occurring organic matter in the sediments, the groundwaters are anoxic and rich in iron. In rural groundwater samples from private small-scale tubewells, contamination levels ranged from 1 to 3050 μg/L, with an average concentration of 159 μg/L arsenic. In a highly affected rural area, the groundwater that is used directly as drinking-water had an average concentration of 430 μg/L. Analysis of raw groundwater pumped from the lower aquifer for the Hanoi water supply yielded arsenic levels of 240–320 μg/L in three of eight treatment plants and 37–82 μg/L in another five plants. Aeration and sand filtration that are applied in the treatment plants for the removal of iron lowered the arsenic content to levels of 25–91 μg/L, but 50% remained above 50 μg/L. The high arsenic concentrations found in tubewells (48% above 50 μg/L and 20% above 150 μg/L) indicate that several million people consuming untreated groundwater might be at a considerable risk for chronic arsenic poisoning.

(e)   Japan

In March 1994, high concentrations of arsenic (> 10 μg/L) were detected in 29 of 67 well-water samples in the southern region of the Fukuoka Prefecture, Japan. The range of arsenic concentrations was 1–293 μg/L: As^V ranged from 11 to 220 μg/L; As^{III} ranged from 15 to 70 μg/L; and MMA and DMA were both < 1 μg/L. The maximum concentration was lower than the figures recorded in Taiwan, China, and India, but higher than those reported in Sendai (range, 1–35 μg/L), Takatsuki (range, 3–60 μg/L) and Kumamoto (range, 5–66 μg/L), Japan (Kondo et al., 1999).

Arsenic concentrations in water from 34 wells in the Niigata Prefecture were measured between 1955 and 1959 as part of a historical cohort study using the Gutzeit method, and ranged from non-detectable to 3000 μg/L: six wells had a non-detectable concentration; 17 wells contained < 1000 μg/L; and 11 wells contained ≥ 1000 μg/L. All wells with arsenic concentrations > 100 μg/L were located within a distance of 500 m from a factory that produced arsenic trisulfide (Tsuda et al., 1995).

(f)   Other

In a recent United Nations Economic and Social Commission for Asia and the Pacific–United Nations International Children's Emergency Fund–World Health Organization (UNESCAP-UNICEF-WHO, 2001) expert group meeting, contamination of groundwater by arsenic was also reported from other countries including Lao People's Democratic Republic, Cambodia, Myanmar and Pakistan. It has also been reported from Nepal (Tandukar et al., 2001; Shreshta et al., 2002).

1.4.5   *Exposure in other countries*

Exposure in other countries is summarized in Table 20.

(*a*)   *Africa (Egypt, Ghana) and the Middle East (Iran)*

In a 1999 study of 100 subjects in Cairo, Egypt, arsenic was measured by HG–AAS (detection limit, 1 µg/L) in hair samples and drinking-water. Levels of arsenic in hair samples ranged from 40 to 1040 µg/kg and levels in drinking-water samples were less than 1 µg/L (Saad & Hassanien, 2001).

Concentrations of arsenic in groundwaters from two areas in Ghana — the Obuasi area in the Ashanti region and the Bolgatanga area of the Upper East region — ranged from < 1 to 64 µg/L [As^{III} range, 6–30 µg/L] and < 1 to 141 µg/L [As^{III} range, < 1–9 µg/L], respectively. Sulfide minerals such as arsenopyrite and pyrite were present in the Birimian basement rocks of both areas and these constitute the dominant sources of arsenic. Concentrations were lowest in the shallowest groundwaters, and increased at greater depths. The lateral and vertical variations in dissolved arsenic concentrations were controlled by ambient pH and redox conditions and by the relative influences of sulfide oxidation and sorption (Smedley, 1996).

Concentrations of arsenic were measured in the scalp hair of three groups of people from a village in western Iran using NAA. One group consisted of healthy subjects, the second of subjects with suspected arsenic poisoning, and the third of subjects with confirmed arsenic poisoning. The arsenic content of water sources used by the inhabitants was also measured. The average arsenic concentration in hair was 200 µg/kg in the healthy group, 4900 µg/kg in the group with suspected poisoning and 5600 µg/kg in the group with arsenic poisoning; arsenic concentrations in water samples varied between 30 µg/L and 1040 µg/L (Pazirandeh *et al.*, 1998).

(*b*)   *Australia*

Australia is a country rich in minerals that present a significant source of natural arsenic contamination to the environment, in addition to anthropogenic sources such as mining activities and pesticide use. In 1991, survey data showed elevated levels of arsenic in the surface water and groundwater in Victoria, particularly around gold mining areas. Concentrations of arsenic in groundwater ranged from < 1 to 300 000 µg/L (*n* = 109) and those in surface water ranged from < 1 to 28 300 µg/L (*n* = 590). In a follow-up study of the same region in the mid-1990s, arsenic concentrations ranged from 1 to 12 µg/L in groundwater samples (*n* = 18), from 1 to 220 µg/L in surface water samples (*n* = 30) and from 1 to 73 µg/L in drinking-water samples (*n* = 170) (Hinwood *et al.*, 1998).

In an investigation of the relationship between environmental exposure to arsenic from contaminated soil and drinking-water and the incidence of cancer in the Victoria region, median arsenic concentrations in groundwater ranged from 1 to 1077 µg/L (total range, 1–300 000 µg/L; *n* = 22 areas) (Hinwood *et al.*, 1999).

Table 20. Concentrations of arsenic (As) in drinking-water in other countries

Country	Population	Date	Sample (no.)	Levels (µg/L)	Source of arsenic	Reference
Africa						
Egypt	Cairo	1999	Tap water (5 districts)	?	NR	Saad & Hassanien (2001)
Ghana	Obuasi area	NR	Groundwater	< 1–64 (total As) 6–30 (As^{III})	Natural	Smedley (1996)
	Bolgatanga area	NR		1–141 (total As) 1–9 (As^{III})		
Middle East						
Iran	West Iran	NR	Spring water (20)	30–1040	Natural	Pazirandeh et al. (1998)
Australia						
Victoria	Victoria	mid-1990s	Groundwater (18) Surface water (30) Drinking-water (170)	1–12 1–220 1–73	Natural anthropogenic (mining, pesticide)	Hinwood et al. (1998)
			Ground- and surface water (22 geographical areas)	1–300 000 (1–1077 medians)		Hinwood et al. (1999)
Europe						
Finland		1993–94 1996	Groundwater (69) Wells (72) Control population	17–980 < 0.05–64 (median, 0.14)	Natural	Kurttio et al. (1998) Kurttio et al. (1999)
Spain	Madrid	1998	353 water supplies Wells (< 2% of population uses wells)	74% < 10 23% 10–50 3.7% > 50	Natural	Aragones Sanz et al. (2001)
Romania (Transylvania)	Bihor and Arad counties	1992–95	Drinking-water	0–176	Natural	Gurzau & Gurzau (2001)
Switzerland	Grisons Canton Valais Canton	1998	Public water supplies (336) 14 000 people	< 10–170 12–50	Natural	Pfeifer & Zobrist (2002)
United Kingdom	South-west		Private supplies (3)	11–80		Farago et al. (1997)

Table 20 (contd)

Country	Population	Date	Sample (no.)	Levels (µg/L)	Source of arsenic	Reference
North America						
Canada	Nova Scotia	NR	Well-water (94 households)	1.5–738	Natural	Meranger et al. (1984)
	Rural areas (Saskatchewan)	1981–85 NR	Communities (121) Private wells and municipality wells (61 wells)	<2–34 <1–117	Natural	Health Canada (1992) Thompson et al. (1999)
USA						
Western USA		NR	Rainwater and snow Rivers Lakes Seawater Groundwater	<0.002–0.59 0.20–264 0.38–1000 0.15–6.0 130–48 000 50–2750 170–3400 80–15 000	Mining area Basin fill deposits Volcanic areas Geothermal area	Welch et al. (1988)
Maine, Michigan, Minnesota, South Dakota, Oklahoma, Wisconsin			Groundwater 17 496 samples	40% > 1 5% > 20		Welch et al. (1999)
National Survey		NR	Surface water (189) Groundwater (239)	68 max 117 max	Natural	Chen & Edwards (1997)
Arizona	Verde Valley Groundwater	1994–2001	Groundwater (41)	10–210 µg/L	Natural	Foust et al. (2000)
Illinois			Deep glacial drift aquifer Shallow glacial drift aquifer	> 5–83 1–28		Warner (2001)
Montana, Wyoming		1988–95 1973–95	Madison River Missouri River	35–370 2–69	Natural	Nimick et al. (1998)
National		mid 1990s	Drinking water supplies (21 120)	6–17% > 5 1–3% > 20		Frey & Edwards (1997)
National survey	36% population of US population	1992–93	Water companies (140 utilities)	56% > 0.5 16% > 2 5% > 5	Natural	Davis et al. (1994)

Table 20 (contd)

Country	Population	Date	Sample (no.)	Levels (μg/L)	Source of arsenic	Reference
Missouri and Iowa		NR	Family wells (11)	34–490	Natural	Korte & Fernando (1991)
New Jersey		1977–79	Groundwater (1064)	1 (median) 1160 (max.)	Natural	Page (1981)
			Surface water (591)	1 (median) 392 (max.)		
Ohio		NR	88 wells	0–96	Natural	Matisoff et al. (1982)
Alaska		1976	Well-water (59)	1–2450		Harrington et al. (1978)
Oregon		1968–74	Tap-water (558)	0–2150 8% > 50		Morton et al. (1976)
New Hampshire		1994	Drinking-water (793)	< 0.01–180		Karagas et al. (1998, 2002)
Utah		1978–79	Community water supplies (88)	0.5–160	Natural	Bates et al. (1995)
Utah		1976–97	151 drinking-water (151)	3.5–620	Natural	Lewis et al. (1999)
National survey	25 states		Groundwater systems	5.3% > 10	Natural	Environmental Protection Agency (2001)
			Surface water systems	0.8% > 10	Natural	

NR, not reported

(c)   Europe (Finland, Romania, Spain, Switzerland, United Kingdom)

Samples of well-water were collected in Finland between July and November 1996. The final study population (144 627 from a register-based cohort) consisted of 61 bladder cancer cases and 49 kidney cancer cases diagnosed between 1981 and 1995, as well as an age- and sex-balanced random sample of 275 subjects (reference cohort). To evaluate the validity of water sampling, two water samples were taken from each of 36 randomly selected wells at two different times (on average 31 days apart; range, 2 h–88 days). The arsenic concentrations in the original samples and field duplicates were not significantly different. The arsenic concentrations in the wells of the reference cohort ranged from < 0.05 to 64 µg/L (median, 0.14 µg/L). Five per cent of the reference cohort had arsenic concentrations > 5 µg/L and 1% (4/275) had consumed well-water containing levels of arsenic that exceed the WHO drinking-water quality guideline value of 10 µg/L (Kurttio *et al.*, 1999). Locally in Finland, drinking-water from privately drilled wells contains high concentrations of arsenic up to 980 µg/L (Kurttio *et al.*, 1998). The arsenic is of geological origin.

In the north-west region of Transylvania, Romania, drinking-water contains arsenic as a result of the geochemical characteristics of the land. The geographical distribution of arsenic in drinking-water in this region, sampled between 1992 and 1995, was heterogeneous, with a mixture of high (mostly in rural areas) and low concentrations in contiguous areas (range, 0–176 µg/L arsenic). Estimates indicated that about 36 000 people were exposed to concentrations of arsenic in the drinking-water ranging from 11 to 48 µg/L, and about 14 000 inhabitants were exposed to arsenic levels exceeding 50 µg/L (Gurzau & Gurzau, 2001).

In 1998 in Madrid, Spain, arsenic concentrations of more than 50 µg/L, the maximum permissible concentration for drinking-water in Spain, were detected in some drinking-water supplies from underground sources. In the initial phase, water samples from 353 Madrid water supplies were analysed. In a second phase, 6 months later, analyses were repeated on those 35 water supplies that were considered to pose a possible risk to public health. Seventy-four per cent of the water supplies studied in the initial phase had an arsenic concentration of less than 10 µg/L, 22.6% had levels of 10–50 µg/L and 3.7% had over 50 µg/L. Most of the water supplies showing arsenic levels greater than 10 µg/L were located in the same geographical area. In the second phase, 26 of the 35 water supplies were in the same range (10–50 µg/L arsenic) as in the first survey; nine had changed category, six of which had less than 10 µg/L and three had more than 50 µg/L. In Madrid, less than 2% of the population drinks water from underground sources (Aragones Sanz *et al.*, 2001).

In Switzerland, areas with elevated levels of arsenic have been found primarily in the Jura mountains and in the Alps. Weathering and erosion of rocks containing arsenic releases this element into soils, sediments and natural waters. The limit for drinking-water (50 µg/L) in Switzerland is not generally exceeded but, in the cantons of Ticino, Grisons and Valais, concentrations of arsenic above 10 µg/L have been found in the drinking-water. The canton of Grisons tested all of the 336 public water supplies in 1998. In 312 drinking-water supplies, arsenic concentrations were below 10 μg/L (93%), while 21 samples had arsenic concentrations between 10 and 50 μg/L (6%). Three samples exceeded the Swiss limit of 50 μg/L (0.9%); the maximum concentration found was 170 μg/L. Ore deposits and sediments in the canton of Valais have also been known for some time to contain arsenic. The drinking-water in this area was not tested for arsenic until 1999. Since then, it has been determined that in this canton approximately 14 000 people live in areas where the drinking-water contains between 12 and 50 μg/L (Pfeifer & Zobrist, 2002).

Although levels of arsenic in public water supplies are low, there is concern about the 20 000–30 000 private well-water supplies in South-West England, particularly those in old mining areas, which undergo limited or no treatment. From limited available data, three private supplies of those tested in Cornwall had arsenic levels above the 5-μg/L detection limit, and contained 11, 60 and 80 μg/L (Farago et al., 1997).

(d)   North America

(i)   Canada

Samples from 61 groundwater sources, including 25 privately owned wells and 36 wells operated by rural municipalities, in Saskatchewan, Canada, were tested for arsenic. For virtually all of the rural municipal wells, no chemical or physical water treatment was performed other than periodic chlorination, whereas approximately half of the private wells underwent some form of water treatment. The most commonly used forms of water treatment included water softening with an ion exchange device, filtration and removal of iron. Arsenic was not detected in 25 samples (10 private wells and 15 rural municipal wells) using a method with a detection limit of 1 μg/L; 34 samples (13 private wells and 21 rural municipal wells) had levels between 1 and 50 μg/L; only two wells (private) had levels greater than 50 μg/L (maximum concentration, 117 μg/L) (Thompson et al., 1999).

In an earlier survey of water supplies from 121 communities in Saskatchewan sampled between 1981 and 1985, arsenic levels were below 10 μg/L in 88% and below 2 μg/L in 42% of the samples taken; the maximum level recorded was 34 μg/L (Health Canada, 1992).

The concentration of total soluble inorganic arsenic (arsenate plus arsenite) was measured in duplicate water samples from the wells of 94 residents in seven communities in Halifax County, Nova Scotia, where arsenic contamination of well-water was suspected. Levels of arsenic exceeded 50 μg/L in 33–93% of wells in each of the communities; in 10% of the wells sampled, concentrations were in the range of 500 μg/L. The total measured levels ranged from 1.5 to 738.8 μg/L (Meranger et al., 1984).

(ii)   USA

The occurrence of arsenic in groundwater has been reported in the USA for areas within the states of Alaska, Arizona, California, Hawaii, Idaho, Nevada, Oregon and Washington. High concentrations are generally associated with one of the following geochemical environments: (a) basin-fill deposits of alluvial-lacustrine origin, particularly in semi-arid areas; (b) volcanic deposits; (c) geothermal systems; and (d) uranium and gold mining areas. Arsenic concentrations ranged from < 0.002 to 0.59 µg/L in rainwater and snow, from 0.20 to 264 µg/L in rivers, from 0.38 to 1000 µg/L in lakes and from 0.15 to 6.0 µg/L in seawater. Maximum observed concentrations of arsenic ranged from 130 to 48 000 µg/L in groundwater from mining areas, from 50 to 2750 in basin-fill deposits, from 170 to 3400 in volcanic areas and from 80 to 15 000 µg/L in geothermal areas. Total inorganic arsenic ranged from 1.1 to 6000 µg/L, arsenite ranged from 0.6 to 4600 µg/L and arsenate ranged from 0 to 4300 µg/L (Welch et al., 1988).

Within the last decade, high concentrations of arsenic exceeding 10 µg/L in groundwater have been documented in many other areas of the USA (Morton et al., 1976; Harrington et al., 1978; Page, 1981; Matisoff et al., 1982; Korte & Fernando, 1991; Davis et al., 1994; Bates et al., 1995; Chen & Edwards, 1997; Frey & Edwards, 1997; Karagas et al., 1998; Nimick et al., 1998; Lewis et al., 1999; Foust et al., 2000; Warner, 2001; Karagas et al., 2002) (Table 20). The US Geological Survey reported that these high concentrations most commonly result from: (a) upflow of geothermal water; (b) dissolution of, or desorption from, iron oxide; and (c) dissolution of sulfide minerals. Overall, analyses of approximately 17 000 groundwater samples in the USA suggest that about 40% of both large and small regulated water supplies have arsenic concentrations greater than 1 µg/L. About 5% of regulated water systems are estimated to have arsenic concentrations greater than 20 µg/L (Welch et al., 1999).

Using a 25-state database of compliance monitoring from community systems, the Environmental Protection Agency (2001) found that 5.3% of groundwater systems and 0.8% of surface water systems had concentrations > 10 µg/L.

In a national retrospective groundwater study of 18 850 drinking-water samples (2262 from community wells and 16 602 from private wells), the US Geological Survey found the 90th percentiles for community wells and private wells to be 8 µg/L and 13 µg/L, respectively (Focazio et al., 2000). A study in New Hampshire found that drinking-water from private wells contained significantly more arsenic than that from community wells. In addition, this study found that deep wells had higher arsenic concentrations than superficial wells and that samples voluntarily submitted to the state for analysis had higher concentrations than randomly selected household water samples (Peters et al., 1999).

1.5   Regulations and guidelines

Arsenic has been a contaminant of concern in drinking-water for several years. For example, in the USA in 1942, a maximum permissible concentration for arsenic was set at 50 µg/L by the Public Health Service. This standard was reaffirmed in 1946 and 1962; however, in 1962, the Public Health Service advised that concentrations in water should not exceed 10 µg/L when "more suitable supplies are or can be made available" (Smith et al., 2002). In 2002, the maximum contaminant level for arsenic in the USA was lowered from 50 µg/L to 10 µg/L (Environmental Protection Agency, 2001). Table 21

Table 21. Regulations and guidelines for arsenic in drinking-water

Region	Guideline/ regulation (μg/L)	Reference
World	10	WHO (1998)
Europe	10	European Commission (1998)
USA	10	Environmental Protection Agency (2001)
Canada	25	Health Canada (2003)
Australia	7	National Health and Medical Research Council and Agriculture and Resource Management Council of Australia and New Zealand (1996)
South-East Asia (Bangladesh, India, Viet Nam, China)	50	WHO (2000)
Laos, Mongolia, Japan and Taiwan	10	WHO (2000); Taiwan Environmental Protection Agency (2000)
Argentina, Bolivia, Brazil and Chile	50	WHO (2000); Penedo & Zigarán (2002); Chilean Institute of National Standards (1984)
Philippines and Indonesia	50	WHO (2000)
Sri Lanka and Zimbabwe	50	WHO (2000)
Namibia	10	WHO (2000)
Bahrain, Egypt, Oman and Saudi Arabia	50	WHO (2000)
Jordan and Syria	10	WHO (2000)

details the various regulations and guidelines that have been established for arsenic in drinking-water.

The WHO (1998) guideline of 10 μg/L is a provisional value. A provisional guideline is established when there is some evidence of a potential health hazard but for which available data on health effects are limited, or when an uncertainty factor greater than 1000 has been used in the derivation of the tolerable daily intake.

The Canadian guideline (Health Canada, 2003) is an interim maximum acceptable concentration, again, due to the limited data on health effects.

Several other countries have also established standards for arsenic, and several developing countries have established a standard for arsenic of 50 μg/L (WHO, 2000).

2. Studies of Cancer in Humans

Major epidemiological studies of cancer in relation to arsenic in drinking-water include ecological studies and fewer case–control and cohort studies. For most other known human carcinogens, the major source of causal evidence arises from case–control and cohort studies, with little, if any, evidence from ecological studies. In contrast, for arsenic in drinking-water, ecological studies provide important information on causal inference, because of large exposure contrasts and limited population migration. As a consequence of widespread exposure to local or regional water sources, ecological measures provide a strong indication of individual exposure. Moreover, in the case of arsenic, the ecological estimates of relative risk are often so high that potential confounding with known causal factors cannot explain the results. Hence, in the review that follows, ecological studies are presented in detail.

2.1   Cancer of the urinary bladder and kidney

The findings of epidemiological studies on arsenic in drinking-water and the risk for cancers of the urinary bladder and kidney are summarized in Table 22.

Historically, several case reports have related cancers of the urinary tract with medicinal arsenic treatments or arsenic-related diseases such as Bowen disease. In 1953, a series of 27 cases with multiple skin cancers attributed to arsenical medicines was reported (Sommers & McManus, 1953). Of these cases, 10 were diagnosed as also having internal cancers at various sites, three of which were urinary tract tumours. Graham and Helwig (1959) first investigated an association between Bowen disease and primary internal cancers. Twenty-eight (80%) of 35 cases had primary internal cancers, two of which were malignant tumours of the bladder and one a tumour of the kidney. Cuzick et al. (1982) examined a cohort of subjects in the United Kingdom who had taken Fowler's solution (potassium arsenite) between 1945 and 1969. After further follow-up of the cohort through 1990 (Cuzick et al., 1992), a threefold increase in mortality from bladder cancer (standardized mortality ratio [SMR], 3.07; 95% confidence interval [CI], 1.01–7.3) was reported, strengthening the evidence on bladder cancer reported previously (Cuzick et al., 1982).

Bergoglio (1964) published the first report of bladder cancer associated with arsenic in drinking-water in the Province of Cordóba in Argentina. He identified 2355 deaths between 1949 and 1959 in nine towns of a highly exposed region and found that cancer was the cause of death of 24%; 11% of these cancer deaths involved cancers of the urinary tract. Biagini (1972) followed 116 patients with arsenic-related skin lesions in the same region and found that 12.5% of the cancer deaths were patients with urinary tract cancers.

Table 22. Summary of epidemiological studies of arsenic in drinking-water and risk for bladder and kidney cancers

Reference	Location	End-point	Exposure	No. of cases		Study outcome	Comments
Ecological studies							
Taiwan							
Chen *et al.* (1985)	84 villages from four neighbouring townships on the SW coast	Mortality 1968–82	Comparison of mortality in an area endemic for Blackfoot disease with general population	*Bladder* Men Women *Kidney* Men Women	Obs/exp. 167/15.2 165/8.2 42/5.4 62/5.5	SMR (95% CI) 11.1 (9.3–12.7) 20.1 (17.0–23.2) 7.7 (5.4–10.1) 11.2 (8.4–14.0)	Reference: national rates
Chen *et al.* (1988a)	Area endemic for Blackfoot disease	Mortality 1973–86	Village of residence; median arsenic levels of well-water samples			SMR	Age-standardized mortality rates per 100 000; 899 811 person–years Reference: world population in 1976
				Bladder Men			
			General population < 300 μg/L 300–590 μg/L ≥ 600 μg/L			3.1 15.7 37.8 89.1	
				Women			
			General population < 300 μg/L 300–590 μg/L ≥ 600 μg/mL			1.4 16.7 35.1 91.5	
				Kidney Men			
			General population < 300 μg/L 300–590 μg/L ≥ 600 μg/L			1.1 5.4 13.1 21.6	
				Women			
			General population < 300 μg/L 300–590 μg/L ≥ 600 μg/L			0.9 3.6 12.5 33.3	

Table 22 (contd)

Reference	Location	End-point	Exposure	No. of cases	Study outcome	Comments
Wu *et al.* (1989)	42 villages in an area endemic for Blackfoot disease	Mortality 1973–86	Arsenic levels: 3 groups based on median level of well-water/village		Age-adjusted mortality rates per 100 000	Reference: world population in 1976
			Group 1 (< 300 µg/L)	*Bladder*		
			Men (248 728)	Men	22.64 ($p < 0.001$)	
			Women (248 728)	Women	25.60 ($p < 0.001$)	
				Kidney		
				Men	8.42 ($p < 0.05$)	
			Group 2 (300–590 µg/L)	Women	3.42 ($p < 0.001$)	
			Men (138 562)	*Bladder*		
			Women (127 502)	Men	61.02 ($p < 0.001$)	
				Women	57.02 ($p < 0.001$)	
				Kidney		
				Men	18.90 ($p < 0.05$)	
			Group 3 (≥ 600 µg/L)	Women	19.42 ($p < 0.001$)	
			Men (79 883)	*Bladder*		
			Women (74 083)	Men	92.71 ($p < 0.001$)	
				Women	111.30 ($p < 0.001$)	
				Kidney		
				Men	25.26 ($p < 0.05$)	
				Women	57.98 ($p < 0.001$)	

Table 22 (contd)

Reference	Location	End-point	Exposure	No. of cases	Study outcome	Comments
Chen & Wang (1990)	314 geographical units (precincts and townships), including 4 townships in the endemic area of Blackfoot disease	Mortality 1972–83	Average arsenic levels in water samples of all geographical units. 73.9% of study precincts or townships had < 5% of wells with > 50 µg/L; 14.7% had 5–14%; 11.5% had ≥ 15%.	**All 314 precincts and townships** *Bladder* Men Women *Kidney* Men Women **170 south-western townships** *Bladder* Men Women *Kidney* Men Women	3.9 (0.5) 4.2 (0.5) 1.1 (0.2) 1.7 (0.2) 3.7 (0.7) 4.5 (0.7) 1.2 (0.2) 1.7 (0.3)	Reference: world population in 1976. Analysis weighted by population in each group. Regression coefficients indicating an increase in age-adjusted mortality/100 000 person–years for every 0.1 µg/L increase in arsenic level (SE)
Chiang et al. (1993)		Incidence 1981–85	Exposure not evaluated Endemic area	*Bladder* Total 140 Men 81 Women 59	Incicence per 100 000 23.5 26.1 21.1	Adjusted for age
			Neighbouring endemic area	Total 13 Men 7 Women 6	4.5 4.7 4.3	
			All Taiwan	Total 2135 Men 1608 Women 527	2.3 3.3 1.2	

Table 22 (contd)

Reference	Location	End-point	Exposure	No. of cases		Study outcome	Comments
Guo et al. (1997)	National survey of 83 656 wells in 243 townships	Incidence of transitional-cell carcinoma 1980–87	Arsenic levels in town of residence (ppm) <0.05 0.05–0.08 0.09–0.16 0.17–0.32 0.33–0.64 >0.64	*Bladder* Men Women	1185 363	0.57 (0.07) 0.33 (0.04)	Estimates of rate difference (per 100 000 person–years) for one unit increase in the predictor and associated standard error for exposure > 0.64 ppm (SE). Results shown for transitional-cell carcinoma
				Kidney Men Women	158 81	0.03 (0.02) 0.142 (0.013)	
Tsai et al. (1999)	4 townships	Mortality 1971–94	Area endemic for Blackfoot disease	**Deaths** *Bladder* Men	312	8.9 (7.96–9.96)	Local reference (Chiayi-Tainan county) National reference (Taiwan)
				Women	295	10.5 (9.4–11.7) 14.1 (12.5–15.8) 17.7 (5.7–19.8)	Local National
				Kidney Men	94	6.7 (5.5–8.3) 6.8 (5.5–8.3)	Local National
				Women	128	8.9 (7.4–10.6) 10.5 (8.8–12.5)	Local National

Table 22 (contd)

Reference	Location	End-point	Exposure	No. of cases		Study outcome	Comments
South America							
Hopenhayn-Rich et al. (1996a, 1998)	26 counties in the Province of Córdoba, Argentina	Mortality 1986–91	Exposure levels	**Deaths** *Bladder*		SMR (95% CI)	SMR for period 1976–92
			Low (690 421)	Men	113	0.8 (0.7–0.96)	
				Women	39	1.2 (0.9–1.7)	
			Medium (406 000)	Men	116	1.3 (1.05–1.5)	
				Women	29	1.4 (0.93–1.99)	
			High (mean arsenic level, 178 µg/L) (273 014)	Men	131	2.1 (1.8–2.5)	
				Women	27	1.8 (1.2–2.6)	
				Kidney			
			Low (690 421)	Men	66	0.9 (0.7–1.1)	
				Women	38	1.0 (0.7–1.4)	
			Medium (406 000)	Men	66	1.3 (1.02–1.7)	
				Women	34	1.4 (0.94–1.9)	
			High (273 014)	Men	53	1.6 (1.2–2.1)	
				Women	27	1.8 (1.2–2.6)	
Rivara et al. (1997)	Chile	Mortality 1950–92	Arsenic-contaminated Region II of northern Chile versus non-contaminated region VIII	Bladder Kidney		SMR (95% CI) 10.2 (8.6–12.2) 3.8 (3.1–4.7)	
Smith et al. (1998)	Chile	Mortality 1989–93	Region II of northern Chile with population-weighted average arsenic concentration up to 569 µg/L compared with rest of Chile; exposure generally < 10 µg/L	*Bladder* Men Women *Kidney* Men Women		SMR (95% CI) 6.0 (4.8–7.4) 8.2 (6.3–10.5) 1.6 (1.1–2.1) 2.7 (1.9–3.8)	Population of about 400 000
Australia							
Hinwood et al. (1999)	Victoria	Incidence 1982–91	Median arsenic concentration in drinking-water ranged 1–1077 µg/L	Bladder Kidney	303 134	SIR (95% CI) 0.9 (0.8–1.1) 1.2 (0.98–1.4)	State rates used as reference

Table 22 (contd)

Reference	Location	End-point	Exposure	No. of cases		Study outcome		Comments
						OR	OR from multiple logistic regression analyses	

Case–control studies

Taiwan

Reference	Location	End-point	Exposure	Community controls	Bladder cancer cases	OR	OR from multiple logistic regression analyses	Comments
Chen *et al.* (1986)	4 neighbour-ing townships in endemic area of Black-foot disease	Mortality 1980–82	Median arsenic content of artesian well-water, 0.78 ppm Years of artesian water consumption					Adjusted for age and sex
			0	136	17	1.0	1.0	
			1–20	131	19	1.2	1.3	
			21–40	50	10	1.6	1.7	
			≥40	51	23	3.9	4.1	

USA

Reference	Location	End-point	Exposure	Controls	Bladder cancer cases	OR (90% CI)		Comments
Bates *et al.* (1995)	10 areas of the USA	Incident cases (aged 21–84 years) dia-gnosed in a 1-year period in the 1970s. Age-, sex- and area-matched controls	Cumulative dose (mg)			All subjects		Adjusted for sex, age, smoking, years of exposure to chlorinated surface water, history of bladder infection, educational level, urbanization of the place of longest lifetime residence, ever employed in a high-risk occupation
			< 19	47	14	1.0		
			19–32	36	21	1.6 (0.8–3.2)		
			33–52	39	17	0.95 (0.4–2.0)		
			≥ 53	38	19	1.4 (0.7–2.9)		
			mg/L × years					
			< 33	42	18	1.0		
			33–52	42	16	0.7 (0.3–1.5)		
			53–73	40	16	0.5 (0.3–1.2)		
			≥ 74	36	21	1.0 (0.5–2.1)		

Table 22 (contd)

Reference	Location	End-point	Exposure	No. of cases	Study outcome	Comments
Bates *et al.* (1995) (contd)			Cumulative dose (mg)		Never smoked	
			< 19		1.0	
			19–32		1.1 (0.4–3.1)	
			33–52		0.7 (0.2–2.3)	
			≥ 53		0.5 (0.1–1.9)	
			mg/L × years			
			< 33		1.0	
			33–52		0.2 (0.1–0.8)	
			53–73		0.3 (0.1–0.9)	
			≥ 74		0.9 (0.3–3.2)	
			Cumulative dose (mg)		Ever smoked	
			< 19		1.0	
			19–32		3.3 (1.0–10.8)	
			33–52		1.9 (0.6–6.2)	
			≥ 53		3.3 (1.1–10.3)	
			mg/L × years			
			< 33		1.0	
			33–52		1.95 (0.7–5.6)	
			53–73		1.2 (0.4–3.7)	
			≥ 74		1.4 (0.5–4.3)	

Table 22 (contd)

Reference	Location	End-point	Exposure	No. of cases		Study outcome	Comments
Europe							
Kurttio *et al.* (1999)	Areas in Finland in which < 10% of population belong to the municipal drinking-water system	Incidence 1981–95	Concentration of arsenic in water	**Bladder**		Relative risk (95% CI)	Case–cohort design Adjusted for age, sex and smoking Short latency: exposure in the 3rd to 9th calendar year prior to the cancer diagnosis Long latency: exposure in the 10th calendar year and earlier prior to the cancer diagnosis
				Short latency			
			< 0.1 μg/L	23		1.0	
			0.1–0.5 μg/L	19		1.5 (0.8–3.1)	
			≥ 0.5 μg/L	19		2.4 (1.1–5.4)	
			Total	61		1.4 (0.95–1.96)	
				Long latency			
			< 0.1 μg/L	26		1.0	
			0.1–0.5 μg/L	18		0.8 (0.4–1.6)	
			≥ 0.5 μg/L	17		1.5 (0.7–3.4)	
			Total	61		0.96 (0.6–1.6)	
			Concentration of arsenic in water	**Kidney**			
				Short latency			
			< 0.1 μg/L	23		1.0	
			0.1–0.5 μg/L	12		0.8 (0.4–1.7)	
			≥ 0.5 μg/L	14		1.5 (0.7–3.3)	
			Total	49		1.2 (0.8–1.7)	
				Long latency			
			< 0.1 μg/L	25		1.0	
			0.1–0.5 μg/L	9		0.3 (0.1–0.8)	
			≥ 0.5 μg/L	15		1.07 (0.5–2.5)	
			Total	49		0.7 (0.4–1.4)	

Table 22 (contd)

Reference	Location	End-point	Exposure	No. of cases		Study outcome	Comments
Cohort studies							
Taiwan							
Chen *et al.* (1988b)	4 neighbour-ing townships in area endemic for Blackfoot disease	Mortality 1968–93	Comparison of mortality with general and endemic population	Cancer deaths		SMR	95% CI calculated by the Working Group General population as reference
				Bladder	15	38.8 [21.7–64.0]	
				Kidney	3	19.5 [4.0–57.0]	
						SMR	Area endemic for Blackfoot disease as reference
				Bladder	15	2.6 [1.4–4.2]	
				Kidney	3	1.6 [0.3–4.7]	
Chiou *et al.* (1995)	4 neighbour-ing townships in area endemic for Blackfoot disease (BFD)		Cumulative index derived for each subject: Σ (Ci × Di). Cumulative exposure (mg/L × year)	BFD patients	263	*Bladder cancer* RR* (95% CI)	*Relative risk after adjustment for age, sex and smoking **Relative risk after adjustment for age, sex, smoking and BFD status Ci, median concentration of arsenic in wells of village; Di, duration of drinking water in that village
				Healthy controls	2295		
			0	Bladder cancer cases	29	1.0	
			0.1–19.9			2.1 (0.6–7.2)	
			≥ 20			5.1 (1.5–17.3)	
						RR** (95% CI)	
			0			1.0	
			0.1–19.9			1.6 (0.4–5.6)	
			≥ 20			3.6 (1.1–12.2)	

Table 22 (contd)

Reference	Location	End-point	Exposure	No. of cases	Study outcome	Comments
Chiou *et al.* (2001)	North-eastern Taiwan	Incidence 1991–94 Person–years of observation	Area endemic for arseniasis		RR (95% CI)	Adjusted for age, sex, smoking and duration of drinking well-water
			Arsenic concentration in well-water (μg/L)		*Urinary tract*	
		7978	< 10.0	3	1.0	
		6694	10.1–50.0	3	1.5 (0.3–8.0)	
		3013	50.1–100.0	2	2.2 (0.4–13.7)	
		5220	> 100.0	7	4.8 (1.2–19.4)	*p* for trend < 0.01
					Transitional-cell carcinoma	
		7978	< 10.0	1	1.0	
		6694	10.1–50.0	1	1.9 (0.1–32.5)	
		3013	50.1–100.0	2	8.2 (0.7–99.1)	
		5220	> 100.0	6	15.3 (1.7–139.2)	*p* for trend < 0.05
Japan						
Tsuda *et al.* (1995)	Niigata prefecture	1959–92		No. of persons exposed at concentration level (1955–59)	SMR (95% CI) Urinary Obs/exp.	
			Arsenic concentration in well-water (ppm) from arsenic-polluted area			
			< 0.05	254	0 (0–12.5) 0/0.3	
			0.05–0.99	76	0 (0–47.1) 0/0.08	
			≥ 1	113	31.2 (8.6–91.8) 3/0.10	
			Total	443	6.3 (1.7–18.4) 3/0.48	

Table 22 (contd)

Reference	Location	End-point	Exposure	No. of cases	Study outcome	Comments
USA						
Lewis *et al.* (1999)	Millard County, UT	Mortality	Index of exposure to arsenic calculated for each cohort member and derived from number of years of residence and median arsenic concentration in the given community		SMR (95% CI)	Confidence intervals not given for exposure categories 4058 members in cohort (2092 men and 1966 women)
					Bladder/other urinary cancers	
			Low exposure (< 1000 ppb–years)	Men	0.36	
				Women	1.18	
			Medium exposure (1000–4999 ppb–years)	Men	–	
				Women	–	
			High exposure (≥ 5000 ppb–years)	Men	0.95	
				Women	1.10	
			Total	Men	0.4 (0.1–1.2)	
				Women	0.8 (0.1–2.9)	
					Kidney cancer	
			Low exposure (< 1000 ppb–years)	Men	2.51	
				Women	2.36	
			Medium exposure (1000–4999 ppb–years)	Men	1.13	
				Women	1.32	
			High exposure (≥ 5000 ppb–years)	Men	1.43	
				Women	1.13	
			Total	Men	1.8 (0.8–3.3)	
				Women	1.6 (0.4–4.1)	

SMR, standardized mortality ratio; CI, confidence interval; SIR, standardized incidence ratio; OR, odds ratio; RR, relative risk; BFD, Blackfoot disease

More systematic studies were conducted in various parts of the world, the most extensive being in Taiwan, China.

2.1.1   Studies in Taiwan, China

There are two areas in Taiwan, China, where exposure to arsenic is endemic. One is located in the south-western coastal area where Blackfoot disease, a unique peripheral vascular disease induced by long-term ingestion of arsenic from artesian well-water, is endemic. Eighty-four villages constitute the four Blackfoot disease-endemic townships of Peimen, Hsuehchia, Putai and Ichu, and artesian wells and patients with Blackfoot disease were also found in the two neighbouring townships of Yensui and Hsiaying (Wu *et al.*, 1989). Residents in the south-western endemic areas in Taiwan drank artesian well-water with high concentrations of arsenic from the early 1910s to the late 1970s. The concentrations of arsenic in artesian well-water, tested by Natelson's method in 1964–66 in 42 villages of the six townships, ranged from 10 to 1752 µg/L, and were mostly above 100 µg/L (Kuo, 1968; Tseng *et al.*, 1968). As well-water was the only source of drinking-water in the endemic area, all residents of a given village consumed the water from a small number of shared wells in their daily home and working environments. Most residents were engaged in fishing, salt production and farming, and the migration rate was low. More than 90% of residents had lived in the study area all their lives. As the study population lived in a small area, they shared similar dietary patterns, lifestyle, socioeconomic status and health care facilities. The piped water supply system using surface water was first implemented in the south-western endemic area, but its coverage was not complete until the late 1970s (Wu *et al.*, 1989).

Another area with exposure to arsenic is located in the Lanyang Basin in north-eastern Taiwan (Chiou *et al.*, 2001), and is comprised of four townships: Chiaohsi (four villages), Chuangwei (seven villages), Wuchieh (three villages) and Tungshan (four villages). Residents in this area of endemic arseniasis used river water for drinking and cooking before the Second World War, and started to use water from tubewells in their houses in the late 1940s. The concentrations of arsenic in the water of 3901 wells in the endemic area was tested from 1991 to 1994 by hydride generation combined with flame atomic absorption spectrometry, and ranged from undetectable (< 0.15 µg/L) to 3.59 ng/L, but were mostly between 10 and 100 µg/L, with a median of 27.3 µg/L. A piped water supply system using surface water was first implemented in the endemic area in late 1997, and its coverage was almost complete in 2001. Residents in the endemic area were engaged in farming, and most of them had lived in the area all their lives.

(a)   Ecological studies

Chen *et al.* (1985) reported an elevation in mortality from cancers of the urinary bladder and kidney during the period 1968–82 in endemic areas of Blackfoot disease (four neighbouring townships comprising 84 villages) compared with the general population of Taiwan. The arsenic content of well-water ranged from 0.35 to 1.14 ppm [mg/L] with a median of 0.78 ppm, while shallow well-water contained arsenic at concentrations ranging from 0 to 0.3 ppm with a median of 0.04 ppm (Chen *et al.*, 1962). The SMRs for bladder cancer and kidney cancer increased with the prevalence of Blackfoot disease. Similarly, the SMRs for cancers of the bladder and kidney were highest in villages where only artesian wells were in use and lowest in those villages using shallow wells. The high SMRs for bladder and kidney cancer were not readily explained by the higher rate of cigarette smoking in the Blackfoot disease-endemic area compared with all of Taiwan (40% versus 32%).

Chen *et al.* (1988a) briefly described a dose–response relationship between median arsenic levels in artesian well-water in the 84 villages studied by Chen *et al.* (1985) and rates of mortality from bladder cancer. The study period (1973–86) covered 899 811 person–years of observation, and exposure was stratified into three categories (< 300, 300–590 and ≥ 600 μg/L arsenic) based on concentrations from a survey of over 83 000 wells, including 313 townships in all of Taiwan, conducted from 1962 to 1964.

Wu *et al.* (1989) examined age-adjusted mortality rates for various cancers in an area of south-western Taiwan that comprised 42 villages in six townships (27 villages studied by Chen *et al.* (1988a) and another 15 villages). The arsenic content of the 155 wells sampled, measured in 1964–66, ranged from 10 to 1750 μg/L. The villages were classified according to median arsenic levels in water into three exposure groups (< 300, 300–590 and ≥ 600 μg/L). Death certificates were used to ascertain cause of death during the period 1973–86. A dose–response relationship was found with concentration of arsenic in water for cancers of the bladder and kidney for both men and women.

Chen and Wang (1990) further investigated cancer mortality rates throughout Taiwan in 1972–83. Of 361 administrative areas, 314 were included in the study following measurements of arsenic contents in well-water in 1974–76. Exposure measurements were derived from a national water survey of over 83 000 wells throughout Taiwan. About 19% of the wells contained levels of arsenic above 50 μg/L, and most of them were in north-western and south-eastern Taiwan. Indices of urbanization and industrialization were included in the analysis to adjust for the possible confounding effect of differing socioeconomic characteristics between the 314 precincts and townships. Mortality data were used to evaluate 21 malignant neoplasms, using population-weighted regression. Results were presented for increases in mortality per 100 000 that were calculated to occur for every 0.1-mg/L increase in arsenic concentration in water.

Chiang *et al.* (1993) showed that the age-adjusted incidence of bladder cancer in the period 1981–85 in the Blackfoot disease-endemic area of Taiwan was higher than that in a neighbouring area of Taiwan and in the country as a whole.

Guo *et al.* (1997) used tumour registry data and the exposure data from the 1974–76 nationwide water-quality survey used by Chen and Wang (1990), which included concentrations of arsenic in drinking-water from 243 townships with about 11.4 million residents. The annual incidence of cancers of the bladder and kidney for townships in 1980–87 and subcategories of those cancer diagnoses were regressed against a model that included six variables for the proportion of wells in each of six categories of arsenic concentration in each township. Sex-specific models were adjusted for age and included an urbanization index and the annual number of cigarettes sold *per capita*. Regression models were weighted by the total population of each township. A total of 1962 bladder, 726 kidney, 170 ureter and 57 urethral cancers were included. The investigators found associations of high arsenic concentrations (more than 0.64 ppm) in both sexes with transitional-cell carcinomas of the bladder, kidney and ureter, and all urethral cancers combined, but they did not present relative risk estimates.

Tsai *et al.* (1999) compared mortality in people aged over 40 years in the Blackfoot disease-endemic area of Taiwan with both local and national references for the period 1971–94. Greater mortality was found for men and women with cancers of the bladder and kidney.

(b)   Case–control study

In a retrospective case–control study using death certificates from 1980–82, Chen *et al.* (1986) examined the relationship between exposure to high concentrations of arsenic in artesian well-water and mortality from internal malignancies, including tumours of the bladder ($n = 69$) in four townships from the Blackfoot disease-endemic area. Controls ($n = 368$) were selected by random sampling from the same geographical areas as the cases and were frequency-matched on age and sex. The response rate was 93% for proxies of cases and 92% for matched controls. Adjustment for age, sex and other variables (smoking, tea drinking, vegetarianism and frequency of consumption of vege-tables and fermented beans) was performed by logistic regression analysis. The results indicated increasing trends in odds ratios with increasing duration of intake of artesian well-water containing arsenic. The highest risks were seen for over 40 years of exposure, with an odds ratio of 4.1 for bladder cancer in a multivariate analysis. Smoking, alcohol consumption and other potential risk factors evaluated in the study did not confound the association between arsenic and cancer.

(c)   Cohort studies

Chen *et al.* (1988b) studied the association between arsenic in artesian well-water in relation to Blackfoot disease and cancer from a multiple risk factor perspective. The study area included the four townships in south-western Taiwan where high rates of Blackfoot disease had been described. Levels of arsenic were reported to be high in water, soil and food, and estimates of ingestion of arsenic by local residents were up to 1 mg per day. The study examined mortality in a cohort of people who had or had since developed Blackfoot disease in 1968, totalling 789 patients and 7578 person–years of observation through 1984. Follow-up started in 1968, since this was the year that registration of deaths in Taiwan was computerized and completeness and quality of death certificate registration was improved. Mortality of persons who had died ($n = 457$) and were not lost to follow-up ($n = 84$) was compared with that of the general population of Taiwan using age- and sex-specific mortality rates from 1968 through 1983. The SMRs for cancers of the bladder and kidney (men and women combined) were 38.8 [95% CI, 21.7–64.0] and 19.5

[95% CI, 4.0–57.0], respectively. The latter result, however, is based on only three deaths. SMRs were also calculated using all residents in the Blackfoot disease-endemic area, which includes people exposed to arsenic. These much lower SMRs were 2.6 [95% CI, 1.4–4.2] for bladder cancer and 1.6 [95% CI, 0.3–4.7] for kidney cancer, indicating that patients with Blackfoot disease had somewhat higher rates for these cancers than the residents in the arsenic-exposed region combined.

Chiou *et al.* (1995) investigated the relationship between incidence of internal cancers and arsenic in relation to Blackfoot disease in 2256 subjects from 1986 to 1993. Patients with Blackfoot disease (*n* = 263) and a referent group of 2293 residents of the same region were followed for 7 years. In contrast to many other studies that evaluate mortality, incident cancer was the outcome of interest. Follow-up occurred many years after exposure to elevated concentrations of arsenic in drinking-water had ended. Information on exposure to other risk factors was gathered by individual interviews. Several measures of exposure were evaluated, including average concentration of arsenic in artesian wells and cumulative exposure to arsenic from drinking artesian well-water. Relative risks were calculated using Cox's proportional hazard regression analysis. After controlling for the effects of age, sex and smoking in the regression analysis, a dose–response relationship was observed between exposure to arsenic from drinking well-water and the incidence of bladder cancer. Patients with Blackfoot disease were found to be at increased risk even after adjustment for cumulative exposure to arsenic.

Chiou *et al.* (2001) studied the incidence of urinary tract cancers among 8102 residents in the arsenic-endemic area in north-eastern Taiwan from 1991 to 1994. Levels of arsenic in the drinking-water ranged from less than 0.15 µg/L (undetectable) to 3590 µg/L. Exposure for each member of the cohort was assessed by measuring concentrations of arsenic in the well associated with the individual's household at one point in time only, although most households had used their current wells for at least 10 years (Chen & Chiou, 2001). Using the general population as referent, the standardized incidence ratio (SIR) for bladder cancer was 1.96 (95% CI, 0.9–3.6), while that for kidney cancer was 2.8 (95% CI, 1.3–5.4). These results were based on 10 subjects with bladder cancer and nine with kidney cancer. A dose–response relationship was observed between urinary tract cancers, particularly transitional-cell carcinoma, after adjusting for age, sex and smoking.

2.1.2   Studies in Japan

A retrospective cohort study was conducted in a small Japanese population, which, between 1955 and 1959, used well-water contaminated with arsenic from a factory producing King's yellow (As_2O_3; Tsuda *et al.*, 1995). The levels of arsenic measured in 34 of 54 wells tested in the area around the factory ranged from undetectable to 3000 µg/L, with 11 wells having levels exceeding 1000 µg/L. A total of 454 residents were enlisted in the cohort. Death certificates, autopsy records and medical records were obtained for the period 1 October 1959 to 30 September 1987. Smoking and occupational histories were ascertained from residents or close relatives. Expected numbers of deaths were based on sex-, age- and cause-specific mortality in Niigata Prefecture from 1960 to 1989. Exposure was grouped into high (≥ 500 μg/L), medium (50–500 μg/L) and low (< 50 μg/L), based on the arsenic content of the well-water. The SMR for urinary-tract cancer in the high-exposure group was 31.2 (95% CI, 8.6–91.8) based on three observed deaths versus 0.1 expected. Two of these were deaths from bladder cancer and one from cancer of the renal pelvis. [Excluding the cancer of the renal pelvis, the SMR for bladder cancer alone would be at least 20.]

2.1.3   Studies in South America

(a)   Argentina

As early as the beginning of the twentieth century, physicians noted an increase in the incidence of clinical skin alterations due to arsenic in well-water in certain areas of the Province of Córdoba, Argentina (Hopenhayn-Rich et al., 1996a). In a study of 2355 deaths in 1949–59 in a highly exposed region, Bergoglio (1964) found that 24% of deaths in the exposed region were due to cancer compared with 15% in the Province of Córdoba. In a 14-year follow-up of 116 patients diagnosed with arsenic-related skin lesions, 30.5% died of cancer, and 12.5% of these deaths were due to cancers of the urinary tract (Biagini, 1972). This was later contrasted with bladder cancer mortality in all of Argentina in 1980, with 2.9% of all cancer deaths attributable to bladder cancer (Hopenhayn-Rich et al., 1996a).

These early reports led to an ecological study on bladder cancer mortality for the period 1986–91 comparing counties categorized as previously having had high, medium and low concentrations of arsenic in water in Córdoba, Argentina (Hopenhayn-Rich et al., 1996a). The majority of reported cases of arsenic-related skin lesions were residents of two counties that were classified as having high exposure since there were extensive reports of elevated concentrations of arsenic in the water there. The average concentration of arsenic in water tested in these counties was 178 μg/L. The medium-exposure group comprised six counties based on some reports of elevated arsenic levels in the water and the occurrence of a few cases of skin lesions. The remaining 16 rural counties were classified as having low exposure. Clear trends in mortality from bladder cancer were observed. Increasing trends were also observed for mortality from kidney cancer as exposure to arsenic increased (Hopenhayn-Rich et al., 1998). No differences were found between the exposure groups for chronic obstructive pulmonary disease, suggesting that the trends for bladder and kidney cancer were not attributable to confounding by smoking.

(b)   Chile

Chile is a long and narrow country divided into geopolitical units called regions (like provinces in other countries), which are numbered sequentially from north to south, starting with Region I. Region II is thus located in the northern part of Chile, in an arid zone where the Atacama Desert is situated. At the time of the 1992 Census, the population in Region II was about 420 000. About 90% of the population live in the cities and towns in this Region, and more than half of the population lives in the city of Antofagasta (Smith *et al.*, 1998). In view of the extremely low level of rainfall and the inability to obtain water from wells, each city and town obtains drinking-water from rivers which originate in the Andes mountains, located on the eastern border of the Region. Many of these rivers are naturally contaminated with inorganic arsenic, some at very high concentrations. This has resulted in widespread exposure of the population of Region II to varying levels of arsenic in the drinking-water. In 1955–70, the majority of the population of Region II was exposed to very high levels of arsenic in drinking-water (see Table 17). Prior to 1955, the drinking-water supply in the main city of Antofagasta had an arsenic concentration of about 90 μg/L. A growing population, and the consequent increased need for water, led to supplementation of the drinking-water supply at Antofagasta with water from the Toconce and Holajar Rivers, which, unknown at the time, had arsenic concentrations of 800 μg/L and 1300 μg/L, respectively. The concentration of arsenic in the drinking-water at Antofagasta, together with that of neighbouring Mejillones, which shared the same supply, increased to an average of 870 μg/L. As shown in Table 17, the other towns in the region, with the exception of Taltal, also had high concentrations of arsenic in the drinking-water for variable periods. The population-weighted average concentration of arsenic in drinking-water for the entire region was about 580 μg/L over a period of approximately 15 years (1955–70). With the introduction in 1970 of a water-treatment plant, the concentration of arsenic in the water at Antofagasta initially dropped to 260 μg/L, and further reductions occurred as a result of improvements to the treatment plant. At present, levels of arsenic in water in Antofagasta are about 40 μg/L. Other cities and towns also implemented water-treatment strategies or used alternative sources that reduced arsenic levels. By the late 1980s, all of the towns with populations over 1000 had concentrations of arsenic in drinking-water of less than 100 μg/L, with the exception of San Pedro de Atacama (population about 3700, some of whom still drink the contaminated water). In contrast, most water sources in the rest of Chile have had low concentrations of arsenic (less than 10 μg/L) (Ferreccio *et al.*, 2000).

Evidence of chronic arsenic toxicity in Region II was noted in the 1960s with the emergence of classic dermatological manifestations (Borgoño & Greiber, 1971; Zaldívar, 1974; Zaldívar *et al.*, 1978). In 1969, in a study of 180 residents in Antofagasta, abnormal skin pigmentation was found in 144 of the participants, 43.7% of whom also had hyperkeratoses (Borgoño *et al.*, 1977). Evidence of effects on the respiratory and cardiovascular system, together with skin lesions, was also reported by Zaldívar *et al.* (1978), who conducted a series of studies concerning the effects of arsenic in Antofagasta during the high-exposure period.

Two ecological mortality studies were conducted on kidney and bladder cancer in Region II. Rivara *et al.* (1997) conducted a study comparing mortality for both sexes combined in Region II with that in Region VIII for the period 1950–92. SMRs for bladder cancer and kidney cancer were 10.2 (95% CI, 8.6–12.2) and 3.8 (95% CI, 3.1–4.7), respectively.

Smith *et al.* (1998) also investigated cancer mortality in Region II for the years 1989–93, using mortality rates in the rest of Chile (excluding Region II) in 1992, a census year, for reference. SMRs were calculated for men and women over the age of 30 years, using 10-year age groupings. The results indicated marked increases in mortality from bladder cancer and kidney cancer in Region II. Data on smoking obtained from a national survey of stratified random samples carried out in 1990, comparing the two largest cities of Antofagasta and Calama with the rest of Chile, were included. No overall increases in mortality from chronic obstructive pulmonary disease were observed in Region II: the SMR for men was 1.0 (0.8–1.1) and mortality among women was lower than expected (SMR, 0.6; 0.4–0.7). In addition, not only did the national survey not find higher rates of smoking in Region II, but in Antofagasta, 76.4% of respondents reported being non-smokers at that time compared with 75.1% of respondents in the rest of Chile. The proportion of people who smoked more than one pack per day was lower in Antofagasta (0.8%) and Calama (1.1%) than in the rest of Chile (1.5%). The SMRs for other causes of death excluding cancers of the bladder, kidney, lung, liver and skin were 1.0 (95% CI, 0.99–1.05) for men and 1.0 (95% CI, 0.97–1.03) for women.

2.1.4   *Studies in the USA*

Some ecological studies have been reported in the USA but they are not informative in view of the relatively small contrasts in exposure between counties.

(a)   *Case–control study*

Bates *et al.* (1995) linked 71 cases of bladder cancer and 160 controls from a sub-sample of residents of Utah from the large national bladder cancer study conducted in 1978 (Cantor *et al.*, 1987) to levels of arsenic in water supplies. Exposures ranged from 0.5 to 160 μg/L, but most concentrations were very low, with only 1.1% of samples having concentrations greater than 50 μg/L. The findings did not provide evidence for an overall increase in the incidence of bladder cancer with the two indices of exposure used. However, among smokers only, there was an increase in risk for the highest category of cumulative dose of arsenic.

(b)   *Cohort study*

Levels of arsenic in drinking-water and mortality were investigated in a cohort of members of the Church of Jesus Christ of Latter-day Saints in Millard County, UT (Lewis *et al.*, 1999). The cohort was assembled from an earlier study that consisted of 2073 participants (Southwick *et al.*, 1983). Most of these individuals had a history of at least 20 years of exposure in their respective places of residence. The cohort was expanded to include all persons who lived for any length of time in the study area, resulting in a total combined cohort of 4058 members. More than 70% of the cohort had reached the age of 60 years at the end of the follow-up period or by the time of their deaths. Approximately 7% of the cohort was lost to follow-up. Arsenic concentrations in the drinking-water supplies were based on measurements maintained by the state of Utah dating back to 1964. The median drinking-water concentrations ranged from 14 to 166 ppb [μg/L], with wide variability in each town. An index of exposure to arsenic was calculated from the number of years of residence and the median concentration of arsenic in drinking-water in a given community, and was categorized as low (< 1000 ppb–years), medium (1000–4999 ppb–years) and high (> 5000 ppb–years). Data on confounding factors were not available; however, Church members are prohibited from using tobacco and consuming alcohol or caffeine. SMRs for kidney cancer were increased in the low- and high-exposure groups among men and in the low- and medium-exposure groups among women. The overall SMRs for cancers of the bladder and other urinary organs were below unity for both sexes, but these results were based on only three and two bladder cancers in men and women, respectively. [The Working Group noted that there were several problems in the interpretation of this study. The exposure assessment was ecological in nature because of relatively low exposures. There was widespread variability in water concentrations, which adds to the uncertainty of the study. Furthermore, the findings are influenced by lower rates of smoking for the cohort compared with all of Utah. This is manifest in the SMRs for non-malignant respiratory disease (SMR for men, 0.7; 95% CI, 0.5–0.9; SMR for women, 0.9; 95% CI, 0.7–1.2) and for mortality from chronic bronchitis, emphysema and asthma (SMR for men, 0.6; 95% CI, 0.4–0.9; SMR for women, 0.5; 95% CI, 0.1–1.2). For these reasons, the study is uninformative with regard to the relationship between exposure to arsenic and mortality from bladder and kidney cancer.]

2.1.5   *Studies in Europe*

In a type of case–control study known as case–cohort, Kurttio *et al.* (1999) investigated the association between low exposure to arsenic in well-water in Finland and the risk for cancers of the bladder and kidney. Cases of bladder and kidney cancer were identified from 1981 to 1995 within a registry-based cohort of the population who had lived at an address outside the municipal drinking-water system during 1967–80. The final study population consisted of 61 cases of bladder cancer and 49 cases of kidney cancer and an age- and sex-matched random sample of 275 subjects from the population register. The daily dose of arsenic in drinking-water was calculated from the concentration of arsenic in well-water and its reported consumption in the 1970s. Cumulative dose was defined as the integral of duration and intensity of exposure to arsenic from well-water. For the shorter latency period of 3–9 years prior to diagnosis of bladder cancer, cumulative dose was estimated from the beginning of use of well-water until 2 years before the diagnosis of cancer. For the longer latency period, the cumulative dose was calculated until 10 years before diagnosis. The concentrations of arsenic in the wells of the reference cohort ranged from less than 0.05 to 64 μg/L (median, 0.14 μg/L). After adjusting for age, sex and smoking, an increasing trend of arsenic in drinking-water and incidence of bladder cancer was observed with shorter latency but not with longer latency, whereas no evidence of an association between kidney cancer and arsenic in well-water was observed.

2.1.6   *Studies in Australia*

Two geographical areas in Victoria, Australia, were selected for study because of reports of concentrations of arsenic in the soil above 100 µg/g and/or concentrations in water above 10 µg/L (Hinwood *et al.*, 1999). Median concentrations of arsenic in water were reported for various towns and showed a wide range up to a median of 1077 µg/L for Ballarat, which had a population of 43 947 in 1986. However, the extent to which contaminated water was used for drinking was not known. The authors noted that "high percentages of the population may be relying on alternative drinking-water sources such as bottled water and tank rain water". Cancer incidence was assessed for the period 1982–91 using the Victorian Cancer Registry. SIRs were estimated for the exposed population in 22 areas of Victoria, using cancer incidence rates for all of the State of Victoria as reference. SIRs for both bladder and kidney cancers were close to unity. [The Working Group noted that no information was presented on the actual use of water contaminated with arsenic for drinking by the population.]

2.2   Liver and lung cancer

2.2.1   *Liver cancer*

A previous IARC monograph on arsenic noted reports of liver angiosarcoma due to medicinal exposure to Fowler's solution (IARC, 1980). A summary of the findings of epidemiological studies on arsenic in drinking-water and risk for liver cancer, mainly hepatocarcinoma, are shown in Table 23.

(*a*)   *Taiwan, China*

(i)   *Ecological studies*

Chen *et al.* (1985) studied the mortality from liver cancer during the period 1968–82 among residents in 84 villages exposed to arsenic of four townships in south-western coastal Taiwan. Increased mortality was observed among both men and women. There was an exposure–response relationship between SMR and prevalence of Blackfoot disease. An exposure–response gradient for mortality from liver cancer was noted in evaluating the risk in areas with shallow wells (presumably with low exposure to arsenic), both shallow and artesian wells (intermediate exposure) and artesian wells only (highest exposure). In villages with artesian wells, the SMR was approximately 2.0 [CI not reported] for liver cancer.

Chen *et al.* (1988a) and Wu *et al.* (1989) reported the age-adjusted mortality rates for liver cancer for men and women in 42 villages in south-western Taiwan and calculated age-adjusted cancer mortality during the period 1973–86 within three groups of villages stratified by exposure concentration (< 300 µg/L, 300–590 µg/L and ≥ 600 µg/L arsenic) tested in 1964–66. Age-adjusted mortality rates (per 100 000 person–years) from liver cancer for residents of all ages in Taiwan (referent) increased with increasing concen-

Table 23. Summary of epidemiological studies on arsenic in drinking-water and risk for liver cancer

Reference	Location	End-point	Exposure	No. of cases	Study outcome	Comments
Ecological studies						
Taiwan						
Chen et al. (1985)	84 villages on the SW coast	Mortality 1968–82, all ages	Endemic area for chronic arsenic toxicity (Blackfoot disease)	Men 305 Women 146	Age- and sex-adjusted SMR (95% CI) 1.7 (1.5–1.9) 2.3 (1.9–2.7)	Mid-year population: 141 733 in 1968, 120 607 in 1982; national rate in 1968–82 used as the standard for estimation of SMR
Chen et al. (1988a)	42 villages on the SW coast	Mortality 1973–86, all ages	Average arsenic (1962–64) General population <300 µg/L 300–590 µg/L ≥ 600 µg/mL General population <300 µg/L 300–590 µg/L ≥ 600 µg/mL	Men Women	Age-adjusted SMR 28.0 32.6 42.7 68.8 8.9 14.2 18.8 31.8	899 811 person-years, rate per 100 000, age-standardized to 1976 world population
Wu et al. (1989)	42 villages on the SW coast	Mortality 1973–86, age ≥ 20	Average arsenic (1962–64) <300 µg/L 300–590 µg/L ≥ 600 µg/mL <300 µg/L 300–590 µg/L ≥ 600 µg/mL	Men 54 42 27 Women 25 16 10	47.8 67.7 86.7 p for trend < 0.05 21.4 24.2 31.8 p for trend < 0.05	Men, 257 935 person-years; women, 234 519 person–years; rate per 100 000, age-standardized to 1976 world population

Table 23 (contd)

Reference	Location	End-point	Exposure	No. of cases	Study outcome	Comments
Chen & Wang (1990)	Taiwan	Mortality 1972–83, all ages	National survey of 83 656 wells (1974–76); average arsenic for each of 314 precincts or townships	Men Women Men Women	*β (SE) from regression* 6.8 (1.3) 2.0 (0.5) *Percentiles of age-adjusted mortality rate/100 000* *person–years* 25th 21.8 50th 27.0 75th 34.1 25th 7.0 50th 8.7 75th 11.6	Regression coefficient (β) estimates increase in age-adjusted mortality per 100 000 per 100 µg/L arsenic increase in water
Tsai et al. (1999)	SW Taiwan, 4 townships	Mortality 1971–94, all ages	Arsenic-exposed area	Men 631 Women 224	SMR (95% CI) 1.8 (1.7–1.98) 1.9 (1.6–2.1)	Men, 1 508 623 person-years; women, 1 404 759 person–years; national rates in 1971–94 used as the standard for estimation of SMR
South America						
Rivara et al. (1997)	Region II and VIII, northern Chile	Mortality 1976–92	Arsenic-contaminated Region II		Relative risk 1.2 (0.99–1.6)	Population: 411 000 in Region II, 1 700 000 in Region VIII. Antofagasta (Region II) versus Region VIII.
Hopenhayn-Rich et al. (1998)	Córdoba Province, Argentina, 26 counties	Mortality 1986–91, age ≥ 20	County group *Men* Low exposure (341 547) Medium exposure (201 546) High exposure (135 209) *Women* Low exposure (348 874) Medium exposure (204 454) High exposure (137 805)	 186 142 98 173 125 90	SMR 1.5 (1.3–1.8) 1.8 (1.5–2.1) 1.8 (1.5–2.2) 1.7 (1.4–1.96) 1.9 (1.6–2.2) 1.9 (1.5–2.4)	National rate in 1989 used as the standard for estimation of SMR

Table 23 (contd)

Reference	Location	End-point	Exposure	No. of cases		Study outcome	Comments
Smith et al. (1998)	Region II, Northern Chile	Mortality 1989–93, age ≥ 30	5-year intervals, 420 µg/L average	Men 48 Women 37		SMR 1.1 (0.8–1.5) 1.1 (0.8–1.5)	National rates in 1991 used as the standard estimation of SMR; arsenic concentration is population-weighted average for major cities or towns in Region II, 1950–74
Australia							
Hinwood et al. (1999)	Victoria	Incidence 1982–91	Median concentration of arsenic in drinking-water ranged 1–1077 µg/L	749		SIR (95% CI) 0.5 (0.3–0.8)	State rates in 1982–91 used as the standard for estimation of SIR
Cohort studies							
Chen et al. (1988b)	SW Taiwan	Mortality	Area endemic for Blackfoot disease	17		SMR: 4.66 (p < 0.001) compared with national standard; 2.48 (p < 0.01) compared with regional standard	789 patients with Blackfoot disease followed from 1968 to 1984. National and regional rates in 1968–83 used as the standard for estimation of SMR
Tsuda et al. (1995)	Niigata Prefecture, Japan	Mortality, 1959–92, all ages	Level of arsenic < 0.05 mg/L 0.05–0.99 mg/L ≥ 1.0 mg/L Total	0 0 2 2		SMR 0.0 (0–4.4) 0.0 (0–15.1) 7.2 (1.3–26.1) 1.5 (0.3–5.5)	113 persons who drank from industrially contaminated wells in 1955–59, then followed for 33 years; rates in Niigata Prefecture in 1960–89 used as the standard for estimation of SMR
Lewis et al. (1999)	Millard County, UT, USA	Mortality	Arsenic in well-water ranged 3.5–620 µg/L	Men 3 Women 7		SMR 0.9 (0.2–2.5) 1.4 (0.6–2.9)	State rates in 1950–92 used as the standard for estimation of SMR.
Nakadaira et al. (2002)	Niigata Prefecture, Japan	Mortality	Industrially contaminated well-water in the town of Nakajo	1		O/E = 0.7	86 patients with chronic arsenic poisoning. National rates in 1959–92 used as the standard for estimation of SMR

Table 23 (contd)

Reference	Location	End-point	Exposure	No. of cases	Study outcome	Comments
Case–control study						
Chen *et al.* (1986)	SW Taiwan, 4 townships	Mortality	Duration of consumption of artesian well-water containing high levels of arsenic	65 cases 368 controls	Age- and gender-adjusted ORs by years of consuming high-arsenic artesian well-water: Never 1.00 1–20 years 0.85 21–40 years 1.24 > 40 years 2.67	ORs calculated using subjects who never consumed artesian well-water as referent Mantel-Haenszel χ^2 value: 9.01 ($p < 0.01$)

SMR, standardized mortality ratio; CI, confidence interval; SIR, standardized incidence ratio; O/E, observed/expected; OR, odds ratio trations of arsenic in water for both men and women in the first study (Chen *et al.*, 1988a), as well as for residents aged 20 years or older in the second study (Wu *et al.*, 1989).

Chen and Wang (1990) analysed nationwide cancer mortality in Taiwan using measurements of arsenic concentrations in water from 83 656 wells located in 314 precincts and townships from 1974 to 1976. Using a multiple regression approach, the authors compared age-adjusted mortality for all ages during the period 1972–83 with arsenic concentrations in these locations. A significant association with concentration of arsenic was found for liver cancer in both men and women. Using multiple linear regression models, a regression coefficient indicating the change in age-adjusted mortality per 100 000 person–years for every 0.1 μg/L increase in arsenic in well-water was calculated, after adjusting for indices of industrialization and urbanization.

Tsai *et al.* (1999) studied mortality from liver cancer in four townships exposed to arsenic in south-western Taiwan during the period 1971–94. SMRs were calculated using two comparison groups: mortality in the whole of Taiwan and mortality in the two counties in which the four townships are located. Although differences in nutrition, socio-economic status or other factors between populations in south-western Taiwan and the remainder of the country may influence their respective cancer rates, Tsai *et al.* (1999) provided evidence that such differences are relatively unimportant. SMRs in both men and women, using both regional and national references, were all close to 1.8.

(ii)   Case–control study

Chen *et al.* (1986) carried out a case–control study on liver cancer and consumption of artesian well-water with high concentrations of arsenic in four townships of south-western Taiwan. A total of 65 cases of liver cancer, identified from death certificates, and 368 healthy controls were studied. Information on consumption of arsenic-contaminated artesian well-water, cigarette smoking, habitual alcohol and tea drinking, and consumption of vegetables and fermented beans was obtained through interview using a standardized questionnaire. Unconditional logistic regression was used to estimate multivariate-adjusted odds ratios for developing liver cancer and various risk factors. There was an exposure–response relationship between the duration of consumption of artesian well-water with high arsenic content and risk for liver cancer.

(b)   Japan

Cohort study

Tsuda *et al.* (1995) found excess mortality from liver cancer among a cohort of 113 persons exposed to levels of arsenic above 1.0 mg/L from industrially contaminated drinking-water in villages of Niigata Prefecture, Japan. The expected number of deaths was based on sex-, age- and cause-specific mortality in Niigata Prefecture in 1960–89. Based on a subgroup of 86 study patients, Nakadaira *et al.* (2002) did not find excess mortality from liver cancer (SMR, 0.7 [95% CI, 0.02–3.9]; one case observed and 1.42 expected). [The small number of liver cancer deaths limited further analysis by severity of chronic arsenic poisoning.] [See complete comment by the Working Group in Section 2.1.2.]

(c)   *Australia*

Hinwood *et al.* (1999) investigated the association between arsenic in drinking-water and liver cancer incidence in Victoria, Australia, in 1982–91. This study included 22 areas where the median arsenic concentration in drinking-water ranged from 14 to 166 µg/L. Using the incidence rate in Victoria, an SIR of 0.5 (95% CI, 0.3–0.8) was observed for liver cancer. [The small number of liver cancer deaths limited further analysis by severity of chronic arsenic poisoning.]

(d)   *South America*

Ecological studies

Rivara *et al.* (1997) compared the mortality from liver cancer in Antofagasta in Region II with that in Region VIII of Chile in 1976–92. The relative risk for liver cancer was 1.2 (95% CI, 0.99–1.6) in arsenic-exposed Region II compared with the control area, Region VIII. [The data source and statistical analysis were not clearly described.]

Smith *et al.* (1998) examined liver cancer mortality during the period 1989–93 among persons 30 years of age and over in Region II of northern Chile. Concentrations of arsenic in drinking-water were well documented and had been high in all major population centres of Region II, especially before 1975. The population-weighted average in the years 1950–74 was 420 µg/L, with a maximum of 870 µg/L in Antofagasta, the largest city, between 1955 and 1969. SMRs for Region II were calculated using the national rate as the standard, and for liver cancer, were 1.1 for both men and women.

Hopenhayn-Rich *et al.* (1998) examined SMRs for liver cancer during the period 1986–91 among residents aged 20 years or older in the 26 counties of Córdoba Province, Argentina. They grouped counties into three strata according to the concentration of arsenic in drinking-water. The low- and intermediate-exposure groups were defined quali-tatively. In the highest exposure group comprising two counties, the concentration of arsenic in drinking-water ranged from 40 to 433 µg/L in the towns of one county and from 50 to 353 µg/L in those of the other. Separate average concentrations in each county were 181 and 174 µg/L. SMRs were calculated using sex- and age-specific rates for Argentina as the referent. Increased mortality from liver cancer was observed for men and women, but SMRs were not related to exposure to arsenic.

(e)   *USA*

Cohort study

Lewis *et al.* (1999) reported the association between arsenic in drinking-water and mortality from liver cancer in a cohort of residents of Millard County, UT, where the median concentration of arsenic in drinking-water ranged from 14 to 166 µg/L. [The limitations of this study are cited in Section 2.1.4.]

2.2.2   *Lung cancer*

A summary of the findings of epidemiological studies on arsenic in drinking-water and risk for lung cancer are shown in Table 24.

> (*a*)   *Taiwan, China*
>
> > (i)   *Ecological studies*

In the study of Chen *et al.* (1985) (described in Section 2.1), increased mortality from lung cancer was observed among men and women in 1968–82 in an area endemic for Blackfoot disease. There was an exposure–response relationship between the SMR and the prevalence of Blackfoot disease. The exposure–response gradient for mortality from lung cancer was noted in evaluating the risk in areas with shallow wells (presumably with low exposure to arsenic), both shallow and artesian wells (intermediate exposure) and artesian wells only (highest exposure). In villages with artesian wells, SMRs were approximately 5.0 [CIs not reported] for lung cancer.

In the studies of Chen *et al.* (1988a) and Wu *et al.* (1989) (described in Section 2.1), age-adjusted mortality rates (per 100 000 person–years) from lung cancer increased with increasing concentrations of arsenic in water for both men and women, for residents of all ages in Taiwan (referent) (Chen *et al.*, 1988a), as well as for residents aged 20 years or older (Wu *et al.*, 1989).

In the analysis of Chen and Wang (1990) (described in Section 2.1.1), regression coefficients (SE) for lung cancer showed a significant association with concentration of arsenic for lung cancer in both men and women.

In the study of Tsai *et al.* (1999) (described in Section 2.1.1) using national and regional rates as standard, SMRs for lung cancer were also increased for both sexes.

> > (ii)   *Cohort study*

Chiou *et al.* (1995) (described in Section 2.1.1) followed 2556 subjects in an area endemic for Blackfoot disease of south-western Taiwan for periods ranging up to approximately 7 years from 1986 to 1993, including 263 patients with Blackfoot disease and 2293 healthy individuals. Results, adjusted for cigarette smoking habits, showed an increased risk for lung cancer in relation to increasing average concentrations of arsenic and to increasing cumulative exposure to arsenic.

> > (iii)   *Case–control study*

Chen *et al.* (1986) (described in Section 2.1.1) studied a total of 76 cases of lung cancer and 368 healthy controls and observed a dose–response relationship between the duration of consumption of artesian well-water containing high levels of arsenic and risk for lung cancer, showing the highest age- and gender-adjusted odds ratio for those who consumed artesian well-water for more than 40 years compared with those who never consumed artesian well-water.

Table 24. Summary of epidemiological studies on arsenic in drinking-water and risk for lung cancer

Reference	Location	End-point	Exposure	No. of cases	Study outcome	Comments
Ecological studies						
Taiwan						
Chen *et al.* (1985)	84 villages on the SW coast	Mortality 1968–82, all ages	Area endemic for chronic arsenic toxicity (Blackfoot disease)		Age- and sex-adjusted SMR (95% CI)	Mid-year population: 141 733 in 1968, 120 607 in 1982; national rate in 1968–82 used as the standard for SMR estimation
				Men 332	3.2 (2.9–3.5)	
				Women 233	4.1 (3.6–4.7)	
Chen *et al.* (1988a)	42 villages on the SW coast	Mortality 1973–1986, all ages	Average arsenic (1964–66)	Men	Age-adjusted SMR	899 811 person–years, rate per 100 000, age-standardized to 1976 world population
			General population		19.4	
			< 300 µg/L		35.1	
			300–600 µg/L		64.7	
			≥ 600 µg/L		87.9	
			General population	Women	9.5	
			< 300 µg/L		26.5	
			300–600 µg/L		40.9	
			≥ 600 µg/L		83.8	
Wu *et al.* (1989)	42 villages on the SW coast	Mortality 1973–86, age ≥ 20	Average arsenic (1964–66)	Men		Men: 257 935 person–years; females, Women: 234 519 person–years; rate per 100 000 age-standardized to 1976 world population
			< 300 µg/L	53	49.16	
			300–600 µg/L	62	100.67	
			≥ 600 µg/L	32	104.08 (*p* for trend < 0.001)	
			< 300 µg/L	Women 43	36.71	
			300–600 µg/L	40	60.82	
			≥ 600 µg/L	38	122.16 (*p* for trend < 0.001)	

Table 24 (contd)

Reference	Location	End-point	Exposure	No. of cases		Study outcome	Comments
Chen & Wang (1990)	Taiwan	Mortality 1972–83, all ages	National survey of 83 656 wells (1974–76); average arsenic for each of 314 precincts or townships	Men Women		β (SE) from regression 5.3 (0.9) 5.3 (0.7) Percentiles of age-adjusted mortality rate/100 000 person–years 25th 11.8 50th 16.2 75th 20.7 25th 5.2 50th 7.4 75th 10.4	Regression coefficient (β) estimates increase in age-adjusted mortality per 100 000 per 100 µg/L arsenic increase in water
				Men Women			
Tsai et al. (1999)	SW Taiwan, 4 townships	Mortality 1971–94, all ages	Arsenic-exposed area	Men Women Men Women Men Women	699 471	SMR (95% CI) 2.6 (2.5–2.8) 3.5 (3.2–3.8) 3.1 (2.9–3.3) 4.1 (3.8–4.5)	Men: 1 508 623 person–years; Women: 1 404 759 person–years National rates in 1971–94 used as the standard for estimation of SMR Regional rates in 1971–94
South America							
Rivara et al. (1997)	Region II and VIII, northern Chile	Mortality 1976–92	Arsenic-contaminated Region II			Relative risk (95% CI) Region II versus region VIII 8.8 (8.1–9.5)	Population: 411 000 in Region II, 1 700 000 in Region VIII. Antofagasta (Region II) versus Region VIII.

Table 24 (contd)

Reference	Location	End-point	Exposure	No. of cases	Study outcome	Comments
Hopenhayn-Rich et al. (1998)	Córdoba Province, Argentina, 26 counties	Mortality 1986–91, age ≥ 20	County group: Men Low exposure Medium exposure High exposure Women Low exposure Medium exposure High exposure	826 914 708 194 138 156	0.92 (0.85–0.98) 1.5 (1.4–1.6) 1.8 (1.6–1.9) 1.2 (1.1–1.4) 1.3 (1.1–1.6) 2.2 (1.8–2.5)	Population: low exposure, 341 547, medium exposure, 201 006; high exposure, 135 209; national rate in 1989 used as the standard for SMR estimation
Smith et al. (1998)	Region II, northern Chile	Mortality 1989–93, age ≥ 30	5-year intervals, 420 μg/L average	Men 544 Women 154	SMR 3.8 (3.5–4.1) 3.1 (2.7–3.7)	National rates in 1991 used as the standard for estimation of SMR; arsenic concentration is population-weighted average for major cities or towns in Region II, 1950–74
Australia						
Hinwood et al. (1999)	Victoria	Incidence 1982–91	Median arsenic concentration in drinking-water ranging 1–1077 μg/L	20	SIR (95% CI) 1.0 (0.9–1.1)	State rates in 1982–91 used as the standard for estimation of SIR
Cohort studies						
Chen et al. (1988b)	SW Taiwan	Mortality 1968–83	Area endemic for Blackfoot disease	28	SMR: 10.49 (p < 0.001) compared with national standard; 2.84 (p < 0.01) compared with regional standard	789 patients with Blackfoot disease followed from 1968 to 1984. National and regional rates in 1968–83 used as the standard for estimation of SMR
Tsuda et al. (1995)	Niigata Prefecture, Japan	Mortality, 1959–92, all ages	Arsenic level: < 0.05 mg/L 0.05–0.99 mg/L ≥ 1.0 mg/L Total	0 1 8 9	SMR 0.0 (0–2.4) 2.3 (0.1–13.4) 15.7 (7.4–31.0) 3.7 (1.8–7.0)	113 persons who drank from industrially contaminated wells in 1955–59, then followed for 33 years; rates in Niigata Prefecture in 1960–89 used as the standard for estimation of SMR

Table 24 (contd)

Reference	Location	End-point	Exposure	No. of cases		Study outcome	Comments
Chiou et al. (1995)	SW Taiwan; 4 neighbouring townships	Incidence 1986–93	Cumulative arsenic exposure (mg/L × year) <0.1 0.1–19.9 ≥20	3 7 7		Relative risk (95% CI) 1.0 3.1 (0.8–12.2) 4.7 (1.2–18.9)	Incidence among a cohort of 2556 subjects (263 Blackfoot disease patients and 2293 healthy individuals) followed for 7 years
			Average arsenic concentration (mg/L) <0.05 0.05–0.70 ≥0.71	5 7 7		1.0 2.1 (0.7–6.8) 2.7 (0.7–10.2)	
Lewis et al. (1999)	Millard County, UT, USA	Mortality	Arsenic in well-water, 3.5–620 µg/L	Men Women	28 6	SMR 0.6 (0.4–0.8) 0.4 (0.2–0.95)	State rates in 1950–92 used as the standard for SMR estimation.
Nakadaira et al. (2002)	Niigata Prefecture, Japan	Mortality	Industrially contaminated well-water in the town of Nakajo	Men Women Total	7 1 8	Poisson probability distribution in men: 9.6 0/E = 11.01	86 patients with chronic arsenic poisoning. National rates in 1959–92 used as the standard for SMR estimation.
Case–control studies							
Chen et al. (1986)	SW Taiwan, 4 townships	Mortality	Duration of consumption of artesian well-water containing high levels of arsenic	76 cases 368 controls		Age- and sex-adjusted OR by years of consuming high-arsenic artesian well-water Never 1.00 1–20 years 1.26 21–40 years 1.52 >40 years 3.39	OR calculated using subjects who never consumed artesian well-water as referent Mantel-Haenszel $\chi 2$ value: 8.49 ($p < 0.01$)

Table 24 (contd)

Reference	Location	End-point	Exposure	No. of cases	Study outcome	Comments
Ferreccio et al. (2000)	Northern Chile	Incidence 1994–96	Individual ≥ 40-year average arsenic concentration from public water supply records during 1930–94 0–10 µg/L 10–29 µg/L 30–49 µg/L 50–199 µg/L 200–400 µg/L	151 cases 419 matched hospital controls	Age- and sex-adjusted OR (95% CI) 1.0 1.6 (0.5–5.3) 3.9 (1.2–12.3) 5.2 (2.3–11.7) 8.9 (4.0–19.6)	OR calculated using subjects with average exposures of 0–10 µg/L as referent

SMR, standardized mortality ratio; CI, confidence interval; SIR, standardized incidence ratio; OR, odds ratio

(b)   *Japan*

Cohort study

Tsuda *et al.* (1995) (described in Section 2.1.2) found excess mortality from lung cancer among a cohort of 113 persons exposed to levels of arsenic above 1.0 mg/L from industrially contaminated drinking-water in villages of Niigata Prefecture, Japan. The expected number of deaths was based on sex-, age- and cause-specific mortality in Niigata Prefecture in 1960–89. Based on a subgroup of 86 study patients, Nakadaira *et al.* (2002) did not find excess mortality from lung cancer.

(c)   *Australia*

Hinwood *et al.* (1999) investigated the association between levels of arsenic in drinking-water and lung cancer incidence in Victoria, Australia, during the period 1982–91. This study included 22 areas where median concentrations of arsenic in drinking-water ranged from 14 to 166 µg/L. Using the incidence rate in Victoria, an SIR of 1.0 was observed for lung cancer.

(d)   *South America*

(i)   *Ecological studies*

Rivara *et al.* (1997) compared the mortality from lung cancer in 1976–92 between Antofagasta in Region II with that in Region VIII of Chile. The relative risk for lung cancer was higher in Antofagasta compared with Region VIII. [The data source and statistical analysis were not clearly described.]

Smith *et al.* (1998) (described in Section 2.2.1) found elevated SMRs of about 3 for lung cancer for both sexes in Region II, using the national rate as standard.

In the study of Hopenhayn-Rich *et al.* (1998) (described in Section 2.2.1), significant increases in the incidence of lung cancer associated with increasing exposure to arsenic were observed for lung cancer.

(ii)   *Case–control study*

Ferreccio *et al.* (2000) conducted a case–control study of incident lung cancer cases in northern Chile. Eligible cases included all lung cancer cases admitted to public hospitals in Regions I, II and III of Chile from November 1994 to July 1996. Eighty to ninety per cent of all cancer patients in the north of Chile are admitted to public hospitals, and a total of 151 cases participated. Controls were selected from all patients admitted to any public hospital in the study region and frequency-matched to cases by age and sex. Two control series were selected: cancers other than lung cancer and non-cancer controls [no response rates were indicated for cases and controls]. Potential biases in control selection were assessed by several approaches including comparisons with geographical distribution of the general population based on census data. Information regarding residential history, socioeconomic status, occupational history (to ascertain employment in copper smelting) and smoking was obtained by questionnaire interview. Historical exposure to arsenic in drinking-water was estimated by linking information on residential history with a database of information on arsenic concentrations in public water supplies collected for the years 1950–94. Arsenic concentrations in the year prior to 1950 were based on concentrations in the 1950s. Average concentration of arsenic in the place of residence was assigned to each subject on a year-by-year basis for the period 1930–94. Population coverage of public water systems in the main cities in Regions I and II was over 90% and was between 80 and 90% in the major cities of Region III. The coverage in smaller cities varied between 64 and 91%. Odds ratios were calculated using unconditional logistic regression, adjusted for age, sex, socioeconomic status, smoking and working in a copper smelter. Results from the analysis based on average exposures during 1930–94 and using all controls showed an increase in the odds ratio with concentration of arsenic. Evidence for a synergistic effect of arsenic in water and smoking was found for those who both smoked and had high concentrations of arsenic in their drinking-water (results not shown).

(e)   *USA*

Cohort study

Lewis *et al.* (1999) reported the association between arsenic in drinking-water and lung cancer mortality in a cohort of residents of Millard County, UT, where the median concentration of arsenic in drinking-water ranged from 14 to 166 µg/L. The SMRs for lung cancer for both men and women were below unity. [Limitations of this study have been cited in Section 2.1.4.]

2.3   Skin cancer

The recognition of arsenic as a carcinogen originally came from case series describing skin cancers following ingestion of arsenical medicine, and exposure to arsenical pesticide residues and arsenic-contaminated drinking-water. Hutchinson (1888) noted skin cancers among patients treated for psoriasis and other ailments with arsenic-containing compounds (e.g. Fowler's Solution containing 1% potassium arsenite). Neubauer (1947) summarized 143 skin cancer cases among arsenic-treated patients. Over 50% of the skin cancers developed in patients treated for 10 years or less and lesions developed after 3–40 years, and on average after 18 years. Clinical reports have described an association of chronic arsenicism with skin cancer in vineyard workers of the Moselle region, Germany (Roth, 1957; Grobe, 1977). Numerous cases of skin cancer have been documented from communities with arsenic-contaminated drinking-water. These include, but are not limited to, case reports from Silesia (Neubauer, 1947), North America (Wagner *et al.*, 1979), Taiwan, China (Yeh, 1973), Argentina (Bergoglio, 1964), Mexico (Cebrían *et al.*, 1983), Chile (Zaldívar, 1974; Zaldívar *et al.*, 1981) and, more recently, Bangladesh (Kurokawa *et al.*, 2001), West Bengal, India (Saha, 2001) and Malaysia (Jaafar *et al.*, 1993). The characteristic arsenic-associated skin tumours include squamous-cell carcinoma arising in keratoses (including Bowen disease) and multiple basal-cell carcinomas (e.g. Neubauer, 1947; Neuman & Schwank, 1960; Yeh *et al.*, 1968). Therefore, this section focuses on these non-melanoma skin cancers. In addition, ecological studies of skin cancer based on mortality rates in areas with low exposure to arsenic such as the USA or exposure imputed from soil levels are not considered here. Findings of epidemiological studies on arsenic in drinking-water and risk for skin cancer are summarized in Table 25.

2.3.1   Taiwan, China

(a)   Ecological studies

(i)   Study based on prevalence of skin cancer

In 1965, Tseng *et al.* (1968) completed a skin cancer prevalence survey on the south-west coast of Taiwan, a region known to have arsenic-contaminated artesian wells that were introduced into the region in 1910–20. A house-to-house examination was conducted of family members from 37 villages: 10 in Chai-yi County, 25 in Tainan County and two in a suburb of Tainan City. The study covered a total population of 40 421 inhabitants. A total of 428 skin cancers were identified, 238 of which were sent for histopathological review. The study was based on clinical diagnoses. In the survey region, 142 water samples from 114 wells were tested for arsenic. Arsenic concentrations ranged from 1 µg/L to 1820 µg/L, and the majority of wells in the endemic region contained between 400 and 600 µg/L arsenic. Skin cancer prevalence was computed according to the median arsenic concentrations per village, categorized as < 300 µg/L, 300–600 µg/L and > 600 µg/L. Villages with either wide-ranging arsenic concentrations or residents who no longer drank the water were deemed to be indeterminate in the analysis. Prevalence rates of skin cancer (based on clinical diagnosis) for inhabitants residing in low- (< 300 µg/L), medium- (300–600 µg/L) and high- (> 600 µg/L) arsenic areas represented over an eightfold difference from the highest to the lowest category.

(ii)   Studies based on incidence of skin cancer

Guo *et al.* (1998, 2001) correlated incidence rates of skin cancer with levels of arsenic in well-water for 243 townships in Taiwan (with a total of about 11.4 million residents). Skin cancers were identified from the National Cancer Registry Program from 1980 to 1987 by a hospital-based registry covering both clinical and pathological diagnoses of skin cancers (reporting is not mandatory by law) (Guo *et al.*, 1998). Arsenic concentrations were provided by a national survey of wells conducted by the Taiwan Provincial Institute of Environmental Sanitation and published in 1977. The investigators modelled the percentage of the population living in townships with drinking-water concentrations of < 50, 50–89, 90–169, 170–329, 330–640 and > 640 µg/L arsenic. The models assume that the same number of individuals drank from each well within a township. The results for all skin cancers combined estimated a relative risk for the highest versus lowest exposure category of about 14 in men and 19 in women, with no excess detected in the other categories (Guo *et al.*, 1998).

Table 25. Summary of epidemiological studies on arsenic in drinking-water and risk for skin cancer

Reference	Location	End-point	Exposure	No. of cases	Study outcome	Comments
Ecological studies						
Taiwan						
Tseng et al. (1968)	40 421 residents from 37 villages (SW)	Prevalence ≥ 20 years of age	Median arsenic concentrations of wells in village of residence (µg/L) <300 300–600 >600	428	Prevalence (per 1000) 2.6 10.1 21.4	Prevalence based on clinical examination of all households. Excludes villages with wells no longer in use or with variations in arsenic concentration (range, 1–1820 µg/L; most wells contained 400–600 µg/L arsenic)
Chen et al. (1985)	4 neighbouring townships on the SW coast	Mortality 1968–82	Areas hyperendemic (21 villages), endemic (25 villages) and not endemic (38 villages) for Blackfoot disease, corresponding to high, medium and low exposure	46 men 49 women	SMR (95% CI) 534 (379–689) 652 (469–835)	Mortality rates in all Taiwan as standard
Chen et al. (1988a)	Region endemic for Blackfoot disease (SW)	Mortality 1973–86	Median arsenic concentrations of well-water (µg/L) <300 300–600 >600	 Men Women Men Women Men Women	SMR 1.6 1.6 10.7 10.0 28.0 15.1	Age-standardized to the 1976 world standard population

Table 25 (contd)

Reference	Location	End-point	Exposure	No. of cases	Study outcome	Comments
Wu et al. (1989)	42 villages in region endemic for Blackfoot disease (SW)	Mortality 1973–86	Median arsenic concentrations of well-water in village of residence (µg/L) in 1964–66 < 300 300–600 > 600 < 300 300–600 > 600	19 men 17 women Men Women	SMR 2.03 14.01 32.41 (p < 0.001) 1.73 14.75 18.66 (p < 0.001)	Age-standardized to the 1976 world standard population
Chen & Wang (1990)	314 precincts and townships	Mortality 1972–83	Average arsenic concentrations	NS Men Women	Increase (β) in mortality rate per 100 000 per 0.1 µg/L increase: β (SE) = 0.9 (0.2) β (SE) = 1.0 (0.2)	Multiple regression adjusted for age and indices of urbanization and industrialization. Mortality rates standardized to the 1976 world standard population
Guo et al. (1998)	243 townships, 11.4 million residents	Incidence 1980–87	Arsenic concentration in wells Exposure categories: > 50, 50–89, 90–169, 170–329, 330–640 and > 640 µg/L	952 men 595 women	Risk difference of 0.34/100 000 (p < 0.01) associated with a 1% increase in arsenic concentrations > 640 µg/L Relative risk of highest versus lowest exposure category: 14.21 in men; 19.25 in women No excess risk for other categories	Rates standardized using the 1976 world standard population. Model assumes that same number of individuals use each well.
Tsai et al. (1999)	Four townships (SW)	Mortality 1971–94	Area endemic for Blackfoot disease	66 men 68 women	Age- and sex-adjusted SMR (95% CI) 4.8 (3.7–6.2) 5.97 (4.6–7.6) 5.7 (4.4–7.2) 6.8 (5.3–8.6)	Local standard National standard Local standard National standard

Table 25 (contd)

Reference	Location	End-point	Exposure	No. of cases	Study outcome	Comments
Guo et al. (2001)	243 townships, 11.4 million residents	Incidence 1980–89	Concentration of arsenic in well-water: Exposure categories (µg/L) arsenic	2369 (1415 men, 954 women)	Rate difference association with a 1% increase in residents with categories of arsenic (µg/L):	Cancers identified through National Cancer Registry. Models include age and urbanization index. Models assume same number of individuals use each well. BCC, basal-cell carcinoma SCC, squamous-cell carcinoma
			Intercept	764 BCC Men	0.779	$*p < 0.05$
				Women	−0.002	$**p < 0.01$
			50–89	Women	0.004	
				Women	−0.012	
			90–169	Men	−0.017	
				Women	0.018	
			170–329	Men	0.006	
				Women	0.004	
			330–640	Men	−0.024	
				Women	0.016	
			> 640	Women	0.128**	
				Men	0.027	
			Intercept	736 SCC Women	0.821	
				Men	1.488	
			50–89	Women	0.024	
				Men	−0.006	
			90–169	Women	−0.026	
				Men	0.006	
			170–329	Women	0.073**	
				Men	0.016	
			330–640	Women	−0.100**	
				Men	−0.064*	
			> 640	Women	0.155**	
					0.212**	
				182 melanoma	No increase associated with melanoma	

Table 25 (contd)

Reference	Location	End-point	Exposure	No. of cases	Study outcome	Comments
Guo et al. (2001) (contd)			Three categories of township: (1) No well with arsenic > 40 μg/L; (2) Some wells > 40 μg/L but none > 640 μg/L; (3) More wells > 640 μg/L than between 320 and 640 μg/L		Dose–response relationship between basal-cell and squamous-cell skin cancer in both men and women and in all age categories (except for basal-cell < 30 years of age, which had few subjects). No consistent increase in melanoma incidence by exposure category	
Mexico						
Cebrián et al. (1983)	Two rural populations in Lagunera region; 2486 residents	Prevalence (time frame not specified)	Town of El Salvador de Arriba: high exposure to arsenic (410 μg/L); town of San Jose del Vinedo: low exposure (5 μg/L)	4	High exposure: 1.4% (4 cases in 57 households and 296 individuals); low exposure: 0% (0 cases in 68 households and 318 individuals)	Epidermoid or basal-cell carcinomas detected on physical exam of every 3rd household
Chile						
Zaldivar et al. (1974)	City of Antofagasta	Incidence of cutaneous lesions of chronic arsenic poisoning, 1968–71	Concentration of arsenic fell from 580 μg/L in 1968–69 to 8 μg/L in 1971		Incidence rates: Men: 145.5/100 000 in 1968–69, 9.1/100 000 in 1971; women: 168.0/100 000 in 1968–69 and 10.0/100 000 in 1971	
Rivara et al. (1997)	Regions II and VIII	Mortality 1976–92	Exposed group: Antofagasta in region II (arsenic concentration in drinking-water, 40–860 μg/L; 1950–92) Unexposed group: region VIII, no arsenic contamination (reference)	NS	SMR (95% CI) 3.2 (2.1–4.8)	
Smith et al. (1998)	Region II, northern Chile	Mortality 1989–93, age ≥ 30	Annual average arsenic concentrations ranging 43–569 μg/L in 1950–94	20 men 7 women	SMR (95% CI) 7.7 (4.7–11.9) 3.2 (1.3–6.6)	Age-standardized to the national rates of Chile in 1991

Table 25 (contd)

Reference	Location	End-point	Exposure	No. of cases	Study outcome	Comments
USA						
Berg & Burbank (1972)		Mortality 1950–67	Trace metals in water supplies from 10 basins throughout the USA; concentration of arsenic in water, Oct. 1962–Sept. 1967		No correlation	
Morton et al. (1976)	Lane County, OR	Incidence 1958–71	Mean arsenic concentration in municipal water system and single-family systems	3039	Correlation of arsenic content in drinking-water: squamous-cell carcinoma: 0.151 for men and −0.20 for women; basal-cell carcinoma: −0.064 for men and 0.10 for women	Non-melanoma cases identified by review of pathology reports
Wong et al. (1992)	Four counties in Montana	Incidence 1980–86	Two contaminated counties (copper smelter and copper mines); two control counties	Around 2300 in the 4 counties	Age-adjusted skin cancer incidence higher in control counties	Overall incidence rates for exposed counties within range observed for other US locations
Cohort studies						
Taiwan						
Chen et al. (1988b)	Four townships (SW)	Mortality 1968–83, all ages	Diagnosis of Blackfoot disease as a surrogate for high exposure to arsenic	7	SMR 28.46 ($p < 0.01$) (national standard) 4.51 ($p < 0.05$) (local standard)	871 people who developed Blackfoot disease after 1968 were followed for 15 years. National standard used for the age- and sex-standardized rates of the general Taiwanese population.
Hsueh et al. (1997)	Three villages in Putai township (SW)	Incidence 1989–92, age, ≥ 30 years	Duration of residence in area endemic for Blackfoot disease (years) ≥ 33 34–43 44–53 > 53	1 4 8 20	Relative risk (95% CI) 1.0 5.01 (0.5–48.1) 4.9 (0.6–41.6) 6.8 (0.9–53.7) (p for trend = 0.07)	654 subjects (275 men and 379 women) without skin cancer followed with dermatological examinations. Total of 2239 person–years. Relative risk adjusted for age, sex and level of education

Table 25 (contd)

Reference	Location	End-point	Exposure	No. of cases	Study outcome	Comments
Hsueh et al. (1997) (contd)			Duration of consumption of artesian well-water (years)			OR adjusted for age, sex, cumulative exposure to arsenic, serum cholesterol and triglyceride levels, cigarette smoking and alcohol drinking
			0	1	1.0	Incidence 14.74/1000 person–years
			1–15	1	1.2 (0.4–19.7)	
			16–25	8	3.9 (0.5–32.1)	
			> 25	23	8.9 (1.1–72.9)	
					(p for trend < 0.05)	
			Average concentration of arsenic in drinking-water (mg/L)			
			0	1	1.0	
			0.01–0.70	12	3.3 (0.4–35.8)	
			0.71–1.10	13	8.7 (1.1–65.5)	
					(p for trend < 0.05)	
			Unknown	7	4.8 (0.6–40.4)	
			Cumulative exposure to arsenic (mg/L–years)			
			0	1	1.0	
			0.1–10.6	2	2.8 (0.3–31.9)	
			10.7–17.7	5	2.6 (0.3–22.9)	
			> 17.7	18	7.6 (0.95–60.3)	
					(p for trend = 0.06)	
			Unknown	7	5.1 (0.6–44.4)	
			Level of serum β-carotene (µg/mL)	16 cases (61 controls)	OR (95% CI)	
			≤ 0.14		1.0	
			0.15–0.18		0.4 (0.1–2.9)	
			> 0.18		0.01 (0.0–0.4) (p for trend < 0.01)	

Table 25 (contd)

Reference	Location	End-point	Exposure	No. of cases	Study outcome	Comments
Case–control studies						
USA						
Karagas *et al.* (2001b, 2002)	New Hampshire	Incidence 1993–96	Concentration of arsenic in toenails (µg/g)		OR (95% CI) *Squamous-cell carcinoma*	OR adjusted for age and sex 284 cases, 524 controls
			0.009–0.089	155	1.0	
			0.090–0.133	64	0.9 (0.6–1.3)	
			0.134–0.211	33	0.98 (0.6–1.6)	
			0.212–0.280	14	1.1 (0.55–2.2)	
			0.281–0.344	5	1.0 (0.3–3.0)	
			0.345–0.81	13	2.1 (0.9–4.7)	
					Basal-cell carcinoma	587 cases, 524 controls
			0.009–0.089	281	1.0	
			0.090–0.133	156	1.01 (0.8–1.4)	
			0.134–0.211	92	1.06 (0.7–1.5)	
			0.212–0.280	22	0.7 (0.4–1.3)	
			0.281–0.344	10	0.8 (0.3–1.8)	
			0.345–0.81	26	1.4 (0.7–2.8)	
Nested case–control study						
Taiwan						
Hsueh *et al.* (1995)	Three villages in Putai Township (SW)	Prevalence	Duration of residence in area endemic for Blackfoot disease (years)		OR (95% CI)	OR adjusted for age and sex; 1081 residents (468 men, 613 women) underwent a physical examination.
			≤ 45	2	1.0	
			46–49	11	5.2 (1.1–25.8)	
			≥ 50	53	8.5 (1.96–37.2)	
					p for trend < 0.05	

Table 25 (contd)

Reference	Location	End-point	Exposure	No. of cases	Study outcome	Comments
Hsueh et al. (1995) (contd)			Duration of drinking artesian well-water (years)			
			≤ 13	2	1.0	
			14–25	15	5.1 (1.03–24.98)	
			≥ 60	52	6.4 (1.4–27.9)	
					p for trend < 0.05	
			Average exposure to arsenic (ppm)			
			0	2	1.0	
			0–0.70	20	3.5 (0.7–17.0)	
			> 0.71	30	5.0 (1.1–23.8)	
			Cumulative exposure to arsenic (ppm–years)			
			≤ 4	1	1.0	
			5–24	22	8.9 (1.1–73.8)	
			≥ 25	28	13.7 (1.7–111.6)	
			Chronic hepatitis B carrier and liver function status:			
			Non-carrier with normal liver function	41	1.0	
			HBsAg carrier with normal liver function	13	1.1 (0.56–2.2)	
			Non-carrier with liver dysfunction	3	2.1 (0.54–7.7)	
			HBsAg carrier with liver dysfunction	4	8.4 (2.37–29.9)	
					p for trend < 0.05	

SMR, standardized mortality ratio; CI, confidence interval; NS, not specified; OR, odds ratio

In a second study, the incidence rates of each major histological type of skin cancer (basal-cell, squamous-cell and melanoma) were modelled using the previously defined exposure categories (Guo *et al.*, 2001). The model included the percentage of residents in categories of age (30–49, 50–69 and > 69 years) and urbanization index as covariates. In this analysis, they found a statistically significant increase in squamous-cell skin cancer in the highest category (> 640 µg/L arsenic) in both men and women. Similarly, for basal-cell carcinoma, there was also a positive association in the highest exposure category. Melanoma was unrelated to any category of exposure to arsenic. In a post-hoc analysis, the investigators created three exposure groups: townships with no well containing levels of arsenic above 40 µg/L, townships with some wells containing more than 40 µg/L arsenic but none above 640 µg/L and townships with more wells containing more than 640 µg/L than between 320 and 640 µg/L. A dose–response relation between basal-cell and squamous-cell skin cancer rates was observed in both men and women and in all age categories (except for basal-cell cancer below age 30 years, which had few subjects). Again, no association was seen for melanoma. [The Working Group noted that the reporting of incident skin cancer may have been incomplete.]

(iii)   *Studies based on mortality from skin cancer*

A series of studies conducted in Taiwan, China (Chen *et al.*, 1985, 1988a; Wu *et al.*, 1989; Chen & Wang, 1990; Tsai *et al.*, 1999) (described in Section 2.1.1) analysed skin cancer mortality in relation to levels of arsenic in well-water. The overall SMRs in the four counties were 534 in men and 652 in women, with 100 as the referent value. There was also a gradient of increasing SMRs for skin cancer in areas not endemic or hyperendemic for Blackfoot disease (Chen *et al.*, 1985). A subsequent report (Chen *et al.*, 1988a) presented SMRs for skin cancer grouped by the median levels of arsenic in well-water measured in 1962–64 into < 300, 300–600 and > 600 µg/L. The age-standardized mortality rates for skin cancer increased in the three categories for both genders (Chen *et al.*, 1988a). Wu *et al.* (1989) published age-adjusted mortality rates from skin cancer for the years 1973–86 for the four townships endemic for Blackfoot disease plus 15 villages in the townships of Yensui and Hsiaying. Using the same data on well-water contents and the classification described above, skin cancer mortality rates also increased in men and women with increasing median concentrations of arsenic. Subsequently, Chen and Wang (1990) used data on wells measured from 1974 to 1976 from 314 precincts and townships throughout Taiwan. Based on a multiple regression analysis (adjusted for urbanization and age), a 0.1-µg/L increase in arsenic corresponded to an increase of 0.9 (SE, 0.2) and 1.0 (SE, 0.2) per 100 000 in skin cancer mortality in men and women, respectively. More recently, Tsai *et al.* (1999) computed SMRs for the period 1971–94 for the four townships endemic for Blackfoot disease by applying both local (Chiayi County) and national (Taiwan) rates as the standards. The SMRs for skin cancer were approximately 5.0 for men and women, using both local and national standards.

(b)   *Cohort studies*

Chen *et al.* (1988b) conducted a retrospective cohort study of 871 patients who met clinical criteria of Blackfoot disease after January 1968. From 1968 to 1983, seven death certificates identified the underlying cause of death as skin cancer. SMRs for skin cancer were computed using both the area endemic for Blackfoot disease and the whole of Taiwan as the population standards. The age-standardized observed versus expected number of skin cancer deaths was around 28 using Taiwan as the standard and 4.5 using the Blackfoot disease townships as the standard. [Use of the national population as the standard may provide a better estimate of excess risk than use of the local population for whom arsenic concentrations in well-water were elevated.]

Hsueh *et al.* (1997) established a cohort of residents living in three villages (Homei, Fuhsin and Hsinming) in Putai Township, one of the regions of the south-west coast of Taiwan endemic for Blackfoot disease. Arsenic-contaminated well-water had existed in these villages for over 50 years, with reported values ranging from 700 to 930 µg/L. The government introduced a new water system in the 1960s, with relatively low penetration until the 1970s. Well-water remained in use for agriculture and aquaculture. The 1571 residents aged 30 years or older who lived for at least 5 days a week in the villages were recruited to take part in the study. Of these, 1081 (68.8%) participated in a physical examination (468 men and 613 women). The 1015 study subjects who did not have a prevalent skin cancer comprised the cohort. Of these, 275 men and 379 women (64%) underwent regular examinations for dermatological conditions. Thirty-three incident skin cancers developed during the follow-up period from September 1989 through December 1992, with a total of 2239 person–years and a rate of 14.74/1000 person–years. Public health nurses interviewed cohort members on length of residence, history of drinking well-water and other potentially confounding factors. An index for cumulative exposure to arsenic was derived for each subject using concentrations of arsenic in well-water for each village of residence measured in the 1960s. These data were multiplied by the duration of consumption of well-water in each village of residence and totalled for each residence. Dermatologists diagnosed clinically the skin cancers that occurred in the cohort, and 91% of clinically diagnosed carcinomas were confirmed histologically. Fasting blood samples and urine samples were collected at the time of interview. Serum was analysed for β-carotene and urine for arsenic metabolites from 16 skin cancer cases [48%] and 61 age- and sex-matched controls. Cox proportional hazard regression was used to analyse data on exposure to arsenic in relation to incidence of skin cancer adjusted for age, sex and educational level. Conditional logistic regression analysis was used to assess the effects of β-carotene and urinary metabolites on the risk for skin cancer. Risk for skin cancer was significantly related to duration of living in the area endemic for Blackfoot disease, duration of consumption of artesian well-water, average concentration of arsenic and index for cumulative exposure to arsenic. There was evidence of a reduced risk for skin cancer in the highest two tertiles of β-carotene versus the lowest tertile after adjustment for multiple potential confounders. Also, compared with controls, cases of skin cancer had higher total urinary concentration of arsenic, percentage of MMA and ratio of MMA to inorganic arsenic and a lower percentage of DMA and ratio of DMA to MMA (not shown in Table 25).

Hsueh *et al*. (1995) conducted a nested case–control study within the cohort study of Hsueh *et al*. (1997). A total of 66 prevalent skin cancers were identified after clinical examination. Age- and sex-adjusted prevalence odds ratios were computed for the same variables of exposure to arsenic included in the cohort analysis: duration of living in the area endemic for Blackfoot disease, duration of drinking artesian well-water, and average and cumulative exposure to arsenic. Significant increases in risk for each exposure group were shown. In addition, a significantly elevated risk for skin cancer was observed among chronic hepatitis B carriers with liver dysfunction.

2.3.2 *Mexico*

Ecological study based on prevalence of skin cancer

Cebrían *et al*. (1983) reported results from a household survey of two rural Mexican towns in the Region of Lagunera. Towns were selected on the basis of levels of arsenic in drinking-water, one town with a high level of arsenic (average, 410 µg/L) and the other with a low level of arsenic (average, 5 µg/L). The towns are located 37 km apart, and their populations were of comparable size (1488 and 998 inhabitants, respectively) and socio-economic and environmental conditions. A questionnaire and physical examination were administered to all family members of every third household in each community. Seventy-five per cent of the residents had lived in these communities since birth. After clinical diagnosis, prevalence of epidermoid or basal-cell carcinoma (referred to as ulcerative lesions) was 1.4% in the exposed town, whereas no case was observed in the control town. The 20 water samples tested from the exposed town between 1975 and 1978 had arsenic concentrations ranging from 160 to 590 µg/L (standard deviation [SD], 114 µg/L), indicating significant variability. The control town showed little variability (SD, 7 µg/L) in the 18 samples collected over the same period.

2.3.3 *Chile*

Ecological study based on incidence of cutaneous lesions

Zaldívar *et al*. (1981) investigated the incidence of cutaneous lesions (leukoderma, melanoderma, hyperkeratosis, and squamous-cell carcinoma) in residents of Antofagasta in arsenic-contaminated Region II from 1968 to 1971. Among 457 patients, about 70% were children aged 0–15 years. Incidence rates decreased from 1968–69 to 1971 due to a filter plant which started operation in 1970.

Ecological studies based on mortality from skin cancer

In an ecological analysis, Rivara *et al*. (1997) compared mortality rates from skin cancer in 1976–92 between Antofagasta and the unexposed control Region VIII of Chile.

The SMR for skin cancer was 3.2 (95% CI, 2.1–4.8). In a later study, Smith *et al.* (1998) compared sex- and site-specific mortality for the years 1989–93 in Region II of Chile with national mortality rates. The SMR for skin cancer was 7.7 (95% CI, 4.7–11.9) among men and 3.2 (95% CI, 1.3–6.6) among women.

2.3.4   USA

(a)   Ecological study

Berg and Burbank (1972) showed no correlation between trace metals from 10 water basins throughout the USA and mortality rates in 1950–67, using concentrations of arsenic measured in 1962–67.

Morton *et al.* (1976) studied the incidence of histologically confirmed basal-cell and squamous-cell skin cancers (*in situ* and invasive) for the period 1958–71 in Lane County, OR, USA. They identified skin cancers by reviewing the pathology records of facilities serving residents of the county and the biopsy files of two of five dermatologists. Water samples were tested from selected points throughout the county from public water systems and from a number of single-family systems. Single-family systems were reported to have been over-sampled in regions suspected of having a problem with arsenic. Within a given region, arsenic values ranged from undetectable to 2150 µg/L. The correlation between census tract estimates of arsenic in drinking-water and incidence of squamous-cell carcinoma was 0.151 in men and –0.20 in women; for the incidence of basal-cell carcinoma, the correlation was –0.064 in men and 0.10 in women. [A major weakness of the study is the misclassification of exposure because of widely varying concentrations of arsenic within a census tract.]

Wong *et al.* (1992) studied approximately 2300 incident cases of skin cancer in four counties in Montana (two contaminated and two controls) in 1980–86. Contamination arose from copper smelters and mines. No difference in incidence rates was observed between exposed counties and the rest of the country.

(b)   Case–control studies

Karagas *et al.* (1998, 2001b, 2002) designed a case–control study of basal-cell and squamous-cell skin cancers in the population of New Hampshire to evaluate the effects of low to moderate levels of exposure to arsenic. About 40% of the population relied on private, unregulated water systems; over 10% of the private supplies contained levels of arsenic above the WHO recommended level of 10 µg/L and 1% of supplies overall contained > 50 µg/L. A biomarker of internal dose was chosen to determine exposure levels. Earlier studies had indicated the reliability of concentrations of arsenic in toenails as a measure of exposure > 1 µg/L arsenic through drinking-water (Karagas *et al.*, 1996, 2000) and reproducibility of concentrations over a period of 3–6 years (Garland *et al.*, 1993; Karagas *et al.*, 2001a). Cases of basal-cell and squamous-cell skin cancer diagnosed from 1 July 1993 to 30 June 1995 were identified through a statewide network of dermatologists, dermatopathologists and pathologists, with participation rates of over 90% (Karagas *et al.*,

2001b). Because of the high incidence of basal-cell carcinoma, incident cases of basal-cell carcinoma were randomly selected in a 2:1 ratio to cases of squamous-cell carcinoma. To minimize detection bias, cases of squamous-cell carcinoma were restricted to invasive disease only (cases of in-situ carcinomas were excluded). A 2:1 ratio of controls to squamous-cell carcinoma cases was randomly selected from population lists (driver's licence files for cases < 65 years and Medicare enrollment lists for cases ≥ 65 years), frequency-matched to the combined distribution of the basal-cell and squamous-cell carcinoma cases. To be eligible to participate, subjects were required to speak English and have a working telephone. Of the 1143 potential case subjects, 896 took part in the study (78%) and, of the 820 potential controls, 540 (66%) enrolled. The analysis included the 587 cases of basal-cell carcinoma, 284 cases of squamous-cell carcinoma and 524 controls (97% of subjects) who contributed a toenail sample for arsenic analysis. Study participants underwent a personal interview to obtain information on confounding factors such as exposure and sensitivity to sun, history of radiation treatment and other medical and lifestyle factors. Age- and sex-adjusted odds ratios were computed using logistic regression analysis according to percentiles of arsenic concentrations in toenails based on the control distribution. In this categorical analysis, concentrations of arsenic appeared to be unrelated to risk for squamous-cell and basal-cell carcinomas except for the highest category (the top 97th percentile; concentrations above 0.344 μg/g) versus concentrations below the median.

An analysis using continuous exposure variables was presented separately (Karagas et al., 2002). A quadratic and two-segment linear model fitted the data for both squamous-cell carcinoma and basal-cell carcinoma. The point at which the dose–response appeared to increase was at 0.105 μg/g (95% CI, 0.093–0.219 μg/g) for squamous-cell carcinoma using a maximum likelihood estimation of the change point for the two-segment linear model. After the change point, a 1% increase in arsenic concentration in toenails was related to a 0.61% increase in risk for squamous-cell carcinoma. The quadratic model for both squamous-cell carcinoma and basal-cell carcinoma produced a consistent nadir or change point of 0.088 and 0.091, respectively. However, it was not possible to estimate a two-segment model for basal-cell carcinoma because of sparse data at the extremes. Based on a regression analysis of concentrations of arsenic in water and toenails, a change point of 0.105 μg/g in toenails translated to 1–2 μg/L in water, with the 95% confidence interval ranging from < 1 to 10–20 μg/L.

2.4   Other organ sites

Studies on cancer at other organ sites are summarized in Table 26.

Neubauer (1947) summarized cancers that had been reported in patients treated with medicinal arsenic. His report of 143 published cases included patients who developed cutaneous tumours and other malignancies such as cancers of the stomach (one case), tongue (two cases, one who also had cancer of oral mucosa), oesophagus (two cases), uterus (one case) and urethra (two cases including one papilloma of the ureter). Among patients who had not developed cutaneous tumours, other reported malignancies included

Table 26. Summary of epidemiological studies of arsenic in drinking-water and risk for other cancers

Reference	Location	End-point	Exposure	Site	No. of cases	Study outcome	Comments
Ecological studies							
Taiwan							
Chen et al. (1985)	84 villages on the SW coast	Mortality 1968–82, all ages	Endemic area for chronic arsenic toxicity (Blackfoot disease)		Men	Age- and sex-adjusted SMR (95% CI)	SMRs for a total of 11 sites. Prostate not investigated; mid-year population: 141 733 in 1968, 120 607 in 1982; national rate in 1968–82 used as the standard for estimation of SMR
				Colon	54	1.6 (1.2–2.0)	
				Small intestine	17	2.98 (0.6–3.4)	
				Leukaemia	45	1.4 (1.0–1.8)	
					Women		
				Colon	61	1.7 (1.3–2.1)	
				Small intestine	5	0.97 (0.1–1.8)	
				Leukaemia	22	0.9 (0.5–1.3)	
Chen et al. (1988a)	42 villages on the SW coast	Mortality 1973–86, all ages	Median level of arsenic in drinking-water grouped into 3 strata, 1962–64 General population < 300 µg/L 300–600 µg/L > 600 µg/L	Prostate		Age-adjusted SMR 1.5 0.5 5.8 8.4	899 811 person–years, rate per 100 000, age-standardized to 1976 world population

Table 26 (contd)

Reference	Location	End-point	Exposure	Site	No. of cases	Study outcome	Comments
						SMR	
Wu et al. (1989)	42 villages (SW)	Mortality 1973–86, age ≥ 20 years	Median level of arsenic in drinking-water grouped into 3 strata, 1962–64				Age-adjusted mortality. All results are non-significant. Men, 257 935 person–years; women, 234 519 person–years; rate per 100 000, age-standardized to 1976 world population
			< 30 µg/L	Prostate	9 M	0.95	
				Leukaemia	11 M	4.87	
					7 F	3.03	
				Nasopharynx	11 M	3.58	
					7 F	1.59	
				Oesophagus	15 M	7.62	
					4 F	1.83	
				Stomach	46 M	25.6	
					21 F	6.71	
				Colon	17 M	7.94	
					21 F	9.05	
				Uterine cervix	6 F	0.91	
			300–600 µg/L	Prostate	9 M	9.00	
				Leukaemia	11 M	6.52	
					7 F	4.55	
				Nasopharynx	11 M	8.16	
					7 F	5.81	
				Oesophagus	15 M	9.37	
					4 F	3.64	
				Stomach	46 M	17.82	
					21 F	18.72	
				Colon	17 M	8.30	
					21 F	8.16	
				Uterine cervix	6 F	5.46	

Table 26 (contd)

Reference	Location	End-point	Exposure	Site	No. of cases	Study outcome		Comments
Wu et al. (1989) (contd)			> 600 μg/L	Prostate	9 M	9.18		
				Leukaemia	11 M	2.69		
					7 F	0.00		
				Nasopharynx	11 M	8.58		
					7 F	4.89		
				Oesophagus	15 M	6.55		
					4 F	0.00		
				Stomach	46 M	56.42		
					21 F	5.98		
				Colon	17 M	12.5		
					21 F	17.21		
				Uterine cervix	6 F	3.92		

						Percentiles of age-adjusted mortality rate/100 000 person–years		
Chen & Wang (1990)	42 villages on the SW coast	Mortality 1972–83, all ages	National survey of 83 656 wells (1974–76); average arsenic content for each of 314 precincts or townships		*Men*	*25th*	*50th*	*75th*
				Oesophagus		3.6	6.0	9.2
				Stomach		14.8	10.2	28.8
				Small intestine		0.6	1.1	1.9
				Colon		4.2	5.6	7.2
				Rectum		1.9	2.7	3.9
				Pancreas		1.3	2.1	3.0
				Nasal cavity		1.1	1.3	2.6
				Larynx		1.1	1.7	2.8
				Bone/cartilage		1.1	1.8	2.9
				Prostate		0.9	1.4	2.3
				Brain		0.7	1.1	1.8
				Leukaemia		1.3	2.1	2.7
					Women	*25th*	*50th*	*75th*
				Oesophagus		1.1	1.8	2.8
				Stomach		7.2	10.0	13.7
				Small intestine		0.6	0.9	1.6
				Colon		3.6	5.5	6.9

Table 26 (contd)

Reference	Location	End-point	Exposure	Site	No. of cases	Study outcome	Comments
Chen & Wang (1990) (contd)				Rectum		1.5　2.3　3.3	
				Pancreas		1.4　2.1　2.7	
				Nasal cavity		0.6　1.0　1.6	
				Larynx		0.5　0.9　1.5	
				Bone/cartilage		1.0　1.7　2.5	
				Breast		2.7　4.4　6.2	
				Cervix uteri		3.8　6.2　8.3	
				Ovary		0.8　1.4　2.0	
				Brain		1.0　1.5　2.2	
				Leukaemia		1.1　1.7　2.4	
Tsai *et al.* (1999)	Four townships (SW)	Mortality 1971–94, all ages	Endemic area for chronic arsenic toxicity			SMR (95% CI)	SMRs with national reference, unless otherwise stated; *SMRs with local reference Men, 1 508 623 person–years; women, 1 404 759 person–years; national rates in 1971–94 used as the standard for SMR estimation
				Men			
				Pharynx	24	1.1 (0.7–1.7)	
				Oesophagus	69	1.7* (1.3–2.1)	
				Stomach	195	1.4* (1.2–1.5)	
				Intestine	15	2.1 (1.2–3.5)	
				Colon	91	1.4 (1.1–1.7)	
				Rectum	46	1.2 (0.9–1.7)	
				Nasal cavity	40	3.7 (2.6–5.0)	
				Larynx	30	1.8 (1.2–2.5)	
				Bone	41	2.3 (1.7–3.2)	
				Prostate	48	1.96 (1.4–2.6)	
				Brain	19	1.1 (90.7–1.8)	
				Lymphoma	56	1.4 (1.1–1.8)	
				Leukaemia	67	1.3 (1.04–1.7)	
				Women			
				Pharynx	10	2.2 (1.1–4.1)	
				Oesophagus	12	0.8 (0.4–1.4)	
				Stomach	111	1.4* (1.2–1.7)	
				Intestine	8	1.3 (0.5–2.5)	
				Colon	83	1.4* (1.1–1.8)	
				Rectum	33	1.5* (1.03–2.11)	
				Nasal cavity	29	5.1 (3.4–7.3)	
				Larynx	13	3.8 (2.0–6.4)	
				Bone	34	2.2 (1.5–3.1)	
				Brain	21	1.8* (1.1–2.7)	
				Lymphoma	35	1.4 (1.0–2.0)	
				Leukaemia	40	1.1 (0.8–1.4)	

Table 26 (contd)

Reference	Location	End-point	Exposure	Site	No. of cases	Study outcome	Comments
Chile							
Rivara et al. (1997)	Regions II and VIII, northern Chile	Mortality 1950–92	Arsenic-contaminated Region II	Larynx		Relative risk (Region II versus Region VIII), 3.4 (95% CI, 1.3–8.6)	Population: 411 000 in Region II, 1 700 000 in Region VIII. Antofagasta (Region II) versus Region VIII.
USA							
Berg & Burbank (1972)	10 water basins	Mortality 1950–67	Trace metals in water supplies (As, Be, Cd, Cr, Co, Fe, Pb, Ni)	Larynx Eye Myeloid leu-kaemia		Probability of a positive association 0.024 0.009 0.042	
Australia							
Hinwood et al. (1999)	Victoria	Incidence 1982–91	Median arsenic concentration in drinking-water ranging 1–1077 µg/L	Prostate Melanoma Breast Chronic myeloid	*Obs. no.* 619 477 762 40	SIR (95% CI) 1.1 (1.05–1.2) 1.4 (1.2–1.5) 1.1 (1.03–1.2) 1.5 (1.1–2.1)	State rates in 1982–91 used as the standard for estimation of SIR

Table 26 (contd)

Reference	Location	End-point	Exposure	Site	No. of cases	Study outcome	Comments
Case–control study							
Canada							
Infante-Rivard *et al.* (2001)	Québec province	Incidence 1980–93	Trihalomethanes, metals (As, Cd, Cr, Pb, Zn) and nitrates in drinking-water during prenatal and postnatal periods	Childhood acute lymphocytic leukaemia			Adjusted for maternal age and level of education
			Arsenic exposure index Average level (> 95th versus ≤ 95th percentile [5 μg/L])		*Exposed cases* Prenatal, 18 Postnatal, 20	OR (95% CI) *Prenatal period* 0.9 (0.5–1.8)	*Postnatal period* 1.4 (0.7–2.8)
			Cumulative exposure (> 95th versus ≤ 95th percentile		Prenatal, 20 Postnatal, 19	0.7 (0.4–1.3)	1.1 (0.6–2.2)
Cohort studies							
Japan							
Tsuda *et al.* (1989)	Nakajo-machi town, Niigata Prefecture, 281 residents	Mortality 1959–87	High dose of arsenic contamination (through a factory) of well-water used for drinking (1955–59)	Uterus	17 deaths from all cancers in the entire cohort	Among the residents in high-exposure areas (low, < 0.05 ppm; medium, 0.05–0.5 ppm; high, ≥ 0.5 ppm), significant excess mortality from cancer of the uterus over expected value based on mortality for Niigata Prefecture and for all Japan	

Table 26 (contd)

Reference	Location	End-point	Exposure	Site	No. of cases	Study outcome				Comments
Tsuda *et al.* (1995)	454 residents living in Namikicho and Nakajo-machi in Niigata Prefecture	Mortality 1959–92	High-dose arsenic contamination (through a factory) of well-water used for drinking (1955–59)	Uterus	0 0 2 2	Arsenic concentration (ppm) < 0.05 0.05–0.99 ≥ 1 Total			SMR (95% CI) 0.0 (0.0–8.0) 0.0 (0–37.6) 13.5 (2.4–48.6) 3.0 (0.5–11.1)	113 persons who drank from industrially contaminated wells in 1955–59, then followed for 33 years; rates in Niigata Prefecture in 1960–89 used as the standard for estimation of SMR
USA										
Garland *et al.* (1996)	Nested case–control in the Nurses' Health Study Cohort, USA	Incidence 1984–86	Exposure to 5 metals (As, Cu, Cr, Fe, Zn) through any route measured	Breast, diagnosed from 1984 to 1986, 459 matched controls	Total, 433 54 67 56 62 69	Cut-point (µg/g) < 0.059 0.059–0.078 0.079–0.103 0.104–0.138 > 0.138	Quintile 1 2 3 4 5	Odds ratio 1.0 1.2 1.01 1.1 1.1	95% CI 0.7–1.98 0.6–1.7 0.7–1.9 0.7–1.9	Multivariate logistic regression models controlled for age, date of nail return, smoking, age at first birth, parity, history of benign breast disease, history of breast cancer in mother or sister, age at menarche, menopausal status, body mass index and alcohol consumption
Lewis *et al.* (1999)	Millard County, UT	Mortality	Exposure to arsenic in drinking-water	Prostate	50	All men (µg/L–years) Low (< 1000) Medium (1000–4999) High (≥ 5000)			SMR 1.5 (95% CI, 1.07–1.9) 1.07 1.70 (*p* < 0.05) 1.65	Exposure index relies on ecological measures of arsenic concentration, median value for the community. State rates in 1950–92 used as the standard for estimation of SMR

SMR, standardized mortality ratio; CI, confidence interval; M, male; F, female; SIR, standardized incidence ratio; OR, odds ratio cancers of the breast (two cases, one case who had keratoses present), pancreas (one case who had keratoses present) and mouth (one patient who had treatment to the mouth for syphilis). A report of seven cases of cancer following treatment with Fowler's solution (Jackson & Grainge, 1975) included one case of bilateral breast cancer and one case of colon cancer. Multiple skin cancers were present in both cases. The literature also includes cases of meningioma and intestinal malignancies associated with ingestion of arsenic (IARC, 1987). [Case reports have helped to identify the role of ingestion of arsenic in the occurrence of skin cancer and could provide leads to occurrences of other malignancies. However, without an appropriate comparison group, it is unclear whether any of the cases represent an excess over the norm.]

2.4.1   Taiwan, China

Ecological studies

Chen *et al.* (1985) (described in Section 2.1.1) reported SMRs in the four-county region in South-West Taiwan that is endemic for Blackfoot disease using mortality rates for the whole country as the standard. The analyses were based on mortality data obtained from the Department of Health for the period 1968–82. To estimate dose, SMRs were computed for regions that were hyperendemic (21 villages), endemic (25 villages) and not endemic (38 villages) for Blackfoot disease. Age-standardized mortality rates for the four counties were elevated compared with national rates for cancer of the colon in both men and women. Age- and sex-standardized mortality rates for colon cancer were higher in endemic areas than in non-endemic areas, but were lower in the hyperendemic areas compared with other areas. SMRs for leukaemia were of borderline statistical significance in men and close to unity in women. The SMR for cancer of the small intestine also was increased in men but not in women, and was not statistically significant in either sex [Prostate cancer was not investigated in this report.] (Chen *et al.*, 1985). A subsequent report used the median concentrations of arsenic in well-water measured in 1962–64 grouped into levels of < 300, 300–600 and ≥ 600 µg/L. A dose-related gradient in age-adjusted mortality was noted for prostate cancer. In the ≥ 600-µg/L group, the age-standardized mortality rate for prostate cancer was 5.6 times higher than that of the general population of Taiwan (Chen *et al.*, 1988a) (study described in Section 2.1.1).

Wu *et al.* (1989) (study described in Section 2.1.1) published age-adjusted mortality rates for the years 1973–86 in 27 townships in the four counties endemic for Blackfoot disease together with 15 additional villages in the townships of Yensui and Hsiaying. Using the same data on well-water content and the classification scheme described above (Chen *et al.*, 1988a), a dose-related trend in mortality from prostate cancer was again observed (Mantel-Haenszel test for trend, $p < 0.05$). The authors also noted dose-related increases in mortality rates from nasopharyngeal and colon cancer in men, which, however, were not statistically significant. In this analysis, leukaemia and oesophageal, stomach or uterine cancers did not appear to be related to levels of arsenic (Wu *et al.*, 1989).

In a later study, Chen and Wang (1990) (described in Section 2.1.1) used data on arsenic in well-water measured from 1974 to 1976 in 314 precincts and townships throughout Taiwan. Based on a multiple regression analysis (adjusted for urbanization and age), mortality rates from prostate cancer significantly increased with higher average level of arsenic. Age-adjusted mortality rates for cancers of the nasal cavity also correlated with average arsenic concentration in men and women for all precincts and townships and for the south-western townships. Cancers of the oesophagus, stomach, small intestine, colon, rectum, pancreas, larynx, bone and cartilage, breast, cervix, ovary and brain and leukaemia were not significantly correlated. [The regression estimates were only presented for statistically significant results.]

Based on data from death certificates for the period 1971–94, Tsai *et al.* (1999) (described in Section 2.1.1) computed SMRs for the four townships that are endemic for Blackfoot disease using local (Chiayi County) and national (Taiwan) rates as the standard. Applying national rates, SMRs were elevated for cancers of the intestine, colon and prostate in men, cancers of the nasal cavity and larynx in both men and women, lymphoma in men and women, leukaemia in men and cancers of the pharynx and bone in women. These SMRs were similarly elevated using the local mortality rates as the standard. In addition, using the local rates as the standard, higher SMRs were found for cancers of the oesophagus in men, cancers of the stomach in men and women and cancers of the colon, rectum and brain in women for the four townships.

[Data from South-West Taiwan indicate a consistent pattern of increases in mortality from prostate cancer in areas with high contamination by arsenic, and there is evidence of a dose-related effect. These studies do not specifically address the issue of dose of exposure, nor do they raise issues of latency and duration. These issues cannot be addressed using mortality as an end-point for prostate cancer since the disease has a low case-fatality rate. One possible source of bias is that prostate cancer often goes undetected, and a higher mortality rate in regions with known exposure to arsenic could occur if screening for cancer deaths is enhanced in the region.]

2.4.2   Chile

Ecological study

Rivara *et al.* (1997) compared the mortality rates for various cancers in 1976 between Antofagasta in Region II and the non-contaminated Region VIII. An elevated risk of cancer of the larynx was observed in the arsenic-exposed Region II. Among the various other cancers (17 sites) investigated, no other elevated SMRs were reported.

2.4.3   Japan

Cohort studies

From about 1945 to 1959, wells in the small town of Nakajo-machi in Niigata Prefecture became contaminated with arsenic (up to 3000 µg/L) from a factory producing

King's yellow (As$_2$O$_3$). Two uterine cancer deaths occurred from 1959 to 1992 in the highest exposure category (≥ 500 μg/L), with an SMR for uterine cancer of around 3.0 using the age- and cause-specific mortality rates from Niigata Prefecture (Tsuda *et al.*, 1989, 1995). [No other cancer sites were discussed.]

2.4.4   *North America*

(*a*)   *Cohort studies*

Garland *et al.* (1996) conducted a nested case–control study in the USA to investigate the relationship between concentrations of arsenic and other trace elements in toenails and the incidence of breast cancer. Cases and controls were selected from the Nurses' Health Study cohort comprising 121 700 nurses aged 30–55 years living in 11 states of the USA. Toenail samples were obtained from 72% of 94 115 cohort members in 1982. A total of 62 641 women provided toenail clippings and did not have a diagnosis of breast cancer as of 1982. The nested study was based on the 433 women who had a diagnosis of breast cancer reported in mailed questionnaires in 1984 and 1986 and an age-matched control group. Compared with the quintile, the adjusted odds ratio for breast cancer was 1.1 (95% CI, 0.7–1.9) for the highest quintile of arsenic.

Lewis *et al.* (1999) reported SMRs for various cancers using a retrospective cohort of residents from Millard County, UT, USA. Of the multiple types of cancers examined, mortality was only elevated for prostate cancer [The limitations of this study have been cited in Section 2.1.4.].

(*b*)   *Case–control studies*

Infante-Rivard *et al.* (2001) conducted a population-based case–control study of childhood leukaemia (occurring before the age of 9 years) in Québec Province, Canada. Cases included individuals who were newly diagnosed with childhood leukaemia from 1980 to 1993. Of 510 eligible cases, 491 participated (96.3%) and, of 588 controls, 493 took part (83.8%). The investigators sought drinking-water test data from 1970 onward through a postal questionnaire that yielded usable data from 112 of 202 municipalities (55%). Additional data was provided from the Ministry of Municipal Distribution Systems in 1986. Further analysis of tap-water was performed by the study investigators and covered 103 of the municipalities. These data were linked to information on residential history derived from interviews with the subjects' parents, and the value used for exposure to arsenic was that of the subjects' municipalities of residence for the closest year when data were available. Values from multiple test results were averaged over a given year, and individuals with private water systems were assigned the arsenic value of their municipality of residence. Separate exposure variables were computed for subjects' pre- and postnatal periods and included average arsenic levels and cumulative exposure. A logistic regression analysis included maternal age and level of schooling. The odds ratio for childhood leukaemia above versus less than or equal to the 95th percentile (5 μg/L arsenic) was 0.9 (95% CI, 0.5–1.8) for prenatal exposure and 1.4 (95% CI, 0.7–2.7) for postnatal exposure.

For cumulative exposure (above versus less than or equal to the 95th percentile), the odds ratios were 0.7 (95% CI, 0.4–1.3) for prenatal exposure and 1.1 (95% CI, 0.6–2.2) for post-natal exposure. [The study does not provide evidence of a link between childhood leukaemia and exposure to arsenic through drinking-water either prenatally or postnatally. However, the drinking-water concentrations were relatively low (95% were below 5 µg/L) and the estimates were imprecise. Furthermore, the exposure estimates are subject to mis-classification, but there is no discussion on the variability of arsenic concentrations within municipalities and the fraction of residents that used private systems that were assumed to contain the same concentrations of arsenic as the public systems.]

2.4.5 *Australia*

Ecological study

In an ecological study using incidence data from the Victoria Cancer Registry for the years 1982–91, Hinwood *et al.* (1999) calculated SMRs for 14 cancer sites (in addition to liver, lung, bladder and kidney) in 22 postcode areas characterized by a level of arsenic in water > 0.01 mg/L in most areas. Incidence rates for all of Victoria were used as the standard. Cancer sites that showed elevated rates with 95% CIs that excluded 1.0 were prostate, melanoma, breast and chronic myeloid leukaemia. [The Working Group noted that no information was presented on the actual use of water contaminated with arsenic for drinking by the population.]

3. Studies of Cancer in Experimental Animals

Previous evaluation

Various inorganic arsenic compounds were tested for carcinogenicity by oral adminis-tration, skin application, inhalation and/or intratracheal administration, subcutaneous and/or intramuscular administration, intravenous administration and other experimental systems in mice, rats, hamsters, dogs or rabbits. Arsenic trioxide produced lung adenomas in mice after perinatal treatment (Rudnay & Börzsönyi, 1981) and in hamsters after its intratracheal instillation (Ishinishi *et al.*, 1983; Pershagen *et al.*, 1984). It induced a low incidence of adenocarcinomas at the site of its implantation into the stomach of rats (Katsnelson *et al.*, 1986). A higher incidence of lung carcinomas was induced in rats following a single intratracheal instillation of a Bordeaux pesticide mixture (copper sulfate and calcium oxide in a concentration of 1–2%) containing calcium arsenate (IARC, 1987). Intratracheal instillations of calcium arsenate into hamsters resulted in a borderline increase in the incidence of lung adenomas, while no such effect was observed with arsenic trisulfide (Pershagen & Björklund, 1985). These studies provide *limited evidence* for carcinogenicity of inorganic arsenics (IARC, 1980, 1987).

No adequate data on the carcinogenicity of organic arsenic compounds were available to the previous working group (IARC, 1980, 1987).

3.1   Oral administration

3.1.1   *Mouse*

Groups of 24 male A/J mice, 6 weeks of age, were given tap-water (control) or a solution of 50, 200 or 400 ppm [μg/mL] dimethylarsinic acid (DMA^V) as drinking-water for 25 (10 mice per group) or 50 weeks (14 mice per group). The incidences of lung tumours were 2/10 (20%), 3/10 (30%), 4/10 (40%) and 3/10 (30%) in control, 50-, 200- and 400-ppm groups, after 25 weeks, with average numbers of tumours/mouse of 0.2 ± 0.42, 0.3 ± 0.48, 0.5 ± 0.71 and 0.4 ± 0.70, respectively; no significant differences were apparent, nor did average tumour size vary significantly (0.9, 0.5, 1.4 and 1.1 mm, respectively). After 50 weeks, a non-significant increase in the incidence of lung tumours (50, 71.4, 64.3 and 78.6% in 14 animals per group, respectively), a significant increase in multiplicity (0.5 ± 0.52, 1.07 ± 1.0, 1.07 ± 1.07 and 1.36 ± 1.01, respectively; $p < 0.05$ for the 400-ppm group) and an increase in average diameter (1.0, 1.2, 1.4 and 1.5 mm, respectively) were observed. The numbers of mice with papillary lung adenoma and/or adenocarcinoma at 50 weeks were two, five, seven and 10 ($p = 0.002$ for the 200- and 400-ppm groups) and increased with increasing dose of DMA^V. In animals that received 0, 50, 200 or 400 ppm DMA^V, the number of alveolar adenomas ranged from 3/14 to 5/14 per treatment group (Hayashi *et al.*, 1998).

Groups of 90 female C57BL/6J mice and 140 female metallothionein heterozygous mice (MT$^{-/-}$), aged 4–5 weeks, were given drinking-water containing sodium arsenate (500 μg/L arsenic) *ad libitum* for up to 26 months. Groups of 60 control mice were given tap-water. Preliminary findings indicate that tumours were observed in the lung (C57BL/6J, 17.8%; MT$^{-/-}$, 7.1%), gastrointestinal tract (14.4%; 12.9%), liver (7.8%; 5.0%), spleen (3.3%; 0.7%), reproductive organs (3.3%; 5.0%), skin (3.3%; 1.4%), bone (2.2%; 0%) and eye (1.1%; 0%) of treated animals. No tumours were observed in the control groups (Ng *et al.*, 1999). [The Working Group decided that this study was preliminary because no histopathological findings were reported.]

Groups of 20–30 K6/ODC transgenic mice, 7 weeks of age, were administered DMA^V at either 10 or 100 ppm [μg/mL] in their drinking water or sodium arsenite at 10 ppm for 5 months. The incidence of squamous skin tumours was 0% in the controls, 8 and 22% in the 10- and 100-ppm DMA groups, respectively, and 15% in the arsenite groups (Chen *et al.*, 2000).

Groups of 29 or 30 male p53$^{+/-}$ heterozygous or p53$^{+/+}$ mice (C57BL/6J background) were exposed to 0, 50 or 200 ppm [μg/mL] DMA^V in the drinking-water for 80 weeks. In p53$^{+/+}$ mice, a significant increase in the incidence (control, 10%; 50-ppm, 30% [$p < 0.05$]; and 200-ppm, 30% [$p < 0.05$] in 30 animals per group) and multiplicity (0.2, 0.6 [$p < 0.02$] and 0.6 [$p < 0.02$] tumours per mouse, respectively) of total tumours was observed at the terminal killing, but with no dose dependence. In the heterozygotes, a non-significant increase in incidence was observed (control, 14/29 [48.3%]; 50-ppm, 18/29 [62.1%]; and 200-ppm, 19/30 [63.3%]), but the number of tumours per mouse was significantly increased at 200 ppm ($p < 0.05$) (control, 0.8; 50-ppm, 1.1; 200-ppm, 1.2). No effects were observed in either heterozygous or p53$^{+/+}$ mice regarding the number of tumours per tumour-bearing animal (control, 1.6; 50-ppm, 1.8; 200-ppm, 1.9 in hetero-zygous mice; control, 2; 50 ppm, 1.9; 200 ppm, 2 in p53$^{+/+}$ mice). No significant influence on tumour development was noted in any particular organ or tissue site. The tumours induced in the p53$^{+/-}$ heterozygous mice were mainly malignant lymphomas or leukaemia (control, 8/29 [28%]; 50-ppm, 13/29 [45%]; 200-ppm, 10/30 [33%]), fibrosarcomas (5/29 [17%], 8/29 [28%], 10/30 [33%]) and osteosarcomas (3/29 [10%], 2/29 [8%], 4/30 [13%]), with lower incidences of other types of tumours such as hepatocellular carci-nomas, thyroid follicular carcinomas, squamous-cell carcinomas of the skin and lung ade-nomas. In p53$^{+/+}$ mice, tumours were generally malignant lymphomas or leukaemia (2/30 [7%], 9/30 [30%], 9/30 [30%]) with very low incidences of the other types of tumour. No fibrosarcomas or osteosarcomas were detected in p53$^{+/+}$ mice. Tumour latency curves in DMA^V-treated p53$^{+/-}$ heterozygous and p53$^{+/+}$ mice showed a dose-dependently signifi-cant shift towards early induction ($p < 0.03$) in comparison with untreated controls (Salim et al., 2003).

3.1.2   Rat

Groups of 36 male Fischer 344/DuCrj rats, 10 weeks of age, received 0, 12.5, 50 and 200 ppm DMA^V [µg/mL] (100% pure) in the drinking-water for 104 weeks. There was no significant difference in body weight or survival (25, 28, 28 and 24 animals) among the groups at week 104. At week 97, the first tumour in the urinary bladder was observed in one animal of the 200-ppm DMA^V group. Effective numbers were considered to be the numbers of animals alive at week 97. Incidences of urinary bladder tumours were 0/28, 0/33, 8/31 (26%; two papillomas and six carcinomas; $p < 0.01$, Fisher's exact probability test) and 12/31 (39%; two papillomas and 12 carcinomas; $p < 0.001$, Fisher's exact proba-bility test), respectively, and were multiple in two animals given the highest dose. Histo-pathologically, the carcinomas were transitional-cell carcinomas. Urinary pH did not differ significantly between groups during the experiment. Bladder calculi were not observed in any of the rats (Wei et al., 1999, 2002). In a more exhaustive examination of the urinary bladder in the same animals, preneoplastic lesions (papillary or nodular hyperplasia) were observed in 0/28, 0/33, 12/31 (39%; $p < 0.01$) and 14/31 (45%, $p < 0.01$) animals in the 0-, 12.5-, 50- and 200-ppm groups, respectively. The incidences of tumours, other than those of the urinary bladder, in all DMA^V-treated groups were not different from those of controls (Wei et al., 2002).

3.2   Transplacental exposure

Mouse: Groups of 10 pregnant C3H mice were given drinking-water containing 0, 42.5 and 85 ppm [mg/mL] *ad libitum* from day 8 to 18 of gestation. Offspring were weaned at 4 weeks and then divided into separate groups of 25 males and 25 females. The offspring received no additional treatment with arsenic for the next 74 (males) or 90 (females) weeks. Transplacental exposure to arsenic did not reduce body weight in any group of offspring over the course of the experiment. In male offspring, there was a marked increase in the incidence of hepatocellular carcinomas (control, 3/24 [12%]; 42.5-ppm, 8/21 [38%]; 85-ppm, 14/23 [61%]; p for trend = 0.00006, two-sided chi-square test) and multiplicity per mouse (control, $0.13 - 0.07$; 42.5-ppm, $0.42 - 0.13$; 85-ppm, $1.30 - 0.28$; p for trend = 0.003) in a dose-related fashion. There was also a dose-related increase in the incidence of adrenal cortical adenomas (control, 9/24 [37.5%]; 42.5-ppm, 14/21 [66.7%]; 85-ppm, 21/23 [91.3%]; p for trend = 0.001) and multiplicity (control, $0.71 - 0.20$; 42.5-ppm, $1.10 - 0.22$; 85-ppm, $1.57 - 0.32$; p for trend = 0.016). In female offspring, there was a strong, dose-related increase in the incidence of ovarian tumours. Total tumour (benign and malignant) incidence was control, 2/25 (8%); 42.5-ppm, 6/23 (26%); and 85-ppm, 9/24 (38%) (p for trend = 0.015). Controls had one adenoma and one benign granulosa-cell tumour. The 42.5-ppm treatment group had three adenomas, one adenocarcinoma, one benign granulosa-cell tumour and one malignant granulosa-cell tumour. The 85-ppm treatment group developed seven adenomas, one luteoma and one haemangiosarcoma. Lung carcinomas developed (control, 0/25 [0%]; 42.5-ppm, 1/23 [4%]; 85-ppm, 5/24 [21%]; p for trend = 0.0086) in a dose-dependent manner. There were significant increases in the number of mice bearing at least one tumour (control, 11/24; 42.5-ppm, 17/21; 85-ppm, 22/23; p for trend = 0.0006 in males; control, 12/25; 42.5-ppm, 17/23; 85-ppm, 16/24; $p < 0.172$ in females) and in mice bearing at least one malignant tumour (control, 3/24; 42.5-ppm, 9/21; 85-ppm, 14/23; $p < 0.0001$ in males; control, 2/25; 42.5-ppm, 9/23; 85-ppm, 8/24; $p < 0.042$ in females) with both doses of arsenic. Exposure to arsenic also increased the incidence of hyperplasia of the uterus and oviduct. In this experiment, four of the organs that developed tumours or hyperplasia were endocrine-responsive organs: adrenal gland, liver, ovary and uterus (Waalkes *et al.*, 2003).

3.3   Intratracheal administration

Hamster: Groups of 30 (20 for a control) male Syrian golden hamsters, 8 weeks of age, were given arsenic trioxide, calcium arsenate or arsenic trisulfide by intratracheal instillation once a week for 15 weeks. Each compound contained 0.25 mg arsenic suspended in 0.1 mL phosphate buffer solution. The control group received buffer solution alone. All hamsters were kept during their entire lifespan. Numbers of survivors after 15 instillations were 18/30 (60%) in the arsenic trioxide-treated group, 27/30 (90%) in the calcium arsenate-treated group, 23/30 (77%) in the arsenic trisulfide-treated group and 22/22 (100%) in the control group, showing a similar tendency in survival rates of all groups. All hamsters had died by day 794 (arsenic trioxide group), day 806 (calcium arsenate group), day 821 (arsenic trisulfide group) and day 847 (control group) after the initial instillation. Incidences of lung tumours were 1/17 (5.8%; adenocarcinoma) arsenic trioxide-treated, 7/25 (28.0%; one adenocarcinoma, six adenomas; p value, significant versus controls) calcium arsenate-treated, 1/22 (4.5%; adenoma) arsenic trisulfide-treated and 1/21 (4.8%; adenosquamous carcinoma) control animals. No tumours of the upper respiratory tract including the trachea were observed in any group. Besides lung tumours, one adrenal adenoma and one liver haemangiosarcoma in the arsenic trioxide-treated group, two adrenal adenocarcinomas and one leukaemia in the calcium arsenate-treated group, one nephroblastoma and one adrenal adenoma in the arsenic trisulfide-treated group and one adrenal adenocarcinoma and one adrenal adenoma in the control group were found (Yamamoto *et al.*, 1987).

3.4   Administration with known carcinogens

3.4.1   *Mouse*

Groups of 30–32 female Swiss mice, 21–24 days of age, received concentrations of 0, 10, 50 or 100 μg/L sodium arsenate or sodium arsenite in the drinking-water for 15 weeks. At week 3, the animals were administered a single intraperitoneal injection of 1.5 mg/kg bw urethane in saline. When killed at week 15, the numbers of lung adenoma per mouse were 29.0 ± 5.4, 21.4 ± 2.6, 15.7 ± 1.8 and 16.0 ± 2.1 (p for trend = 0.0185) in the 0-, 10-, 50- and 100-μg/mL arsenate-treated groups, respectively, and 20.1 ± 1.8, 25.7 ± 4.0, 19.5 ± 2.1 and 10.8 ± 1.6 (p for trend = 0.00082) in 0-, 10-, 50- and 100-μg/mL arsenite-treated groups, respectively [suggesting that both forms of arsenic exerted an inhibitory effect]. Both arsenate and arsenite at 100 μg/mL caused a significant reduction in tumour size (0.64 ± 0.01, 0.65 ± 0.02; $p < 0.05$) compared with control animals (0.74 ± 0.01, 0.71 ± 0.02) (Blakley, 1987).

In a two-stage protocol, groups of 9–13 male ddY mice, 6 weeks of age, were given a single subcutaneous injection of 10 mg/kg bw 4-nitroquinoline 1-oxide (4NQO) and then received tap-water, a 5% glycerol solution or a 200- or 400-ppm [μg/mL] DMA^V solution in drinking-water for 25 weeks. The incidences of lung tumour-bearing mice were 2/9 (22%), 5/10 (50%), 8/13 (62%) and 10/13 (77%), respectively, while the numbers of tumours per mouse were 0.22 ± 0.15, 1.40 ± 0.62, 3.92 ± 1.79 and 4.38 ± 1.07 ($p < 0.05$ Cochran-Cox t-test). Thus, DMA^V promoted lung tumorigenesis initiated by 4NQO (Yamanaka *et al.*, 1996). [The authors described a shift from papillary-type adenomas to adenosquamous carcinomas but no quantitative data were provided.]

Groups of 60 male and female C57BL6J outbred mice [sex distribution unspecified], 2 months of age, were fed a diet containing 10% lipids and were given either 0.01% arsenic trioxide in the drinking-water for 28 weeks or 3 mg per mouse benzo[*a*]pyrene in 0.2 mL corn oil by gavage once a week for 3 weeks or both arsenic trioxide and benzo[*a*]pyrene. No significant differences between groups given benzo[*a*]pyrene with or without arsenic trioxide were observed at the end of the 28-week experimental period with regard to focal localized hyperplasia, ulcers, focal multiple hyperplasia, papillomatosis or papillomas of the forestomach (total number per mouse, 4.20 ± 0.39 in the benzo[a]pyrene-treated group and 5.40 ± 0.89 in the benzo[a]pyrene–arsenic-treated group) (Silva et al., 2000).

Groups of 10–11 female Hos:HR-1 hairless mice, 6 weeks of age, were administered 0, 400 or 1000 ppm [μg/mL] DMA^V in the drinking-water and irradiated twice weekly with 2 kJ/m^2 ultraviolet B (UVB) rays for 25 weeks. DMA^V had no effect on body weight gain. The number of skin tumours per mouse was significantly increased by 1000 ppm DMA^V compared with the 0-ppm value from weeks 13 to 19, and incidence of tumour-bearing mice was significantly increased at weeks 12 and 13 [exact data not clear as there were no tabulations]. No differences were noted at later time points up to week 25 (100% incidence in all groups by week 16), showing a shift towards early tumour induction following treatment with DMA^V. Malignant tumours were observed in only two animals in the 1000-ppm group (Yamanaka et al., 2000).

Groups of 15 female Crl: SK1-hrBR hairless mice, 21 days of age, received 0 or 10 mg/L sodium arsenite in the drinking-water and were irradiated with solar lamps at a dose of 1.7 J/m^2 (lamp output: 85% in the UVB range, < 1% UVC and 4% UVA, and the remainder visible) three times weekly. The UVR dose was chosen to be approximately half of the minimal erythemic dose. Two control groups of five mice received sodium arsenite only or no treatment. Sodium arsenite did not influence body weight gain. No skin tumours were observed with arsenite alone or in the untreated controls. The first tumours were noted after 8 weeks with arsenite and UVR but after 12 weeks with UVR alone (significantly earlier appearance). All UVR-treated animals had at least one tumour at 26 weeks; however, after 19 weeks of exposure to UVR, incidences were 100% for UVR and arsenite and 33% for UVR alone. The total number of tumours in the group treated with UVR alone (15 animals) was 53 and that in the group treated with UVR and arsenite (15 animals) was 127. In UVR and arsenite-treated animals, 64/127 (50.4%) tumours were highly invasive squamous-cell carcinomas, whereas in UVR alone-treated animals, 14/53 (26.4%) were highly invasive squamous-cell carcinoma ($p = 0.003$) (Rossman et al., 2001)

Groups of 10 female Hos:HR-1 hairless mice, 6 weeks of age, were treated with a single topical application of 200 nmol 7,12-dimethylbenz[a]anthracene (DMBA) dissolved in acetone and were then administered 0, 400 or 1000 ppm [μg/mL] DMA^V in the drinking-water and/or irradiated twice weekly with 0.3 kJ/m^2 UVB. All mice were killed after 50 weeks. Skin tumours occurred faster in the DMA^V-treated group. DMA^V without UVB increased the incidence of skin tumours ($p < 0.05$ at 20–22 weeks) but not dose-dependently. Greater effects were seen in combination with UVB, particularly at 1000 ppm [exact data not clear as there was no tabulation]. Incidences of papillomas in animals treated with DMA^V without UVB were 1/10, 9/10 and 7/10 in the 0-, 400- and 1000-ppm groups, respectively, and those of squamous-cell carcinomas were 2/10, 0/10 and 0/10, respectively. Incidences of papillomas in animals treated with DMA^V and UVB were 0/10, 7/10 and 7/10, and those of squamous-cell carcinomas were 0/10, 3/10 and 1/10 in the 0-, 400- and 1000-ppm groups, respectively (Yamanaka *et al.*, 2001).

3.4.2 *Transgenic mouse*

Groups of 7–8 female *keratin (K6)/ODC* transgenic mice, 10–14 weeks of age, received two weekly applications of 3.6 mg DMA^V in neutral cream or 5 µg 12-*O*-tetradecanoyl-phorbol 13-acetate (TPA) in 200 µL acetone 1 week after initiation with 50 µg DMBA in 200 µL acetone. A significantly accelerated development of skin tumours (first tumour after 8 weeks in DMA^V-treated animals and after 11 weeks in controls) was observed following treatment with DMA^V; 20 weeks after initiation, the numbers of tumours (average, 19.4 ± 10.2 per mouse compared with 9.7 ± 3.5 in controls) were increased. Promoting activity was similar to that achieved with application of 5 µg TPA (20.7 ± 8.4) twice weekly. Microscopically, most of the tumours were squamous papillomas, although squamous carcinomas occurred in some DMA^V- and some TPA-treated animals (Morikawa *et al.*, 2000).

3.4.3 *Rat*

Sodium arsenite has been reported to enhance the incidence of renal tumours induced in rats by intraperitoneal injection of *N*-nitrosodiethylamine (NDEA) (Shirachi *et al.*, 1983). A subsequent re-evaluation of the study indicated that not only sodium arsenite but also sodium arsenate enhanced NDEA-induced kidney tumours (Smith *et al.*, 1992).

Groups of 20 male Fischer 344/DuCrj rats, 6 weeks of age, received a single intraperi-toneal injection of 100 mg/kg bw NDEA, followed by intraperitoneal injections of 20 mg/kg bw *N*-methyl-*N*-nitrosourea on days 5, 8, 11 and 14 and subcutaneous injections of 50 mg/kg bw 1,2-dimethylhydrazine chloride on days 18, 22, 26 and 30. At the same time, animals received 0.05% *N*-butyl-*N*-(4-hydroxybutyl)nitrosamine (BBN) in the drinking-water for the first 2 weeks then 0.1% *N*-bis(2-hydroxypropyl)nitrosamine for the next 2 weeks (so-called DMBDD model treatment). After a 2-week interval, the animals received 0, 50, 100, 200 or 400 ppm [µg/mL] DMA^V in the drinking-water from weeks 6 to 30, at which time they were killed. DMA^V significantly enhanced tumour induction in the urinary bladder (both papillomas and transitional-cell carcinomas), kidney (both adenomas and adenocarcinomas), liver (hepatocellular carcinomas) and thyroid (adeno-mas) (see Table 27). Values for preneoplastic lesions such as papillary or nodular hyper-plasia in the urinary bladder, atypical tubules in the kidney and altered hepatocyte foci in the liver were also significantly increased. No promoting effects were noted in the lungs or the nasal cavity (Yamamoto *et al.*, 1995).

To confirm the above-mentioned results and to evaluate low-dose effects of DMA in urinary bladder and liver carcinogenesis, the following two studies were conducted. Groups of 20 male Fischer 344 rats, 6 weeks of age, received 0.05% BBN in drinking-water for 4 weeks followed by 0, 2, 10, 25, 50 or 100 ppm [µg/mL] DMA^V for 32 weeks.

Table 27. Incidence of preneoplastic and neoplastic lesions in various organs of Fischer 344/Du Crj rats treated with DMA^V after initiation with DMBDD treatment

Organ and finding	DMA^V					Two-tailed Cochran-Armitage analysis[a]
	0 ppm $n = 20$ (%)	50 ppm $n = 20$ (%)	100 ppm $n = 19$ (%)	200 ppm $n = 20$ (%)	400 ppm $n = 20$ (%)	
Urinary bladder						
Papillary or nodular hyperplasia	4 (20)	13 (65)[c]	14 (73.7)[d]	11 (55)[b]	11 (55)[b]	NE
Papilloma	1 (5)	12 (60)[d]	12 (63.2)[d]	11 (55)[d]	7 (35)[b]	NE
Transitional-cell carcinoma	1 (5)	10 (50)[c]	11 (57.9)[d]	12 (60)[d]	13 (65)[d]	NE
No. of tumour-bearing animals	2 (10)	17 (85)[d]	16 (84.2)[d]	17 (85)[d]	16 (80)[d]	NE
Kidney						
Adenoma	1 (5)	3 (15)	1 (5.2)	7 (35)[b]	3 (15)	$p < 0.01$
Adenocarcinoma	0	0	2 (10.5)	1 (5)	7 (35)[c]	$p < 0.05$
Nephroblastoma	4 (20)	0	4 (12.1)	6 (30)	9 (45)	$p < 0.001$
No. of tumour-bearing animals	5 (25)	3 (15)	6 (31.6)	13 (65)[b]	13 (65)[b]	$p < 0.001$
Liver						
Altered cell foci						
Clear-cell foci	10 (50)	12 (60)	14 (73.7)	19 (95)[c]	20 (100)[d]	$p < 0.001$
Basophilic foci	1 (5)	2 (10)	3 (15.8)	10 (50)[c]	17 (85)[d]	$p < 0.001$
Eosinophilic foci	1 (5)	2 (10)	8 (42.1)[c]	15 (75)[d]	16 (80)[d]	$p < 0.001$
Hyperplastic nodule	0	0	2 (10.5)	9 (45)[d]	7 (35)[c]	$p < 0.001$
Hepatocellular carcinoma	0	2 (10)	0	8 (40)[c]	8 (40)[c]	$p < 0.001$
Cholangioma	0	0	0	1 (5)	1 (5)	
Haemangioma	0	0	0	1 (5)	0	
No. of tumour-bearing animals	0	2 (10)	2 (10.5)	17 (85)[d]	13 (65)[d]	$p < 0.001$

Table 27 (contd)

Organ and finding	DMA^V 0 ppm $n = 20$ (%)	50 ppm $n = 20$ (%)	100 ppm $n = 19$ (%)	200 ppm $n = 20$ (%)	400 ppm $n = 20$ (%)	Two-tailed Cochran-Armitage analysis[a]
Thyroid gland						
Hyperplasia	3 (15)	4 (20)	2 (10.5)	13 (65)[c]	13 (65)[c]	$p < 0.001$
Adenoma	2 (10)	1 (5)	3 (15.8)	1 (5)	6 (30)	
Adenocarcinoma	1 (5)	1 (5)	5 (26.3)	5 (25)	4 (20)	
No. of tumour-bearing animals	3 (15)	2 (10)	8 (42.1)	6 (30)	9 (45)[b]	$p < 0.05$

From Yamamoto *et al.* (1995)

DMA^V, dimethylarsinic acid; DMBDD, *N*-nitrosodiethylamine + *N*-methyl-*N*-nitrosourea + 1,2-dimethylhydrazine chloride + *N*-butyl-*N*-(4-hydroxybutyl)nitrosamine + *N*-bis(2-hydroxypropyl)nitrosamine; NE, not exam ned

[a] The significance of differences in the incidence of lesions between groups was assessed using the Fisher's exact probability test. To evaluate the dose–response relationships of the incidences in lesions in the kidney, liver and thyroid gland, two-tailed Cochran-Armitage analysis was used.

Significantly different from 0 ppm at [b] $p < 0.05$, [c] $p < 0.01$ and [d] $p < 0.001$

Development of preneoplastic lesions and tumours of the bladder (papillary or nodular hyperplasia, papillomas and carcinomas) was enhanced in a dose-dependent manner (see Table 28). Doses of 25, 50 and 100 ppm increased the incidences (%) and multiplicities (number per rat) of bladder papillomas and carcinomas. A significant increase in multiplicity of total tumours (papillomas plus carcinomas) was observed with doses as low as 10 ppm DMA^V ($p < 0.05$): 0 ppm, 0.20; 2 ppm, 0.20; 10 ppm, 0.55; 25 ppm, 1.47; 50 ppm, 2.30; 100 ppm, 2.40. Compared with controls, doses of 50 or 100 ppm significantly increased the incidence of papillary or nodular hyperplasia (Wanibuchi *et al.*, 1996).

Groups of 10 male Fischer 344 rats, 6 weeks of age, were given a single intraperitoneal injection of 0 (control) or 200 mg/kg bw NDEA in saline and 2 weeks later received 0, 25, 50 or 100 ppm [µg/mL] DMA^V in the drinking-water for 6 weeks. Partial hepatectomy was performed on all animals at the end of week 3. Final body weights were decreased dose-dependently but not significantly. No significant variation in relative liver weights was noted. Dose-dependent significant increases in both numbers and areas of glutathione *S*-transferase placental form (GST-P)-positive foci in the liver were observed after initiation with NDEA; the significance was evident at doses of 50 and 100 ppm ($p < 0.01$) for numbers and at all three doses ($p < 0.05$ or $p < 0.01$) for areas [exact values were not listed because of figure]. No GST-P-positive foci were observed in groups not initiated with NDEA (Wanibuchi *et al.*, 1997).

Groups of eight male NCI-Black-Reiter rats (which lack α_{2u}-globulin), 9–14 weeks of age, received 0.05% BBN in the drinking-water for 4 weeks followed by 0 or 100 ppm [µg/mL] DMA^V for 32 weeks. When killed at the end of week 36, the incidence and multiplicity of papillary or nodular hyperplasia in the bladder was significantly increased in DMA^V-treated rats (6/8 [75%]; $p < 0.05$; number per rat, 1.1 ± 1.0, $p < 0.05$) compared with rats receiving BBN alone (0/8 [0%]; number per rat, 0). A 38% incidence of bladder papillomas or carcinomas was observed in DMA^V-treated but not in control animals (Li *et al.*, 1998).

Groups of 20 male Fischer 344 rats, 10 weeks of age, received a single intraperitoneal injection of 200 mg/kg bw NDEA followed 2 weeks later by DMA^V, monomethylarsonic acid (MMA) or trimethylarsine oxide (TMAO) at a dose of 100 ppm in the drinking-water for 6 weeks. Numbers of GST-P-positive foci in the liver were significantly increased in rats treated with MMA, DMA^V and TMAO compared with the controls. Areas of GST-positive foci were also significantly increased in rats treated with MMA, DMA^V and TMAO compared with the controls (Nishikawa *et al.*, 2002).

Table 28. Induction of urinary bladder lesions in Fischer 344 rats treated with BBN followed by DMA^V at various doses

DMA^V (ppm)	No. of rats examined	Papillary or nodular hyperplasia		Papilloma		Carcinoma	
		Incidence (%)	No./rat	Incidence (%)	No./rat	Incidence (%)	No./rat
0 (control)	20	14 (90)	1.05 ± 95[a]	3 (15)	0.15 ± 0.37	1 (5)	0.05 ± 0.22
2	20	13 (65)	1.30 ± 1.30	2 (10)	0.10 ± 0.31	2 (10)	0.10 ± 0.31
10	20	14 (70)	1.55 ± 1.47	7 (35)	0.40 ± 0.60	3 (15)	0.15 ± 0.37
25	19	18 (95)	2.37 ± 1.17[b]	11 (58)[c]	1.05 ± 1.18[b]	7 (37)[d]	0.42 ± 0.61[e]
50	20	20 (100)[d]	2.95 ± 1.88[b]	13 (65)[f]	1.50 ± 1.36[b]	10 (50)[f]	0.80 ± 0.95[c]
100	20	20 (100)[d]	4.10 ± 3.02[b]	17 (85)[g]	1.70 ± 1.17[b]	12 (60)[g]	0.70 ± 0.66[b]

From Wanibuchi et al. (1996)

BBN, N-butyl-N-(4-hydroxybutyl)nitrosamine; DMA^V, dimethylarsinic acid

[a] Mean ± SD

[b] $p < 0.001$ (significantly different from control, Student's t-test); [c] $p < 0.01$; [d] $p < 0.05$; [e] $p < 0.05$; [f] $p < 0.01$; [g] $p < 0.001$ (significantly different from control, Fisher's exact probability test)

4. Other Data Relevant to an Evaluation of Carcinogenicity and its Mechanisms

Extensive reviews of the metabolism of arsenic have been published recently (National Research Council, 1999, 2001; WHO, 2001). This section focuses on data relevant for the evaluation of carcinogenic effects. Thus, it is not a complete review on all published data. The oxidation state of arsenic and its metabolites are given if reported. If speciation of oxidation state has not been performed, the metabolites are given as monomethylarsonic acid (MMA) and dimethylarsinic acid (DMA).

4.1 Absorption, distribution, metabolism and excretion

4.1.1 Absorption

Arsenic in drinking-water is easily absorbed in the gastrointestinal tract. About 70–90% of a single dose of dissolved arsenite (As^{III}) or arsenate (As^V) was absorbed from the gastrointestinal tract of humans and experimental animals (Pomroy et al., 1980; Vahter & Norin, 1980; Freeman et al., 1995). A high rate of gastrointestinal absorption is also supported by the fact that people whose main fluid intake consists of drinking-water with elevated arsenic concentrations have very high concentrations of arsenic in their urine.

With regard to absorption of arsenic through the skin, a few experimental studies indicate a low degree of systemic absorption. Application of water solutions of radiolabelled arsenate in vivo to the skin of rhesus monkey and in vitro to human cadaver skin showed that about 2–6% and 0.98% of the applied arsenic was absorbed within 24 h, respectively (Wester et al., 1993). Similar in-vitro studies using dorsal skin of mice showed a much higher absorption: 33–62% of the applied dose of radiolabelled arsenate in aqueous solution was absorbed through the skin within 24 h. However, most of it, on average 60–90%, was retained in the skin (Rahman et al., 1994). These authors suggested that absorption of arsenic through the skin may be species-specific. For many chemicals, mouse skin is more permeable in vitro than human cadaver skin (Bronaugh et al., 1982). In-vitro studies with human keratinocytes showed that 1–8% of the applied arsenic dose was retained per hour (Bernstam et al., 2002). Morphological changes, cytotoxicity and inhibition of DNA and protein syntheses were found with in-vitro doses of As^{III} as low as 10 µg/L. Thus, it seems probable that inorganic arsenic can be absorbed from the exterior, leading to a breakdown in skin barrier function.

In-vitro studies of skin absorption of DMA^V (10 µg in 20–100 µL water) by application to dorsal skin from adult mice mounted in flow-through diffusion cells showed that 16–25% was retained in the receptor fluid (Hanks balanced salt solution), about 15% in the skin and the remainder in the wash after 24 h (Hughes et al., 1995). After exposure for 1 h only, essentially all of the applied dose was washed away. Less than 1% of the applied dose was absorbed.

A low degree of systemic skin absorption of inorganic arsenic is supported by studies showing that people in Fairbanks, AK, who used tap-water containing about 345 µg/L arsenic for washing, but only bottled water (which did not contain arsenic) for drinking, had about the same low concentrations of arsenic metabolites in urine (on average about 40 µg/L) as people with less than 50 µg/L in their tap-water (Harrington *et al.*, 1978). The concentration of arsenic in hair was clearly elevated in the group drinking bottled water (5.74 µg/g compared with 0.43 µg/g in the low-arsenic group), which shows that arsenic is bound to hair and probably also to skin during washing with water rich in arsenic.

4.1.2   *Distribution*

Following its absorption, arsenate is rapidly reduced to As^{III}; the distribution of its metabolites in the body are therefore very similar to that following exposure to As^{III}. However, studies in mice given arsenite or arsenate (0.4 mg/kg bw) intravenously showed that the concentrations in stomach and intestines were higher after exposure to As^{III} than after exposure to As^{V}, while incorporation in bone was higher following exposure to As^{V} (Lindgren *et al.*, 1982). The differences are less marked in the case of oral exposure, probably due to faster methylation that occurs when the absorbed arsenic passes directly to the liver. After exposure to toxic doses at which methylation capacity is exceeded or inhibited, the differences in distribution patterns for the two forms are greater (Vahter & Norin, 1980).

Absorbed arsenic is transported, mainly bound to SH groups in proteins and low-molecular-weight compounds such as glutathione (GSH) and cysteine, to different organs in the body (National Research Council, 1999, 2001). Complexation of trivalent arsenical compounds with GSH, probably mainly in the form of As(GS)$_3$, has been demonstrated, but As^{III} is easily transferred to binding sites of higher affinity, especially vicinal dithiols, such as lipoic acid and dimercaptosuccinic acid (Cullen & Reimer, 1989; Delnomdedieu *et al.*, 1993). Studies on serum arsenic in dialysis patients showed the presence of inorganic arsenic, partly bound to proteins, and DMA (Zhang *et al.*, 1997, 1998a,b; De Kimpe *et al.*, 1999). Transferrin was the main carrier protein, but the extent to which this occurs in healthy individuals is not known. Most of the arsenic in blood is rapidly cleared, following a three-exponential clearance curve (Mealey *et al.*, 1959; Pomroy *et al.*, 1980). The majority of arsenic in blood is cleared with a half-time of about 2 or 3 h. The half-times of the second and third phases are about 168 and 240 h, respectively (Mealey *et al.*, 1959; National Research Council, 1999).

In experimental studies on mammals exposed to inorganic arsenic, the tissues with the longest retention of arsenic, depending on species, were skin, hair, liver, kidney, blood, squamous epithelium of the upper gastrointestinal tract, epididymis, thyroid, skeleton and lens. Arsenic does not readily cross the blood–brain barrier, and concentrations in the brain are generally low compared with most other tissues (Lindgren *et al.*, 1982; Vahter *et al.*, 1982; Lindgren *et al.*, 1984; Yamauchi & Yamamura, 1985).

In human subjects exposed chronically to arsenic and also at background environmental concentrations, the hair and nails generally show the highest concentrations (0.02–10 mg/kg dry wt; Hindmarsh, 2002). Thus, arsenic appears to concentrate in tissues with a high content of cysteine-containing proteins. In areas of West Bengal and Bangladesh that have high concentrations of arsenic in the drinking-water, maximal concentrations of arsenic in hair, nail and skin exceeding 40 mg/kg have been reported (Guha Mazumder *et al.*, 1988; Chowdhury *et al.*, 2001; Basu *et al.*, 2002). Very few studies have been carried out on the distribution of arsenic in human tissues. Postmortem analysis of human tissues confirm that arsenic is widely distributed in the body after long-term exposure, with highest concentrations in the skin and lungs (0.01–1 mg/kg dry wt), as well as hair and nails (Liebscher & Smith, 1968; Cross *et al.*, 1979; Dang *et al.*, 1983). In people exposed to high concentrations (0.2–2 mg/L) of arsenic in drinking-water, the concentration in liver was 0.6–6 mg/kg dry wt compared with 0.16 mg/kg in unexposed people (Guha Mazumder *et al.*, 1988). In a case of acute intoxication by arsenic, the liver and kidneys showed the highest concentrations of total arsenic with values 350- and 63-fold higher than those in blood, respectively. In all organs, As^{III} was the predominant species, and MMA occurred at higher concentrations than DMA. MMA and DMA were more prevalent in lipid-rich organs (49% and 45% of total arsenic in cerebellum and in brain, respectively) compared with other organs (~ 20% of total arsenic). As^V was found in small quantities in the liver, kidneys and blood (2% of total arsenic) (Benramdane *et al.*, 1999).

Dang *et al.* (1983) used neutron activation analysis (NAA) to measure total arsenic in tissues of people [age and sex not specified] dying in accidents in Mumbai, India (Table 29). Concentrations in the brain were generally low compared with most other tissues. Thus, it appears that arsenic does not readily cross the blood–brain barrier. Notably, there was a large variation in tissue concentrations of arsenic among individuals, similar to that reported in earlier studies (Liebscher & Smith, 1968; Larsen *et al.*, 1974).

Table 29. Levels of arsenic in human tissues obtained from traffic accident victims in the Mumbai area of India

Tissue	No. of samples	Mean concentration (± SD) of arsenic (mg/kg wet wt)
Brain	12	3.9 ± 1.0
Blood	8	5.9 ± 3.9
Kidney	13	12.4 ± 20.7
Liver	19	14.5 ± 6.9
Spleen	18	15.2 ± 16.6
Lung	13	19.9 ± 22.7

From Dang *et al.* (1983)

Few studies have examined the distribution of arsenic metabolites in tissues, owing to analytical difficulties. Marafante *et al.* (1982) reported predominantly inorganic arsenic in ultrafiltrates of rat and rabbit liver and kidney 1 h after intraperitoneal injection of 50 µg/kg bw [^{74}As] as sodium arsenite, using ion-exchange chromatography with radiometric detection. The fraction present as MMA was generally less than one-tenth that of inorganic arsenic. De Kimpe *et al.* (1996) studied the tissue distribution of arsenic metabolites up to 120 h after intraperitoneal injection of a trace amount of [^{74}As]-arsenate in male Flemish giant rabbits, also using ion-exchange chromatographic separation of ultrafiltrates with radiometric detection. The predominant metabolite present in tissues was DMA, followed by inorganic arsenic species and low concentrations of MMA. The percentage of DMA increased steadily over time in bone marrow, heart, liver, muscle, pancreas, small intestine and spleen, but levelled off or declined in kidney and lung.

Yamauchi and Yamamura (1985) studied the tissue distribution over time of arsenic metabolites in male Syrian golden hamsters given a single oral dose of 4.5 mg/kg bw arsenic trioxide. Speciation of arsenic metabolites was carried out by hybrid generation–atomic absorption spectrophotometry (HG–AAS) with a cold trap after alkaline digestion. The predominant form of arsenic present in all tissues up to 120 h after dosing was inorganic arsenic. In contrast to other studies, the concentrations of MMA in tissue were two- to fourfold higher than those of DMA at all time-points, while much more DMA (22% of the dose in 5 days) than MMA (2.5% of the dose) was excreted in urine. The highest concentrations of MMA were found in lungs and spleen at 12–24 h, and those of DMA in liver, lung and kidney at 24 h.

Yamauchi *et al.* (1988) reported data on the time-course tissue distribution in hamsters given a single oral dose of 50 mg/kg bw MMA. Peak MMA concentrations were achieved within 6–120 h after dosing and were highest in the kidney, followed by spleen, lung, skin, liver, muscle and brain. MMA itself accumulated in the kidney and levels declined very slowly. DMA was also detected in several tissues, with highest levels occurring in the lung, followed by kidney and liver. Trimethylated arsenic was not detected in any tissues.

The fate of DMA has been studied in mice administered [^{74}As] or [^{14}C]DMA intravenously. The highest levels of radioactivity were present in kidney at all time-points (5–60 min after injection). Tissues that retained arsenic for the longest time (24 h) were the lungs, intestinal walls, thyroid and lens (Vahter *et al.*, 1984; Hughes *et al.*, 2000). Yamauchi and Yamamura (1984) studied the tissue distribution of DMA in hamsters administered a single oral dose of 50 mg/kg bw DMA. Concentrations were elevated in all tissues examined, including the brain, indicating that DMA is widely distributed in the body and that it passes the blood–brain barrier, although not to a large extent. Concentrations of DMA peaked at 6 h in all tissues examined except hair, with the highest levels in lung, followed by kidney, spleen, liver, skin, muscle and brain. Part of the DMA was found to be methylated further to trimethylarsenic (TMA) *in vivo*. Concentrations of TMA peaked at 6 h in all tissues except skin and hair in which none was detected. The highest concentrations of TMA were found in lung, and were equivalent to about half of those of DMA. It is notable that the peak concentrations of DMA and TMA in the lung were over fourfold higher than those in the liver and kidney, but at 120 h after dosing, both had declined to control levels. The authors presumed that the TMA biosynthesized by hamsters is more likely to behave as an organic arsenic compound, such as arsenobetaine which is present in seafoods, than to be a very toxic substance such as trimethylarsine.

4.1.3 Metabolism

(a) Methylation of arsenic

For several decades, it has been known that inorganic arsenic is metabolized via methylation in microorganisms, aquatic organisms, birds and mammals. The methylation occurs through alternating reductive and oxidative methylation reactions, that is, reduction of pentavalent to trivalent arsenic followed by addition of a methyl group (Figure 3). In certain microorganisms, the methylation of inorganic arsenic may proceed to trimethylated metabolites. In humans, the relative amounts of species in urine are generally 10–30% inorganic arsenic, 10–20% MMA_{total} and 60–80% DMA_{total} (Hopenhayn-Rich et al., 1993; National Research Council, 1999; Vahter, 1999a) (see below for further discussions of variations). The main metabolites excreted in the urine of humans exposed to inorganic arsenic are mono- and dimethylated arsenic acids, together with some unmetabolized inorganic arsenic. The major urinary methylated metabolites of arsenic are MMA^V and DMA^V, with arsenic in its pentavalent oxidation state. However, recent studies have demonstrated MMA^V reductase activity in different tissues that gives rise to the presence of both monomethylarsonous acid (MMA^{III}) and/or dimethylarsinous acid (DMA^{III}) in hamster liver (Sampayo-Reyes et al., 2000) and of MMA^{III} in the bile of various experimental

Figure 3. Biotransformation of inorganic arsenic

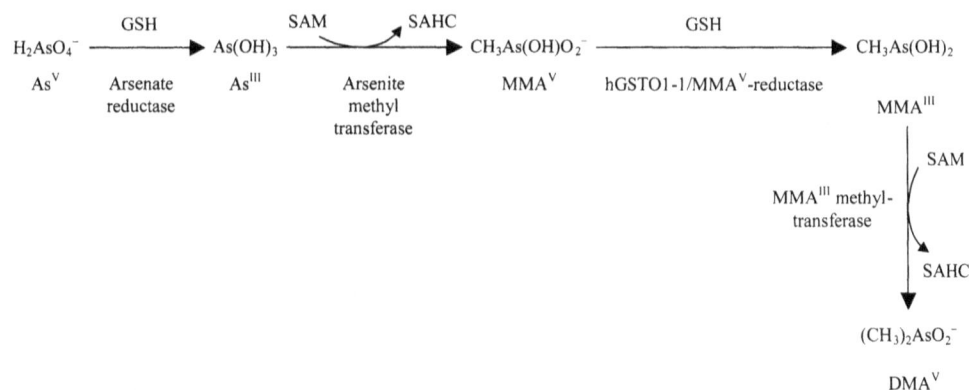

Adapted from Zakharyan et al. (2001)
The conjugate acids and bases of the several forms of arsenic that are thought to predominate at physiological pH are shown.
SAM, S-adenosyl-L-methionine; SAHC, S-adenosyl-L-homocysteine; hGSTO1-1, human glutathione-S-transferase omega 1-1 (which is identical to MMA^V-reductase)

animal species (Csanaky & Gregus, 2002), as well as DMA[III] and/or MMA[III] in human urine (Aposhian *et al.*, 2000a; Le *et al.*, 2000b; Del Razo *et al.*, 2001a; Mandal *et al.*, 2001) after exposure to inorganic arsenic.

Following exposure of mice, hamsters, rats and humans to DMA[V], further methylation to trimethylarsine oxide (TMAO) has been observed (Marafante *et al.*, 1987; Yoshida *et al.*, 1997; Kenyon & Hughes, 2001). This probably occurs via DMA[III] or the DMA-complex observed in urine. About 5% of urinary arsenic was in the form of TMAO. Of the species studied, demethylation of DMA to inorganic arsenic was detected only in rats, in which inorganic arsenic constituted less than 10% of the total excreted (Yoshida *et al.*, 1997). Inorganic arsenic appeared in urine during the first 24–48 h after administration, whereas the highest rate of excretion of unchanged DMA had occurred by 6 h and that of TMAO between 6 and 24 h after dosing. The authors suggested that demethylation of DMA was effected by intestinal flora. TMAO was not found in an in-vitro study after incubation of inorganic arsenic with rat liver cytosol. Indeed, it was shown that the methylation of MMA to DMA was inhibited by increasing concentrations of As[III] preventing the formation of TMAO (Buchet & Lauwerys, 1985, 1988).

It should be noted that there are pronounced species differences in the metabolism of arsenic (National Research Council, 1999; Vahter, 1999b). Most experimental animals excrete very little MMA in urine compared with humans (Vahter, 1999b), and some animals, in particular guinea-pigs and several species of non-human primates (Vahter, 1999b; Wildfang *et al.*, 2001), are unable to methylate inorganic arsenic at all. In addition, rats show different kinetics of arsenic metabolism with a pronounced accumulation of DMA[III] in red blood cells (Shiobara *et al.*, 2001) and greater biliary excretion of arsenic (Klaassen, 1974; Gregus *et al.*, 2000), compared with humans. The unique disposition of arsenic in rats may be due to the pronounced biliary excretion of MMA[III] and blood cell uptake of DMA[III] (Gregus *et al.*, 2000; Shiobara *et al.*, 2001). Thus, it is difficult to evaluate human metabolism of arsenic based on much of the available experimental animal data. Studies in hamsters and rabbits seem to be the most useful because their metabolism is most similar to that in humans (National Research Council, 1999). This phenomenon has been taken into consideration here, and data from rats are not included, except for some information from in-vitro studies with rat hepatocytes, which has been used for the purpose of adding mechanistic information, where appropriate.

Compared with inorganic arsenic, the methylated metabolites containing pentavalent arsenic (MMA[V] and DMA[V]) are less cytotoxic, less reactive with tissue constituents and more readily excreted in urine (for review see National Research Council, 1999, 2001; Vahter & Concha, 2001). This has been taken as evidence that methylation of arsenic is an efficient detoxification process. In general, trivalent arsenic is more toxic than the pentavalent form. Recent studies, however, show that the trivalent methylated metabolites are considerably more toxic than inorganic As[III] (National Research Council, 1999, 2001; Thomas *et al.*, 2001; Sections 4.2 and 4.4). Thus, their presence in tissues and body fluids implies that the metabolism of inorganic arsenic involves important bioactivation processes, and that the toxicity of inorganic arsenic probably depends on its metabolism, especially the capacity of cells to produce methylated intermediates that react with tissue constituents. It should be noted that there may be other mechanisms of transport out of tissues to urine. Excretion of arsenic in chimpanzees was found to be more rapid than that in humans, although methylation of arsenic does not occur in chimpanzees (Vahter *et al.*, 1995a). A more complete understanding of the mechanisms of the metabolism of arsenic will provide further insight into the factors determining susceptibility to its toxicity. It should be noted that it is difficult to evaluate the tissue concentrations of MMA^{III} and DMA^{III} based on the amounts detected in urine.

A few studies have indicated a slightly larger fraction of urinary MMA and a smaller fraction of DMA in people with arsenic-related health effects, including skin lesions (Del Razo *et al.*, 1997; Yu *et al.*, 2000) and chromosomal aberrations (Mäki-Paakkanen *et al.*, 1998). Similarly, there are indications that a relatively large amount of MMA in urine is associated with greater retention of arsenic in the body. Evaluation of data from a number of experimental studies on humans receiving specified doses of inorganic arsenic indicates that a higher percentage of DMA in urine is associated with greater overall excretion, while a higher percentage of inorganic arsenic and MMA is associated with slower excretion of total arsenic metabolites (Vahter, 2002). It should also be noted that other mammals that excrete little (rat, rabbit, hamster, beagle and mouse) or no MMA (marmoset, chimpanzee and guinea-pig) in the urine, that is, most experimental animals, show a rapid overall excretion of arsenic (Vahter, 1999b). They also seem to be less susceptible than humans to arsenic-induced toxicity, including cancer (National Research Council, 1999).

(b)   Mechanism of methylation of arsenic

The mechanism of methylation of arsenic in humans has not been elucidated, but *S*-adenosylmethionine (SAM) seems to be the main methyl donor. In experimental studies, inhibition of SAM-dependent methylation pathways (by periodate-oxidized adenosine [PAD] or *S*-adenosylhomocysteine [SAH]) resulted in a marked decrease in methylation of arsenic (Marafante *et al.*, 1985; De Kimpe *et al.*, 1999; Csanaky & Gregus, 2001). Rabbits fed diets with a low content of methyl groups (low in methionine, protein or choline) methylated arsenic to a lesser degree (Vahter & Marafante, 1987). In-vitro studies using rat liver preparations have confirmed the requirement of SAM and thiols (reduced GSH) in the formation of MMA and DMA from As^{III} (Buchet & Lauwerys, 1988; Styblo *et al.*, 1996; Healy *et al.*, 1999; Thomas *et al.*, 2001).

As shown in Figure 3, the methyl groups are transferred from SAM to arsenic in its trivalent form. GSH or other thiols serve as reducing agents for As^{V} and MMA^{V} (National Research Council, 1999, 2001) and are required for the methylation of arsenic. Complexation of trivalent arsenic with GSH, probably mainly in the form of As(GS)$_3$, has been demonstrated. Depletion of hepatic GSH by buthionine sulfoximine in rats and hamsters has been shown to decrease the methylation of inorganic arsenic (Buchet & Lauwerys, 1988; Hirata *et al.*, 1990). Although As^{V} may be reduced non-enzymatically by GSH, enzyme-catalysed reduction seems to predominate. Studies with mice, rabbits, and marmoset monkeys showed that a substantial fraction of absorbed arsenate (As^V) is rapidly reduced, probably mainly in the blood, to As^{III} (Marafante *et al.*, 1985; Vahter & Marafante, 1989; Vahter, 2002). Arsenate and pentavalent methylated metabolites may also be reduced to the corresponding trivalent form in tissues. Arsenate reductase activity has been detected in human liver (Radabaugh & Aposhian, 2000) and MMA^V reductase in human activity and rabbit liver (Zakharyan & Aposhian, 1999; Zakharyan *et al.*, 2001) and various hamster tissues (Sampayo-Reyes *et al.*, 2000). There is evidence that human MMA^V reductase is identical to glutathione-*S*-transferase class omega 1-1 (GSTO1-1) (Zakharyan *et al.*, 2001). Based on studies on rabbit liver, it appears that reduction of MMA^V is the rate-limiting step in the metabolism of arsenic (Zakharyan & Aposhian, 1999). In male hamsters, MMA^V reductase activity was found to vary considerably among tissues: the highest activity was found in the brain followed by urinary bladder, spleen, liver, lung, heart, skin, kidney and testis. The activity in the testis was only about 10% of that in the brain (Sampayo-Reyes *et al.*, 2000).

Experimental studies conducted in rabbits have indicated that the liver is the main site of arsenic methylation, especially following ingestion, when the absorbed arsenic initially passes through the liver, the only organ in which DMA is present 1 h after administration of the parent compound (Marafante *et al.*, 1985). This is supported by studies showing a marked improvement in the methylation of arsenic following liver transplantation in patients with end-stage liver disease (Geubel *et al.*, 1988). In-vitro studies have shown that the methylating capacity of different tissues may vary considerably. Investigation of the methylating activity of arsenic in male mice showed that the highest activity occurred in the testes, followed by kidney, liver and lung (Healy *et al.*, 1998). The situation in female animals or humans remains to be elucidated. DMA was the main excretory metabolite of rat and human hepatocytes exposed to inorganic arsenic *in vitro* (Styblo *et al.*, 1999). In addition to arsenic methyltransferase activity, the tissue in which arsenic is methylated may also depend on the cellular uptake of its different forms. Experimental studies show a several-fold higher uptake of As^{III} and MMA^{III} than of As^V and MMA^V in liver cells (National Research Council, 1999; Styblo *et al.*, 1999, 2000; National Research Council, 2001). In contrast, arsenate is readily taken up in the kidneys, after which it can be reduced and excreted in the urine, partly in methylated form. Whether this also occurs with MMA^V and DMA^V is not known. The uptake of MMA^V and DMA^V in most other tissues of mice seems to be low (Hughes & Kenyon, 1998; National Research Council, 2001).

The methyltransferases involved in arsenic methylation have not been fully characterized, although enzymes from liver cells of rats, rabbits, hamsters and rhesus monkeys have been partially characterized (Zakharyan *et al.*, 1995, 1996; Wildfang *et al.*, 1998; Lin *et al.*, 2002) as cytosolic enzymes of 46–60 kDa, the activity of which requires both SAM and a thiol. As^{III} and MMA^V methylating activities seem to involve the same protein. Arsenate, selenate, selenite and selenide were not methylated by the purified enzyme preparations (Zakharyan *et al.*, 1995). Arsenite and $MAs^{III}O$ (methylarsine oxide) methyltransferase activities have been detected in primary cultures of human hepatocytes (Styblo *et al.*, 1999). The mRNA for As^{III} methyltransferase, purified from liver cytosol of male rats, was found to be similar to Cyt19, a putative methyltransferase expressed in human and mouse tissues (Lin *et al.*, 2002), and was detected in rat tissues and in HepG2 cells, a human cell line that was reported to methylate arsenic, but not in UROtsa cells, an immortalized human urothelial cell line that does not methylate arsenite.

(c)   *Variation in arsenic metabolism*

Although a number of studies have shown that the average relative distribution of arsenic metabolites in the urine is 10–30% inorganic arsenic, 10–20% MMA_{total} and 60–70% DMA_{total} (calculated percentages) (National Research Council, 1999), there is a wide variation among individuals (Vahter, 1999a,b). In one study, interindividual variation was found to exceed intra-individual variation considerably; the efficiency of arsenic methylation of an individual is remarkably stable over time (Concha *et al.*, 2002). Also, a few recent studies in which the trivalent and pentavalent metabolite forms have been speciated indicate a considerable variation in urinary excretion of MMA^{III} and DMA^{III} (Aposhian *et al.*, 2000a; Del Razo *et al.*, 2001b; Mandal *et al.*, 2001). Differences between population groups have also been reported. In one study, indigenous people living in the Andes, mainly Atacameños, excreted less MMA in urine (often only a few per cent) (Vahter *et al.*, 1995b); people living in certain areas of Taiwan, China, however, seem to have an unusually high percentage of MMA in urine (20–30% on average) (Chiou *et al.*, 1997b; Hsueh *et al.*, 1998). These findings indicate that the influence of genetic polymorphims is more important than environmental factors for the variation in arsenic methylation. Recently, it was reported that the methylation pattern among 11 families in Chile correlated more strongly between siblings than between father–mother pairs (Chung *et al.*, 2002), supporting a genetic basis for the variation in arsenic methylation.

A few human studies indicate that high doses of arsenic may influence its methylation in humans. In humans acutely intoxicated by inorganic arsenic, there is a marked delay in the urinary excretion of DMA which exceeds all other metabolites (Mahieu *et al.*, 1981; Foà *et al.*, 1984). Only after 1 or 2 weeks did the fraction of DMA in urine reach 70–80%, a level commonly seen after lower exposures. In people exposed to arsenic via drinking-water, the ratio of DMA to MMA in urine decreased somewhat with increasing level of exposure (Hopenhayn-Rich *et al.*, 1993). This is probably related to inhibition of methyl-transferase, especially in the second methylation step, by high concentrations of arsenite, as demonstrated in experimental studies *in vitro* (Buchet & Lauwerys, 1988; Styblo *et al.*, 1996). In people exposed to high concentrations of arsenic in drinking-water (several hundred micrograms per litre), there is a slight decrease in the percentage of DMA and a corresponding decrease in the percentage of MMA (Hepenhayn-Rich *et al.*, 1993). In some studies, women tend to methylate arsenic more efficiently than men (Hopenhayn-Rich *et al.*, 1996c), which may in part be related to the observed increase in arsenic methylation during pregnancy (Concha *et al.*, 1998c).

4.1.4   *Placental transfer*

Studies in experimental animals and humans show that both inorganic arsenic and methylated metabolites cross the placenta to the fetus (Concha *et al.*, 1998c). In women exposed to arsenic in drinking-water (about 200 µg/L), the concentrations of arsenic in umbilical cord blood were about as high as those in maternal blood in late gestation (about 10 µg/L) (Concha *et al*, 1998c). Placentas also had elevated concentrations of arsenic (median, 34 µg/kg wet wt; range, 17–54 µg/kg; *n* = 11). More than 90% of the arsenic in urine and plasma of both newborns and their mothers (at the time of delivery) was in the form of DMA (compared with about 70% in non-pregnant women), indicating an increase in arsenic methylation during pregnancy. The authors suggested that the DMA is much less toxic to the embryo and fetus than inorganic arsenic; the increased arsenic methylation during pregnancy could be highly protective for the developing organism.

Studies on women living in north-western Argentina indicated a low degree of arsenic excretion in human breast milk (Concha *et al.*, 1998a). The average concentration of arsenic in milk was 2 µg/kg fresh wt, compared with 10 µg/L in maternal blood and 320 µg/L in maternal urine. Breastfeeding of newborns from highly exposed areas decreased the levels of arsenic in their urine (which were elevated directly after birth) because of the low concentration in maternal breast milk (compared with formula milk prepared from the local water, which would provide about 200 µg arsenic/day) (Concha *et al.*, 1998c). A study of 36 German women showed that the average concentration of total arsenic in breast milk was less than 0.3 µg/L (Sternowsky *et al.*, 2002). The few women who reported a high intake of seafood showed increased arsenic levels in breast milk, indicating that organic arsenic compounds of marine origin, e.g. arsenobetaine, are excreted in milk.

4.1.5   *Excretion*

In humans, the major route of excretion of most arsenic compounds is via the urine. The biological half-time of inorganic arsenic is about 4 days, but is slightly shorter following exposure to arsenate than to arsenite (Crecilius, 1977; Yamauchi & Yamamura, 1979; Tam *et al.*, 1979; Pomroy *et al.*, 1980; Buchet *et al.*, 1981). In six human subjects who ingested radiolabelled [^{74}As]arsenate, 38% of the dose was excreted in the urine within 48 h and 58% within 5 days (Tam *et al.*, 1979). The results of another study indicate that the data were best fit to a three-compartment exponential model, with 66% excreted with a half-time of 2.1 days, 30.4% with a half-time of 9.5 days and 3.7% with a half-time of 38.4 days (Pomroy *et al.*, 1980). In three subjects who ingested 500 µg arsenic in the form of arsenite in water, about 33% of the dose was excreted in the urine within 48 h and 45% within 4 days (Buchet *et al.*, 1981).

The administration of sodium 2,3-dimercapto-1-propane sulfonate (DMPS), a chelating agent, to humans chronically exposed to inorganic arsenic in the drinking-water resulted in increased urinary excretion of arsenic (Aposhian *et al.*, 2000b; Guha Mazumder *et al.*, 2001a). In particular, there was a marked increase in urinary excretion of MMA^{III} and

MMA^V, while the concentration and percentage of urinary DMA decreased (Aposhian *et al.*, 2000b). Experimental studies supported the hypothesis that DMPS competes with endogenous ligands for MMA^{III}, forming a DMPS–MMA complex that is not a substrate for the MMA^{III} methyltransferase enzyme. This may explain the decrease in the conversion of the MMA^{III} to DMA. The DMPS–MMA complex is readily excreted in urine. Interestingly, MMA^{III} was excreted in the urine only after administration of DMPS (Aposhian *et al.*, 2000b).

4.2 Toxic effects

4.2.1 *Humans*

(*a*) *Acute and subacute toxicity*

Acute effects caused by the ingestion of inorganic arsenic compounds, mainly As^{III} oxide, are well documented. The major lesion is profound gastrointestinal damage, resulting in severe vomiting and diarrhoea, often with blood-tinged stools — symptoms that resemble cholera. Other acute symptoms and signs include muscular cramps, facial oedema and cardiac abnormalities; shock can develop rapidly as a result of dehydration. Subacute effects mainly involve the respiratory, gastrointestinal, cardiovascular, nervous and haematopoietic systems (WHO, 1981).

(*b*) *Chronic toxicity*

Most of the reports on chronic exposure to arsenic in humans focus attention on skin manifestations because of their diagnostic specificity. However, data derived from population-based studies and clinical case series and reports relating to intake of inorganic arsenic through drinking-water, medications or occupational and environmental exposure show that chronic exposure to arsenic adversely affects multiorgan systems. The clinical appearance of non-cancerous manifestations of arsenic intoxication in humans is insidious in onset and is dependent on the magnitude of the dose and the time course of exposure.

(i) *Cutaneous manifestations*

The specific cutaneous lesions of chronic arsenic toxicity are characterized by pigmentation and keratosis. These have been reported from different regions of the world including Argentina, Bangladesh, Chile, China, India (West Bengal), Japan, Mexico and Taiwan, China, where the content of arsenic in drinking-water is elevated (Zaldívar, 1974; Borgoño *et al.*, 1977; Cebrián *et al.*, 1983; Saha, 1984; Chakraborty & Saha, 1987; Guha Mazumder *et al.*, 1988, 1992; Ahmad *et al.*, 1997; Guha Mazumder *et al.*, 1998b; Biswas *et al.*, 1998; Mandal *et al.*, 1998; Ahmad, S.A. *et al.*, 1999; Milton & Rahman, 1999; Guo *et al.*, 2001). The magnitude of dose and the time frame of exposure to arsenic needed to induce the hyperpigmentation and hyperkeratosis characteristic of chronic arsenic intoxication have been investigated to a limited extent.

Among the population exposed to arsenic in drinking-water in the Antofagasta region of Chile, where levels reached 0.8 mg/L, cases of cutaneous arsenicosis, including both hyper-pigmentation and hyperkeratosis, have been described in children as young as 2 years of age (Rosenberg, 1974). In a cohort of 40 421 inhabitants of south-western Taiwan, China, inves-tigated by Tseng *et al.* (1968), the youngest subjects found to have hyperpigmentation and hyperkeratosis were reported to be aged 3 and 4 years, respectively; in a later investigation, the youngest subjects were aged 5 and 15 years (Tseng, 1977). The amount of arsenic consumed by these children was not specified. In a clinical evaluation conducted among 296 residents of Region Lagunera in northern Mexico, where ingested groundwater contained a mean arsenic concentration of approximately 0.4 mg/L, the shortest time of exposure associated with hypopigmentation was 8 years, increasing to 12 years for hyperpigmen-tation and palmoplantar keratosis (Cebrián *et al.*, 1983).

The hyperpigmentation of chronic arsenic poisoning commonly appears in a finely freckled, raindrop pattern that is particularly pronounced on the trunk and extremities, and distributed bilaterally symmetrically, but can also involve mucous membranes such as the undersurface of the tongue or buccal mucosa (Yeh, 1973; Tay, 1974; Saha, 1984; Guha Mazumder, 1988, 1992; Saha, 1995; Guha Mazumder *et al.*, 1998b; Ahmad, S.A. *et al.*, 1999; Milton & Rahman, 1999). Although less common, other patterns include diffuse hyperpigmentation (melanosis) (Tay, 1974; Saha, 1984), localized or patchy pigmen-tation, particularly affecting skinfolds (Tay, 1974; Szuler *et al.*, 1979), and so-called leukodermia or leukomelanosis (Saha, 1984; Mandal *et al.*, 1996, 1997; Chowdhury *et al.*, 2000a,b) in which the hypopigmented macules take a spotty, white appearance.

Arsenical hyperkeratosis appears predominantly on the palms of the hands and on the plantar aspect of the feet, although involvement of the dorsum of the extremities and the trunk have also been described. Occasionally, lesions may be larger and have a nodular or horny appearance. In severe cases, the hands and soles present diffuse verrucous lesions. Cracks and fissures may be severe on the soles (Sommers & McManus, 1953; Black, 1967; Tseng *et al.*, 1968; Yeh, 1973; Tay, 1974; Zaldívar, 1974; Borgoño *et al.*, 1977; Cebrián *et al.*, 1983; Saha, 1984; Guha Mazumder *et al.*, 1988; Ahmad *et al.*, 1997; Guha Mazumder *et al.*, 1998b; Saha & Chakraborti, 2001). Histological examination of the lesions typically reveals hyperkeratosis with or without parakeratosis, acanthosis and enlargement of the rete ridges. In some cases, there may be evidence of cellular atypia (mitotic figure) in large vacuolated epidermal cells (Tay, 1974). Yeh (1973) classified arsenical keratosis into two types: a benign type A (further subgrouped into those with no cellular atypia and those with mild cellular atypia); and a malignant type B (intra-epidermal carcinoma or carcinoma *in situ*, basal-cell carcinoma or squamous-cell carci-noma). Type B arsenical keratosis is histologically similar to but not indistinguishable from Bowen disease. Skin cancer can arise in the hyperkeratotic areas or appear on non-keratotic areas of the trunk, extremities or hands (Sommers & McManus, 1953; Yeh, 1973). In epidemiological studies in West Bengal (India) and Bangladesh, a higher pre-valence of arsenical skin lesions was observed in men compared with women, with a clear dose–response relationship (Guha Mazumder *et al.*, 1998c; Rahman *et al.*, 1999a; Tondel *et al.*, 1999).

Early studies provided estimates for the dose–response relationship of arsenic-induced skin lesions. Tseng *et al.* (1968) and Yeh (1973) evaluated 40 421 inhabitants of south-western Taiwan, China, where the drinking-water supply (artesian well-water) had been contaminated with arsenic for more than 50 years. The concentration of arsenic in the water supply varied from 0.01 to 1.82 mg/L, and most well-water in the endemic area had a range of 0.4–0.6 mg/L. The entire population at risk numbered 103 154. Of the people surveyed and examined clinically, characteristic arsenic-induced hyperpigmentation was diagnosed in 18.4%, keratosis in 7.1%, skin cancer in 1.1% and invasive skin cancer in 0.4%. Of the 428 people with clinically diagnosed skin cancer, 71.7% also had keratosis and 89.7% had hyperpigmentation. Ninety-nine per cent of the people with skin cancer had multiple skin cancers; 74.5% of the malignant lesions were on unexposed areas (Tseng *et al.*, 1968). Yeh (1973) studied 303 samples of skin cancers histologically: 57 were squamous-cell carcinomas, 45 were basal-cell carcinomas, 176 were intra-epidermal carcinomas (including 23 type B arsenical keratoses and 153 Bowen disease) and 25 were combined forms. The study in Taiwan lacked individual data on exposure to arsenic, since the levels were reported by village. In general, however, the incidence of hyperpigmentation, keratoses and skin cancer increased with increasing content of arsenic in the drinking-water and with age and length of exposure. The youngest patient with skin cancer was 2 years old. No case of melanosis, keratosis or skin cancer was identified in the nearby control population.

Guha Mazumder *et al.* (1998c) carried out the first population survey with individual data on exposure to arsenic among 7683 participants in West Bengal, India, to ascertain the prevalence of keratoses and hyperpigmentation. The arsenic content of their current water source ranged up to 3.4 mg/L, although 80% of participants consumed water containing < 0.5 mg/L arsenic. Of 4093 female and 3590 male participants, 48 and 108 had keratotic lesions and 127 and 234 had hyperpigmentation, respectively. Clear exposure–response relationships were found for levels of arsenic in water and the prevalence of these arsenic-induced skin effects. Men were affected more than women. Subjects who had body weights below 80% of the standard for their age and sex had a 1.6-fold and 1.2-fold increase in the prevalence of keratosis and hyperpigmentation, respectively. However, the survey examined only the participants' primary current drinking-water source. A similar cross-sectional study was conducted in Bangladesh by Tondel *et al.* (1999) who interviewed and examined 1481 subjects ≥ 30 years of age. A total of 430 subjects had skin lesions. Individual exposure assessment could only be estimated by present levels. Concentrations of arsenic in water ranged from 0.01 to 2.04 mg/L and the crude overall prevalence rate for skin lesions was 29/100. This study also showed a higher prevalence rate of arsenic-related skin lesions in men than in women, with a clear dose–response relationship.

Haque *et al.* (2003) recently completed a nested case–control study of a previous study (Guha Mazumder *et al.*, 1998c) to examine the dose–response relationship between concentrations below 0.5 mg/L in drinking-water and arsenic-induced skin lesions using a detailed exposure assessment that incorporated data on arsenic concentrations from current and past water sources used in households and work sites. A subset of 158 participants (69 cases and 89 controls) had complete histories of water concentrations. No case of a skin lesion was found with peak water concentrations of arsenic less than 0.1 mg/L. All of the eight cases (four men aged 31–75 years, four women aged 21–66 years), who currently had skin lesions and had ingested peak arsenic concentrations between 0.1 and 0.19 mg/L, had hyperpigmentation, and four also had keratoses.

Skin cancers are frequently associated with hyperkeratotic lesions (Yeh, 1973). Hyperkeratosis occurs more commonly and earlier in arsenic-exposed populations than skin cancer. A dose–response analysis of hyperkeratotic lesions may therefore allow the observation of a potential carcinogenic response to exposures lower than those used for skin cancer.

(ii)   *Respiratory disease*

The possible role of chronic ingestion of arsenic in the genesis of non-malignant pulmonary disease has been suggested in a few case series describing medical problems among individuals chronically exposed to increased concentrations of arsenic in the drinking-water. Among a total cohort of 180 residents of Antofagasta, Chile, exposed to drinking-water containing 0.8 mg/L arsenic, 38.8% of 144 subjects with abnormal skin pigmentation complained of chronic cough, compared with 3% of 36 subjects with normal skin (Borgoño *et al.*, 1977). In autopsies of four children and one adolescent from the Antofagasta region with an antecedent history of cutaneous arsenicosis and postmortem findings of extensive (non-pulmonary) vascular disease, two of the subjects were noted to have chronic bronchitis, slight bronchiectasis and slight diffuse interstitial fibrosis of the lung (Rosenberg, 1974). Symptoms of chronic lung disease were present in 89 (57%) of 156 cases of chronic arsenic toxicity caused by drinking arsenic-contaminated water in West Bengal, India (Guha Mazumder *et al.*, 1998b). Lung function tests carried out on 17 patients showed features of restrictive lung disease in nine (53%) and combined obstructive and restrictive lung disease in seven (41%) cases.

To investigate the relationship between non-malignant respiratory disease and ingested arsenic, Guha Mazumder *et al.* (2000) analysed data from the cross-sectional survey of 7683 participants who were examined clinically and interviewed, and measured the arsenic content in their current primary drinking-water source. Arsenic concentrations ranged from < 0.003 mg/L to 3.4 mg/L. Because there were few smokers, analyses were confined to nonsmokers (n = 6864 participants). Study subjects had arsenic-associated skin lesions, such as hyperpigmentation and hyperkeratosis, and were also highly exposed at the time of the survey (concentration of arsenic in water ≥ 0.5 mg/L). Individuals with normal skin and low concentration of arsenic in water (< 0.05 mg/L) were used as the referent group. Participants with skin lesions had age-adjusted prevalence odds ratio estimates for cough, crepitations and shortness of breath of 7.8 (95% CI, 3.1–19.5), 9.6 (95% CI, 4.0–22.9) and 23.2 (95% CI, 5.8–92.8) in women and 5.0 (95% CI, 2.6–9.9), 6.9 (95% CI, 3.1–15.0) and 3.7 (95% CI, 1.3–10.6) in men, respectively.

The effect of chronic exposure to arsenic on the respiratory system was studied in 218 individuals (94 exposed to arsenic [0.136–1 mg/L] and 124 control cases), most of whom were non-smokers, in Bangladesh (Milton *et al.*, 2001). The overall crude prevalence (or risk) of chronic cough and chronic bronchitis among exposed subjects was three times that in controls. Women were reported to be affected more than men.

The occurrence of chronic respiratory disease in the form of chronic cough or chronic bronchitis due to constant ingestion of arsenic through drinking-water has also been reported (Hotta, 1989; Chowdhury *et al.*, 1997; Kilburn, 1997; Chakraborti *et al.*, 1998; Ahmad, S.A. *et al.*, 1999; Ma *et al.*, 1999; Chowdhury *et al.*, 2000b).

(iii)   *Gastrointestinal system*

Chronic arsenic toxicity has been reported to produce various gastrointestinal symptoms. Hotta (1989) reported gastrointestinal impairment in 76% of subjects exposed to environmental arsenic at Torku, Japan. The symptoms were not serious in most patients, as they had possibly been afflicted with the initial stage of disease. Gastroenteritis was reported in a study of 1447 cases of chronic arsenicosis caused by drinking arsenic-contaminated water (0.05–1.8 mg/L) in the Inner Mongolian Autonomous region of China (Ma *et al.*, 1999). Of patients suffering from chronic arsenicosis after drinking arsenic-contaminated water (0.05–14.2 mg/L) in West Bengal, India, gastrointestinal symptoms characterized by dyspepsia were present in 60/156 (38.4%) cases studied (Guha Mazumder *et al.*, 1998b). Many investigators variously reported symptoms such as nausea, diarrhoea, anorexia and abdominal pain in cases of chronic arsenic toxicity (Rosenberg, 1974; Zaldívar, 1974; Borgoño *et al.*, 1977; Cebrián *et al.*, 1983; Guha Mazumder *et al.*, 1988; Ahmad *et al.*, 1997). However, in an epidemiological study carried out in the affected population in West Bengal, there was no difference in the incidence of abdominal pain among people drinking arsenic-contaminated water (0.05–3.4 mg/L) and the control population (< 0.05 mg/L) (27.84% versus 31.81%) (Guha Mazumder *et al.*, 2001b).

(iv)   *Liver and spleen*

Exposure to inorganic arsenic compounds has been associated with the development of chronic pathological changes in the liver. Several authors have reported cases of liver damage following treatment with trivalent inorganic arsenic (Morris *et al.*, 1974; Cowlishaw *et al.*, 1979; Szuler *et al.*, 1979; Nevens *et al.*, 1990). A common finding in these reports was portal hypertension without signs of liver cirrhosis. All patients had been given arsenic as a medication, mostly Fowler's solution, for several years. Typical cutaneous signs of long-term exposure to arsenic were also observed in some of the patients. In addition, there have been case reports on liver cirrhosis following medication with inorganic arsenic compounds (Franklin *et al.*, 1950; Rosenberg, 1974).

Datta *et al.* (1979) reported portal hypertension associated with periportal fibrosis in nine patients who were found to have high levels of arsenic in their liver in Chandigarh, India, two of whom had been drinking arsenic-contaminated water (0.549 and 0.360 mg/L). Guha Mazumder *et al.* (1988) reported hepatomegaly in 62/67 (92.5%)

members of families who had drunk arsenic-contaminated water (0.2–2 mg/L) in West Bengal, India, but in only 6/96 (6.25%) people from the same area who had drunk un-contaminated water (< 0.05 mg/L). Thirteen arsenic-exposed patients who had hepato-megaly were further investigated in hospital. All showed varying degrees of portal zone expansion and liver fibrosis histologically. Four of the five patients who had spleno-megaly showed evidence of increased intrasplenic pressure (30–36 cm saline), suggesting portal hypertension. Splenoportography of these cases showed evidence of intrahepatic portal vein obstruction. Although routine liver function tests were normal in all these cases, the bromosulphthalin retention test was abnormal in three. The level of arsenic in liver tissue, estimated by neutron activation analysis, was found to be elevated in 10/13 cases (0.5–6 mg/kg dry wt versus 0.16 ± 0.04 mg/kg dry wt in controls). Santra et al. (1999) and Guha Mazumder (2001a) subsequently reported hepatomegaly in 190/248 cases (76.6%) of chronic arsenicosis investigated in the same hospital. Evidence of non-cirrhotic portal zone fibrosis of the liver was found histologically in 63/69 cases (91.30%) of hepatomegaly. Liver function tests carried out on 93 such patients showed evidence of elevated levels of alanine aminotransferase (ALT), aspartase aminotransferase and alkaline phosphatase in 25.8%, 61.3% and 29% of cases, respectively. Serum globulin was found to be high (> 3.5 g/dL) in 19 (20.7%) cases.

Liver enlargement has been reported in cases of chronic arsenic toxicity caused by drinking arsenic-contaminated water (Saha, 1984; Chakraborty & Saha, 1987; Ahmad, S.A. et al., 1997, 1999; Ma et al., 1999; Saha & Chakribori, 2001).

(v)   *Chronic cardiovascular effects*

Ingested inorganic arsenic has been related to an increased incidence of cardio-vascular disease, especially ischaemic heart disease. This has been reviewed extensively (WHO, 1981; Engel & Smith, 1994; Chen et al., 1997; National Research Council, 1999, 2001).

Arsenic has been well documented as one of the major risk factors for Blackfoot disease, a unique peripheral arterial disease characterized by severe systemic arterio-sclerosis, as well as dry gangrene and spontaneous loss of affected extremities at end-stages. Histologically, Blackfoot disease can be divided into two reaction groups: arteriosclerosis obliterans and thromboangiitis obliterans. The prevalence of Blackfoot disease was reported to be 8.9/1000 among 40 421 inhabitants studied by Tseng et al. (1968) in Taiwan, China. The villages surveyed were arbitrarily divided according to the arsenic content of the well-water into low (< 0.3 mg/L), medium (0.3–0.6 mg/L) and high (> 0.6 mg/L) exposure. The prevalence of Blackfoot disease revealed a clear-cut ascendency gradient from low to medium to high exposure for both sexes and the three different age groups studied (Tseng, 1977). Atherogenicity and carcinogenicity of high levels of arsenic in artesian well-water was examined by Chen et al. (1988b). The lifetable method used to analyse cancer mortality of 789 patients with Blackfoot disease followed for 15 years showed a significantly higher mortality from cardiovascular and peripheral vascular disease among these patients com-pared with the general population in Taiwan and residents in the area endemic for Blackfoot disease. Whether fluorescent humic substances isolated from artesian well-water play an etiological role in Blackfoot disease has not been ascertained by epidemiological or animal studies (Van Duuren *et al.*, 1986; Lu *et al.*, 1990). A causal role for arsenic in the induction of Blackfoot disease offers the best explanation for the observations in Taiwan (Engel *et al.*, 1994).

Tsai *et al.* (1999) conducted a study in Taiwan, China, to analyse mortality from all causes in areas endemic for Blackfoot disease. They calculated standardized mortality ratios (SMRs) for cancer and non-cancer diseases, by sex, during the period 1971–94 and compared them with the local reference group (Chiayi-Tainan County) and the national reference group (population of Taiwan). The results revealed marked differences in SMR for the two reference groups. With respect to non-cancer disease, mortality was greater for men and women in the endemic area who had vascular disease, ischaemic heart disease, hypertension, diabetes mellitus and bronchitis than for the local reference groups. Mortality from other diseases including cancers of the rectum, stomach and oesophagus and cerebrovascular disease was higher among subjects in the study area than among the local reference group. These results indicated that the hazardous effect of arsenic was systemic.

Comparable peripheral vascular disorders with varying degrees of severity including Raynaud syndrome, acrocyanosis and gangrene of the feet have also been reported among people drinking arsenic-contaminated water (Rosenberg, 1974; Zaldívar, 1974; Borgoño *et al.*, 1977; Tseng *et al.*, 1996; Ahmad, S.A. *et al.*, 1999; Ma *et al.*, 1999; Guha Mazumder *et al.*, 2001b). It should be emphasized that there are differences in the prevalence of peripheral vascular diseases that cause gangrene and limb loss among different populations exposed to arsenic; the incidence is high in Taiwan, China, but low in Chile, India and Bangladesh, and none has been reported from Mexico or Argentina (Engel *et al.*, 1994).

An epidemiological study reported an increased prevalence of hypertension among residents in an area endemic for Blackfoot disease and a dose–response relationship with ingested inorganic arsenic (Chen *et al.*, 1995). A total of 382 men and 516 women residing in areas of Taiwan, China, endemic for arsenic were studied. A 1.5-fold increase in the age- and sex-adjusted prevalence of hypertension was observed compared with residents in non-endemic areas, and was associated with higher cumulative exposure to arsenic. The dose–response relation remained significant after adjustment for age, sex, diabetes mellitus, proteinuria, body mass index and level of serum triglycerides. Increased prevalence of hypertension was also observed in 6.2% of patients affected with arsenic-induced skin lesions (144) compared with none of those with no skin lesion (36) in Antofagasta, Chile (Borgoño *et al.*, 1977). Rahman *et al.* (1999b) conducted studies on arsenic-exposed people in Bangladesh and demonstrated an association between hypertension and cumulative exposure to arsenic in drinking-water (Rahman & Axelson, 2001; Rahman, 2002).

Significant dose–response relationships between the level of ingested inorganic arsenic and risk for ischaemic heart disease were observed in recent cohort and case–control studies in Taiwan, China (Chen *et al.*, 1994). In an ecological correlational study in Taiwan based on 898 806 person–years and 172 deaths from ischaemic heart disease observed from 1973 to 1986, a dose–response relationship between concentration of arsenic in drinking-water and age-adjusted mortality from ischaemic heart disease was observed. A total of 257 patients with Blackfoot disease and 753 matched healthy controls were recruited and followed-up for more than 7 years. Significantly increased mortality from ischaemic heart disease was observed for patients with Blackfoot disease and matched controls showing SMRs (95% confidence interval [CI]) of 937 (536–1519) and 248 (139–409), respectively, compared with the general population of Taiwan (SMR, 100). Cox's proportional hazard regression analysis also showed a dose–response relationship between mortality from ischaemic heart disease and cumulative exposure to arsenic after adjustment for age, sex, body mass index and disease status for hypertension and diabetes mellitus. A case–control study including 78 patients with electrocardiogram-based ischaemic heart disease and 384 healthy residents was carried out in three villages where Blackfoot disease was endemic. Based on a multiple logistic regression analysis, cumulative exposure to arsenic was found to be associated with ischaemic heart disease in a dose-related manner after adjustment for age, sex, body mass index, disease status for hypertension and diabetes mellitus, ratio between total cholesterol and high-density lipoprotein cholesterol and cumulative alcohol consumption. The occurrence of ischaemic heart disease due to chronic exposure to arsenic has also been reported by other investigators (Rosenberg, 1974; Zaldívar, 1974; Hotta, 1989; Chen et al., 1994, 1995; Ma et al., 1999), as has the occurrence of cardiac arrhythmia (Hotta, 1989; Ma et al., 1999).

Mortality rates from 1973 through 1986 for ischaemic heart disease among residents in 60 villages of an area in Taiwan, China, endemic for arsenicosis were analysed by Chen et al. (1996) to examine their association with the concentration of arsenic in drinking-water. Based on 1 355 915 person–years and 217 deaths from ischaemic heart disease, the cumulative mortality from birth to age 79 years as 3.4%, 3.5%, 4.7% and 6.6%, respectively, for residents who lived in villages in which the median concentrations of arsenic in drinking-water were < 0.1, 0.1–0.34, 0.35–0.59 and ≥ 0.6 mg/L. A cohort of 263 patients with Blackfoot disease and 2293 residents in the endemic area of arsenicosis without Blackfoot disease were recruited and followed up for an average period of 5.0 years. There was a monotonic biological gradient relationship between cumulative exposure to arsenic through drinking artesian well-water and mortality from ischaemic heart disease. The relative risks (95% CI) were 2.5 (0.53–11.37), 4.0 (1.01–15.60) and 6.5 (1.88–22.24), respectively, for those who had cumulative exposures to arsenic of 0.1–9.9, 10.0–19.9 and ≥ 20.0 mg/L–years, compared with those with no known exposure to arsenic, after adjustment for age, sex, cigarette smoking, body mass index, serum cholesterol and triglyceride levels, and disease status for hypertension and diabetes through proportional hazard regression analysis. Patients with Blackfoot disease were found to have a significantly higher mortality from ischaemic heart disease than residents without Blackfoot disease, showing a multivariate-adjusted relative risk of 2.5 (95% CI, 1.14–5.40).

Wang et al. (2002) reported evidence of a dose–response relationship between long-term exposure to arsenic in drinking-water and prevalence of carotid atherosclerosis in the arsenic-exposed area of south-western Taiwan, China. The extent of carotid athero-sclerosis was assessed by duplex ultrasonography among 199 male cases and 264 residents who participated in the study. Three indices of exposure, duration of consumption of artesian well-water, average concentration of arsenic in consumed well-water and cumulative exposure to arsenic, were all significantly associated with an increased prevalence of carotid atherosclerosis with a dose–response relationship. The biological gradient remained significant after adjustment for age, sex, hypertension, diabetes mellitus, cigarette smoking, alcohol consumption, waist-to-hip ratio and serum levels of total and low-density lipoprotein cholesterol. The multivariate-adjusted prevalence odds ratio was 1.8 (95% CI, 0.8–3.8) and 3.1 (95% CI, 1.3–3.4) for those who had a cumulative exposure to arsenic of 0.1–19.9 and ≥ 20 mg/L–years, respectively, compared with those without exposure to arsenic from drinking artesian well-water.

(vi) *Nervous system*

Abnormal electromyographic (EMG) findings suggestive mostly of sensory neuropathy were reported in 10/32 (31.25%) subjects exposed to arsenic by drinking contaminated well-water (range, 0.06–1.4 mg/L) in Canada (Hindmarsh *et al.*, 1977). Paresthaesia was present in 74/156 (47.43%) patients with chronic arsenicosis caused by drinking arsenic-contaminated water (0.05–14.2 mg/L) in West Bengal, India. Objective evaluation of neuronal involvement carried out in 29 patients showed abnormal EMGs in 10 (34.5%) and altered nerve conduction velocity and EMGs in 11 (38%) cases (Guha Mazumder *et al.*, 1997). Evidence of parasthaesia or peripheral neuropathy due to chronic exposure to arsenic through drinking-water has also been reported (Saha, 1984; Hotta, 1989; Kilburn, 1997; Ahmad, S.A. *et al.*, 1999; Ma *et al.*, 1999; Chowdhury *et al.*, 2000a; Rahman *et al.*, 2001, 2003). More sensory than motor neuropathy has also been reported among arsenicosis patients in West Bengal (Basu *et al.*, 1996; Mukherjee *et al.*, 2003).

The relationship between the prevalence of cerebrovascular disease and ingestion of inorganic arsenic in drinking-water was reported by Chiou *et al.* (1997a) in a cross-sectional study in Taiwan, China, that recruited a total of 8102 men and women from 3901 households. The status of cerebrovascular disease of study subjects was identified through personal home interviews and ascertained by review of hospital medical records according to WHO criteria. Information on consumption of well-water, sociodemographic characteristics, cigarette smoking habits and alcohol consumption, as well as personal and family history of disease, was also obtained. The concentration of arsenic in the well-water of each household was determined by HG–AAS. A significant dose–response relationship was observed between concentration of arsenic in well-water and the prevalence of cerebrovascular disease after adjustement for age, sex, hypertension, diabetes mellitus, cigarette smoking and alcohol consumption. The biological gradient was even more prominent for cerebral infarction, showing multivariate-adjusted odds ratios of 1.0, 3.21 (95% CI, 1.51–6.88), 4.37 (95% CI, 1.99–9.60) and 6.58 (95% CI, 2.82–15.28), respectively, for those who consumed well-water with an arsenic content of 0, 0.001–0.05, 0.051–0.299 and

> 0.3 mg/L. Increased incidences of cerebrovascular disease in cases of chronic arsenicosis have been reported elsewhere (Hotta, 1989; Chen *et al.*, 1997; Ma *et al.*, 1999).

Kilburn (1997) reported the occurrence of peripheral neuritis, sleep disturbances, weakness and cognitive and memory impairment in residents of Bryan-College Station, TX, USA, exposed to arsenic in air and water from the use of arsenic trioxide to produce defoliants for cotton at an Atochem plant. Siripitayakunkit *et al.* (1999) reported retardation of intelligence among 529 children (6–9 years of age) living in Thailand who had chronic exposure to arsenic from the environment.

(vii) *Diabetes mellitus*

To examine the association between ingested inorganic arsenic and the prevalence of diabetes mellitus, Lai *et al.* (1994) studied 891 adults residing in villages in southern Taiwan, China. The status of diabetes mellitus was determined by an oral glucose tolerance test and a history of diabetes regularly treated with sulfonylureas or insulin. They observed a dose–response relationship between cumulative exposure to arsenic and prevalence of diabetes mellitus. The relationship remained significant after adjustment for age, sex, body mass index and physical activity level at work by a multiple logistic regression analysis, giving multivariate-adjusted odds ratios of 6.61 (95% CI, 0.86–51.0) and 10.05 (95% CI, 1.30–77.9), respectively, for those who had a cumulative exposure to arsenic of 0.1–15.0 and > 15.1 mg/L–years compared with those who were unexposed.

Rahman *et al.* (1998) reported a significantly increased prevalence of diabetes mellitus in Bangladesh caused by drinking arsenic-contaminated water among subjects with keratosis compared with subjects who did not have keratosis. A significant trend in risk between an approximate, time-weighted exposure to arsenic and the prevalence of diabetes mellitus strengthened the possibility of a causal association. [The lack of comprehensive, systematic, long-term sampling of the water supplies in the study area is a limitation of the study and data on individual exposures measured directly over time would have been more informative. However, these results suggest that chronic exposure to arsenic may induce diabetes mellitus in humans.] A further study regarding glucosuria patients with and without skin lesions in relation to exposure to arsenic in drinking-water reported that the prevalence ratios among the subjects without skin lesions were 0.8 (95% CI, 0.4–1.3), 1.4 (95% CI, 0.8–2.3) and 1.4 (95% CI, 0.7–2.4), after ajustment for age and sex compared with unexposed subjects as reference. The exposure categories were < 0.5, 0.5–1 and > 1 mg/L, respectively. For those with skin lesions, the prevalence ratios were slightly higher; 1.1 (95% CI, 0.5–2.0), 2.2 (95% CI, 1.3–3.8) and 2.6 (95% CI, 1.5–4.6), respectively, in comparison with unexposed subjects (Rahman *et al.*, 1999a). [In this study also, a lack of systematic sampling of water supplies in the study area is a limitation. Furthermore, although glucosuria is a primary indicator of diabetes mellitus, identification of the hyperglycaemic patients among those with glucosuria would have been more informative.]

Tseng *et al.* (2000) reported a cohort study on 446 non-diabetic residents from an arsenic-contaminated area in south-western Taiwan, China. Diabetes mellitus was determined by an oral glucose tolerance test. The age-specific incidence density ratio of diabetes mellitus was between two- and five-fold higher in the exposed cohort than in the unexposed cohort. An exposure–response relationship was observed between incidence of diabetes mellitus and long-term exposure to ingested arsenic from artesian well-water, showing a relative risk of 2.1 (95% CI, 1.1–4.2) for those who had a cumulative exposure to arsenic $\geq$ 17 mg/L–years compared with those who had a lower cumulative exposure (< 17 mg/L–years).

(viii) *Study of oxidative stress in humans*

8-Hydroxy-2′-deoxyguanosine (8-OHdG) is generated by the hydroxyl radical (Kasai & Nishimura, 1984) or singlet oxygen (Devasagayam *et al.*, 1991) or by direct electron transfer, which does not involve the participation of any reactive oxygen species (Kasai *et al.*, 1992). 8-OHdG is considered to be one of the main indicators of oxidative damage to DNA and may cause mutation (G:C→T:A) during DNA replication (Shibutani *et al.*, 1991).

Matsui *et al.* (1999) investigated whether neoplastic and precancerous skin lesions of arsenic-exposed individuals are under oxidative stress using 8-OHdG as a marker. Biopsy samples of arsenic keratosis, arsenic-induced Bowen disease and arsenic-induced Bowen carcinoma arising in areas not exposed or less exposed to the sun were obtained from 28 individuals (aged 26–83 years) living in areas where chronic arsenicism was endemic in either Taiwan, China, Thailand or Japan. The presence of 8-OHdG was studied by immuno-histochemistry using N45.1 monoclonal antibody in the 28 cases of arsenic-related skin neoplasm and arsenic keratosis as well as in 11 cases of Bowen disease unrelated to arsenic. The frequency of 8-OHdG-positive cases was significantly higher in arsenic-related skin neoplasms (22/28; 78%) than in Bowen disease unrelated to arsenic (1/11; 9%) ($p < 0.001$ by chi-square test). 8-OHdG was also detected in normal tissue adjacent to the arsenic-related Bowen disease lesion. Furthermore, arsenic was detected by neutron activation analysis in deparaffined skin tumour samples of arsenic-related disease (four of five, 80%), whereas it was not detected in control samples. The results may suggest the involvement of reactive oxygen species in arsenic-induced human skin cancer.

Wu *et al.* (2001) reported an association of blood arsenic levels with increased reactive oxidants and decreased antioxidant capacity among 64 subjects aged 42–75 years from an arsenic-contaminated area in north-eastern Taiwan, China. The blood level of arsenic determined by HG–AAS ranged from undetectable to 46.5 µg/L. The capacity of subjects to methylate arsenic was determined by speciation of inorganic arsenic and its metabolites in urine using high-performance liquid chromatography (HPLC) linked with HG–AAS. The plasma level of reactive oxidants was determined by a chemiluminescence method using lucigenin as an amplifier for measuring superoxide anion (O_2^-), while the plasma level of antioxidant capacity was measured by the 2,2′-azino-di[3-ethylbenzthiazoline]-sulfonate method. There was a positive association between blood arsenic level and plasma level of reactive oxidants ($r = 0.41$; $p = 0.001$) and a negative association with plasma level of antioxidant capacity ($r = -0.30$; $p = 0.014$). Categorical analysis showed that greater primary capacity for arsenic methylation was correlated with a higher plasma level of anti-oxidant capacity ($p = 0.029$).

(ix)   Other

Generalized weakness and fatigue have been reported in people chronically exposed to arsenic-contaminated drinking-water (Zaldívar & Guillier, 1977; Saha, 1984; Guha Mazumder *et al.*, 1988, 1992; Kilburn, 1997; Guha Mazumder *et al.*, 1998; Guha Mazumder, 2001b; Guha Mazumder *et al.*, 2001b). Conjunctival congestion and non-pitting oedema of the legs and hands have also been reported in patients with chronic arsenic toxicity in West Bengal, India, and Bangladesh (Ahmad *et al.*, 1997; Chowdhury *et al.*, 1997; Guha Mazumder *et al.*, 1998b; Ahmad, S.A. *et al.*, 1999).

García Vargas *et al.* (1994) carried out a detailed study of the urinary excretion pattern of porphyrins in humans chronically exposed to arsenic via drinking-water in Mexico using HPLC. Thirty-six individuals (15 men and 21 women) were selected from a town which had 0.40 mg/L arsenic in the drinking-water. The control group consisted of 31 individuals (13 men and 18 women) whose concentration of arsenic in the drinking-water was 0.02 mg/L. Major abnormalities in the urinary porphyrin excretion pattern observed in arsenic-exposed individuals were (*a*) significant reductions in coproporphyrin III excretion resulting in decreases in the coproporphyrin III/coproporphyrin I ratio and (*b*) significant increases in uroporphyrin excretion. Both alterations were responsible for the decrease in the urinary coproporphyrin/uroporphyrin ratio. No porphyrinogenic response was found in individuals with urinary concentrations below 1 mg arsenic/g creatinine. However, as arsenic concentrations exceeded this value, the excretion of porphyrins (except copropor-phyrin III) increased proportionally, and most of the individuals with high urinary arsenic concentrations had alterations in porphyrin ratios and also presented cutaneous signs of chronic arsenic poisoning. The prevalence of clinical signs of arsenicism showed a direct relationship with both concentration of arsenic in urine and time-weighted exposure to arsenic. A direct relationship between time-weighted exposure to arsenic and alterations in urinary porphyrin excretion ratios was also observed. These alterations in arsenic-exposed individuals are compatible with a lower activity of uroporphyrinogen decarboxylase, the enzyme that converts the substrate uroporphyrinogen to a coproporphyrinogen product.

Except for anaemia, no haematological abnormality (in differential lymphocyte count or in the levels of blood sugar, urea or creatinine) has been described in cases of chronic toxicity caused by drinking arsenic-contaminated water (Guha Mazumder *et al.*, 1988, 1997, 1998b, 1999). Haematological consequences of subacute and chronic arsenic toxicity have been reviewed extensively (National Research Council, 1999).

4.2.2   *Experimental systems*

(a)   *Acute toxicity*

The acute toxicity of arsenic is related to its chemical form and oxidation state. The LD_{50} (50% lethal dose) values of several arsenicals in laboratory animals have been reviewed (Hughes, 2002). In mice, the oral lethal dose of arsenic trioxide varies from 15 to 48 mg/kg bw. In contrast, the lethal dose range of inorganic arsenic in adult humans is estimated at 1–3 mg/kg bw (Hughes, 2002). A basic tenet is that the acute toxicity of trivalent arsenic is greater than that of pentavalent arsenic. For example, in mice, the oral LD_{50} of arsenic trioxide is more than 36-fold lower than that of MMA^V. However, MMA^{III} has been found to be more toxic than trivalent arsenic. When MMA^{III} or sodium arsenite was administered intraperitoneally to hamsters, the LD_{50}s were found to be 29.3 and 112.0 µmol/kg bw, respectively (Petrick et al., 2001).

(b) Chronic toxicity

Many different systems within the body are affected by chronic exposure to inorganic arsenic. Arsenates can replace phosphate in many biochemical reactions because they have similar structure and properties. Arsenate uncouples in-vitro formation of adenosine-5′-triphosphate (ATP) by a mechanism termed arsenolysis. In the substrate, arsenolysis may occur during glycolysis. ATP is generated during glycolysis in the presence of phosphate (substrate phosphorylation), but not arsenate. In the mitochondria, arsenolysis may occur during oxidative phosphorylation. Adenosine-5′-diphosphate (ADP)-arsenate is synthesized by submitochondrial particles from ADP and arsenate, in the presence of succinate. ADP-arsenate hydrolyses easily compared with ADP-phosphate, which is formed during oxidative phosphorylation. In both the substrate and the mitochondria, arsenolysis diminishes in-vitro formation of ATP by the replacement of phosphate with arsenate in the enzymatic reactions. Depletion of ATP by arsenate has been observed in cellular systems: ATP levels are reduced in rabbit and human erythrocytes after in-vitro exposure to arsenate (rabbits, 0.8 mM; humans, 0.01–10 mM) (Delnomdedieu et al., 1994a; Winski & Carter, 1998; Hughes, 2002).

Trivalent arsenic reacts readily in vitro with thiol-containing molecules such as GSH and cysteine (Scott et al., 1993; Delnomdedieu et al., 1994b). In rat red blood cells, As^{III} forms mixed complexes with protein and GSH, and the main protein-binding species is haemoglobin (Winski & Carter, 1995). Binding of MMA^{III} and DMA^{III} to protein in vitro occurs to a greater extent than with the pentavalent organic forms (Styblo et al., 1995). Arsenite has a higher affinity for dithiols than monothiols, as shown by the highly favoured transfer of arsenite from a $(GSH)_3$–arsenic complex to the dithiol 2,3-dimercaptosuccinic acid (Delnomdedieu et al., 1993). The binding of trivalent arsenic to critical thiol groups may inhibit important biochemical events that could lead to toxicity (Hughes, 2002).

(i) Mitochondrial damage

Hepatic phosphate resonances were evaluated by Chen et al. (1986) in vivo by ^{31}P nuclear magnetic resonance spectroscopy following a single intravenous dose of sodium arsenite (10 mg/kg bw) in rats. Acute in-vivo administration of arsenite rapidly decreased intracellular pools of ATP with concomitant increases in inorganic phosphate and phosphomonoesters (phosphocholine and adenosine monophosphate). Glycerolphosphorylcholine and glycerolphosphorylethanolamine were also increased. The data suggest that liver cannot compensate for the rapid loss of nicotinamide-adenine dinucleotide (NAD)-linked substrate oxidation via other metabolic pathways, such as glycolysis, for the production of ATP.

Arsenic fed to laboratory animals is known to accumulate in the mitochondria and has been related to the swelling of this subcellular organelle in a number of tissues, especially the liver (Fowler *et al.*, 1977, 1979). It has been suggested that the effects of arsenic on mitochondrial utilization of pyruvate results from arsenic binding to the lipoic acid and dithiol moieties of the pyruvate dehydrogenase (PDH) complex. The initial step in the mitochondrial metabolism of pyruvate, which is catalysed by the PDH–enzyme complex, involves the formation of acetyl-coenzyme A (CoA) and the generation of CO_2 and hydrogenated NAD (NADH). This complex is composed of three enzymes: pyruvate decarboxylase (PDH), dehydrolipoate transacetylase and dihydrolipoate dehydrogenase. The latter two enzymes contain dithiol moieties. Pyruvate decarboxylase is regulated by inactivation and activation reactions, which are controlled by phosphorylation/dephosphorylation. Phosphorylation and the concomitant inactivation of PDH is catalysed by a Mg-ATP-requiring kinase, and dephosphorylation and concomitant reactivation is catalysed by a Mg^{2+}- and Ca^{2+}-requiring phosphatase. In order to examine whether this phosphorylation/dephosphorylation is a mechanism of action for arsenic, PDH activities, before and after in-vitro activation with Mg^{2+}, were measured in tissue from animals fed arsenic (Schiller *et al.*, 1977). Adult male Charles River CD rats were given deionized drinking-water containing 0, 20, 40, and 85 mg/L arsenic as sodium arsenate (As^V) for 3 and 6 weeks. PDH activity was assayed in the liver tissue obtained from the animals. After 3 weeks, the effects of arsenic at the highest dose were pronounced compared with the basal activity (before activation), with up to 47.5% inhibition of the control values. The total PDH activity (after activation) was inhibited by 13.5, 15.3 and 27.6% of the control values at 20, 40 and 85 mg/L sodium arsenate, respectively. A similar pattern of inhibition of PDH activity was observed at 6 weeks, although the inhibition was lower at the highest dose. This pattern may be indicative of mitochondrial regeneration at the highest dose after 6 weeks (Schiller *et al.*, 1977). The possible metabolic effects of this inhibition are a decrease in acetyl-CoA formation, which leads to a decrease in carbon flow through the tricarboxylic acid cycle, and a decrease in the citrate available to allow mitochondria to supply acetyl-CoA for fatty acid synthesis, which in turn results in fewer storage triglycerides.

Petrick *et al.* (2001) have compared the in-vivo toxicity of MMA[III] and arsenite in hamsters. Groups of six male golden Syrian hamsters, 11–12 weeks old and weighing 100–130 g, were injected intraperitoneally with MMA[III] oxide or sodium arsenite. Inhibition of PDH activity of the hamster kidney or purified porcine heart by MMA[III] or arsenite was determined. To inhibit PDH activity of hamster kidney by 50%, concentrations (mean ± SE) of 59.9 ± 6.5 µM MMA[III] as methylarsine oxide, 62.0 ± 1.8 µM MMA[III] as diiodomethylarsine and 115.7 ± 2.3 µM arsenite were needed. To inhibit in-vitro PDH activity of the purified porcine heart by 50%, concentrations (mean ± SE) of 17.6 ± 4.1 µM MMA[III] as methylarsine oxide and 106.1 ± 19.8 µM arsenite were needed. These data demonstrate that MMA[III] is more toxic than inorganic arsenite, both *in vivo* and *in vitro*.

Brown *et al.* (1976) demonstrated alteration of normal ultrastructure and respiratory ability of proximal renal tubules following administration of arsenate. Groups of male Sprague-Dawley rats weighing 70–150 g were fed laboratory chow. The control groups received deionized water while the experimental groups received 40, 85 or 125 mg/L arsenic as sodium arsenate in deionized water ($n = 28, 11, 7$ and 10 in each group, respectively). After 6 weeks, the rats were killed in pairs of one experimental rat and a control rat matched by weight. The kidneys were excised and the capsule removed; combined oxygen electrode and electron microscopic studies were conducted. Decreased state 3 respiration and respiratory control ratios were observed in kidneys of rats given the 85- and 125-mg/L dose levels. Ultrastructural alterations, which consisted of swollen mitochondria and an increased number of dense autophagic lysosome-like bodies, were confined to proximal tubule cells of animals at all dose levels of arsenic.

Fowler *et al.* (1977) carried out investigations to delineate the subcellular manifestations of arsenic toxicity following chronic exposure using combined ultrastructural and biochemical techniques. Four groups of 18 male Charles River CD rats were fed a casein-based purified diet and had access to deionized drinking-water containing 0, 20, 40 or 85 mg/L arsenic as sodium arsenate (As^v) for 6 weeks. At the end of this period, three animals from each group were killed and the livers removed. Mitochondrial respiration studies were conducted. Extensive in-situ swelling of liver mitochondria and matrix rarification with lipidic vacuolation were the most prominent ultrastructural changes observed at 40- and 80-mg/L As^V dose levels. Mitochondrial respiration studies indicated decreased state 3 respiration and respiratory control ratios for pyruvate/malate- but not succinate-coupled respiration. Specific activity of monoamine oxidase, which is localized on the outer mitochondrial membrane, showed increases of up to 150% of control, and cytochrome *c* oxidase, which is localized on the inner mitochondrial membrane, showed an increase in specific activity of 150–200%. Activity of malate dehydrogenase, which is localized in the mitochondrial matrix, remained unchanged at all dose levels. These studies indicate that decreased mitochondrial respiration is only one aspect of arsenic toxicity to this organelle. Marked arsenic-mediated perturbation of important enzyme systems localized in mitochondria, which participate in the control of respiration and other normal mitochondrial functions (such as haeme synthesis, carbohydrate metabolism and fatty acid synthesis), are also important manifestations of cellular dysfunction.

A positive, quantitative in-vivo correlation between mitochondrial structure and function and their alteration following administration of sodium arsenate has further been demonstrated (Fowler *et al.*, 1979). Two groups of male Charles River CD rats were fed a casein-based semipurified diet for 6 weeks and had access to deionized drinking-water containing 0 or 40 mg/L arsenic as sodium arsenate. Ultrastructural morphometric and biochemical studies were conducted on hepatic mitochondria. Morphometric analysis disclosed an overall 1.2-fold increase in the relative mitochondrial volume density and a 1.4-fold increase in the surface density of the inner mitochondrial membrane plus cristae of arsenate-exposed rats. These observations suggest that arsenate-mediated perturbation of mitochondrial membrane integrity compromises the mechanisms of normal ion transport. These structural changes were associated with a perturbation of mitochondrial protein synthesis as expressed by a 1.5-fold increase in [^{14}C]leucine incorporation into all mitochondrial proteins, which was primarily associated with the acid-insoluble membranous fraction. Mitochondria from arsenate-treated rats showed a marked disruption of normal conformational behaviour with depression of NAD$^+$-linked substrate oxidation and a subsequent approximately two-fold in-vivo increase in the mitochondrial NAD$^+$ to NADH$^+$ ratio. Observed changes in mitochondrial membranes from arsenate exposure also resulted in 1.5–2-fold increases in the specific activities of the membrane marker enzymes monoamine oxidase, cytochrome c oxidase and Mg^{2+}-ATPase, which are localized in both inner and outer mitochondrial membranes. Activity of malate dehydrogenase, which is localized in the mitochondrial matrix, was unchanged.

Larochette *et al.* (1999) investigated whether arsenic compounds act on mitochondria to induce apoptosis. The mechanisms by which arsenic induces apoptosis are not clear. U937 cells transfected with an SFFV.neo-vector containing the human *bcl-2* gene coding for apoptosis-inhibitory protein or the neomycin-resistance gene (*Neo*) only, or 2B4.11T cell hybridoma cells (1–5 × 10^5/mL) were incubated with variable doses of sodium arsenite, sodium arsenate, phenylarsine oxide, *para*-arsanilic acid or 1-(2-chlorophenyl)-*N*-methyl-*N*-(1-methylpropyl)-3-isoquinolinecarboxamide (PK11195) and 100 μM of the caspase inhibitors *N*-benzyloxycarbonyl-Val-Ala-Asp.fluoromethylketone (Z-VAD.fmk) and *tert*-butyloxycarbonyl-Asp.fluoromethylketone (Boc-D.fmk) and 100 μM of the cathepsin inhibitor *N*-benzyloxycarbonyl-Phe-Ala.fluoromethylketone (Z-FA.fmk). Arsenite induced apoptosis accompanied by a loss of mitochondrial transmembrane potential (ΔΨ$_m$). Inhibition of caspases by Z-VAD.fmk and Boc-D.fmk prevented arsenite-induced nuclear DNA loss, but had no effect on the ΔΨ$_m$ dissipation and cytolysis induced by arsenite, suggesting that arsenite might cause necrosis when the caspase pathway is blocked. In contrast, *bcl-2* expression induced by gene transfer prevented all hallmarks of arsenite-induced cell death (such as generation of reactive oxygen species, hypoploidy and loss of viability), including the collapse of ΔΨ$_m$. PK11195, a ligand of the mitochondrial benzodiazepine receptor, neutralized the bcl-2-mediated resistance to arsenite. Mitochondria were required in a cell-free system (isolated nuclei *in vitro*) to mediate arsenite-induced nuclear apoptosis. Arsenite caused the release of an apoptosis-inducing factor from the mitochondrial intermembrane space. This effect was prevented by the permeability transition (PT) pore inhibitor cyclosporin A, as well as by bcl-2, which is known to function as an endogenous PT pore antagonist. Arsenite-permeabilized liposomal membranes contained the purified, reconstituted PT pore complex. Bcl-2 also inhibited the arsenite-triggered opening of the PT pore in the reconstituted system. As a control, a mutant bcl-2 Δα5/6 protein, which had lost its anti-apoptotic function as a result of the deletion of a putative membrane insertion domain, failed to prevent arsenite-induced PT pore opening. Together these data suggest that arsenite could induce apoptosis via a direct effect on the mitochondrial PT pore.

(ii)   *Urinary porphyrins, haeme biosynthetic enzyme activities, haeme metabolism and arsenic*

Arsenic can modify the urinary excretion of porphyrins in animals and humans. It also interferes with the activities of several enzymes of the haeme biosynthetic pathway, such as δ-aminolevulinate (ALA) synthase (ALA-S), porphobilinogen deaminase, uroporphyrinogen III synthase, uroporphyrinogen decarboxylase, coproporphyrinogen oxidase, ferrochelatase and haeme oxygenase (H-O). The urinary porphyrins and several haeme enzymes can be used as early biomarkers of arsenic toxicity (García-Vargas & Hernández-Zavala, 1996).

Rodents exposed for 6 weeks to sodium arsenate in drinking-water showed a substantial increase in the urinary excretion of porphyrins, with excretion of uroporphyrin exceeding that of coproporphyrin (Woods & Fowler, 1978). Groups of 12 male Sprague-Dawley rats (CD strain) (150–200 g) or male C57 BL mice (20–30 g) were given access to laboratory chow and deionized drinking-water containing 0, 20, 40 or 85 mg/L arsenic as sodium arsenate (As^V) for up to 6 weeks. Livers of animals were homogenized and mitochondria and microsomal fractions were then prepared. Continuous prolonged exposure to sodium arsenate resulted in depression of hepatic δ-aminolevulinate synthase and haeme synthase, the first and last enzymes in haeme biosynthesis, respectively, in both rats and mice. ALA-S was maximally depressed to approximately 80% of control values at 40 mg/L in both species, whereas haeme synthase activity was maximally decreased to 63 and 75% of control at 85 mg/L in rats and mice, respectively. Uroporphyrinogen I synthase, the third enzyme in haeme biosynthesis, was increased at all doses in mice, whereas ALA dehydratase, the second haeme biosynthetic pathway enzyme, was unaltered in either species. Concomitantly, urinary uroporphyrin concentrations were increased up to 12-fold and coproporphyrin levels up to 9-fold the control values in rats. Similar patterns of increased porphyrin excretion were seen in mice. In contrast, no changes were observed in the activities of cytochrome oxidase or cytochrome P450, indicators of mitochondrial and microsomal haemoprotein function, respectively. These results demonstrate that prolonged exposure to low levels of arsenic results in selective alteration of hepatic haeme biosynthetic pathway enzymes, with concomitant increases in urinary porphyrin concentrations.

Cebrián *et al.* (1988) demonstrated that sodium arsenite is a potent inducer of H-O, which is the rate-limiting enzyme of haeme degradation. Male Wistar albino rats were fasted for 24 h before treatment and until they were killed. Animals received 0.1 mL sodium chloride (0.9%, w/v), or arsenic salts by subcutaneous injection. The doses of As^{III} were 12.5, 25, 50, 75 and 100 μmol/kg bw and those of As^V were 25, 50, 100, 150 and 200 μmol/kg bw. Animals were killed 16 h after injection. In a subchronic study, animals were exposed to As^{III} in the drinking-water at a concentration of 50 mg/L for periods of 5, 10, 20 or 30 days, and food was withheld for 24 h before sacrifice. The livers were excised, perfused and homogenized, and tryptophan pyrrolase (TP), ALA-S and H-O activities were measured. Cytochrome P450 and *b5* contents were also measured. Acute administration of arsenic produced a decrease in the haeme saturation of TP in rat liver, accompanied by dose-related increased ALA-S and H-O activities, and a corresponding decrease in cytochrome P450 concentration. The decrease in the haeme saturation of TP indicates that arsenic reduced the content of cytosolic haeme in liver cells and that the increase in hepatic ALA-S activity appears to be in response to a reduction in haeme availability. The alteration in the relationship between haeme synthesis and degradation is a result of treatment with arsenic. The magnitude of these effects was related to the oxidation state of arsenic: sodium arsenite (As[III]) was more potent than sodium arsenate (As[V]). These results support the suggestion that haeme saturation of TP is sensitive to treatments that modify liver haeme concentration. The increase in H-O activity produced by arsenic appears to be mediated by a mechanism largely or entirely independent of haeme. Indeed, there were no indications of an increase in the free haeme pool that could trigger a positive feedback on H-O. On the contrary, it appears that one reason for the reduction in haeme saturation was the increase in H-O activity. Moreover, the concomitant increase in ALA-S activity was a further indication of cellular depletion of haeme. The main effects of continuous exposure to As[III] were an initial decrease in the haeme saturation of TP, which remained constant during the period of treatment, and an initial increase in ALA-S activity, which after 10 days of exposure dropped somewhat but remained above control values. No significant effects on H-O or P450 activity were observed. These results were interpreted as being indicative that a new balance between haeme synthesis and degradation had been reached and that an adaptive response to the subchronic effects of As[III] was taking place.

(iii)   *Arsenic and oxidative stress*

Among the various proposed mechanisms by which arsenic induces cancer, oxidative damage may play a role in arsenic-induced carcinogenesis. Exposure to arsenite, arsenic trioxide or arsenate has been reported to result in the generation of reactive oxygen species in laboratory animals or in cultured animal and human cells by many investigators (Wang *et al.*, 1996; Chen *et al.*, 1998; Hei *et al.*, 1998; Ahmad *et al.*, 2000; Lynn *et al.*, 2000; Chouchane & Snow, 2001; Liu, S.X. *et al.*, 2001). The topic has been reviewed by Del Razo *et al.* (2001b), Ercal *et al.* (2001), and Thomas *et al.* (2001).

Arsenic-induced free-radical formation was indicated by Yamanaka *et al.* (1991), who studied cellular response in the lung induced by the administration of DMA[V], and in particular the enzymes that participate directly in protective reactions against active oxygen species, superoxide dismutase, catalase and GSH peroxidase (GPx). Male ICR mice, weighing approximately 25 g, were given an oral dose of 1500 mg/kg bw DMA[V] after fasting for several hours. The activities of mitochondrial superoxide dismutase, GPx and glucose-6-phosphate dehydrogenase (G6PDH) significantly increased at 6 h or longer after dosing, whereas cytosolic superoxide dismutase and catalase were not. Furthermore, the NADPH levels were markedly decreased at 6–9 h after treatment with DMA[V] while NADP[+] levels increased, resulting in a marked reduction in the NADPH/NADP[+] ratio. This change, accompanied by an increase in G6PDH activity, indicates that the pentose-phosphate pathway is activated by the oxidation of reduced GSH with hydrogen peroxide

(H_2O_2). With regard to cellular sulfhydryls, after treatment with DMA^V, levels of GSH and non-protein sulfhydryls were decreased and levels of oxidized GSH (GSSG) remained constant, whereas those of mixed disulfides were significantly increased. These cellular variations suggest that mouse pulmonary cells produced reduced oxygen species, that is, superoxide anion radical, hydrogen peroxide and subsequent radicals in the metabolism of DMA^V, and that these and the dimethylarsenic peroxyl radical were responsible for pulmonary DNA damage, the diethylarsenic peroxyl radical probably being produced from the reaction of molecular oxygene and dimethylarsine (Yamanaka et al., 1990, 2001). The same investigators had demonstrated previously that oral administration of DMA^V, the main metabolite of inorganic arsenic, induces lung-specific DNA damage in mice. An in-vitro experiment indicated that the breaks were not caused directly by DMA^V but by DMA^{III}, a further metabolite of DMA. They hypothesized that this damage was partially due to the active oxygen species produced in the metabolism of DMA^V (Yamanaka et al., 1989). In a further study, the authors had shown that oral administration of DMA to mice significantly enhanced the amounts of 8-oxo-2'-deoxyguanosine (8-oxodG) specifically in target organs of arsenic carcinogenesis (skin, lung, liver and urinary bladder) and in urine. The dimethyl arsenics thus may play an important role in the carcinogenesis of arsenic through the induction of oxidative damage, particularly of base-oxidation (Yamanaka et al., 2001).

Ahmad, S. et al. (1999) investigated the biochemical effects of exposure to DMA in $B6C3F_1$ mice using six biochemical parameters: DNA damage, GSH and GSSG content, cytochrome P450 content, ornithine decarboxylase (ODC) activity in liver and/or lung and ALT activity in serum. GSH was selected as an important constituent for cellular protection against oxidative damage by free radicals and the three enzymes were employed as biological markers of cell proliferation and promotion of carcinogenesis. Groups of 10 or 12 adult female $B6C3F_1$ mice received DMA^V at a dose of 720 mg/kg bw by oral gavage at one of three times (2 h, 15 h or at both 21 and 24 h) before sacrifice. Four or five control mice were run on each of 5 experimental days and received distilled water alone. Significant ($p < 0.05$) decreases in liver GSH and GSSG contents (15–37%) were observed. Pulmonary and hepatic ODC activities were reduced (19–59%) by treatment with DMA^V. A significant decrease in hepatic cytochrome P450 content (21%) was observed only in the group treated at both 21 and 24 h before sacrifice. The mouse serum ALT activity was not reduced after in-vivo administration of DMA^V but the addition of 2.8, 28 and 280 mM DMA^V in vitro reduced ALT activity by 0, 8 and 6.5%, respectively.

Santra et al. (2000) examined the hepatic effects of chronic ingestion (for up to 15 months) of drinking-water containing arsenic (1:1 arsenite to arsenate) at 3.2 mg/L in male BALB/c mice (5–14 experimental animals, 5–10 control animals). Groups of arsenic-exposed mice and unexposed controls were killed at 3, 6, 9, 12 and 15 months for examination of hepatic histology and certain biochemical parameters of oxidative stress. Statistically significant decrements in body weight were observed in the exposed animals at 12 months and 15 months, without significant differences in the amount of food or water consumption between exposed and control groups. No abnormal hepatic morphology was observed by light microscopy during the first 9 months of exposure to arsenic, but at 12 months, 11/14 mice in the experimental group exhibited hepatocellular degeneration and focal mononuclear cell collection. After 15 months, exposed mice displayed evidence of hepatocellular necrosis, intralobular mononuclear cell infiltration, Kupffer cell proliferation and portal fibrosis. Hepatic morphology was normal in all control mice. Biochemical changes, consistent with oxidative stress, preceded the overt histological pathology. Hepatic GSH was significantly reduced after 6 months, in a time-related manner; the hepatic activities of enzymes related to GSH homeostasis, namely G6PDH, GST, GSH reductase, GPx and catalase were also reduced in a time-related manner (at 9, 12 and 15 months). There was a progressive, time-dependent increase in lipid peroxidation, as demonstrated by increased production of malondialdehyde, and concomitant time-dependent damage to hepatocellular plasma membranes, as demonstrated by decreases in membrane Na^+/K^+ ATPase activity. Depletion of GSH may result in the accumulation of free radicals that initiate lipid peroxidation and biochemical damage by covalent binding to macromolecules. Biochemical changes observed in this long-term in-vivo animal feeding experiment suggest that the adverse histological effects of arsenic on the liver may be mediated through oxidative stress.

Ishinishi *et al.* (1980) studied the chronic toxicity of arsenic trioxide in rats with special reference to liver damage. Four groups of male adult Wistar rats were given distilled water containing 0, 0.125, 12.5 or 62.5 ppm arsenic trioxide orally for 7 months and were thereafter given distilled water with no arsenic trioxide for 4 months. Despite no difference in growth among the four groups of rats, chronic exposure to arsenic trioxide induced not only liver injury but also dose-dependent proliferation of the bile duct with chronic angitis. The liver injury was characterized by degenerative changes in hepatocytes, such as cloudy swelling, disordered trabeculae or irregularity of hepatocyte tracts, and spotty coagulative necrosis with infiltration of round cells. Sarin *et al.* (1999) demonstrated hepatic fibrosis and fibrogenesis following chronic ingestion of arsenic in Swiss albino mice fed arsenic daily (120, 240, 360 or 500 mg/L). A significant increase in hepatic collagen and its deposition in the extracellular matrix, an expression of hepatic fibrosis, were seen in arsenic-treated mice compared with controls. Hepatic 4-hydroxy-proline levels, indicative of fibrogenesis, were increased four- to 14-fold with different doses of arsenic compared with controls.

Effects on levels of GSH and some related enzymes in tissues after acute exposure to arsenic were studied in rats by Maiti and Chatterjee (2001). Male Wistar rats, maintained on either an 18% or 6% protein (casein) diet, received an intraperitoneal injection of sodium arsenite at its LD_{50} dose (15.86 mg/kg bw). One hour after exposure to arsenic, the GSH concentration was significantly depleted and lipid peroxidation was increased in both the high- and low-protein diet groups. Acute exposure to arsenic significantly increased GPx activity in the liver in both groups. GST activity was significantly decreased in the liver of the animals fed 18% protein, whereas it increased in the kidneys of both groups. No significant change in GSH reductase or G6PDH activity in the liver and kidneys was observed. In this study, liver as a whole seemed to be more affected in terms of level of

GSH and GST activity. The animals fed 6% protein appeared to be less affected in terms of tissue arsenic concentration, level of GSH, level of lipid peroxidation and GST activity compared with those fed 18% protein. This might be due to the deficiency in tissues of possible target proteins for arsenic binding and a lesser availability of specific amino acid to synthesize different stress proteins in the animals fed 6% protein.

As reviewed by Del Razo *et al.* (2001b), a variety of genes related to base excision repair and oxidative stress are commensurately up-regulated by nanomolar concentrations of inorganic arsenic. Reactive oxygen species induced by low levels of As^{III} or As^{V} increase the DNA-binding activity of activator protein 1 (AP-1) and nuclear factor-κB (NF-κB) in cultured aortic endothelial cells (Barchowsky *et al.*, 1996), human MDA-MB-435 breast cancer and rat H411E hepatoma cells (Kaltreider *et al.*, 1999) and precision-cut lung slices (Wijeweera *et al.*, 2001). This results in stimulation of cell proliferation and up-regulation of gene expression including that of mdm2 protein, which is a key regulator of the critical tumour-suppressor gene *p53* (Germolec *et al.*, 1996; Hamadeh *et al.*, 1999). In contrast, high levels of inorganic arsenic inhibit the activation of NF-κB and cell proliferation and induce apoptosis in human acute myelogenous leukaemia cells, human embryonic kidney (HEK 293) cells and human bronchial epithelial (BEAS 2B) cells (Estrov *et al.*, 1999; Roussel & Barchowsky, 2000). Based on results obtained in NIH 3T3 cells exposed to arsenic, Chen *et al.* (1998) have suggested that apoptosis is triggered by generation of H_2O_2 through the activation of flavoprotein-dependent superoxide-producing enzymes (e.g. NADPH oxidase) and the increase in superoxide levels in cells. The event probably acts as a mediator to induce apoptosis through the release of cytochrome c from the mitochondria to cytosol, the activation of caspase 3 and the degradation of poly(ADP-ribose) polymerase (PARP) leading to DNA fragmentation (Chen *et al.*, 1998).

(iv)   *Stress proteins*

Exposure to arsenicals either *in vitro* or *in vivo* in a variety of model systems has been shown to induce a number of the major stress protein families such as heat-shock proteins. Among them are members with a low molecular weight, such as metallothionein and ubiquitin, and others with masses of 27, 32, 60, 70, 90 and 110 kDa. In most cases, the induction of stress proteins depends on the capacity of the arsenic compound to reach the target, its valence and the type of exposure, with arsenite being the strongest inducer of most heat-shock proteins in several organs and systems. Induction of heat-shock proteins is a rapid dose-dependent response (1–8 h) to acute exposure to arsenite. Thus, the stress response appears to be useful for monitoring toxicity resulting from a single exposure to arsenite. The capacity of arsenic compounds to modulate the expression and/or accumulation of stress proteins has been studied in normal and transformed cell lines by Caltabiano *et al.* (1986), Keyse and Tyrrell (1989), van Wijk *et al.* (1993), Wu and Welsh (1996) and Wijeweera *et al.* (2001) and has been reviewed by Bernstam and Nriagu (2000) and Del Razo *et al.* (2001b).

Metallothionein is a low-molecular-weight, cysteine-rich, metal-binding protein that has been propounded to play an important role in the homeostasis of essential metals, in the detoxication of heavy metals and in the scavenging of free radicals. Moreover, it is a small protein easily induced by heavy metals, hormones, acute stress and a variety of chemicals. Twenty of the 61 amino acid residues in mettalothionein molecules are cysteinyl residues, all of which are involved in metal binding (Sato & Bremner, 1993; National Research Council, 1999).

The induction of metallothionein is observed following oral administration; the doses of organic arsenic compounds (MMA and DMA) required for its induction are one order higher than those of inorganic arsenic compounds (As^{III} and As^V). Only a small portion of the arsenic dose was found to be associated with the metallothionein fraction, which there fore does not protect against arsenic toxicity by binding the metal (Maitani et al., 1987). Rather, because of its high sulfhydryl content, it has also been suggested that metallothionein reacts with organic free radicals and electrophiles (Klaassen & Cagen, 1981). Indeed, metallothionein can serve as a sacrificial scavenger for superoxide and hydroxyl radicals in vitro (Thornalley & Vašák, 1985). It is induced by metal chemicals that produce oxidative stress (Bauman et al., 1993) and has been shown to protect against oxidative damage (Sato & Bremner, 1993).

The effect of various arsenic forms on the tissue concentrations of metallothionein was determined in male CF-1 mice (25–30 g) injected subcutaneously with various doses of As^{III} (55–145 μmol/kg bw), As^V (165–435 μmol/kg bw), MMA (100–7250 μmol/kg bw) or DMA (2750–10 250 μmol/kg bw) (Kreppel et al., 1993). Controls were injected with an equal volume (0.01 mL/g bw) of saline. Metallothionein content in hepatic cytosol was quantified by the cadmium–haemoglobin assay. As^{III} was found to be a potent inducer of hepatic metallothionein, producing a 30-fold increase at a dose of 85 μmol/kg. In comparison, it took three-, 50- and 120-fold higher molar amounts of As^V, MMA and DMA, respectively, to produce a similar effect. MMA produced the largest increase in hepatic metallothionein (80-fold), followed by As^{III} (30-fold), As^V (25-fold) and DMA (10-fold). However, none of the compounds induced metallothionein in mouse primary hepatocyte cultures, suggesting that arsenicals may be considered as indirect inducers of metallothionein. Both metallothionein-I (MT-I) and metallothionein-II (MT-II) protein isoforms were commensurately induced by As^{III}, As^V and MMA. Induction of metallothionein by As^{III} was further characterized following subcutaneous administration of arsenite (85 μmol/kg). Induction of hepatic metallothionein peaked at 24 h. As^{III} also increased metallothionein in kidney, spleen, stomach, intestine, heart and lung and the most marked increase occurred in the liver. MT-I mRNA increased 24-, 52- and 11-fold at 3, 6 and 15 h after administration of As^{III}, respectively. This induction profile is similar to that observed after exposure to zinc or cadmium. This study showed that arsenic compounds are effective inducers of metallothionein in vivo and that their potency and efficacy are dependent on the chemical form of arsenic. As^{III} is a potent inducer of hepatic metallothionein for both MT-I and MT-II and this effect is associated with an increase in metallothionein mRNA, suggesting that the mechanism of this induction appears to be due, at least in part, to increased metallothionein gene transcription.

In a recent study, Liu *et al.* (2000) demonstrated that MT-I/II-null mice are more sensitive than wild-type mice to the hepatotoxic and nephrotoxic effects of chronic oral administration or injection of inorganic arsenicals. Groups of 4–6 male and female MT-I/II-null mice and corresponding wild-type mice, aged 6–8 weeks, were provided drinking-water containing As^{III} at concentrations of 7.5, 22.5 or 45 mg/L, or As^{V} at concentrations of 37.5 or 75 mg/L, or were injected subcutaneously in the dorsal thoracic midline with 10 mL/kg bw saline containing As^{III} at doses of 10 and 30 μmol/kg bw or As^{V} at a dose of 100 μmol/kg bw once daily on 5 days per week for 15 weeks. Control mice received tap-water or were injected with the same volume of saline. Chronic exposure to arsenic produced only modest increased tissue concentrations of metallothionein (two- to fivefold) in wild-type but not in MT-null mice, either following repeated injections or following oral administration. Arsenic by both routes produced damage to the liver (fatty infiltration, inflammation and focal necrosis) and kidney (tubular cell vacuolization, inflammatory cell infiltration, glomerular swelling, tubular atrophy and interstitial fibrosis) in both MT-null and wild-type mice. However, in MT-null mice, the pathological lesions were more frequent and severe compared with those in wild-type mice in either liver or kidney. This was confirmed biochemically, in that, at the higher oral doses of As^{V}, the levels of blood urea nitrogen, an indicator of kidney injury, were increased to a greater extent in MT-null mice (60%) than in wild-type mice (30%). However, As^{III} resulted in elevated levels of blood urea nitrogen in MT-null mice only. Chronic exposure to arsenic produced a two- to 10-fold increase in levels of serum interleukin-1β (IL-1β), interleukin-6 (IL-6) and tumour necrosis factor-α (TNF-α), with greater increases seen after repeated injections than after oral exposure; again, MT-null mice had higher levels of serum cytokines than wild-type mice, following repeated injections of arsenic, but not after oral exposure. Repeated injections of arsenic also decreased hepatic GSH up to 35% but had no effect on hepatic GPx or GSH reductase activities. MT-null mice were more sensitive than wild-type mice to the effect of GSH depletion by As^{V}. Hepatic caspase 3 activity was increased (two- to threefold) in both wild-type and MT-null mice, indicating apoptotic cell death. The study demonstrated that chronic exposure to inorganic arsenic produced injuries to multiple organs, and that MT-null mice are generally more susceptible than wild-type mice to arsenic-induced toxicity regardless of route of exposure, suggesting that metallothionein could be a cellular factor in protecting against chronic arsenic toxicity.

Kato *et al.* (2000) reported from earlier studies the induction and accumulation of heat-shock protein-72 (Hsp72) in the cell nuclei of human alveolar type II (L-132) cells and DNA damage following exposure to DMA^{V} (Kato *et al.*, 1997). They also found that the accumulation of Hsp72 in cell nuclei was related to the suppression of apoptosis (Kato *et al.*, 1999). Referring to reports indicating that Hsp72 might be involved in the tumorigenic process through the function of apoptosis (Jäättelä, 1999), they assumed that Hsp72 induced by dimethylarsenics may play an important role in DNA damage and tumorigenesis. They therefore investigated whether Hsp72 was induced and accumulated in the lung, a target organ for tumorigenesis, following administration of DMA^{V} to mice. Five-week-old male A/J mice were injected intraperitoneally with DMA^{V} (100–600 mg/kg) or arsenite (5 mg/kg bw) and then killed. Lung, kidney, liver and spleen were excised, homogenized and immunoblotting analysis was performed with anti-Hsp72 monoclonal antibody. Hsp72 in lung was also investigated immunohistochemically. Forty-eight hours after exposure to DMA, Hsp72 was observed in the lung and in the kidney, but not in the liver or spleen. Hsp72 was also detected by immunohistochemical analysis in the nuclei of alveolar flat cells containing capillary endothelium, in the lungs of DMA-treated mice. This result may be consistent with those observed in previous studies showing that oral administration of DMA to mice induces a preferential increase in heterochromatin in the vesicular endothelium of the lung, an early morphological change in the development of pulmonary carcinomas (Nakano *et al.*, 1992; Hayashi *et al.*, 1998). Kato *et al.* (2000) suggested that the increase and accumulation of Hsp72 following administration of DMA to mice occur specifically in target organs for the carcinogenesis of arsenic. It appears that arsenic compounds regulate the expression of the major families of heat-shock proteins and that inorganic As[III] is the most potent inducer of these proteins (Del Razo *et al.*, 2001b).

Stress-related gene expression in mice treated with inorganic arsenic has been studied by Liu, J. *et al.* (2001). Adult male 129/Sv mice, aged 6 8 weeks, were injected subcutaneously in the dorsal thoracic midline with 100 µmol/kg bw As[III], 300 µmol/kg bw As[V] or the same volume of saline (10 mL/kg bw). To examine stress-related gene expression, livers were removed 3 h after injection of arsenic to extract RNA and protein. The Atlas Mouse Stress/Toxicology array revealed that the expression of genes related to stress — DNA damage and repair-responsive genes — and metabolism were altered by acute exposure to arsenic. Expression of H-O-1, a hallmark for arsenic-induced stress, was increased 10-fold, together with increases in heat shock protein-60 (Hsp60), the DNA damage-inducible protein GADD45 and the DNA excision repair protein ERCC1 and growth arrest. Down-regulation of certain cytochrome P450 drug-metabolizing enzymes occurred after treatment with arsenic. Because the AP-1 complex is associated with stress-related gene activation, the effect of arsenic on AP-1 complex activation was examined. A multiprobe RNAse protection assay revealed the activation of the c-Jun–AP-1 transcription complex after treatment with arsenic. Western blot analysis further confirmed the enhanced production of arsenic-induced stress proteins such as H-O-1, Hsp70, Hsp90, metallothionein, metal-responsive transcription factor, NF-κB and c-Jun–AP-1. Increases in caspase 1 and cytokines such as TNF-α and macrophage inflammatory protein-2 were also evident. Activation of caspase has been propounded to play a role in arsenic-induced apoptosis (Chen *et al.*, 1998), and induction of inflammatory cytokines is another important aspect of arsenic toxicity. The results of this study profiled gene expression patterns in mice treated with inorganic arsenicals. The altered gene expressions following acute exposure to arsenic *in vivo* include stress-related components —DNA damage and repair-responsive genes — activation of transcription factors such as the AP-1 complex and an increase in proinflammatory cytokines.

Expression of shock proteins is regulated by a complex mechanism that requires the integration of multiple signal pathways. The inter-relationships among stress signalling, cell death and oncogenesis after exposure to arsenic need further research (Del Razo *et al.*, 2001b).

(v)   *Immunotoxicity*

Although many studies have evaluated the immunological effects of environmental toxic substances such as lead, cadmium and mercury, only a few studies on arsenic have been reported.

Yoshida *et al.* (1986) reported immunological effects of arsenic compounds on mouse spleen cells *in vitro*. Spleens from male C57BL/6N mice were removed aseptically, and sterile viable spleen cells were cultured with 20 μL/mL of an arsenic solution (at concentrations of 1–500 ng/mL, 0.01–50 μg/mL and 0.1–500 μg/mL, for sodium arsenite, sodium arsenate and DMA, respectively). Saline (0.9% NaCl) was added to control cultures. For plaque-forming cell (PFC) response, spleen cells (2.5–3.0×10^5 cells/mL) were cultured in triplicate and incubated with 8×10^6 sheep erythrocytes, and the number of direct (immunoglobulin M) PFCs were enumerated. Spleen cells (2.5×10^6 cells/mL) were also cultured for 48 h with or without the mitogens, phytohaemagglutinin P (PHA) or lipopolysaccharide ω. At high doses, the three arsenic compounds (sodium arsenite, sodium arsenate and DMA) suppressed the PFC response to sheep erythrocytes and the proliferative response to mitogens whereas, at low doses, they enhanced both responses. In other studies, the authors have demonstrated that this immunoenhancing effect of arsenic on PFC response to sheep erythrocyte is not attributable to the augmentation of lymphocyte function, but to the cytotoxicity of arsenic against precursors of suppressor T cells (Yoshida *et al.*, 1987). The concentration at which each arsenic compound exerted the modulatory effects on both responses differed, and was correlated to the general toxicity of each compound.

A pilot study on arsenic-exposed humans was carried out by Ostrosky-Wegman *et al.* (1991) to determine the lymphocyte proliferation of kinetics and genotoxic effects. The exposed group comprised 11 individuals (nine women and two men) from Santa Ana, State of Coahuila, Mexico, where the drinking-water contained 0.39 mg/L arsenic (98% in pentavalent form and the rest in trivalent form). The non-exposed group (13 individuals; 11 women and two men) was chosen from Nuevo Leon, State of Coahuila, where levels of arsenic in the drinking-water ranged from 0.019 to 0.026 mg/L during 1987–89; while sampling was performed, the levels rose to 0.060 mg/L because of new piping that linked several towns in the area. Venous blood samples were taken and lymphocyte cultures were rapidly processed. The analysis of chromosomal aberrations and sister chromatid exchange was performed in 100 consecutive first-division metaphases and in 30 consecutive second-division metaphases, respectively, all with 46 centromeres. The proportion of first, second, third and subsequent metaphases was determined in 100 consecutive mitoses to study the kinetics of proliferation. The highly exposed group excreted greater amounts of arsenic in urine; neverthelesss, the *Bacillus subtilis* rec-assay for genetic damage induced by urine samples showed negative results. There was a significant difference in cell-cycle kinetics between the groups: the average generation time was longer in the highly exposed group. The lag in lymphocyte proliferation could mean an impairment of the cellular immune response due to exposure to arsenic.

Because inhibition of lymphocyte proliferation has been used to identify agents that depress the cellular immune response, Gonsebatt *et al.* (1992) investigated *in vitro* the effect of arsenic on human lymphocyte stimulation and proliferation using concentrations of arsenic similar to those found in blood. When human lymphocytes collected from healthy donors (two men, two women) were exposed to arsenite and arsenate (10^{-7}, 10^{-8} and 10^{-9} M) during culture and harvested after 24 h, a dose-related inhibition of proliferation was observed. Cultures were also treated with 10^{-7} M arsenite and arsenate for 2, 6 and 24 h at the beginning of culture in the presence or absence of PHA. Inhibition of PHA stimulation and proliferation was directly related to the length of treatment with arsenic. The results show that, at the concentrations tested, arsenite and arsenate impaired lymphocyte stimulation and proliferation and confirm that chronic exposure to arsenic can affect the proliferation of whole-blood lymphocytes.

A human monitoring study was subsequently carried out by Gonsebatt *et al.* (1994) to explore the effect on lymphocyte proliferation of chronic exposure to arsenic via drinking-water. Blood and urine samples were taken from 33 volunteers from a town where levels of arsenic in the drinking-water averaged 412 μg/L and from 30 subjects from a matched group with similar socioeconomic status, who drank water with an average level of 37.2 μg/L arsenic. Exposure was assessed by questionnaire and by determining the levels of arsenic in urine and water samples. Peripheral blood lymphocyte proliferation was evaluated at different culture times using labelling (radioactive thymidine incorporation), mitotic and replication indexes as end-points. No significant differences were seen for either labelling or mitotic indices, except for mitotic index in 72-h cultures (higher in the exposed group) and for labelling index (lower) in men and women with skin lesions versus those without lesions. Significant decreases in replication index were seen for exposed women but not for men. Correlations between labelling and mitotic indices showed that progression from the initial S- to M-phase is altered in exposed individuals. The results obtained corroborate the slower cell kinetics found previously in the pilot study by Ostrosky-Wegman *et al.* (1991).

From the preceding reports, it appears that inorganic arsenic is immunotoxic, but the mechanism of immune suppression is not clear. Harrison and McCoy (2001) showed that arsenite inhibits the enzymatic activity of lysosomal protease cathepsin L (CathL) in cultures of the murine antigen-presenting B-cell line TA3 and in lysates from unexposed TA3 cells *in vitro*. Arsenite also significantly inhibits purified CathL. This enzyme plays an important role in antigen processing, the mechanism by which antigen-presenting cells cleave foreign protein antigens to peptides to stimulate a T-cell response. Deficient proteolysis may lead to diminished immune responses. Arsenite suppressed enzymatic activity within TA3 cells after 4 h of exposure without affecting cell viability. Kinetic analyses revealed that arsenite was a reversible, partially noncompetitive inhibitor of CathL with a Ki of 90 μM for TA3-derived and 120 μM for the purified enzyme. Indeed, upon addition of excess dithiotheitol, the enzyme activity of CathL was restored; the value of Ki was comparable to that of the arsenite concentration that maximally decreased CathL in viable TA3 cells after 4 h of exposure. However, an 18-h exposure to arsenite triggered massive cell death at concentrations that were substantially lower than those required for enzymatic inhibition. Morphological analysis (chromatin condensation, cell shrinkage) and annexin V staining showed that arsenite-exposed TA3 cells underwent apoptosis within 18 h and early stages of apoptosis began within 4 h, indicating that arsenic causes apoptosis independent of CathL. Although whether in-vivo exposure to arsenic causes apoptosis in lymphoid organs has not been assessed, these findings suggest that apoptosis could be a major mechanism of arsenic-induced immunosuppression.

4.3 Reproductive and developmental effects

4.3.1 *Humans*

In a case–control study, Zierler *et al.* (1988) compared 270 cases of infants born with congenital heart disease and 665 controls from Massachusetts (USA). The proportional odds ratio, adjusted for all measured contaminants, source of water and maternal education, was not elevated for any congenital heart disease in relation to exposure to arsenic above the detection limit of 0.8 μg/L. However, for a specific malformation, coarctation of the aorta, there was a significant proportional odds ratio of 3.4 (95% CI, 1.3–8.9). The exposure was low, the 90th percentile level being 1 μg/L.

In a case–control study, Aschengrau *et al.* (1989) examined 286 women who experienced spontaneous abortions and 1391 controls from Boston, MA (USA), in relation to the content of their water supplies. An adjusted odds ratio of 1.5 was found for the group with the highest arsenic concentrations. [However, this exposure group had low levels of arsenic in water (1.4–1.9 μg/L), close to or lower than laboratory analytical detection limits, and the possibility of chance or unaccounted confounders could not be discounted.]

An ecological study in an area of south-east Hungary with exposure to arsenic from drinking-water examined the rates of spontaneous abortions and stillbirths for the period 1980–87. Two populations were compared: one from an area with levels of arsenic in drinking-water > 100 μg/L (*n* = 25 648 people) and one control area with low levels of arsenic (*n* = 20 836). [No information on analytical method, timing or frequency of sampling was available.] The incidences of both outcomes were significantly higher in the exposed groups, with a 1.4-fold increase in spontaneous abortions (*p* = 0.007) and a 2.8-fold increase in stillbirths (*p* = 0.028) (Borzsonyi *et al.*, 1992). [Although both populations were stated to have several similar characteristics, such as smoking, lifestyle, occupation and socioeconomic status, no data were provided, and other important factors such as maternal age were not considered. Furthermore, no mention was made of other potential environmental exposures.]

An ecological study conducted in the USA investigated mortality from vascular diseases in the 30 counties with the highest average levels of arsenic in drinking-water for the period 1968–84. The arsenic levels ranged up to 92 μg/L in Churchill County, NV. SMRs were based on comparison with the population of the USA. When counties were grouped into three arsenic-exposure categories, defined as 5–10, 10–20 and > 20 μg/L, there appeared to be an increase in mortality from congenital anomalies of the heart only for females in the highest exposure group (SMR, 1.3; 95% CI, 1.0–1.8) and for both sexes for congenital anomalies of the circulatory system (female SMR, 2.0; 95% CI, 1.0–3.4; male SMR, 1.3; 95% CI, 0.7–2.4) (Engel & Smith, 1994).

A retrospective ecological study examined infant mortality rates in three Chilean cities over a 46-year period (1950–96). Antofagasta, in northern Chile, experienced very high levels of arsenic in drinking-water for a period of 12 years. In 1958, a new water source, which contained arsenic concentrations of around 800 μg/L, was introduced as the main supply of public water. In 1970, because of the overt signs of arsenicism observed in several studies, a plant for the removal of arsenic was installed, and levels decreased initially to around 110 μg/L, and then gradually over time to around 40 μg/L (see Table 18). The changes in late fetal, neonatal and post-neonatal mortality rates over time in Antofagasta were compared with those in Valparaiso, another Chilean city with similar demographic characteristics but with low levels of arsenic. A temporal relationship was observed between the period of high arsenic contamination and a rise in neonatal mortality rates, in particular in Antofagasta, whereas the other city had a fairly steady decline in infant mortality (Hopenhayn-Rich et al., 2000). [Data on other contaminants or factors related to infant mortality were not presented, but the temporal relationship suggests a role for exposure to arsenic.]

A retrospective survey in Bangladesh compared several outcomes in women exposed to high (mean, 240 μg/L; n = 96) and low (< 20 μg/L; n = 96) concentrations of arsenic in drinking-water. Rates of spontaneous abortions, stillbirths and pre-term births were 2.9 (p = 0.08), 2.24 (p = 0.046) and 2.54 (p = 0.018) times higher, respectively, in the high-exposure group than in the low-exposure group. The groups were comparable in terms of age, socioeconomic status, level of education and age at marriage (Ahmad et al., 2001). [This study was based on recall of previous pregnancies, however, and ascertainment of the outcomes was not clearly defined.]

4.3.2   Experimental systems

(a)   Developmental toxicity

(i)   In vivo

Inorganic arsenic is toxic to mouse and hamster embryos and fetuses after oral or intraperitoneal administration to the dams, with arsenite being three- to 10-fold more potent than arsenate. The embryos and fetuses of hamsters are more sensitive to this effect than those of mice. The toxicity is characterized by decreases in fetal weight, crown–rump length, embryo protein content and the number of somites and by growth retardation and lethality (Baxley et al., 1981; Hood & Harrison, 1982; Hood & Vedel-Macrander, 1984; Carpenter, 1987; Domingo et al., 1991; Wlodarczyk et al., 1996).

Sodium arsenite was given by gavage to CD-1 mice on one of days 8–15 of gestation at doses of 20, 40 or 45 mg/kg bw. The lowest dose had no effect. The two highest doses produced 19 and 36% incidences of maternal deaths, respectively, and also decreased fetal weight and increased the incidence of resorptions. Arsenite-induced lethality was dependent on dose and day of gestation (Baxley *et al.*, 1981).

In hamsters, sodium arsenite administered orally (20–25 mg/kg) caused less fetal mortality than parenteral dosing (2.5–5 mg/kg) (Hood & Harrison, 1982).

Nemec *et al.* (1998) evaluated the developmental toxicity of arsenate administered by oral gavage to CD-1 mice and New Zealand white rabbits. Rabbits received doses of 0, 0.19, 0.75 or 3.0 mg/kg bw per day on gestation days 6–18 and mice received 0, 7.5, 24 or 48 mg/kg per day on gestation days 6–15. Increased fetal resorptions and decreased fetal weight were observed only at exposure levels resulting in maternal toxicity (severely decreased weight gain, mortality).

A single intravenous administration of MMA^V (disodium salt) or DMA^V (sodium salt) on day 8 of gestation at dose levels of 20–100 mg/kg elicited a low resorption rate ($\leq$ 10%) in pregnant hamsters (Willhite, 1981). Higher doses of DMA (sodium salt, 900–1000 mg/kg) administered intraperitoneally to pregnant hamsters on one of days 8–12 of gestation induced higher resorption rates, ranging from 30–100% of the litters. MMA^V (500 mg/kg) was less toxic than DMA after intraperitoneal administration, with 6–21% of the litters resorbed. Fetal growth was retarded after administration of MMA on days 9, 10 or 12 of gestation (Hood *et al.*, 1982).

DMA^V administered orally to pregnant mice (200–600 mg/kg per day) and rats (7.5–60 mg/kg per day) on days 7–16 of gestation resulted in significant fetal mortality in mice at 600 mg/kg per day and rats at 50–60 mg/kg per day. A significant decrease in fetal weight gain was observed in mice at 400–600 mg/kg and rats at 40–60 mg/kg (Rogers *et al.*, 1981).

Inorganic arsenic elicits teratogenic effects in mice (Hood & Bishop, 1972; Baxley *et al.*, 1981; Morrissey & Mottet, 1983; Wlodarczyk *et al.*, 1996), rats (Fisher, 1982) and hamsters (Hood & Harrison, 1982; Carpenter, 1987) at levels of tens of milligrams per kilogram body weight after oral or intraperitoneal administration. In these studies, the major teratogenic effect induced is cephalic axial dysraphic disorder or neural tube defect. The defect is characterized by exencephaly and encephalocele, which are characterized by non-closure and partial closure of the cephalic neural folds, respectively. Other malformations that occur to a minor extent include fused ribs, renal agenesis, micromelia, facial malformations, twisted hindlimb, microphthalmia and anophthalmia. The malformations are dose- and gestational age-dependent. Sodium arsenite is more potent than sodium arsenate in inducing a teratogenic response, and intraperitoneal administration of arsenic is more effective than oral administration.

Histological studies of the developing urogenital system in rat embryos after intraperitoneal administration of arsenate to pregnant rats revealed that the first observable change is a retardation in the growth of the mesonephric duct. This retardation led to the absence of the ureteric bud (which arises from the mesonephric duct) and resulted in the absence of the vas deferens, seminal vesicle and part of the epididymis (Burk & Beaudoin, 1977).

Administration of inorganic arsenic to mice on days 7–9 of gestation results in neural tube defects in the developing organism. The time most sensitive to arsenate in mouse embryos is when the dams are administered the chemical on day 8. Of the fetuses that survived a single dose of sodium arsenate (45 mg/kg) administered intraperitoneally to dams on day 8, 65% or more were exencephalic. After administration of a similar dose of arsenate on day 7 or 9, 3% or less of the surviving fetuses were exencephalic (Morrissey & Mottet, 1983).

The neural tube defects seem to result from an apparent arsenic-induced arrest or delay in neural-fold apposition. Takeuchi (1979) examined the changes induced by an embryo-lethal dose of arsenate (30 mg/kg) administered to pregnant rats intraperitoneally on day 9 of gestation. At 4 h after exposure, some cellular necrosis was seen in the neuro-ectoderm and mesoderm of the embryos. By 12 h, abnormal mitotic and interphase cells were observed in both tissues, and necrotic cells and debris from these cells were also present. By 24 h, neurulation had stopped, as evidenced by the presence of the V-shaped neural fold that is normally closed by this time.

In studies by Morrissey and Mottet (1983), pregnant mice were killed 6–21 h after intraperitoneal administration of sodium arsenate (45 mg/kg) on day 8 of gestation. Neural folds were widely separated and not positioned for closure in the prospective hindbrain. Necrotic debris was also found primarily in the neuroepithelium of the pros-pective forebrain and sometimes in the mesenchyme, but it was not clear if this was the main lesion associated with exencephaly.

Fisher (1982) examined the effect on the development of embryos of sodium arsenate (45 mg/kg bw) administered intraperitonally to pregnant rats on day 10 of gestation. These rats were killed 4 h or 24 h after injection. The embryos were removed and the macro-molecule levels were determined immediately, or at 24 h or 42 h after being placed in culture media. In-utero exposure to arsenate for 4 h did not affect the macromolecule levels. A 24-h in-utero exposure to arsenate resulted in a significant decrease in DNA, RNA and protein accumulation at the beginning of cultivation and after 24 h in culture. However, after 42 h in culture, protein levels had recovered. After 24 h in culture, morphological changes in the 24-h exposed embryos included a failure to rotate to a ventroflexed position, failure of closure of the anterior neuropore, no establishment of visceral yolk sac circulation, and no fusion of the allantoic sac in placental formation. The latter effect may reflect problems in the formation of the urogenital system.

Nemec et al. (1998) observed no teratogenic effects in mice or rabbits receiving daily oral administrations of 0–48 or 0–3 mg/kg bw arsenate on gestation days 6–15 or 6–18, respectively.

MMA^V (disodium salt, 20–100 mg/kg) and DMA^V (sodium salt, 20–100 mg/kg) induced a low percentage of fetal malformations ($\leq$ 6%) after intravenous administration on day 8 of gestation to pregnant hamsters. The effects were characterized by fused ribs, renal agenesis or encephalocele, with the latter anomaly was observed only with DMA. Neither MMA nor DMA caused maternal toxicity (Willhite, 1981).

The effect of continuous oral exposure of pregnant mice (200, 400, 600 mg/kg per day) and rats (7.5–60 mg/kg per day) to DMA^V during days 7–16 of gestation was examined by Rogers et al. (1981). In mouse fetuses, cleft palate was the major teratogenic response to DMA and was observed at the two highest doses. There was also a significant decrease in the incidence of supernumerary ribs. In the mid-dose group, four mouse fetuses had irregular palatine rugae. In rats, the average number of sternal and caudal ossifications was decreased at the two highest doses and the percentage of irregular palatine rugae increased significantly with dose. An increase in fetal lethality occurred at the highest dose in mice (39.8%) and at the two highest doses in rats (32.9 and 65.4%).

(ii)   In vitro

Muller et al. (1986) examined the effect of sodium arsenite on mouse embryos at the two-cell pre-implantation stage, which is approximately 30–32 h after conception. Arsenite-induced lethality occured at a concentration of 100 μmol/L. After implantation, arsenite and arsenate are toxic (decreases in crown–rump length, number of somites, protein content, head length, yolk sac diameter) and lethal to embryos of mice (Chaineau et al., 1990; Tabacova et al., 1996) and rats (Mirkes & Cornel, 1992; Mirkes et al., 1994). Tabacova et al. (1996) observed that as gestational age at which the mouse embryos were isolated and exposed to arsenic increased, so did resistance to toxicity or lethality. As in the in-vivo studies, arsenite was more potent than arsenate.

Inorganic arsenic is teratogenic to cultured mouse embryos (day 8), with sodium arsenite (1–4 μmol/L) being approximately 10-fold more effective than sodium arsenate (10–40 μmol/L) after a 48-h incubation. The most sensitive in-vitro effect of arsenic is hypoplasia of the prosencephalon. Other effects include failure of neural tube closure and development of limb buds and sensory placode, somite abnormalities and, in arsenate-exposed embryos, hydropericardium (Chaineau et al., 1990).

Arsenite inhibits chondrogenesis in chick limb bud mesenchymal cells, with complete inhibition at 25 μmol/L. Arsenate was ineffective at concentrations up to 200 μmol/L but, when added with arsenite, gave an apparent dose-dependent additive effect (Lindgren et al., 1984).

Sodium arsenite (50 μmol/L) induces dysmorphology in rat embryos (10 days old) after a 2.5-h exposure followed by a 21.5-h incubation period without arsenic. This effect is characterized by hypoplastic prosencephalon, mild swelling of the rhombencephalon and abnormal somites and flexion of the tail (Mirkes & Cornel, 1992; Mirkes et al., 1994).

Tabacova et al. (1996) examined the teratogenicity of arsenite (1–30 μmol/L) and arsenate (5–100 μmol/L) in mouse embryos isolated from pregnant dams on day 9 of gestation. The embryos were incubated with various concentrations of arsenic, for different lengths of time and at various stages of somite development. Treatment with arsenic led to non-closure of the neural tube, collapsed neural folds, prosencephalic hypoplasia, anophthalmia, pharyngeal arch defects and abnormal somites. The malformation rates were dependent on the dose and oxidation state of arsenic. Arsenite was generally three to four times more potent than arsenate in inducing these effects. As the age of the embryos advanced, a higher dose of arsenic was required to elicit the effect. The developmental effects most sensitive to inorganic arsenic were forebrain growth, neural tube closure, eye differentiation, axial rotation (dorso- to ventroflexion) and pharyngeal arch development, which were induced by a 1-h exposure to inorganic arsenic.

(b)   Gene expression

Wlodarczyk *et al.* (1996) examined the expression of several transcription factors from embryos isolated from pregnant mice administered sodium arsenate intraperitoneally at 30–45 mg/kg, an approximately lethal dose. Expression of several genes was altered by arsenate administered on day 9 of gestation. This day corresponds to the progression of neural tube closure, which is delayed in embryos exposed to inorganic arsenic. In the neuroepithelium of arsenate-exposed embryos, there was significant down-regulation of *Hox 3.1* and up-regulation of *Pax3*, *Emx-1* and *creb*. Both *Hox 3.1* and *Pax3* play a role in the regulation of neural cellular adhesion molecules, a glycoprotein that affects neural crest cell migration and ultimately neural tube closure (Rutishauser *et al.*, 1988).

(c)   Induction of heat-shock proteins

Arsenic induces the biosynthesis in embryos of several heat-shock proteins that protect cells from its detrimental effects. However, its has been proposed that induction of a heat-shock protein response could alter the normal gene programme for organogenesis (German, 1984).

(i)   In vivo

Pregnant mice were administered sodium arsenite (0.5 mg/mouse, approximately 17 mg/kg bw) intraperitoneally on days 9–11 of gestation. Two proteins that were induced were isolated from the embryos and had molecular weights between 45 and 66.2 kDa. Heat-shock treatment of pregnant mice induced one embryonic protein with a molecular weight between 45 and 66.2 kDa and a second with a molecular weight between 66.2 and 92.5 kDa (German *et al.*, 1986). In mice administered sodium arsenite (19 mg/kg) intraperitoneally on day 8 and killed 1 day later, the levels of two proteins, Hsp70 and Hsp105, which are produced constitutively, were increased throughout the embryo. There was a high concentration of these proteins in the neuroepithelial tissue of the embryos after treatment with heat shock or arsenite (Honda *et al.*, 1992).

(ii)   *Animal embryos* in vitro

Four proteins with molecular weights of 27, 35, 73 and 89 kDa and their mRNA were induced in chick embryo cells by sodium arsenite (50 µmol/L) or heat shock in a dose- and time-dependent manner. For example, the 35-kDa protein was induced at a concentration of 5 µmol/L sodium arsenite, but the 73- and 89-kDa proteins were minimal at this concentration. Only the 27-kDa protein was still induced 24–48 h after treatment (Johnston *et al.*, 1980). In chick embryo fibroblasts (10–12 days old), arsenite induced the synthesis of Hsp70A and 70B (Wang & Lazarides, 1984).

Mouse embryo cells (gestation day 11) were exposed to either sodium arsenite (50 μmol/L) for 3 h or heat shock for 10 min, and proteins from cell extracts were analysed by two-dimensional gel electrophoresis. The synthesis of Hsp73 and Hsp105 was increased by both exposures (Honda *et al.*, 1992).

In rat embryos (gestation day 10), exposure for 2.5–5 h to an embryotoxic level of sodium arsenite (50 μmol/L) resulted in the induction of three heat-shock proteins (Mirkes & Cornel, 1992). A monoclonal antibody specific for Hsp72 recognized one of the proteins induced by arsenite. Levels of mRNA for these heat-shock proteins were also increased in the embryos after exposure to arsenite. Hsp72 was detected 10 h after exposure, and maximal levels were observed at 24 h. However, Hsp72 was not detected at 48 h, which indicates that this protein is turned over (Mirkes *et al.*, 1994).

(iii) *Human fetal tissue* in vitro

German *et al.* (1986) treated human fetal tissue (gestational age, 77–84 days) with either sodium arsenite (50 μmol/L) for 2 h or heat shock for 6 min. The cells were then examined for induction of heat-shock proteins. Several proteins were induced by both treatments, and two with molecular weights < 45 kDa were induced only by exposure to arsenite.

Honda *et al.* (1992) treated human chorionic villus cells (gestational age, 70–119 days) with sodium arsenite (50 μmol/L) for 3 h or with heat shock for 10 min. In unstressed tissue, Hsp70, Hsp73, Hsp85 and Hsp105 were synthesized constitutively, but their levels were increased after exposure to sodium arsenite or heat.

4.4 Genetic and related effects

The genetic effects of arsenic compounds have recently been reviewed extensively (National Research Council, 1999; Basu *et al.*, 2001; Gebel, 2001; National Research Council, 2001; WHO, 2001). In this section, the genotoxicity of arsenic in humans and in experimental animals is dealt with comprehensively. Relevant studies on single and combined mammalian genotoxicity have been included. Data on fungi, plants and *Drosophila* have not been reviewed.

4.4.1 *Humans*

Several studies have investigated the genotoxic effects of arsenic after long-term ingestion via drinking-water, but few studies of occupational exposure to arsenic are available. Exposures were mainly to inorganic arsenic, but since arsenic is methylated in humans, mixed internal exposures to inorganic arsenic and methylated arsenic metabolites predominate. Although MMA and DMA (as sodium salts) have been used in pesticides, this use is currently decreasing and no study was available on the monitoring of human biological effects after occupational exposure to these compounds.

In a pilot study in Mexico, nine women and two men exposed to well-water containing high levels of arsenic (390 µg/L, presumably > 10 years) did not show a significantly higher frequency of chromosomal aberrations or sister chromatid exchange than controls exposed to lower levels of arsenic (11 women and two men; 19–60 µg/L arsenic in well-water). The age range for both groups was 21–62 years. Mutant frequencies at the *HPRT* locus were elevated but not significantly in the high-exposure group (Ostrosky-Wegman *et al.*, 1991). In a more recent study, 35 Mexican individuals exposed to well-water containing 408 µg/L arsenic (presumably > 10 years) were compared with 34 controls (well-water concentration, 29.9 µg/L arsenic). The mean age of the two groups was 40.6 years (exposed) and 39.0 years (control), and sex distribution was said to be similar [exact data not supplied]. In the high-exposure group, chromosomal aberrations were significantly elevated, with 0.08 (exposed) versus 0.03 (control) chromosomal aberrations per cell. Moreover, the frequency of micronuclei in buccal and urothelial cells was significantly elevated (average/1000 cells, 2.21 versus 0.56 and 2.22 versus 0.48, respectively) (Gonsebatt *et al.*, 1997). Among the exposed individuals, men showed more chromosomal aberrations and higher frequency of micronuclei than women. This difference could be attributed to the fact that men drank more water; in this study country, men work in the fields and, because of the dry climate, drink more water than women. The proportion of smokers was similar in the two groups: 29% of the exposed and 33% of the controls; smoking was not significantly associated with a higher incidence of chromosomal aberrations or micronuclei. People occupationally exposed to putative genotoxins or those who underwent medical treatment were excluded from the study.

No differences in sister chromatid exchange (98 exposed subjects versus 83 controls) or chromosomal aberration (104 exposed versus 86 controls) frequencies were found in the peripheral lymphocytes of subjects exposed to moderate quantities of arsenic in the drinking-water in Nevada (USA). Drinking-water with mean concentrations of 109 µg/L arsenic had been consumed for at least 5 years; control subjects had drunk water containing 12 µg/L arsenic (Vig *et al.*, 1984). In the statistical evaluation, sex, age, smoking and putative occupational exposures were controlled for. The population studied was exposed to much lower levels of arsenic than the current study population and arsenic has not been shown to be associated with cancer in blood-forming tissue.

In a more recent study in Nevada (USA), 18 people (mean exposure from drinking-water, 1312 µg/L arsenic > 1 year) showed elevated frequencies of micronuclei in exfoliated bladder cells (2.79/1000 cells) in comparison with 18 control subjects exposed to low levels of arsenic (exposure from drinking-water, 16 µg/L arsenic; 1.57/1000 cells) matched for age, sex and smoking status (Warner *et al.*, 1994). Occupation was included as a confounding variable. In contrast, there was no increase in micronucleated buccal cells associated with such high levels of arsenic.

The frequencies of chromosomal aberrations were determined in the peripheral lymphocytes of 32 Finnish subjects (age, 15–83 years; mean, 52 years) after long-term ingestion of drinking-water containing a median concentration of 410 µg/L arsenic (Mäki-Paakkanen et al., 1998) and were compared with those of eight controls (age, 37–76 years; mean, 50 years) from the same village who consumed drinking-water containing < 1 µg/L arsenic. Estimated cumulative median doses of arsenic were 455 and 7 mg per lifetime, respectively. Smoking habits, sex, seafood consumption and residential history were included as confounders in the evaluation. The crude study results did not show elevated frequencies of chromosomal aberrations in arsenic-exposed subjects (6.9 in exposed versus 8.6 in controls) or smokers (6.0 in ex- and current smokers versus 6.9 in never-smokers). However, in the crude and adjusted linear regression analyses, numbers of chromosomal aberrations were significantly associated with levels of arsenic in urine of current users ($r^2 = 0.25$; $p = 0.08$ and $r^2 = 0.27$; $p = 0.04$, respectively).

In a pilot study in Inner Mongolia, 19 residents exposed to arsenic via drinking-water (527.5 µg/L) for 17 years (group average) were compared with 13 control subjects exposed to a low concentration of 4.4 µg/L arsenic (Tian et al., 2001). Data on smoking habits, occupation, diet, demographic factors, age and medical status were collected. Frequencies of micronuclei were significantly (3.4-fold) higher in cells from the buccal mucosa and sputum collected from airway epithelium. The increase observed for bladder cells was smaller: 2.7-fold over control for all subjects and 2.4-fold over control for nonsmokers. When smokers were excluded from high-exposure and control groups, the effects of arsenic were greater, although only in buccal and sputum cells, in which sixfold increases in micronuclei frequency occurred.

A nested case–control study was performed in an area endemic for Blackfoot disease in Taiwan, China (Liou et al., 1999). A cohort of 686 residents was assembled and, after 4 years, 31 people had developed cancer. Twenty-two blood samples obtained from these subjects at the beginning of the cohort study were successfully processed. A control comparison group was selected from among members of the cohort who had not developed cancer, matched on sex, age, history of residence (residential village) or of drinking artesian well-water and smoking. No differences were found in overall frequencies of sister chromatid exchange. The frequency of chromosomal aberrations was significantly higher among cases, which was due to the induction of chromosome-type but not chromatid-type aberrations. [The Working Group noted that there was no difference in exposure to arsenic (mean duration of drinking artesian well-water) among cases and controls.]

A study in West Bengal, India, compared 45 subjects with cutaneous signs of arsenicism (368 µg/L arsenic in drinking-water) with 21 healthy individuals considered as controls residing in two unaffected districts (5.50 µg/L arsenic in drinking-water) (Basu et al., 2002). The frequency of micronuclei was significantly higher in the oral mucosal cells (5.15 versus 0.77 per 1000 cells), urothelial cells (5.74 versus 0.56 per 1000 cells) and peripheral lymphocytes (6.40 versus 0.53 per 1000 cells) of exposed subjects compared with control subjects. The age distribution and socioeconomic status was reported to be similar in the two groups. Exposure of exposed subjects to arsenic via drinking-water had probably been for a mean of 11 years.

In another study, the mean frequency of sister chromatid exchange/cell in human peripheral lymphocytes was not found to be affected by voluntary ingestion of 0.15 g potassium arsenite or poisoning from 1, 10 or 20 g arsenic trioxide. At 20 g arsenic trioxide, the mean frequency of sister chromatid exchange was significantly elevated (Hantson *et al.*, 1996). Doses of 10 and 20 g arsenic trioxide significantly increased the number of cells with a high sister chromatid exchange frequency and produced a shift in the distribution of the cells according to frequency of sister chromatid exchange.

Few studies have dealt with the induction of genetic damage in workers exposed to arsenic. Moreover, these subjects were exposed to other genotoxic agents. In the peripheral lymphocytes of nine smelter workers exposed to arsenic and other compounds, a significant increase in chromosomal damage was found, with 87 aberrations per 819 mitoses compared with 13 per 1012 in controls (Beckman *et al.*, 1977). In this preliminary report, no data on duration of exposure or age of the workers were given. In a further study, 33 male copper smelter workers (aged 20–62 years) exposed to arsenic and other toxic compounds were studied to determine chromosomal aberrations in peripheral lymphocytes (Nordenson & Beckman, 1982). Internal exposures to arsenic were analysed in urine, but the analytical method was not given. The frequencies of chromosomal aberrations were not associated with age, smoking or degree of exposure to arsenic. Significantly increased frequencies of chromosomal aberrations were found in comparison with 15 male employees (aged 26–60 years) without known occupational exposure to arsenic or other toxic agents: 5.4 aberrations versus 2.1 per 100 cells for gaps and 1.4 aberrations versus 0.1 per 100 cells for chromosome breaks ($p < 0.001$). Chromatid breaks showed a lower significance level (1.3 versus 0.6 per 100 cells [$p < 0.05$]).

Some studies investigated whether arsenic-mediated chromosomal damage *in vivo* is caused by an aneugenic or clastogenic effect (Dulout *et al.*, 1996; Moore, L.E. *et al.*, 1996, 1997a). Both types of damage were induced, but clastogenicity predominated with high exposure to arsenic (Moore, L.G. *et al.*, 1996, 1997a).

Apart from the pilot study of Ostrosky-Wegman *et al.* (1991), no induction of *HPRT* mutation was found in a further study of 15 male Chilean copper-roasting-plant workers (aged 24–66 years), who were categorized according to job type as being exposed to arsenic at low, medium or high levels. Their mean duration of employment in the factory was 43 months. The individual exposure was ascertained by analysing levels of arsenic in the urine. In the very highly exposed workers (internal dose, 260 µg/L arsenic in urine), no induction of *HPRT* mutations in peripheral lymphocytes was demonstrated. The authors concluded that the *HPRT* assay seems to have a low sensitivity for the detection of the genotoxicity of arsenic *in vivo* (Harrington-Brock *et al.*, 1999).

Another study of 70 Chilean men with long-term exposure to 600 µg/L arsenic in drinking-water and 55 frequency-matched control subjects (15 µg/L arsenic in drinking-water) determined micronuclei in bladder cells (Biggs *et al.*, 1997; Moore, L.E. *et al.*, 1997a). Matching criteria were age, smoking status, time of local residence (average high exposure, 19.3 years), education and ethnicity. An exposure-related increase in the frequency of micronuclei was found in the exposure quintiles 2–4 (urinary arsenic,

54–729 µg/L), but not in the 5th quintile (urinary arsenic > 729 µg/L). The prevalence of centromer-positive micronuclei increased 3.1-fold in quintile 4 (95% CI, 1.4–6.6), and the prevalence of centromer-negative micronuclei increased 7.5-fold in quintile 3 (95% CI, 2.8–20.3), suggesting that chromosome breakage was the major cause of formation of micronuclei. An intervention study was carried out on a subset of 34 of the arsenic-exposed Chilean men of this investigation. The arsenic-contaminated drinking-water supply (600 µg/L) was changed to water containing 45 µg/L arsenic. After 8 weeks, the prevalence of micronuclei in bladder cells decreased from 2.63/1000 cells before the intervention to 1.80/1000 cells after the intervention for all individuals. The frequencies of micronuclei in exfoliated bladder cells had significantly decreased from 4.45/1000 cells before the inter-vention to 1.44/1000 cells after the intervention in smokers but not in nonsmokers (2.05/1000 cells versus 1.90/1000 cells), suggesting that the bladder cells of smokers could be more susceptible to genotoxic damage caused by arsenic (Moore, L.E. *et al.*, 1997b).

The frequency of micronuclei in 12 Andean women and 10 children with lifetime current exposure to 200 µg/L arsenic in the drinking-water was compared with that in 10 women and 12 children exposed to 0.7 µg/L arsenic. Putative confounding variables such as smoking, consumption of alcohol and coca leaves were included in the evaluation. It was shown that the frequencies of micronuclei per 1000 binucleated cells in peripheral lymphocytes were significantly elevated in the arsenic-exposed groups as compared with controls (women, 41 versus 8.5; children, 35 versus 5.6, respectively) (Dulout *et al.*, 1996). Moreover, the frequency of aneuploidy was significantly elevated (0.21% versus 0%; 12 exposed versus 17 controls). In contrast, the frequencies of sister chromatid exchange in the arsenic-exposed group were not affected (5.7 versus 5.5 per cell in exposed and control women and 4.4 versus 4.6 per cell in exposed and control children, respectively), nor were specific chromosome translocations.

Induction of sister chromatid exchange was found in peripheral lymphocytes of subjects after 20 years of exposure to arsenic in well-water (> 130 µg/L) in Argentina (Lerda, 1994). Putative exposures to other genotoxic compounds were reported to be taken into account in the study. The mean frequency of sister chromatid exchange was 10.50 per cell in exposed men and women (282 nonsmokers) versus 7.50 per cell in 155 control subjects (volunteer men and women) drinking water that contained less than 20 µg/L arsenic for more than 20 years. Exposed subjects were significantly older than the control group (mean age, 56.71 versus 38.90). In a further evaluation, to homogenize the age of the exposed group, participants older than 50 years were excluded from the analysis. In the younger subset, no correlation between sister chromatid exchange and sex, or sister chro-matid exchange and age was found. Sister chromatid exchange was induced by concen-trations as low as 100 µg/L arsenic for the younger subset. Moreover, the arsenic content in drinking-water was associated with the frequency of sister chromatid exchange in both sexes but was not affected by sex. [The Working Group noted that the value of the study is reduced because the statistical evaluation was not reported in detail. Moreover, arsenic in urine was quantified by an insensitive colorimetric method of analysis.]

4.4.2   *Experimental systems* (see Table 30 for details and references)

(*a*)   *In-vitro studies*

The methylated forms of trivalent arsenic are the only arsenic species that cause DNA damage *in vitro* (Mass *et al.*, 2001; Nesnow *et al.*, 2002).

Arsenic (sodium arsenite) did not induce tryptophan revertants in *Escherichia coli* or ouabain- or 6-thioguanine-resistant mutants in Chinese hamster lung (V79), Chinese hamster ovary or Syrian hamster embryo cells (Lee *et al.*, 1985a). Moreover, induction of SOS repair by sodium arsenite was not detected in *E. coli PQ37*. However, sodium arsenite and sodium arsenate were mutagenic in mouse lymphoma L5178Y cells, inducing trifluorothymidine-resistant mutants.

Sodium arsenite induced a significantly increased frequency of sister chromatid exchange in Chinese hamster ovary and Syrian hamster embryo cells. Sodium arsenate was one order of magnitude less potent in inducing sister chromatid exchange than sodium arsenite. It induced the formation of micronuclei in Chinese hamster ovary and V79 cells in the cytokinesis-block micronucleus test using cytochalasin B as well as in the absence of cytochalasin B in V79 cells and also induced chromosomal aberrations in mammalian cells.

Sodium arsenite significantly elevated the frequency of sister chromatid exchange and significantly enhanced micronucleus formation in isolated human peripheral lymphocytes as well as in whole blood after cytokinesis block through cytochalasin B. It induced chromosomal aberrations as chromatid gaps, fragmentation, endoreduplication and chromosomal breaks in human leukocytes, lymphocytes and primary umbilical cord fibroblasts. Moreover, induction of aneuploidy was observed in human peripheral lymphocytes treated with sodium arsenite *in vitro*, suggesting that this clastogenic agent may exhibit some weak aneuploidogenic properties.

There is some evidence that human, mouse and rat leukocytes are more sensitive to the induction of micronuclei after treatment with arsenite than guinea-pig leukocytes (Peng *et al.*, 2002). This difference in the induction of micronuclei by arsenic could not be explained by a species-dependent variability in arsenite methylation. The leukocytes of all four species were able to ethylate arsenic but there was no clear correlation between the ability to methylate arsenic and the induction of micronuclei.

In assays with mouse lymphoma L5178Y cells, arsenate (As^V), MMA^V and DMA^V induced mutations at the *Tk* locus, chromosomal aberrations and micronuclei. Arseno-betaine, the major arsenic compound in seafood, did not induce neoplastic transformation in mouse fibroblast BALB/3T3 cells.

Significant increases in chromosomal aberrations were induced in human umbilical cord fibroblasts by arsenate, MMA^V, DMA^V, trimethylarsine oxide, arsenosugar, arseno-choline, arsenobetaine and tetramethylarsonium iodide. The higher potency of induction of chromosomal aberrations by DMA^V in comparison with MMA^V was probably caused by contamination of DMA^V sample by inorganic arsenic (Eguchi *et al.*, 1997). Nevertheless,

Table 30. Genetic and related effects of arsenic and arsenic compounds

Test system	Results[a]	Dose[b] (LED or HID)	Reference
Arsenate (As^V)			
Gene mutation, mouse lymphoma L5178Y cells, *Tk* locus, *in vitro*	+	10	Moore, M.M. *et al.* (1997)
Gene mutation, Syrian hamster embryo cells, ouabain resistance, *in vitro*	–	31	Lee *et al.* (1985a)
Gene mutation, Syrian hamster embryo cells, 6-thioguanine resistance, *in vitro*	–	31	Lee *et al.* (1985a)
Sister chromatid exchange, Syrian hamster embryo cells *in vitro*	+	3.1	Lee *et al.* (1985a)
Micronucleus formation, mouse lymphoma L5178Y cells *in vitro*	+	10	Moore, M.M. *et al.* (1997)
Chromosomal aberrations, mouse lymphoma L5178Y cells *in vitro*	+	10	Moore, M.M. *et al.* (1997)
Chromosomal aberrations, Syrian hamster embryo cells *in vitro*	+	20	Lee *et al.* (1985a)
Cell transformation, Syrian hamster embryo cells	+	5	Lee *et al.* (1985a)
Chromosomal aberrations, primary human umbilical cord fibroblasts *in vitro*	+	5	Oya-Ohta *et al.* (1996)
Chromosomal aberrations, human lymphocytes *in vitro*	–	1	Nordenson *et al.* (1981)
Chromosomal aberrations, human leukocytes *in vitro*	+	2.25 (0.6 ppm as As)	Nakamuro & Sayato (1981)
Arsenite (As^{III})			
Escherichia coli, gene mutation (tryptophan revertant selection) *in vitro*	–	3250	Rossman *et al.* (1980)
Escherichia coli, *LacZ* gene induction (SOS chromotest) *PQ37 in vitro*	–	105	Lantzsch & Gebel (1997)
Gene mutation, Chinese hamster ovary cells, ouabain resistance, *in vitro*	–	0.65	Rossman *et al.* (1980); Lee *et al.* (1985b)
Gene mutation, Chinese hamster ovary cells, 6-thioguanine resistance, *in vitro*	–	13	Rossman *et al.* (1980)
Gene mutation, Chinese hamster ovary cells, 6-thioguanine resistance, *in vitro*	–	1.3	Lee *et al.* (1985b)
Gene mutation, mouse lymphoma L5178Y cells, *Tk* locus, *in vitro*	+	1	Moore, M.M. *et al.* (1997b)
Gene mutation, Syrian hamster embryo cells, ouabain resistance, *in vitro*	–	1.3	Lee *et al.* (1985a)
Gene mutation, Syrian hamster embryo cells, 6-thioguanine resistance, *in vitro*	–	1.3	Lee *et al.* (1985a)
Sister chromatid exchange, Chinese hamster ovary cells *in vitro*	+	0.65	Lee *et al.* (1985b)

Table 30 (contd)

Test system	Results[a]	Dose[b] (LED or HID)	Reference
Sister chromatid exchange, Syrian hamster embryo cells in vitro	+	0.1	Lee et al. (1985a)
Micronucleus formation, Chinese hamster ovary cells in vitro	+	5.21[c]	Wang et al. (1997)
Micronucleus formation, Chinese hamster V79 cells in vitro	+	0.325	Gebel (1998)
Micronucleus formation, mouse lymphoma L5178Y cells in vitro	+	1.5	Moore, M.M. et al. (1997)
Chromosomal aberrations, mouse lymphoma L5178Y cells in vitro	+	1.5	Moore, M.M. et al. (1997)
Chromosomal aberrations, Syrian hamster embryo cells in vitro	+	0.8	Lee et al. (1985a)
Cell transformation, Syrian hamster embryo cells	+	0.20	Lee et al. (1985a)
Sister chromatid exchange, human lymphocytes in vitro	+	0.03	Gebel et al. (1997); Rasmussen & Menzel (1997); Nordenson et al. (1981)
Micronucleus formation, human lymphocytes in vitro	+	0.06	Schaumlöffel & Gebel (1998)
Chromosomal aberrations, primary human umbilical cord fibroblasts in vitro	+	0.5	Oya-Ohta et al. (1996)
Chromosomal aberrations, human lymphocytes in vitro	+	0.09	Nordenson et al. (1981)
Chromosomal aberrations, human leukocytes in vitro	+	0.31	Nakamuro & Sayato (1981)
Aneuploidy, human lymphocytes in vitro	+	0.4	Eastmond & Tucker (1989)
Aneuploidy, human lymphocytes in vitro	+	0.31 ng/mL	Ramírez et al. (1997)
Single-cell gel assay (comet), Swiss albino mouse leukocytes in vivo	+	0.13 mg/kg po	Saleha Banu et al. (2001)
LacZ gene mutation, Muta™ mouse lung, kidney, bladder, bone marrow in vivo	–	7.6 mg/kg ip × 5	Noda et al. (2002)
Micronucleus formation, BALB/c mouse bone marrow in vivo	+	10 mg/kg (24 h) or 0.5 mg/kg (30 h) ip	Deknudt et al. (1986)
Micronucleus formation, BALB/c/CBA/C57BL mouse bone marrow in vivo	+	5 mg/kg ip	Tinwell et al. (1991)
Micronucleus formation, B6C3F1 mouse bone marrow in vivo	+	5 mg/kg po × 4	Tice et al. (1997)

Table 30 (contd)

Test system	Results[a]	Dose[b] (LED or HID)	Reference
Micronucleus formation, Muta™ mouse peripheral blood reticulocytes *in vivo*	+	7.6 mg/kg ip × 5	Noda *et al.* (2002)
Chromosomal aberrations, Swiss mouse bone marrow *in vivo*	+	0.1 mg/kg sc × 4	Roy Choudhury *et al.* (1996)
Chromosomal aberrations, Swiss mouse bone marrow *in vivo*	+	2.5 mg/kg po	Biswas *et al.* (1999)
Dominant lethal mutation, Balb/c mouse *in vivo*	–	5 mg/kg ip	Deknudt *et al.* (1986)
Monomethylarsonic acid (MMA^V)			
Gene mutation, mouse lymphoma L5178Y cells, *Tk* locus, *in vitro*	+	2500	Moore, M.M. *et al.* (1997)
Micronucleus formation, mouse lymphoma L5178Y cells *in vitro*	+	4000	Moore, M.M. *et al.* (1997)
Chromosomal aberrations, mouse lymphoma L5178Y cells *in vitro*	+	4000	Moore, M.M. *et al.* (1997)
Chromosomal aberrations, primary human umbilical cord fibroblasts *in vitro*	+	196	Oya-Ohta *et al.* (1996)
Dimethylarsinic acid (DMA^V)			
Gene mutation, mouse lymphoma L5178Y cells, *Tk* locus, *in vitro*	+	5000	Moore, M.M. *et al.* (1997)
Micronucleus formation, mouse lymphoma L5178Y cells *in vitro*	–	10 000	Moore, M.M. *et al.* (1997)
Chromosomal aberrations, mouse lymphoma L5178Y cells *in vitro*	+	8000	Moore, M.M. *et al.* (1997)
Chromosomal aberrations, primary human umbilical cord fibroblasts *in vitro*	+	96.6	Oya-Ohta *et al.* (1996)
DNA strand break, ICR CD-1 mouse lung *in vivo*	+	1500 mg/kg	Yamanaka *et al.* (1989); Yamanaka & Okada (1994)
DNA strand break, ICR CD-1 mouse liver, kidney and spleen *in vivo*	–	1500 mg/kg	Yamanaka *et al.* (1989); Yamanaka & Okada (1994)
LacZ gene mutation, Muta™ mouse lung, kidney, bladder, bone marrow *in vivo*	–	10.6 mg/kg ip × 5	Noda *et al.* (2002)
Micronucleus formation, Muta™ mouse peripheral blood reticulocytes	–	10.6 mg/kg ip × 5	Noda *et al.* (2002)
Aneuploidy, CD-1 mouse bone marrow *in vivo*	+	300 mg/kg ip	Kashiwada *et al.* (1998)

Table 30 (contd)

Test system	Results[a]	Dose[b] (LED or HID)	Reference
Trimethylarsine oxide (TMAO)			
Chromosomal aberrations, primary human umbilical cord fibroblasts *in vitro*	+	503	Oya-Ohta *et al.* (1996)
Arsenocholine			
Chromosomal aberrations, primary human umbilical cord fibroblasts *in vitro*	+	4950	Oya-Ohta *et al.* (1996)
Arsenobetaine			
Cell transformation, mouse BALB/3T3 cells	–	89	Sabbioni *et al.* (1991)
Chromosomal aberrations, primary human umbilical cord fibroblasts *in vitro*	+	1958	Oya-Ohta *et al.* (1996)
Tetramethylarsonium iodide			
Chromosomal aberrations, primary human umbilical cord fibroblasts *in vitro*	+	4978	Oya-Ohta *et al.* (1996)
Arsenosugar (2′,3′-Dihydroxypropyl-5-deoxy-5-dimethylarsinoyl-β-D-riboside)			
Chromosomal aberrations, primary human umbilical cord fibroblasts *in vitro*	–	4860	Oya-Ohta *et al.* (1996)
Methylarsonous acid (MAs[III])			
Single-cell gel (comet) assay, human lymphocytes *in vitro*	+	2.12[c]	Mass *et al.* (2001)
Dimethylarsinous acid (DMAs[III])			
Single-cell gel (comet) assay, human lymphocytes *in vitro*	+	1.22[c]	Mass *et al.* (2001)

[a] +, positive; –, negative; without exogenous metabolic system
[b] LED, lowest effective dose; HID, highest ineffective dose unless otherwise stated; in-vitro tests, μg/mL; in-vivo tests, mg/kg bw/day; po, oral; ip, intraperitoneal
[c] Estimated from graph in paper

Eguchi *et al.* (1997) had shown that pure DMA^V but not MMA^V had induced tetraploids in Chinese hamster V79 cells.

MMA^{III} and DMA^{III} were investigated using human lymphocytes in the single-cell gel assay. At low micromolar doses, these methylated trivalent arsenicals showed a comet-like tail corresponding to DNA damage. In this study, neither As^{III}, As^V, nor the methylated pentavalent arsenicals produced significant nicking, strand breaks or alkali labile lesions in DNA compared with the methylated trivalent arsenicals.

Both hypomethylation and hypermethylation of DNA were associated with exposure to arsenic in cultures of human lung A549 cells and in human kidney UOK cells (Zhong & Mass, 2001). This could be consistent with the proposal that changes in DNA methylation can activate some genes and repress others in response to exposure to arsenite.

(b)   In-vivo studies

Swiss Albino mice administered arsenic trioxide (also called arsenite, As^{III}) orally showed a significantly increased DNA tail-length in leukocytes in the single-cell gel (comet) assay at the lowest dose tested.

Induction of DNA single-strand breaks was detected in the lung, but not in liver, kidney or spleen of ICR (CD-1) mice 12 h after oral administration of DMA^V. The DNA damage was completely repaired after a further 12-h interval.

No significant mutagenesis of the *lacZ* gene was observed in male transgenic Muta^{TM} mouse lung, kidney, bladder or bone marrow after five daily intraperitoneal injections of arsenite (As^{III}) or DMA^V. However, arsenite significantly increased the frequencies of micronucleated reticulocytes in peripheral blood, whereas DMA^V had no effect. [The Working Group noted that, in comparison with other studies using DMA^V, the dose tested was more than one order of magnitude lower.]

Sodium arsenite dissolved in water and administered intraperitoneally to CBA, BALB/c and C57BL mice resulted in a significant induction of micronuclei in the polychromatic erythrocytes, as did oral administration to B6C3F$_1$ mice. Potassium arsenite tested only in C57BL mice was also positive in the micronucleus test in polychromatic erythrocytes. Arsenic sulfide (called orpiment) did not induce micronuclei to any quantifiable extent, presumably because of its low solubility and bioavailability, a reflection of elevated blood levels of arsenic in orpiment-treated animals. After oral or subcutaneous administration of sodium arsenite for either 1, 6 or 30 consecutive days, elevated frequencies of chromosomal aberrations were found in the bone-marrow cells of Swiss albino mice.

Significantly elevated numbers of aneuploid cells were detected in bone-marrow cells of ICR (CD-1) mice treated intraperitoneally with DMA^V.

In an assay to detect point mutations caused by arsenic, virgin C57BL/6J mice and female metallothionein knock-out null mice (MT$^{-/-}$) were exposed to drinking-water containing 500 µg/L arsenic for up to 26 months (Ng *et al.*, 2001). Nine of 12 (75%) virgin C57BL/6J and 8/11 (72.72%) MT$^{-/-}$ mice developed one or multiple mutations in exon 5 of the *p53* gene. The most prominent mutation (mutation hot spot) appeared in codon 163 of exon 5, in 9/12 (75%) and 10/14 (71.4%) of the tissues tested in C57BL/6J and MT$^{-/-}$ mice, respectively.

C57BL/6J mice fed methyl-deficient diets were administered arsenite in the drinking-water at doses of 0, 2.6, 4.3, 9.5 or 14.6 mg/kg bw per day for 130 days. Arsenite treatment increased genomic hypomethylation in a dose-dependent manner and reduced the frequency of methylation at several cytosine sites within the promoter region of the Ha-*ras* gene (Okoji *et al.*, 2002).

Co-mutagenicity/co-genotoxicity of arsenic

Trivalent arsenic was demonstrated to act as a synergistic co-mutagen in combination with many genotoxic agents including ultraviolet (UV) light.

For instance, when Chinese hamster ovary cells were treated simultaneously with UV light and sodium arsenite, chromatid and chromosomal aberrations as well as *Hprt* mutations were increased synergistically. An additive effect in the induction of sister chromatid exchange was observed with a combined treatment of low doses of UV and As^{III} but not with a combined treatment of higher doses of UV and arsenic (Lee *et al.*, 1985b). Treatment of Chinese hamster ovary cells with sodium arsenite after incubation with the DNA-alkylating agent methyl methanesulfonate also enhanced clastogenicity and *Hprt* mutagenicity synergistically (Lee *et al.*, 1986a). However, pretreatment with sodium arsenite resulted in a reduction in the mutagenicity of methyl methanesulfonate. Furthermore, post-treatment of Chinese hamster ovary cells with sodium arsenite was shown to increase UV- and alkylating agent-induced chromosomal aberrations (Huang *et al.*, 1986) and the clastogenicity of DNA-cross-linking agents (Lee *et al.*, 1986b). In the presence of sodium arsenite, γ-ray-induced chromosomal aberration frequency was potentiated in human peripheral lymphocytes (Jha *et al.*, 1992). In human UV-irradiated VH16 fibroblasts, micronuclei (but not sister chromatid exchange) were induced synergistically by post-treatment with sodium arsenite (Jha *et al.*, 1992). According to the authors, the lack of synergistic effect on UV-induced sister chromatid exchange in this study may be because sodium arsenite was washed off before the cells were seeded for division.

4.5   Mechanistic considerations

Several different mechanisms of arsenic-induced carcinogenicity have been proposed, and the trivalent species are implicated in most of these mechanisms (National Research Council, 1999, 2001; Simeonova & Luster, 2000; Kitchin, 2001; Hughes, 2002). It should be noted, however, that the trivalent species are formed *in vivo* after exposure to penta-valent arsenic. Methylated trivalent arsenic is more toxic, and genotoxic, than trivalent inorganic arsenic; in contrast, methylated pentavalent arsenic is less toxic, and genotoxic, than pentavalent inorganic arsenic.

4.5.1 Genotoxicity

Arsenic induces chromosomal aberrations, micronuclei, aneuploidy, endoreduplication and gene amplification. These may play a role in the genomic instability that can result from treatment with arsenic. Arsenic appears to have little if any ability to induce point mutations (National Research Council, 1999, 2001). The methylated trivalent molecules of arsenic are potent forms for the induction of DNA damage in cells *in vitro*, and they are the only forms of arsenic that cause DNA breakage *in vitro*, a reaction that is mediated by reactive oxygen species (Yamanaka & Okada, 1994; Nesnow *et al.*, 2002; Kitchin & Ahmad, 2003).

4.5.2 Altered DNA repair

Trivalent arsenic (As^{III}) inhibits nucleotide-excision repair of UVC-induced DNA damage in human fibroblasts by interacting with distinct steps of the repair process. It impaired the incision step at low concentrations and the ligation step at higher concentrations (Hartwig *et al.*, 1997).

As^{III} inhibits several DNA-repair enzymes including DNA ligases I and II (Li & Rossman, 1989; Lee-Chen *et al.*, 1992), and zinc-finger proteins bearing covalent disulfide linkages seem to be potential targets of this metal. The activity of PARP, one of the zinc-finger DNA-repair enzymes, is inhibited in a human T-cell lymphoma-derived Molt-3 cell line and HeLa cells by low concentrations of arsenic (5 μM and 10 nM, respectively) (Yager & Wiencke, 1997; Hartwig *et al.*, 2003). However, other zinc-finger DNA-repair enzymes such as mammalian xeroderma pigmentosum group A protein and bacterial formamido-pyrimidine-DNA glycosylase are not inhibited by As^{III} (Asmuss *et al.*, 2000).

4.5.3 Induction of oxidative stress

Exposure to arsenic results in the generation of reactive oxygen species both *in vitro* and *in vivo*. There is evidence that these may be involved in the DNA-damaging activities of As^{III}, MMA^{III} and DMA^{III}. Arsenic species, particularly DMA^{III}, release iron from ferritin (Ahmad *et al.*, 2000); this free iron can produce reactive oxygen species via Fenton and/or Haber-Weiss type reactions. Reactive oxygen species are detected in human–hamster hybrid cells exposed to arsenite (As^{III}) (Liu, S.X. *et al.*, 2001) and in ϕX174 DNA incubated *in vitro* with MMA^{III} or DMA^{III} (Nesnow *et al.*, 2002). They are also involved in stress responses that may alter DNA and gene expression. For example, 8-OHdG formation and cyclooxygenase Cox-2 expression, most commonly used as a marker for the evaluation of oxidative DNA damage, are increased in the urinary bladder cancers of rats treated with dimethyl arsenite (Wei *et al.*, 2002). The DMA^{III} produced *in vivo* in the urine of rats treated with DMA^{V} (Cohen *et al.*, 2002) and subsequent generation of reactive oxygen species may be important factors in the arsenic-induced bladder cancer observed in these animals (Wei *et al.*, 2002).

4.5.4   Altered DNA methylation

The alteration of DNA methylation by arsenic may also play a role in the development of cancer. In-vitro and in-vivo studies indicate that the carcinogenicity of arsenic may be mediated by alterations in the methylation status of DNA either by hypermethylation or hypomethylation (Mass & Wang, 1997; Zhao *et al.*, 1997; Okoji *et al.*, 2002).

4.5.5   Cell transformation

Arsenic induces cell transformation in Syrian hamster embryo cells, BALB/3T3 cells and in the rat liver cell line TRL1215. Inoculation of the latter cells into nude mice gave rise to malignant tumours (fibrosarcoma and metastases to the lung) (Lee *et al.*, 1985a; Bertolero *et al.*, 1987; Zhao *et al.*, 1997).

4.5.6   Altered cell proliferation

Increased cell proliferation has been demonstrated directly or indirectly in various experimental systems after exposure to arsenic (Germolec *et al.*, 1997; Kitchin, 2001; Hughes, 2002).

Increases in ODC activity, a biomarker of cell proliferation, have been observed in the kidney or liver of rats treated with arsenic (Yamamoto *et al.*, 1995; Brown & Kitchin, 1996). Stimulation of cell proliferation had been shown in normal human epidermal kera-tinocytes treated *in vitro* by arsenic (Germolec *et al.*, 1997).

Hyperplasia has been observed in the bladder of rats treated with DMA^V (Cohen *et al.*, 2002).

4.5.7   Altered cell signalling

Arsenic stimulates the activity of Jun kinases, which belong to the mitogen-activated protein kinase family, and increases the DNA binding of transcriptional factor AP-1. Arsenic also induces the expression of proto-oncogenes such as C-*JUN*, C-*FOS*, C-*MYC* and tumour growth factor-α (Cavigelli *et al.*, 1996; Germolec *et al.*, 1998; Simeonova *et al.*, 2000; Chen *et al.*, 2001). A reduction in p53 protein levels concomitant with an increase in mdm$_2$ protein levels were also observed in a keratinocyte (HaCaT) cell line treated with arsenic. The disruption of *P53-MDM$_2$* loop-regulating cell-cycle arrest as a model for arsenic-related skin carcinogenesis has been proposed (Hamadeh *et al.*, 1999).

4.5.8   Altered steroid receptor binding and gene expression

Arsenic inhibited steroid binding to glycocorticoid receptors but had no effect on the binding of ligands to androgen, estrogen, mineral corticoid or progesterone receptors. This specific inhibition may provide a method of using arsenic to block glucocorticoid receptors selectively in assays of the progesterone receptor content of breast cancer tissues

(Lopez *et al.*, 1990). In MCF-7 cells, arsenite blocked the binding of estradiol to oestrogen receptor-α (ER-α) (Stoica *et al.*, 2000).

Moreover, arsenic inhibited expression of ER-α but had no effect on expression of ER-β in breast cancer cell lines (Chen *et al.*, 2002). Thus, the authors concluded that the role of arsenic in the expression of ER-α provides a novel therapeutic approach for ER-α-positive breast cancer (Chen *et al.*, 2002).

4.5.9 *Gene amplification*

Arsenic enhanced the amplification of the dihydrofolate reductase (*DHFR*) gene in mouse 3T6 cells and gene amplification has been suggested as a possible mechanism of the carcinogenicity of arsenic (Lee *et al.*, 1988).

5. Summary of Data Reported and Evaluation

5.1 Exposure data

Exposure of high levels of arsenic in drinking-water has been recognized for many decades in some regions of the world, notably in China, Taiwan (China) and some countries in Central and South America. More recently, it has been discovered that a number of other regions have drinking-water that is highly contaminated with arsenic. In most of these regions, the drinking-water source is groundwater, naturally contaminated from arsenic-rich geological formations. The primary regions where high concentrations of arsenic have been measured in drinking-water include large areas of Bangladesh, China and West Bengal (India) and smaller areas of Argentina, Australia, Chile, Mexico, Taiwan (China), the USA and Viet Nam. In some areas of Japan, Mexico, Thailand and other countries, mining, smelting and other industrial activities have contributed to elevated concentrations of arsenic in local water sources.

Levels of arsenic in affected areas may range from tens to hundreds or even thousands of micrograms per litre, whereas in unaffected areas levels are typically only a few micrograms per litre. The WHO guideline recommends that levels of arsenic in drinking-water should not exceed 10 μg/L. Arsenic occurs in drinking-water primarily as arsenate (As^V), although in reducing environments significant concentrations of arsenite (As^{III}) have also been reported. Trace amounts of methylated arsenic species are typically found in drinking-water, and higher levels are found in biological systems. Inorganic arsenic (arsenate plus arsenite) is the predominant form of arsenic in drinking-water.

In many areas where contamination of drinking-water by arsenic has been reported, current exposures have been reduced by various interventions.

5.2   Human carcinogenicity data

In previous monographs, the evidence of carcinogenicity to humans of exposure to arsenic and arsenic compounds, such as medical treatment with Fowler's solution and inhalation exposure of mining and smelting workers, was evaluated as *sufficient*.

Informative epidemiological studies of cancer in relation to arsenic in drinking-water include ecological studies and fewer case–control and cohort studies. For most other known human carcinogens, the major source of causal evidence derives from case–control and cohort studies, with little evidence from ecological studies. In contrast, for arsenic in drinking-water, ecological studies provide important information for causal inference. The reasons include large exposure contrasts and limited population migration. As a consequence of widespread exposure to local or regional water sources, ecological measurements provide a strong indication of individual exposure. Moreover, in the case of arsenic, the ecological estimates of relative risk are often so high that potential confounding with known causal factors cannot explain the results. Hence, in the reviews that follow, ecological studies are presented in detail.

Urinary bladder cancer

The Working Group evaluated ecological studies in Taiwan (China), Chile, Argentina and Australia, cohort studies from Taiwan, Japan and the USA and case–control studies in Taiwan, the USA and Finland.

There is extensive evidence of increased risks for urinary bladder cancer associated with arsenic in drinking-water. All studies that involved populations with high long-term exposures found substantial increases in the risk for bladder cancer. Key evidence derives from ecological studies in Taiwan and Chile. In Taiwan, the evidence is supported by case–control studies and cohort studies within the exposed communities that demonstrate evidence of dose–response relationships with levels of arsenic in drinking-water. The evidence of increased mortality from bladder cancer in Chile comes from a large population with exposure to arsenic in all major cities and towns of the contaminated region.

There is also evidence of increased risks for bladder cancer from a small cohort study in Japan of persons drinking from wells that had been highly contaminated with arsenic wastes from a factory and an ecological study from Argentina with moderate exposure to arsenic in well-water. Two case–control studies that investigate low exposure to arsenic found increased risks with increasing exposure in one or more subgroups.

Considered overall, the findings cannot be attributed to chance or confounding, and they are consistent, with strong associations found in populations with high exposure. There is evidence of dose–response relationships within exposed populations.

Lung cancer

The Working Group evaluated ecological studies using mortality data in Taiwan (China), Chile, Argentina and Australia, cohort studies in Taiwan, Japan and the USA and case–control studies in Taiwan and Chile.

Increased risk for lung cancer was consistently observed in ecological, case–control and cohort studies in Taiwan, Japan, Chile and Argentina. Evidence for a dose–response relationship between arsenic in drinking-water and risk for lung cancer was also observed in ecological studies in Taiwan and Argentina, in cohort studies in south-western and north-eastern Taiwan and Japan and in case–control studies in south-western Taiwan and Chile. The potential confounding effect of cigarette smoking was ruled out by direct and indirect evidence in studies from Taiwan and Chile.

Considered overall, the findings cannot be attributed to chance or confounding, are consistent and demonstrate strong associations in populations with high exposure. There is evidence of a dose–response relationship.

Skin cancer

The Working Group evaluated ecological studies from Taiwan (China), Mexico, Chile and the USA, cohort studies from Taiwan and a case–control study from the USA.

The recognition that arsenic was potentially carcinogenic arose from occurrences of skin cancer after ingestion of medicinal arsenic, arsenical pesticide residues and arsenic-contaminated drinking-water. Skin cancer is a commonly observed malignancy related to contamination of drinking-water with arsenic. The characteristic arsenic-associated skin tumours include keratinocytic malignancies (non-melanoma skin cancers), in particular squamous-cell carcinomas, including Bowen disease, and multiple basal-cell carcinomas.

Ecological studies, largely from the south-west of Taiwan, indicate substantially elevated incidence of, prevalence of and mortality rates for skin cancer associated with drinking-water highly contaminated with arsenic, with evidence of a dose–response relationship. Findings in ecological studies were substantiated in two cohort studies in the region of Taiwan that is endemic for arsenic. Increased mortality from skin cancer was found in Chile. A high prevalence of skin lesions, including skin cancers, was found in rural regions of Mexico. An excess risk for skin cancer was observed in a case–control study in the USA conducted in an area with lower concentrations of arsenic in drinking-water. A cohort study from the south-west of Taiwan reported that differences in the levels of serum β-carotene and urinary arsenic metabolites may modify the risk for arsenic-induced skin cancers.

Liver cancer

The Working Group evaluated ecological studies using mortality data in Taiwan (China), Chile, Argentina and Australia, cohort studies in Taiwan, Japan and the USA and a case–control study in Taiwan of liver cancer cases identified from death certificates.

Increased mortality from liver cancer was observed in the ecological studies involving a large population with high exposure to arsenic in Taiwan. Evidence for a dose–response relationship between arsenic in drinking-water and liver cancer mortality was observed in both ecological and case–control studies in Taiwan. Increased risks were also found in small cohort studies in Taiwan and Japan. Findings on mortality from liver cancer observed in ecological studies in Chile are inconsistent.

The interpretation of these findings is limited by the small number of liver cancer cases, questionable accuracy of the diagnosis of liver cancer on death certificates and potential confounding or modifying effects of chronic hepatitis virus infection or other factors.

Kidney cancer

The Working Group evaluated ecological studies in Taiwan (China), Chile, Argentina and Australia, and cohort studies from Taiwan and the USA.

All studies that involved populations with high long-term exposures to arsenic found increased risks for kidney cancer. Key evidence comes from ecological studies in Taiwan and Chile. In Taiwan, the evidence is supplemented by a small cohort study of patients with Blackfoot disease. The evidence of increased mortality from kidney cancer in Chile comes from a large population with exposure to arsenic in all major cities and towns of the region. There is also evidence of increased risk for kidney cancer in populations in Argentina with moderate exposure to arsenic in well-water.

Relative risk estimates for kidney cancer were generally lower than those for urinary bladder cancer, and no studies have reported dose–response relationships on the basis of individual exposure assessment.

Other cancers

The Working Group evaluated ecological studies from Taiwan (China), Chile and the USA, cohort studies from Japan and the USA and one case–control study each from Canada and the USA.

Excess mortality from prostate cancer was found in south-west Taiwan. Inconsistent findings were reported for other cancer sites.

5.3   Animal carcinogenicity data

Dimethylarsinic acid was tested for carcinogenicity by administration in drinking-water in mice and rats. It was also tested in two-stage initiation–promotion studies in mice and rats. Complete carcinogenicity was observed in the urinary bladder of rats and lungs of mice. Dimethylarsinic acid exerted its carcinogenic effect on spontaneous development of tumours in p53$^{+/-}$ and p53$^{+/+}$ mice. Dimethylarsinic acid is a tumour promoter in the skin and lung of mice, and in the liver, urinary bladder, kidney and thyroid gland of rats.

After perinatal treatment, arsenic trioxide induced lung adenomas in mice and, after intratracheal instillation to hamsters, it induced lung adenomas in two of three studies. Calcium arsenate administered to hamsters by intratracheal instillation induced lung adenomas. Sodium arsenate induced tumours at various organ sites in metallothionein knockout mice. Transplacental exposure of mice to sodium arsenite induced liver and lung carcinomas, ovarian tumours (benign and malignant) and adrenal cortical adenomas. Sodium arsenite promoted skin carcinogenesis in mice. Arsenic trisulfide was negative for carcinogenicity when tested in hamsters by intratracheal instillation.

5.4   Other relevant data

Arsenic in drinking-water is well absorbed in the gastrointestinal tract. The trivalent species of arsenic are formed *in vivo* after exposure to pentavalent arsenic. Arsenic is metabolized by a series of reductions and oxidations and by methylation reactions. Methylated trivalent arsenic is more toxic and less genotoxi than trivalent inorganic arsenic; in contrast, methylated pentavalent arsenic is less toxic and less genotoxic than pentavalent inorganic arsenic. There is a large variation in metabolism between animal species, population groups and individuals. Both inorganic arsenic and its methylated metabolites are excreted in urine.

Acute effects due to ingestion of arsenic are characterized by severe vomiting and diarrhoea with features of shock, muscle cramps and cardiac abnormalities. Subacute exposures affect primarily the respiratory, gastrointestinal, cardiovascular, nervous and haematopoietic systems.

Most reports of chronic arsenic toxicity focus on skin manifestations such as pigmentation, with depigmentation affecting trunks and limbs and keratosis affecting hands and feet. Chronic lung disease, peripheral neuropathy, hepatomegaly and peripheral vascular disease have frequently been reported in cases of chronic exposure to arsenic. Exposure to arsenic has been associated with an increased risk for diabetes mellitus. Other systemic manifestations include cardiovascular effects, abdominal pain, anorexia, nausea, diarrhoea, cerebrovascular disease, non-pitting oedema of hands, feet or legs, anaemia and generalized weakness. In a study in Taiwan (China), significantly higher mortality from cardiovascular and peripheral vascular disease was reported among patients with Blackfoot disease compared with the general population of Taiwan or unaffected residents in endemic areas of Blackfoot disease.

The acute toxicity of trivalent arsenic is greater than that of the pentavalent form. The 50% lethal dose of arsenic trioxide in mice by the oral route varies from 15 to 48 mg/kg bw, whereas the acute lethal dose in humans varies from 1 to 3 mg/kg bw. In chronic toxicity studies, arsenic inhibits mitochondrial respiration and induces apoptosis accompanied by a loss of the mitochondrial transmembrane potential. Metallothionein is thought to have a protective effect against the toxicity of arsenic.

Arsenic can modify the urinary excretion of porphyrins in animals and humans. It also interferes with the activities of several enzymes of the haeme biosynthetic pathway. The major abnormalities in urinary porphyrin excretion in chronically exposed humans are (*a*) significant reductions in coproporphyrin III excretion, resulting in a decrease in the ratio of coproporphyrinogen oxidase III to coproporphyrinogen oxidase I and (*b*) significant increases in uroporphyrin excretion.

Exposure to arsenite or arsenate results in generation of reduced oxygen species in laboratory animals and human cells. Exposure to arsenicals either *in vivo* or *in vitro* in a variety of model systems causes induction of a number of major stress-protein families such as heat-shock proteins. Recent studies in animals demonstrated altered gene expression following acute treatment with arsenic that included DNA repair genes, activation of transcription factors, such as activator protein 1, and an increase in pro-inflammatory cytokines. All of these events could play a role in the toxicity of arsenic.

Few studies have been conducted on the immunotoxicity of arsenic. All arsenic compounds evaluated in mouse spleen cells suppressed plaque-forming cell responses to sheep erythrocytes and proliferative response to mitogens. Furthermore, arsenic impaired stimulation and proliferation of human lymphocytes *in vitro*. Recent studies suggest that apoptosis may be an important mechanism for arsenic-induced immunosuppression.

Experimental animal studies have demonstrated the developmental toxicity of trivalent and pentavalent arsenic, monomethylarsonic acid and dimethylarsinic acid. Limited human data suggest that exposure to high concentrations of arsenic in drinking-water during pregnancy may increase fetal and neonatal mortality.

The genotoxicity of arsenic is due largely to the trivalent arsenicals. In humans, arsenic is a chromosomal mutagen (an agent that induces mutations involving more than one gene, typically large deletions or rearrangements). Arsenic appears to have limited ability to induce point mutations. Elevated frequencies of micronuclei, chromosomal aberrations and aneuploidy were detected in the peripheral lymphocytes or urothelial cells, or both, of people exposed to elevated levels of arsenic. *In vitro*, arsenic was not a point mutagen in bacteria. In mammalian cells, arsenic caused various types of chromosomal mutations and aneuploidy. In combination with many genotoxic agents, including ultraviolet light, arsenic was a synergistic co-mutagen. *In vitro*, arsenite was genotoxic at micromolar concentrations. Arsenate was approximately one order of magnitude less genotoxic than arsenite, dimethylarsinic acid and monomethylarsonic acid induced genotoxicity at millimolar concentrations.

Methylarsenous acid and dimethylarsinous acid are intermediary metabolites in the methylation of arsenic. Their genotoxicity has not been fully established, but recent results implicate a major role for these metabolites and reduced (reactive) oxygen species in the induction of urinary bladder cancer in rats.

5.5   Evaluation

There is *sufficient evidence* in humans that arsenic in drinking-water causes cancers of the urinary bladder, lung and skin.

There is *sufficient evidence* in experimental animals for the carcinogenicity of dimethylarsinic acid.

There is *limited evidence* in experimental animals for the carcinogenicity of sodium arsenite, calcium arsenate and arsenic trioxide.

There is *inadequate evidence* in experimental animals for the carcinogenicity of sodium arsenate and arsenic trisulfide.

Taken together, the studies on inorganic arsenic provide *limited evidence* for carcinogenicity in experimental animals.

Overall evaluation

Arsenic in drinking-water is *carcinogenic to humans (Group 1)*.

6.   References

Acharyya, S.K. (2002) Arsenic contamination in groundwater affecting major parts of southern West Bengal and parts of western Chhattisgarh: Source and mobilization process. *Current Sci.*, **82**, 740–744

Agahian, B., Lee, J.S., Nelson, J.H. & Johns, R.E. (1990) Arsenic levels in fingernails as a biological indicator of exposure to arsenic. *Am. ind. Hyg. Assoc. J.*, **51**, 646–651

Ahmad, S.A., Bandaranayake, D., Khan, A.W., Hadi, S.K., Uddin, G. & Halim, M.A. (1997) Arsenic contamination in ground water and arsenicosis in Bangladesh. *Int. J. environ. Health Res.*, **7**, 271–276

Ahmad, S., Anderson, W.L. & Kitchin, K.T. (1999) Dimethylarsinic acid effects on DNA damage and oxidative stress related biochemical parameters in B6C3F1 mice. *Cancer Lett.*, **139**, 129–135

Ahmad, S.A., Sayed, M.H.S.U., Hadi, S.A., Faruquee, M.H., Khan, M.H., Jalil, M.A., Ahmed, R. & Khan, A.W. (1999) Arsenicosis in a village in Bangladesh. *Int. J. environ. Health Res.*, **9**, 187–195

Ahmad, S., Kitchin, K.T. & Cullen, W.R. (2000) Arsenic species that cause release of iron from ferritin and generation of activated oxygen. *Arch. Biochem. Biophys.*, **382**, 195–202

Ahmad, S.A., Sayed, M.H.S.U., Barua, S., Khan, M.H., Faruquee, M.H., Jalil, A., Hadi, S.A. & Talukder, H.K. (2001) Arsenic in drinking water and pregnancy outcomes. *Environ. Health Perspect.*, **109**, 629–631

Alam, M.G.M., Allinson, G., Stagnitti, F., Tanaka, A. & Westbrooke, M. (2002) Arsenic contamination in Bangladesh groundwater: A major environmental and social disaster. *Int. J. environ. Health Res.*, **12**, 236–253

Alam, M.G.M., Snow, E.T. & Tanaka, A. (2003) Arsenic and heavy metal contamination of vegetables grown in Samta village, Bangladesh. *Sci. total Environ.*, **308**, 83–96

Alonso, H. (1992) Arsenic enrichment in superficial waters. II Region northern Chile. In: *Proceedings of an International Seminar on Arsenic in the Environment and its Incidence on Health*, Santiago, Universidad de Chile, pp. 101–108

Aposhian, H.V., Gurzau, E.S., Le, X.C., Gurzau, A., Healy, S.M., Lu, X., Ma, M., Yip, L., Zakharyan, R.A., Maiorino, R.M., Dart, R.C., Tircus, M.G., Gonzalez-Ramirez, D., Morgan, D.L., Avram, D. & Aposhian, M.M. (2000a) Occurrence of monomethylarsonous acid in urine of humans exposed to inorganic arsenic. *Chem. Res. Toxicol.*, **13**, 693–697

Aposhian, H.V., Zheng, B., Aposhian, M.M., Le, X.C., Cebrian, M.E., Cullen, W., Zakharyan, R.A., Ma, M., Dart, R.C., Cheng, Z., Andrewes, P., Yip, L., O'Malley, G.F., Maiorino, R.M., Van Voorhies, W., Healy, S.M. & Titcomb, A. (2000b) DMPS–arsenic challenge test. II. Modulation of arsenic species, including monomethylarsonous acid (MMA[III]), excreted in human urine. *Toxicol. appl. Pharmacol.*, **165**, 74–83

Aragones Sanz, N., Palacios Diez, M., Avello de Miguel, A., Gomez Rodriguez, P., Martinez Cortes, M. & Rodriguez Bernabeu, M.J. (2001) [Nivel de arsenico en abastecimientos de agua de consumo de origen subterraneo en la comunidad de Madrid.] *Rev. esp. Salud. publica*, **75**, 421–432 (in Spanish)

Arguello, R.A., Cenget, D.D. & Tello, E.E. (1938) [Cancer and arsenicism in the endemic region of Córdoba.] *Rev. Argentina Dermatofisiol.*, **22**, 461–470 (in Spanish)

Armienta, M.A., Rodriguez, R. & Cruz, O. (1997) Arsenic content in hair of people exposed to natural arsenic polluted groundwater at Zimapán, México. *Bull. environ. Contam. Toxicol.*, **59**, 583–589

Aschengrau, A., Zierler, S. & Cohen, A. (1989) Quality of community drinking water and the occurrence of spontaneous abortion. *Arch. environ. Health*, **44**, 283–290

Asmuss, M., Mullenders, L.H.F., Eker, A. & Hartwig, A. (2000) Differential effects of toxic metal compounds on the activities of Fpg and XPA, two zinc finger proteins involved in DNA repair. *Carcinogenesis*, **21**, 2097–2104

Ballantyne, J.M. & Moore, J.N. (1988) Arsenic geochemistry in geothermal systems. *Geochim. cosmochim. Acta*, **52**, 475–483

Barchowsky, A., Dudek, E.J., Treadwell, M.D. & Wetterhahn, K.E. (1996) Arsenic induces oxidant stress and NF-κB activation in cultured aortic endothelial cells. *Free rad. Biol. Med.*, **21**, 783–790

Basu, D., Dasgupta, J., Mukherjee, A. & Guha Mazumder, D.N. (1996) Chronic neuropathy due to arsenic intoxication from geo-chemical source — A five-year follow-up study. *JANEI*, **1**, 45–48

Basu, A., Mahata, J., Gupta, S. & Giri, A.K. (2001) Genetic toxicology of a paradoxical human carcinogen, arsenic: A review. *Mutat. Res.*, **488**, 171–194

Basu, A., Mahata, J., Roy, A.K., Sarkar, J.N., Poddar, G., Nandy, A.K., Sarkar, P.K., Dutta, P.K., Banerjee, A., Das, M., Ray, K., Roychaudhury, S., Natarajan, A.T., Nilsson, R. & Giri, A.K. (2002) Enhanced frequency of micronuclei in individuals exposed to arsenic through drinking water in West Bengal, India. *Mutat. Res.*, **516**, 29–40

Bates, M.N., Smith, A.H. & Cantor, K.P. (1995) Case–control study of bladder cancer and arsenic in drinking water. *Am. J. Epidemiol.*, **141**, 523–530

Bauman, J.W., Liu, J. & Klaassen, C.D. (1993) Production of metallothionein and heat-shock proteins in response to metals. *Fundam. appl. Toxicol.*, **21**, 15–22

Daxley, M.N., Hood, R.D., Vedel, G.C., Harrison, W.P. & Szczech, G.M. (1981) Prenatal toxicity of orally administered sodium arsenite in mice. *Bull. environ. Contam. Toxicol.*, **26**, 749–756

Beckman, G., Beckman, L. & Nordenson, I. (1977) Chromosome aberrations in workers exposed to arsenic. *Environ. Health Perspect.*, **19**, 145–146

Benramdane, L., Accominotti, M., Fanton, L., Malicier, D. & Vallon, J.-J. (1999) Arsenic speciation in human organs following fatal arsenic trioxide poisoning — A case report. *Clin. Chem.*, **45**, 301–306

Berg, J.W. & Burbank, F. (1972) Correlations between carcinogenic trace metals in water supplies and cancer mortality. *Ann. N.Y. Acad. Sci.*, **199**, 249–264

Berg, M., Tran, H.C., Nguyen, T.C., Pham, H.V., Schertenleib, R. & Giger, W. (2001) Arsenic contamination of groundwater and drinking water in Vietnam. A human health threat. *Environ. Sci. Technol.*, **35**, 2621–2626

Bergoglio, R.M. (1964) [Mortality in zones with arsenical woter in the Province of Córdoba, Argentine Republic.] *Pren. méd. Argent.*, **51**, 994–998 (in Spanish)

Bernstam, L. & Nriagu, J. (2000) Molecular aspects of arsenic stress. *J. Toxicol. environ. Health*, **B3**, 293–322

Bernstam, L., Lan, C.-H., Lee, J. & Nriagu, J.O. (2002) Effects of arsenic on human keratinocytes: Morphological, physiological, and precursor incorporation studies. *Environ. Res.*, **A89**, 220–235

Bertolero, F., Pozzi, G., Sabbioni, E. & Saffiotti, U. (1987) Cellular uptake and metabolic reduction of pentavalent to trivalent arsenic as determinants of cytotoxicity and morphological transformation. *Carcinogenesis*, **8**, 803–808

Bhattacharya, P., Chatterjee, D. & Jacks, G. (1997) Occurrence of arsenic-contaminated groundwater in alluvial aquifers from delta plains, eastern India: Options for safe drinking water supply. *Int. J. Water Resour. Dev.*, **13**, 79–92

Bhattacharya, P., Larsson, M., Leiss, A., Jacks, G., Sracek, A. & Chatterjee, D. (1998) Genesis of arseniferous groundwater in the alluvial aquifers of Bengal delta plains and strategies for low-cost remediation. In: *International Conference on Arsenic Pollution of Ground Water in Bangladesh: Causes, Effects and Remedies*, Dhaka, Dhaka Community Hospital Trust, pp. 120–123

Biagini, R.E. (1972) [Chronic hydroarsenicism and mortality from malignant cancers.] *Semana med.*, **141**, 812–816 (in Spanish)

Biggs, M.L., Kalman, D.A., Moore, L.E., Hopenhayn-Rich, C., Smith, M.T. & Smith, A.H. (1997) Relationship of urinary arsenic to intake estimates and a biomarker of effect, bladder cell micronuclei. *Mutat. Res.*, **386**, 185–195

Biswas, B.K., Dhar, R.K., Samanta, G., Mandal, B.K., Chakraborti, D., Faruk, I., Islam, K.S., Chowdhury, M.M., Islam, A. & Roy, S. (1998) Detailed study report of Samta, one of the arsenic-affected villages of Jessore District, Bangladesh. *Current Sci.*, **74**, 134–145

Biswas, S., Talukder, G. & Sharma, A. (1999) Prevention of cytotoxic effects of arsenic by short-term dietary supplementation with selenium in mice in vivo. *Mutat. Res.*, **441**, 155–160

Black, M.M. (1967) Prolonged ingestion of arsenic. *Pharma J.*, **9 December**, 593–597

Blackwell, R.Q., Yang, T.H. & Ai, I. (1961) Preliminary report on arsenic level in water and food from the endemic Blackfoot area. *J. formos. med. Assoc.*, **60**, 1139–1140

Blakley, B.R. (1987) The effect of arsenic on urethan-induced adenoma formation in Swiss mice. *Can. J. vet. Res.*, **51**, 240–243

Borgoño, J.M. & Greiber, R. (1971) [Epidemiological study of arsenic poisoning in the city of Antofagasta.] *Rev. méd. Chile*, **99**, 702–707 (in Spanish)

Borgoño, J.M., Vicent, P., Venturino, H. & Infante, A. (1977) Arsenic in the drinking water of the city of Antofagasta: Epidemiological and clinical study before and after the installation of a treatment plant. *Environ. Health Perspect.*, **19**, 103–105

Borgoño, J.M., Venturino, H. & Vicent, P. (1980) [Clinical and epidemiologic study of arsenicism in northern Chile.] *Rev. méd. Chile*, **108**, 1039–1048 (in Spanish)

Borzsonyi, M., Bereczky, A., Rudnai, P., Csanady, M. & Horvath, A. (1992) Epidemiological studies on human subjects exposed to arsenic in drinking water in Southeast Hungary. *Arch. Toxicol.*, **66**, 77–78

Boyle, R.W. & Jonasson, I.R. (1973) The geochemistry of arsenic and its use as an indicator element in geochemical prospecting. *J. geochem. Explor.*, **2**, 251–296

Braman, R.S. & Foreback, C.C. (1973) Methylated forms of arsenic in the environment. *Science*, **182**, 1247–1249

Bronaugh, R.L., Stewart, R.F. & Congdon, E.R. (1982) Methods for *in vitro* percutaneous absorption studies. II. Animal models for human skin. *Toxicol. appl. Pharmacol.*, **62**, 481–488

Brown, J.L. & Kitchin, K.T. (1996) Arsenite, but not cadmium, induces ornithine decarboxylase and heme oxygenase activity in rat liver: Relevance to arsenic carcinogenesis. *Cancer Lett.*, **98**, 227–231

Brown, M.M., Rhyne, B.C. & Goyer, R.A. (1976) Intracellular effects of chronic arsenic adminis-tration on renal proximal tubule cells. *J. Toxicol. environ. Health*, **1**, 505–514

Buchet, J.P. & Lauwerys, R. (1985) Study of inorganic arsenic methylation by rat liver in vitro: Relevance for the interpretation of observations in man. *Arch. Toxicol.*, **57**, 125–129

Buchet, J.P. & Lauwerys, R. (1988) Role of thiols in the *in vitro* methylation of inorganic arsenic by rat liver cytosol. *Biochem. Pharmacol.*, **37**, 3149–3153

Buchet, J.P., Lauwerys, R. & Roels, H. (1981) Comparison of the urinary excretion of arsenic metabolites after a single dose of sodium arsenite, monomethylarsonate or dimethylarsinate in man. *Int. Arch. occup. environ. Health*, **48**, 71–79

Burk, D. & Beaudoin, A.R. (1977) Arsenate-induced renal agenesis in rats. *Teratology*, **16**, 247–259

Caceres, L., Gruttner, E. & Contreras, R. (1992) Water recycling in arid regions — Chilean case. *Ambio*, **21**, 138–144

Calderon, R.L., Hudgens, E., Le, X.C., Schreinemachers, D. & Thomas, D.J. (1999) Excretion of arsenic in urine as a function of exposure to arsenic in drinking water. *Environ. Health Perspect.*, **107**, 663–667

Calderón, J., Navarro, M.E., Jimenez-Capdeville, M.E., Santos-Diaz, M.A., Golden, A., Rodriguez-Leyva, I., Borja-Aburto, V. & Diaz-Barriga, F. (2001) Exposure to arsenic and lead and neuropsychological development in Mexican children. *Environ. Res.*, **A85**, 69–76

Caltabiano, M.M., Koestler, T.P., Poste, G. & Greig, R.G. (1986) Induction of 32- and 34-kDa stress proteins by sodium arsenite, heavy metals, and thiol-reactive agents. *J. biol. Chem.*, **261**, 13381–13386

Cantor, K.P., Hoover, R., Hartge, P., Mason, T.J., Silverman, D.T., Altman, R., Austin, D.F., Child, M.A., Key, C.R., Marrett, L.D., Myers, M.H., Narayana, A.S., Levin, L.I., Sullivan, J.W., Swanson, G.M., Thomas, D.B. & West, D.W. (1987) Bladder cancer, drinking water source, and tap water consumption: A case–control study. *J. natl Cancer Inst.*, **79**, 1269–1279

Cao, S. (1996) [Current status of inorganic arsenic contamination in mainland China.] *Chinese J. public Health*, **15** (Suppl. 3), S6–S10 (in Chinese)

Carpenter, S.J. (1987) Developmental analysis of cephalic axial dysraphic disorders. *Anat. Embryol.*, **176**, 345–365

Cavigelli, M., Li, W.W., Lin, A., Su, B., Yoshioka, K. & Karin, M. (1996) The tumor promoter arsenite stimulates AP-1 activity by inhibiting a JNK phosphatase. *EMBO J.*, **15**, 6269–6279

Cebrián, M.E., Albores, A., Aguilar, M. & Blakely, E. (1983) Chronic arsenic poisoning in the north of Mexico. *Hum. Toxicol.*, **2**, 121–133

Cebrián, M.E., Albores, A., Connelly, J.C. & Bridges, J.W. (1988) Assessment of arsenic effects on cytosolic heme status using tryptophan pyrrolase as an index. *J. biochem. Toxicol.*, **3**, 77–86

Cebrián, M.E., Albores, A., García-Vargas, G., Del Razo, L.M. & Ostrosky-Wegman, P. (1994) Chronic arsenic poisoning in humans: The case of Mexico. In: O'Nriagu, J., ed., *Arsenic in the Environment*, Part II, *Human Health and Ecosystem Effects*, New York, John Wiley & Sons, pp. 93–107

Chaineau, E., Binet, S., Pol, D., Chatellier, G. & Meininger, V. (1990) Embryotoxic effects of sodium arsenite and sodium arsenate on mouse embryos in culture. *Teratology*, **41**, 105–112

Chakraborti, D., Samanta, G., Mandal, B.K., Roy Chowdhury, T., Chanda, C.R., Biswas, B.K., Dhar, R.K., Basu, G.K. & Saha, K.C. (1998) Calcutta's industrial pollution: Groundwater arsenic contamination in a residential area and sufferings of people due to industrial effluent discharge — An eight-year study report. *Current Sci.*, **74**, 346–355

Chakraborti, D., Biswas, B.K., Basu, G.K., Chowdhury, U.K., Roy Chowdhury, T., Lodh, D., Chanda, C.R., Mandal, B.K., Samanta, G., Chakraborti, K., Rahman, M.M., Paul, K., Roy, S., Kabir, S., Ahmed, B., Das, R., Salim, M. & Quamruzzaman, Q. (1999a) Possible arsenic contamination of free groundwater source in Bangladesh. *J. surface Sci. Technol.*, **15**, 180–188

Chakraborti, D., Biswas, B.K., Roy Chowdhury, T., Basu, G.K., Mandal, B.K., Chowdhury, U.K., Mukherjee, S.C., Gupta, J.P., Chowdhury, S.R. & Rathore, K.C. (1999b) Arsenic groundwater contamination and sufferings of people in Rajnandgaon district, Madhya Pradesh, India. *Current Sci.*, **77**, 502–504

Chakraborti, D., Rahman, M.M., Paul, K., Chowdhury, U.K., Sengupta, M.K., Lodh, D., Chanda, C.R., Saha, K.C. & Mukherjee, S.C. (2002) Arsenic calamity in the Indian subcontinent — What lessons have been learned? *Talanta*, **58**, 3–22

Chakraborti, D., Mukherjee, S.C., Pati, S., Sengupta, M.K., Rahman, M.M., Chowdhury, U.K., Lodh, D., Chanda, C.R., Chakraborti, A.K. & Basu, G.K. (2003a) Arsenic groundwater contamination in Middle Ganga Plain, Bihar, India: A future danger. *Environ. Health Perspect.*, **111**, 1194–1201

Chakraborti, D., Rahman, M.M., Paul, K., Chowdhury, U.K. & Quamruzzaman, Q. (2003b) Groundwater arsenic contamination. In: *The Encyclopedia of Water Science*, New York, Marcel Dekker, pp. 324–329

Chakraborty, A.K. & Saha K.C. (1987) Arsenical dermatosis from tubewell water in West Bengal. *Indian J. med. Res.*, **85**, 326–334

Chatterjee, A., Das, D. & Chakraborti, D. (1993) A study of ground water contamination by arsenic in the residential area of Behala, Calcutta due to industrial pollution. *Environ. Pollut.*, **80**, 57–65

Chatterjee, A., Das, D., Mandal, B.K., Roy Chowdhury, T., Samanta, G. & Chakraborti, D. (1995) Arsenic in ground water in six districts of West Bengal, India: The biggest arsenic calamity in the world, Part I. Arsenic species in drinking water and urine of the affected people. *Analyst*, **120**, 645–650

Chatterjee, A., Shibata, Y., Yoshinaga, J. & Morita, M. (2000) Determination of arsenic compounds by high-performance liquid chromatography-ultrasonic nebulizer-high power nitrogen-microwave-induced plasma mass spectrometry: An accepted coupling. *Anal. Chem.*, **72**, 4402–4412

Chávez, A., Pérez Hidalgo, C., Tovar, E. & Garmilla, M. (1964) [Studies in a community with chronic endemic arsenicism.] *Salud públic Méx.*, **6**, 435–441 (in Spanish)

Chen, C.-J. & Chiou, H.-Y. (2001) Chen and Chiou Respond to 'Arsenic and cancer of the urinary tract' by Cantor. *Am. J. Epidemiol.*, **153**, 422–423

Chen, H.-W. & Edwards, M. (1997) Arsenic occurrence and speciation in United States drinking waters: Implications for water utilities. In: *Proceedings of the Water Quality Technology Conference*, Denver, CO, American Water Works Association, pp. 468–484

Chen, C.-J. & Wang, C.-J. (1990) Ecological correlation between arsenic level in well water and age-adjusted mortality from malignant neoplasms. *Cancer Res.*, **50**, 5470–5474

Chen, K.P., Wu, H.Y. & Wu, T.C. (1962) Epidemiologic studies on blackfoot disease in Taiwan. 3. Physiochemical characteristics of drinking water in endemic blackfoot disease areas. *Mem. Coll. Med. natl Taiwan Univ.*, **8**, 115–129

Chen, C.-J., Chuang, Y.-C., Lin, T.-M. & Wu, H.-Y. (1985) Malignant neoplasms among residents of a blackfoot disease-endemic area in Taiwan: High-arsenic artesian well water and cancers. *Cancer Res.*, **45**, 5895–5899

Chen, B., Burt, C.T., Goering, P.L., Fowler, B.A. & London, R.E. (1986) In vivo ^{31}P nuclear magnetic resonance studies of arsenite induced changes in hepatic phosphate levels. *Biochem. biophys. Res. Commun.*, **139**, 228–234

Chen, C.-J., Chuang, Y.-C., You, S.-L., Lin, T.-M. & Wu, H.-Y. (1986) A retrospective study on malignant neoplasms of bladder, lung and liver in blackfoot disease endemic area in Taiwan. *Br. J. Cancer*, **53**, 399–405

Chen, C.-J., Kuo, T.-L. & Wu, M.-M. (1988a) Arsenic and cancers. *Lancet*, **i**, 414–415

Chen, C.-J., Wu, M.-M., Lee, S.-S., Wang, J.-D., Cheng, S.-H. & Wu, H.-Y. (1988b) Atherogenicity and carcinogenicity of high-arsenic artesian well water. Multiple risk factors and related malignant neoplasms of blackfoot disease. *Arteriosclerosis*, **8**, 452–460

Chen, C.-J., Lin, L.-J., Hsueh, Y.-M., Chiou, H.-Y., Liaw, K.-F., Horng, S.-F., Chiang, M.-H., Tseng, C.-H. & Tai, T.-Y. (1994) Ischemic heart disease induced by ingested inorganic arsenic. In: Chappell, W.R., Abernathy, C.O. & Cothern, C.R., eds, *Arsenic Exposure and Health*, Northwood, IL, Science and Technology Letters, pp. 83–90

Chen, C.-J., Hsueh, Y.-M., Lai, M.-S., Shyu, M.-P., Chen, S.-Y., Wu, M.-M., Kuo, T.-L. & Tai, T.-Y. (1995) Increased prevalence of hypertension and long-term arsenic exposure. *Hypertension*, **25**, 53–60

Chen, C.-J., Chiou, H.-Y., Chiang, M.-H., Lin, L.-J. & Tai, T.-Y. (1996) Dose–response relationship between ischemic heart disease mortality and long-term arsenic exposure. *Arterioscler. Thromb. vasc. Biol.*, **16**, 504–510

Chen, C.-J., Chiou, H.-Y., Huang, W.-I., Chen, S.-Y., Hsueh, Y.-M., Tseng, C.-H., Lin, L.-J., Shyu, M.-P. & Lai, M.-S. (1997) Systemic non-carcinogenic effects and developmental toxicity of inorganic arsenic. In: Abernathy, C.O., Calderon, R.L. & Chappell, W.R., eds, *Arsenic Exposure and Health Effects*, London, Chapman & Hall, pp. 124–134

Chen, Y.-C., Lin-Shiau, S.-Y. & Lin, J.-K. (1998) Involvement of reactive oxygen species and caspase 3 activation in arsenite-induced apoptosis. *J. cell. Physiol.*, **177**, 324–333

Chen, K.-L.B., Amarasiriwardena, C.J. & Christiani, D.C. (1999) Determination of total arsenic concentrations in nails by inductively coupled plasma mass spectrometry. *Biol. trace Element Res.*, **67**, 109–125

Chen, Y., Megosh, L.C., Gilmour, S.K., Sawicki, J.A. & O'Brien, T.G. (2000) K6/ODC transgenic mice as a sensitive model for carcinogen identification. *Toxicol. Lett.*, **116**, 27–35

Chen, H., Liu, J., Zhao, C.Q., Diwan, B.A., Merrick, B.A. & Waalkes, M.P. (2001) Association of c-*myc* overexpression and hyperproliferation with arsenite-induced malignant transformation. *Toxicol. appl. Pharmacol.*, **175**, 260–268

Chen, G.-C., Guan, L.-S., Hu, W.-L. & Wang, Z.-Y. (2002) Functional repression of estrogen receptor α by arsenic trioxide in human breast cancer cells. *Anticancer Res.*, **22**, 633–638

Chen, Y.-C., Amarasiriwardena, C.J., Hsueh, Y.-M. & Christiani, D.C. (2002) Stability of arsenic species and insoluble arsenic in human urine. *Cancer Epidemiol. Biomarkers Prev.*, **11**, 1427–1433

Chiang, J., Hermosilla, A.P., Rojas, H. & Henríquez, C. (1990) [Determination of arsenic in individuals exposed to high pollution levels.] *Rev. Chil. Nutr.*, **18**, 39–50 (in Spanish)

Chiang, H.S., Guo, H.R., Hong, C.L., Lin, S.M. & Lee, E.F. (1993) The incidence of bladder cancer in the black foot disease endemic area in Taiwan. *Br. J. Urol.*, **71**, 274–278

Chilean Institute of National Standards (1984) [*Standards for Potable Water*], Santiago (in Spanish)

Chiou, H.-Y., Hsueh, Y.-M., Liaw, K.-F., Horng, S.-F., Chiang, M.-H., Pu, Y.-S., Lin, J.S.-N., Huang, C.-H. & Chen, C.-J. (1995) Incidence of internal cancers and ingested inorganic arsenic: A seven-year follow-up study in Taiwan. *Cancer Res.*, **55**, 1296–1300

Chiou, H.-Y., Hsueh, Y.-M., Hsieh, L.-L., Hsu, L.-I., Hsu, Y.-H., Hsieh, F.-I., Wei, M.-L., Chen, H.-C., Yang, H.-T., Leu, L.-C., Chu, T.-H., Chen-Wu, C., Yang, M.-H. & Chen, C.-J. (1997b) Arsenic methylation capacity, body retention, and null genotypes of glutathione *S*-transferase M1 and T1 among current arsenic-exposed residents in Taiwan. *Mutat. Res.*, **386**, 197–207

Chiou, H.-Y., Huang, W.-I., Su, C.-L., Chang, S.-F., Hsu, Y.-H. & Chen, C.-J. (1997a) Dose–response relationship between prevalence of cerebrovascular disease and ingested inorganic arsenic. *Stroke*, **28**, 1717–1723

Chiou, H.Y., Chiou, S.T., Hsu, Y.H., Chou, Y.L., Tseng, C.H., Wei, M.L. & Chen, C.J. (2001) Incidence of transitional cell carcinoma and arsenic in drinking water: A follow-up study of 8102 residents in an arseniasis-endemic area in northeastern Taiwan. *Am. J. Epidemiol.*, **153**, 411–418

Choprapawon, C. & Porapakkham, Y. (2001) Occurrence of cancer in arsenic contaminated area, Ronpibool District, Nakorn Srithmmarat Province, Thailand. In: Chappell, W.R., Abernathy, C.O. & Calderon, R.L., eds, *Arsenic Exposure and Health Effects IV*, New York, Elsevier Science, pp. 201–206

Choprapawon, C. & Rodcline, A. (1997) Chronic arsenic poisoning in Ronpibool Nakhon Sri Thammarat, the Southern Province of Thailand. In: Abernathy, C.O., Calderon, R.L. & Chappell, W.R., eds, *Arsenic Exposure and Health Effects*, London, Chapman & Hall, pp. 69–77

Chouchane, S. & Snow, E.T. (2001) In vitro effect of arsenical compounds on glutathione-related enzymes. *Chem. Res. Toxicol.*, **14** , 517–522

Chowdhury, T.R., Mandal, B.K., Samanta, G., Basu, G.K., Chowdhury, P.P., Chanda, C.R., Karan, N.K., Lodh, D., Dhar, R.K., Das, D., Saha, K.C. & Chakraborti, D. (1997) Arsenic in groundwater in six districts of West Bengal, India: The biggest arsenic calamity in the world: The status report up to August, 1995. In: Abernathy, C.O., Calderon, R.L. & Chappell, W.R., eds, *Arsenic Exposure and Health Effects*, London, Chapman & Hall, pp. 93–111

Chowdhury, U.K., Biswas, B.K., Roy Chowdhury, T., Samanta, G., Mandal, B.K., Basu, G.C., Chanda, C.R., Lodh, D., Saha, K.C., Mukherjee, S.K., Roy, S., Kabir, S., Quamruzzaman, Q. & Chakraborti, D. (2000a) Groundwater arsenic contamination in Bangladesh and West Bengal, India. *Environ. Health Perspect.*, **108**, 393–397

Chowdhury, U.K., Biswas, B.K., Roy Chowdhury, T., Mandal, B.K., Samanta, G., Basu, G.K., Chanda, C.R., Lodh, D., Saha, K.C., Chakraborti, D., Mukherjee, S.C., Roy, S., Kabir, S. & Quamruzzaman, Q. (2000b) Arsenic groundwater contamination and sufferings of people in West Bengal-India and Bangladesh. In: Roussel, A.M., Anderson, R.A. & Favrier, A.E., eds, *Trace Elements in Man and Animals 10*, New York, Kluwer Academic/Plenum, pp. 645–650

Chowdhury, U.K., Rahman, M.M., Mondal, B.K., Paul, K., Lodh, D., Biswas, B.K., Basu, G.K., Chanda, C.R., Saha, K.C., Mukherjee, S.C., Roy, S., Das, R., Kaies, I., Barua, A.K., Palit, S.K., Quamruzzaman, Q. & Chakraborti, D. (2001) Groundwater arsenic contamination and human suffering in West Bengal, India and Bangladesh. *Environ. Sci.*, **8**, 393–415

Chowdhury, U.K., Rahman, M.M., Sengupta, M.K., Lodh, D., Chanda, R.C., Roy, S., Quamruzzaman, Q., Tokunaga, H., Ando, M. & Chakraborti, D. (2003) Pattern of excretion of arsenic compounds [arsenite, arsenate, MMA (V), DMA(V)] in urine of children compared to adults from an arsenic exposed area in Bangladesh. *Environ. Sci. Health*, **A38**, 87–113

Chung, J.S., Kalman, D.A., Moore, L.E., Kosnett, M.J., Arroyo, A.P., Beeris, M., Guha Mazumder, D.N., Hernandez, A.L. & Smith, A.H. (2002) Family correlations of arsenic methylation patterns in children and parents exposed to high concentrations of arsenic in drinking water. *Environ. Health Perspect.*, **110**, 729–733

Cohen, S.M., Arnold, L.L., Uzvolgyi, E., Cano, M., St John, M., Yamamoto, S., Lu, X. & Le, X.C. (2002) Possible role of dimethylarsinous acid in dimethylarsinic acid-induced urothelial toxicity and regeneration in the rat. *Chem. Res. Toxicol.*, **15**, 1150–1157

Colbourn, P., Alloway, B.J. & Thornton, I. (1975) Arsenic and heavy metals in soils associated with regional geochemical anomalies in South-West England. *Sci. total Environ.*, **4**, 359–363

Committee Constituted by Government of West Bengal (1994) *Report on Arsenic Pollution in Groundwater in West Bengal* (PHE-1/716/3D-1/88), Bengal, PHE Department

CONAMA (2000) [*Technical Information Sheet: Analysis of Human Exposure to Arsenic in Large Cities*] (Study No. 21–0022–002), Santiago, Ed. P. Pino Z. (in Spanish)

Concha, G., Vogler, G., Nermell, B. & Vahter, M. (1998a) Low-level arsenic excretion in breast milk of native Andean women exposed to high levels of arsenic in the drinking water. *Int. Arch. occup. environ. Health*, **71**, 42–46

Concha, G., Nermell, B. & Vahter, M.V. (1998b) Metabolism of inorganic arsenic in children with chronic high arsenic exposure in northern Argentina. *Environ. Health Perspect.*, **106**, 355–359

Concha, G., Vogler, G., Lezcano, D., Nermell, B. & Vahter, M. (1998c) Exposure to inorganic arsenic metobolites during early human development. *Toxicol. Sci.*, **44**, 185–190

Concha, G., Vogler, G., Nermell, B. & Vahter, M. (2002) Intra-individual variation in the metabolism of inorganic arsenic. *Int. Arch. occup. environ. Health*, **75**, 576–580

COSUDE (2000) [*Determination of National Baseline Content of Arsenic in Respirable Particulate Material*] (Proyecto CONAMA No. 22-0023-09), Santiago (in Spanish)

Cowlishaw, J.L., Pollard, E.J., Cowen, A.E. & Powell, L.W. (1979) Liver disease associated with chronic arsenic ingestion. *Aust. N.Z. J. Med.*, **9**, 310–313

Crecelius, E.A. (1977) Changes in the chemical speciation of arsenic following ingestion by man. *Environ. Health Perspect.*, **19**, 147–150

Crecelius, E.A. (1978) Modification of the arsenic speciation technique using hydride generation. *Anal. Chem.*, **50**, 826–827

Cross, J.D., Dale, I.M., Leslie, A.C.D. & Smith, H. (1979) Industrial exposure to arsenic. *J. radioanal. Chem.*, **48**, 197–208

Cruz, A.C., Fomsgaard, I.S. & Lacayo, J. (1994) Lead, arsenic, cadmium and copper in Lake Asososca, Nicaragua. *Sci. total Environ.*, **155**, 229–236

Csanaky, I. & Gregus, Z. (2001) Effect of phosphate transporter and methylation inhibitor drugs on the disposition of arsenate and arsenite in rats. *Toxicol. Sci.*, **63**, 29–36

Csanaky, I. & Gregus, Z. (2002) Species variations in the biliary and urinary excretion of arsenate and arsenite and their metabolites. *Comp. Biochem. Physiol.*, **C131**, 355–365

Cullen, W.R. & Reimer, K.J. (1989) Arsenic speciation in the environment. *Chem. Rev.*, **89**, 713–764

Cuzick, J., Evans, S., Gillman, M. & Price Evans, D.A. (1982) Medicinal arsenic and internal malignancies. *Br. J. Cancer*, **45**, 904–911

Cuzick, J., Sasieni, P. & Evans, S. (1992) Ingested arsenic, keratoses, and bladder cancer. *Am. J. Epidemiol.*, **136**, 417–421

Dang, H.S., Jaiswal, D.D. & Somasundaram, S. (1983) Distribution of arsenic in human tissues and milk. *Sci. total Environ.*, **29**, 171–175

Das, D., Chatterjee, A., Samanta, G., Mandal, B., Roy Chowdhury, T., Samanta, G., Chowdhury, P.P., Chanda, C., Basu, G., Lodh, D., Nandi, S., Chakraborty, T., Mandal, S., Bhattacharya, S.M. & Chakraborti, D. (1994) Arsenic contamination in groundwater in six districts of West Bengal, India: The biggest arsenic calamity in the world. *Analyst*, **119**, 168N–170N

Das, D., Chatterjee, A., Mandal, B.K., Samanta, G., Chakraborti, D. & Chanda, B. (1995) Arsenic in ground water in six districts of West Bengal, India: The biggest arsenic calamity in the world. Part 2. Arsenic concentration in drinking water, hair, nails, urine, skin-scale and liver tissue (biopsy) of the affected people. *Analyst*, **120**, 917–924

Das, D., Samanta, G., Mandal, B.K., Roy Chowdhury, T., Chanda, C.R., Chowdhury, P.P., Basu, G.K. & Chakraborti, D. (1996) Arsenic in groundwater in six districts of West Bengal, India. *Environ. Geochem. Health*, **18**, 5–15

Datta, D.V. (1976) Arsenic and non-cirrhotic portal hypertension (Letter to the Editor). *Lancet*, **i**, 433

Datta, D.V. & Kaul, M.K. (1976) Arsenic content of tubewell water in villages in northern India. A concept of arsenicosis. *J. Assoc. Phys. India*, **24**, 599–604

Datta, D.V., Mitra, S.K., Chhuttani, P.N. & Chakravarti, R.N. (1979) Chronic oral arsenic intoxication as a possible aetiological factor in idiopathic portal hypertension (non-cirrhotic portal fibrosis) in India. *Gut*, **20**, 378–384

Davis, M.K., Reich, K.D. & Tikkanen, M.W. (1994) Nationwide and California arsenic occurrence studies. In: Chappell, W.R., Abernathy, C.O. & Cothern, C.R., eds., *Arsenic Exposure and Health Effects*, Northwood, IL, Science and Technology Letters, pp. 31–40

De Kimpe, J., Cornelis, R., Mees, L. & Vanholder, R. (1996) Basal metabolism of intraperitoneally injected carrier-free ^{74}As-labeled arsenate in rabbits. *Fundam. appl. Toxicol.*, **34**, 240–248

De Kimpe J., Cornelis R. & Vanholder, R. (1999) In vitro methylation of arsenite by rabbit liver cytosol: Effect of metal ions, metal chelating agents, methyltransferase inhibitors and uremic toxins. *Drug chem. Toxicol.*, **22**, 613–628

Deknudt, G., Léonard, A., Arany, J., Jenar-Du Buisson, G. & Delavignette, E. (1986) *In vivo* studies in male mice on the mutagenic effects of inorganic arsenic. *Mutagenesis*, **1**, 33–34

Del Razo, L.M., Arellano, M.A. & Cebrián, M.E. (1990) The oxidation states of arsenic in well-water from a chronic arsenicism area of northern Mexico. *Environ. Pollut.*, **64**, 143–153

Del Razo, L.M., Garcia-Vargas, G.G., Vargas, H., Albores, A., Gonsebatt, M.E., Montero, R., Ostrosky-Wegman, P., Kelsh, M. & Cebrián, M.E. (1997) Altered profile of urinary arsenic metabolites in adults with chronic arsenicism. A pilot study. *Arch. Toxicol.*, **71**, 211–217

Del Razo, L.M., Aguilar, C., Sierra-Santoyo, A. & Cebrián, M.E. (1999) Interference in the quantitation of methylated arsenic species in human urine. *J. anal. Toxicol.*, **23**, 103–107

Del Razo, L.M., Styblo, M., Cullen, W.R. & Thomas, D.J. (2001a) Determination of trivalent methylated arsenicals in biological matrices. *Toxicol. appl. Pharmacol.*, **174**, 282–293

Del Razo, L.M., Quintanilla-Vega, B., Brambila-Colombres, E., Calderón-Aranda, E.S., Manno, M. & Albores, A. (2001b) Stress proteins induced by arsenic. *Toxicol. appl. Pharmacol.*, **177**, 132–148

Delnomdedieu, M., Basti, M.M., Otvos, J.D. & Thomas, D.J. (1993) Transfer of arsenite from glutathione to dithiols: A model of interaction. *Chem. Res. Toxicol.*, **6**, 598–602

Delnomdedieu, M., Basti, M.M., Styblo, M., Otvos, J.D. & Thomas, D.J. (1994a) Complexation of arsenic species in rabbit erythrocytes. *Chem. Res. Toxicol.*, **7**, 621–627

Delnomdedieu, M., Basti, M.M., Otvos, J.D. & Thomas, D.J. (1994b) Reduction and binding of arsenate and dimethylarsinate by glutathione: A magnetic resonance study. *Chem.-biol. Interact.*, **90**, 139–155

Devasagayam, T.P.A., Steenken, S., Obendorf, M.S.W., Schulz, W.A. & Sies, H. (1991) Formation of 8-hydroxy(deoxy)guanosine and generation of strand breaks at guanine residues in DNA by singlet oxygen. *Biochemistry*, **30**, 6283–6289

Dhaka Community Hospital Trust and School of Environmental Studies (1998) *International Conference on Arsenic Pollution of Groundwater in Bangladesh: Causes, Effects and Remedies*, Dhaka

Dhar, R.K., Biswas, B.K., Samanta, G., Mandal, B.K., Chakraborti, D., Roy, S., Jafar, A., Islam, A., Ara, G., Kabir, S., Khan, A.W., Ahmed, S.A. & Hadi, S.A. (1997) Groundwater arsenic calamity in Bangladesh. *Current Sci.*, **73**, 48–59

Diáz-Barriga, F., Santos, M.A., Mejía, J.J., Batres, L., Yáñez, L., Carrizales, L., Vera, E., Del Razo, L.M. & Cebrián, M.E. (1993) Arsenic and cadmium exposure in children living near a smelter complex in San Luis Potosí, Mexico. *Environ. Res.*, **62**, 242–250

Domingo, J.L., Bosque, M.A. & Piera, V. (1991) *meso*-2,3-Dimercaptosuccinic acid and prevention of arsenite embryotoxocity and teratogenicity in the mouse. *Fundam. appl. Toxicol.*, **17**, 314–320

Dulout, F.N., Grillo, C.A., Seoane, A.I., Maderna, C.R., Nilsson, R., Vahter, M., Darroudi, F. & Natarajan, A.T. (1996) Chromosomal aberrations in peripheral blood lymphocytes from native Andean women and children from northwestern Argentina exposed to arsenic in drinking water. *Mutat. Res.*, **370**, 151–158

Durham, N.F. & Kosmus, W. (2003) Analytical considerations of arsenic contamination in water. *Asian environ. Technol.*, **7**(2), April/May

Eastmond, D.A. & Tucker, J.D. (1989) Identification of aneuploidy-inducing agents using cytokinesis-blocked human lymphocytes and an antikinetochore antibody. *Environ. mol. Mutag.*, **13**, 34–43

Eguchi, N., Kuroda, K. & Endo, G. (1997) Metabolites of arsenic induced tetraploids and mitotic arrest in cultured cells. *Arch. environ. Contam. Toxicol.*, **32**, 141–145

Engel, R.R. & Smith, A.H. (1994) Arsenic in drinking water and mortality from vascular disease: An ecologic analysis in 30 counties in the United States. *Arch. environ. Health*, **49**, 418–427

Engel, R.R., Hopenhayn-Rich, C., Receveur, O. & Smith, A.H. (1994) Vascular effects of chronic arsenic exposure: A review. *Epidemiol. Rev.*, **16**, 184–209

Ercal, N., Gurer-Orhan, H. & Aykin-Burns, N. (2001) Toxic metals and oxidative stress. Part I: Mechanisms involved in metal-induced oxidative damage. *Current Top. med. Chem.*, **1**, 529–539

Erickson, B.E. (2003) Field kits fail to provide accurate measure of arsenic in groundwater. *Environ. Sci. Technol.*, **37**, 35A–38A

Estrov, Z., Manna, S.K., Harris, D., Van, Q., Estey, E.H., Kantarjian, H.M., Talpaz, M. & Aggarwal, B.B. (1999) Phenylarsine oxide blocks interleukin-1β-induced activation of the nuclear transcription factor NF-κB, inhibits proliferation, and induces apoptosis of acute myelogenous leukemia cells. *Blood*, **94**, 2844–2853

European Commission (1998) *European Union Directive on the Quality of Water Intended for Human Consumption* (Council Directive 98/83/EC), Brussels

Farago, M.E., Thornton, I., Kavanagh, P., Elliott, P. & Leonardi, G.S. (1997) Health aspects of human exposure to high arsenic concentrations in soil in south-west England. In: Abernathy, C.O., Calderon, R.L. & Chappell, W.R., eds, *Arsenic Exposure and Health Effects*, New York, Chapman & Hall, pp. 210–226

Feldmann, J., Lai, V.W.-M., Cullen, W.R., Ma, M., Lu, X. & Le, X.C. (1999) Sample preparation and storage can change arsenic speciation in human urine. *Clin. Chem.*, **45**, 1988–1997

Ferreccio, C., González, C., Solari, J. & Noder, C. (1996) [Lung cancer in workers exposed to arsenic: A case–control study.] *Rev. méd. Chile*, **124**, 119–123 (in Spanish)

Ferreccio, C., González, C., Milosavjlevic, V., Marshall, G., Sancha, A.M. & Smith, A.H. (2000) Lung cancer and arsenic concentrations in drinking water in Chile. *Epidemiology*, **11**, 673–679

Fisher, D.L. (1982) Cultured rat embryo accumulation of DNA, RNA, and protein following maternal administration of sodium arsenate. *Environ. Res.*, **28**, 1–9

Flynn, H.C., McMahon, V., Diaz, G.C., Demergasso, C.S., Corbisier, P., Meharg, A.A. & Paton, G.I. (2002) Assessment of bioavailable arsenic and copper in soils and sediments from the Antofagasta region of northern Chile. *Sci. total Environ.*, **286**, 51–59

Foà, V., Colombi, A., Maroni, M., Buratti, M. & Calzaferri, G. (1984) The speciation of the chemical forms of arsenic in the biological monitoring of exposure to inorganic arsenic, *Sci. total Environ.*, **34**, 241–259

Focazio, M.J., Welch, A.H., Watkins, S.A., Helsel, D.R. & Horn, M.A. (2000) *A Restrospective Analysis on the Occurrence of Arsenic in Ground-water Resources of the United States and Limitations in Drinking-water-supply Characterizations* (Water-Resources Investigations Report 99–4279), Reston, VA, Office of Water Quality, US Geological Survey

Foust, R.D., Jr, Mohapatra, P., Compton, A.-M. & Reifel, J. (2000) Groundwater arsenic in the Verde Valley in Central Arizona. In: Bhattacharya, P. & Welch, A.H., eds., *Arsenic in Ground-water of Sedimentary Aquifers, 31st International Geological Congress, Rio de Janeiro, Brazil*, pp. 43–46

Fowler, B.A., Woods, J.S. & Schiller, C.M. (1977) Ultrastructural and biochemical effects of pro-longed oral arsenic exposure on liver mitochondria of rats. *Environ. Health Perspect.*, **19**, 197–204

Fowler, B.A., Woods, J.S. & Schiller, C.M. (1979) Studies of hepatic mitochondrial structure and function. Morphometric and biochemical evaluation of *in vivo* perturbation by arsenate. *Lab. Invest.*, **41**, 313–320

Francesconi, K.A. & Kuehnelt, D. (2002) Arsenic compounds in the environment. In: Frankenberger, W.T., Jr, ed., *Environmental Chemistry of Arsenic*, New York, Marcel Dekker, pp. 51–94

Franklin, M., Bean, W.B. & Hardin, R.C. (1950) Fowler's solution as an etiologic agent in cirrhosis. *Am. J. med. Sci.*, **219**, 589–596

Freeman, G.B., Schoof, R.A., Ruby, M.V., Davis, A.O., Dill, J.A., Liao, S.C., Lapin, C.A. & Bergstrom, P.D. (1995) Bioavailability of arsenic in soil and house dust impacted by smelter activities following oral administration in cynomolgus monkeys. *Fundam. appl. Toxicol.*, **28**, 215–222

Frey, M.M. & Edwards, M.A. (1997) Surveying arsenic occurrence. *J. Am. Water Works Assoc.*, **89**, 105–117

Gallagher, P.A., Schwegel, C.A., Wei, X. & Creed, J.T. (2001) Speciation and preservation of inorganic arsenic in drinking water sources using EDTA with IC separation and ICP–MS detection. *J. environ. Monit.*, **3**, 371–376

Garai, R., Chakraborty, A.K., Dey, S.B. & Saha, K.C. (1984) Chronic arsenic poisoning from tube-well water. *J. Indian med. Assoc.*, **82**, 34–35

García-Vargas, G.G. & Hernández-Zavala, A. (1996) Urinary porphyrins and heme biosynthetic enzyme activities measured by HPLC in arsenic toxicity. *Biomed. Chromatogr.*, **10**, 278–284

García-Vargas, G.G., Del Razo, L.M., Cebrián, M.E., Albores, A., Ostrosky-Wegman, P., Montero, R., Gonsebatt, M.E., Lim, C.K. & De Matteis, F. (1994) Altered urinary porphyrin excretion in a human population chronically exposed to arsenic in Mexico. *Hum. exp. Toxicol.*, **13**, 839–847

Garland, M., Morris, J.S., Rosner, B.A., Stampfer, M.J., Spate, V.L., Baskett, C.J., Willett, W.C. & Hunter, D.J. (1993) Toenail trace element levels as biomarkers: Reproducibility over a 6-year period. *Cancer Epidemiol. Biomarkers Prev.*, **2**, 493–497. Erratum in: *Cancer Epidemiol. Biomarkers Prev.*, **3**, 523 (1994)

Garland, M., Morris, J.S., Colditz, G.A., Stampfer, M.J., Spate, V.L., Baskett, C.K., Rosner, B., Speizer, F.E., Willett, W.C. & Hunter, D.J. (1996) Toenail trace element levels and breast cancer: A prospective study. *Am. J. Epidemiol.*, **144**, 653–660

Gebel, T. (1998) Suppression of arsenic-induced chromosome mutagenicity by antimony. *Mutat. Res.*, **412**, 213–218

Gebel, T.W. (2001) Genotoxicity of arsenical compounds. *Int. J. Hyg. environ Health*, **203**, 249–262

Gebel, T., Christensen, S. & Dunkelberg, H. (1997) Comparative and environmental genotoxicity of antimony and arsenic. *Anticancer Res.*, **17**, 2603–2607

Geiszinger, A., Goessler, W. & Kosmus, W. (2002) Organoarsenic compounds in plants and soil on top of an ore vein. *Appl. organometal. Chem.*, **16**, 245–249

George, G.M., Frahm, L.J. & McDonnell, J.P. (1973) Dry ashing method for determination of total arsenic in animal tissues: Collaborative study. *J. Am. off. anal. Chem.*, **56**, 793–797

German, J. (1984) Embryonic stress hypothesis of teratogenesis. *Am. J. Med.*, **76**, 293–301

German, J., Louie, E. & Banerjee, D. (1986) The heat-shock response in vivo: Experimental induction during mammalian organogenesis. *Teratog. Carcinog. Mutag.*, **6**, 555–562

Germolec, D.R., Yoshida, T., Gaido, K., Wilmer, J.L., Simeonova, P.P., Kayama, F., Burleson, F., Dong, W., Lange, R.W. & Luster, M.I. (1996) Arsenic induces overexpression of growth factors in human keratinocytes. *Toxicol. appl. Pharmacol.*, **141**, 308–318

Germolec, D.R., Spalding, J., Boorman, G.A., Wilmer, J.L., Yoshida, T., Simeonova, P.P., Bruccoleri, A., Kayama, F., Gaido, K., Tennant, R., Burleson, F., Dong, W., Lang, R.W. & Luster, M.I. (1997) Arsenic can mediate skin neoplasia by chronic stimulation of keratinocyte-derived growth factors. *Mutat. Res.*, **386**, 209–218

Germolec, D.R., Spalding, J., Yu, H.-S., Chen, G.S., Simeonova, P.P., Humble, M.C., Bruccoleri, A., Boorman, G.A., Foley, J.F., Yoshida, T. & Luster, M.I. (1998) Arsenic enhancement of skin neoplasia by chronic stimulation of growth factors. *Am. J. Pathol.*, **153**, 1775–1785

Geubel, A.P., Mairlot, M.C., Buchet, J.P., Dive, C. & Lauwerys, R. (1988) Abnormal methylation capacity in human liver cirrhosis. *Int. J. clin. pharmacol. Res.*, **8**, 117–122

Goessler, W. & Kuehnelt, D. (2002) Analytical methods for the determination of arsenic and arsenic compounds in the environment. In: Frankenberger, W.T., Jr, ed., *Environmental Chemistry of Arsenic*, New York, Marcel Dekker, pp. 27–50

Gómez-Arroyo, S., Armienta, M.A., Cortés-Eslava, J. & Villalobos-Pietrini, R. (1997) Sister chromatid exchanges in *Vicia faba* induced by arsenic-contaminated drinking water from Zimapan, Hidalgo, Mexico. *Mutat. Res.*, **394**, 1–7

Gong, Z., Lu, X., Cullen, W.R. & Le, X.C. (2001) Unstable trivalent arsenic metabolites, monomethylarsonous acid and dimethylarsinous acid. *J. anal. at. Spectrom.*, **16**, 1409–1413

Gonsebatt, M.E., Vega, L., Herrera, L.A., Montero, R., Rojas, E., Cebrián, M.E. & Ostrosky-Wegman, P. (1992) Inorganic arsenic effects on human lymphocyte stimulation and proliferation. *Mutat. Res.*, **283**, 91–95

Gonsebatt, M.E., Vega, L., Montero, R., Garcia-Vargas, G., Del Razo, L.M., Albores, A., Cebrian, M.E. & Ostrosky-Wegman, P. (1994) Lymphocyte replicating ability in individuals exposed to arsenic via drinking water. *Mutat. Res.*, **313**, 293–299

Gonsebatt, M.E., Vega, L., Salazar, A.M., Montero, R., Guzmán, P., Blas, J., Del Razo, L.M., García-Vargas, G., Albores, A., Cebrián, M.E., Kelsh, M. & Ostrosky-Wegman, P. (1997) Cytogenetic effects in human exposure to arsenic. *Mutat. Res.*, **386**, 219–228

Gonzalez Amarales, E. (1970) [Arsenicism in Antofagasta.] *Bol. Inst. bacteriol. Chile*, 30–38 (in Spanish)

Graham, J.H. & Helwig, E.B. (1959) Bowen's disease and its relationship to systemic cancer. *Am. med. Assoc. Arch. Dermatol.*, **80**, 133–159

Gregus, Z., Gyurasics, A. & Csanaky, I. (2000) Biliary and urinary excretion of inorganic arsenic: Monomethylarsonous acid as a major biliary metabolite in rats. *Toxicol. Sci.*, **56**, 18–25

Grobe, J.W. (1977) Expert-testimony and therapeutic findings and observations in wine-dressers of the Mosel-region with late sequelae of arsenic intoxication. *Berufsdermatosen*, **25**, 124–130

Guha Mazumder, D.N. (2001a) Arsenic and liver disease. *J. Indian med. Assoc.*, **99**, 311–320

Guha Mazumder, D.N. (2001b) Clinical aspects of chronic arsenic toxicity. *J. Assoc. Physicians India*, **49**, 650–655

Guha Mazumder, D.N., Chakraborty, A.K., Ghose, A., Gupta, J.D., Chakraborty, D.P., Dey, S.B. & Chattopadhyay, N. (1988) Chronic arsenic toxicity from drinking tubewell water in rural West Bengal. *Bull. World Health Org.*, **66**, 499–506

Guha Mazumder, D.N., Das Gupta, J., Chakraborty, A.K., Chatterjee, A., Das, D. & Chakraborti, D. (1992) Environmental pollution and chronic arsenicosis in south Calcutta. *Bull. World Health Org.*, **70**, 481–485

Guha Mazumder, D.N., Das Gupta, J., Santra, A., Pal, A., Ghose, A., Sarkar, S., Chattopadhaya, N. & Chakraborti, D. (1997) Non-cancer effects of chronic arsenicosis with special reference to liver damage. In: Abernathy, C.O., Calderon, R.L. & Chappell, W.R., eds, *Arsenic Exposure and Health Effects*, London, Chapman & Hall, pp. 112–123

Guha Mazumder, D.N., De, B.K., Santra, A., Das Gupta, J., Ghosh, A., Pal, A., Roy, B., Pal, S. & Saha, J. (1998a) Clinical manifestations of chronic arsenic toxicity: Its natural history and therapy: Experience of study in West Bengal, India. In: *International Conference on Arsenic Pollution of Ground Water in Bangladesh: Causes, Effects and Remedies*, Dhaka, Dhaka Community Hospital Trust, pp. 91–93

Guha Mazumder, D.N., Das Gupta, J., Santra, A., Pal, A., Ghose, A. & Sarkar, S. (1998b) Chronic arsenic toxicity in West Bengal — The worst calamity in the world. *J. Indian med. Assoc.*, **96**, 4–7

Guha Mazumder, D.N., Haque, R., Ghosh, N., De, B.K., Santra, A., Chakraborty, D. & Smith, A.H. (1998c) Arsenic levels in drinking water and the prevalence of skin lesions in West Bengal, India. *Int. J. Epidemiol.*, **27**, 871–877

Guha Mazumder, D.N., De, B.K., Santra, A., Dasgupta, J., Ghosh, N., Roy, B.K., Ghoshal, U.C., Saha, J., Chatterjee, A., Dutta, S, Haque, R., Smith, A.H., Chakraborty, D., Angle, C.R. & Centeno, J.A. (1999) Chronic arsenic toxicity: Epidemiology, natural history and treatment. In: Chappell, W.R., Abernathy, C.O. & Calderon, R.L., eds, *Arsenic Exposure and Health Effects*, Amsterdam, Elsevier Science, pp. 335–347

Guha Mazumder, D.N., Haque, R., Ghosh, N., De, B.K., Santra, A., Chakraborti, D. & Smith, A.H. (2000) Arsenic in drinking water and the prevalence of respiratory effects in West Bengal, India. *Int. J. Epidemiol.*, **29**, 1047–1052

Guha Mazumder, D.N., De, B.K., Santra, A., Ghosh, N., Das, S., Lahiri, S. & Das, T. (2001a) Randomized placebo-controlled trial of 2,3-dimercapto-1-propanesulfonate (DMPS) in therapy of chronic arsenicosis due to drinking arsenic-contaminated water. *Clin. Toxicol.*, **39**, 665–674

Guha Mazumder, D.N., Ghosh, N., De, B.K., Santra, A., Das, S., Lahiri, S., Haque, R., Smith, A.H. & Chakraborti, D. (2001b) Epidemiological study on various non-carcinomatous manifestations of chronic arsenic toxicity in a district of West Bengal. In: Chappell, W.R., Abernathy, C.O. & Calderon, R.L., eds, *Arsenic Exposure and Health Effects IV*, Amsterdam, Elsevier, pp. 153–164

Guo, H.-R., Chiang, H.-S., Hu, H., Lipsitz, S.R. & Monson, R.R. (1997) Arsenic in drinking water and incidence of urinary cancers. *Epidemiology*, **8**, 545–550

Guo, H.-R., Lipsitz, S.R., Hu, H. & Monson, R.R. (1998) Using ecological data to estimate a regression model for individual data: The association between arsenic in drinking water and incidence of skin cancer. *Environ. Res.*, **A79**, 82–93

Guo, H.-R., Yu, H.-S., Hu, H. & Monson, R.R. (2001) Arsenic in drinking water and skin cancers: Cell-type specificity (Taiwan, R.O.C). *Cancer Causes Control*, **12**, 909–916

Guo, X., Fujino, Y., Kaneko, S., Wu, K., Xia, Y. & Yoshimura, T. (2001) Arsenic contamination of groundwater and prevalence of arsenical dermatosis in the Hetao plain area, Inner Mongolia, China. *Mol. cell. Biochem.*, **222**, 137–140

Gurzau, E.S. & Gurzau, A.E. (2001) Arsenic in drinking water from groundwater in Transylvania, Romania: An overview. In: Chappell, W.R., Abernathy, C.O. & Calderon, R.L., eds, *Arsenic Exposure and Health Effects IV*, New York, Elsevier Science, pp. 181–184

Hale, M. (1981) Pathfinder applications of arsenic, antimony and bismuth in geochemical exploration. *J. geochem. Explor.*, **15**, 307–323

Hamadeh, H.K., Vargas, M., Lee, E. & Menzel, D.B. (1999) Arsenic disrupts cellular levels of p53 and mdm2: A potential mechanism of carcinogenesis. *Biochem. biophys. Res. Commun.*, **263**, 446–449

Hantson, P., Verellen-Dumoulin, C., Libouton, J.M., Leonard, A., Leonard, E.D. & Mahieu, P. (1996) Sister chromatid exchanges in human peripheral blood lymphocytes after ingestion of high doses of arsenicals. *Int. Arch. occup. environ. Health*, **68**, 342–344

Haque, R., Guha Mazumder, D.N., Samanta, S., Ghosh, N., Kalman, D., Smith, M.M., Mitra, S., Santra, A., Lahiri, S., Das, S., De, B.K. & Smith, A.H. (2003) Arsenic in drinking water and skin lesions: Dose–response data from West Bengal, India. *Epidemiology*, **14**, 174–182

Harrington, J.M., Middaugh, J.P., Morse, D.L. & Housworth, J. (1978) A survey of a population exposed to high concentrations of arsenic in well water in Fairbanks, Alaska. *Am. J. Epidemiol.*, **108**, 377–385

Harrington-Brock, K., Cabrera, M., Collard, D.D., Doerr, C.L., McConnell, R., Moore, M.M., Sandoval, H. & Fuscoe, J.C. (1999) Effects of arsenic exposure on the frequency of *HPRT*-mutant lymphocytes in a population of copper roasters in Antofagasta, Chile: A pilot study. *Mutat. Res.*, **431**, 247–257

Harrison, M.T. & McCoy, K.L. (2001) Immunosuppression by arsenic: A comparison of cathepsin L inhibition and apoptosis. *Int. Immunopharmacol.*, **1**, 647–656

Hartwig, A., Gröblinghoff, U.D., Beyersmann, D., Natarajan, A.T., Filon, R. & Mullenders, L.H.F. (1997) Interaction of arsenic(III) with nucleotide excision repair in UV-irradiated human fibroblasts. *Carcinogenesis*, **18**, 399–405

Hartwig, A., Pelzer, A., Asmuss, M. & Bürkle, A. (2003) Very low concentrations of arsenite suppress poly(ADP-ribosyl)ation in mammalian cells. *Int. J. Cancer*, **104**, 1–6

Hasegawa, H., Sohrin,Y., Matsui, M., Hojo, M. & Kawashima, M. (1994) Speciation of arsenic in natural waters by solvent extraction and hydride generation atomic absorption spectrometry. *Anal. Chem.*, **66**, 3247–3252

Hasegawa, H., Matsui, M., Okamura, S., Hojo, M., Iwasaki, N. & Sohrin, Y. (1999) Arsenic speciation including 'hidden' arsenic in natural waters. *Appl. organometal. Chem.*, **13**, 113–119

Hayashi, H., Kanisawa, M., Yamanaka, K., Ito, T., Udaka, N., Ohji, H., Okudela, K., Okada, S. & Kitamura, H. (1998) Dimethylarsinic acid, a main metabolite of inorganic arsenics, has tumorigenicity and progression effects in the pulmonary tumors of A/J mice. *Cancer Lett.*, **125**, 83–88

Health Canada (1992) *Guidelines for Canadian Drinking Water Quality — Supporting Documentation for Arsenic*, Ottawa

Health Canada (2003) *Summary of Guidelines for Canadian Drinking Water Quality*, Ottawa

Healy, S.M., Casarez, E.A., Ayala-Fierro, F. & Aposhian, H.V. (1998) Enzymatic methylation of arsenic compounds. V. Arsenite methyltransferase activity in tissues of mice. *Toxicol. appl. Pharmacol.*, **148**, 65–70

Healy, S.M., Wildfang, E., Zakharyan, R.A. & Aposhian, H.V. (1999) Diversity of inorganic arsenite biotransformation. *Biol. trace Elem. Res.*, **68**, 249–266

Hei, T.K., Liu, S.X. & Waldren, C. (1998) Mutagenicity of arsenic in mammalian cells: Role of reactive oxygen species. *Proc. natl Acad. Sci. USA*, **95**, 8103–8107

Hernández-Zavala, A., Del Razo, L.M., García-Vargas, G.G., Aguilar, C., Borja, V.H., Albores, A. & Cebrián, M.E. (1999) Altered activity of heme biosynthesis pathway enzymes in individuals chronically exposed to arsenic in Mexico. *Arch. Toxicol.*, **73**, 90–95

Hickey, C.W., Roper, D.S. & Buckland, S.J. (1995) Metal concentrations of resident and transplanted freshwater mussels *Hyridella menziesi* (Unionacea: Hyriidae) and sediments in the Waikato River, New Zealand. *Sci. total Environ.*, **175**, 163–177

Hindmarsh, J.T. (2002) Caveats in hair analysis in chronic arsenic poisoning. *Clin. Biochem.*, **35**, 1–11

Hindmarsh, J.T., McLetchie, O.R., Heffernan, L.P.M., Hayne, O.A., Ellenberger, H.A., McCurdy, R.F. & Thiebaux, H.J. (1977) Electromyographic abnormalities in chronic environmental arsenicalism. *J. anal. Toxicol.*, **1**, 270–276

Hinwood, A., Bannister, R., Shugg, A. & Sim, M. (1998) Environmental arsenic in rural Victoria: An update. *Water*, **July/August**, 34–36

Hinwood, A.L., Jolley, D.J. & Sim, M.R. (1999) Cancer incidence and high environmental arsenic concentrations in rural populations: Results of an ecological study. *Int. J. environ. Health Res.*, **9**, 131–141

Hirata, M., Tanaka, A., Hisanaga, A. & Ishinishi, N. (1990) Effects of glutathione depletion on the acute nephrotoxic potential of arsenite and on arsenic metabolism in hamsters. *Toxicol. appl. Pharmacol.*, **106**, 469–481

Honda, K.-I., Hatayama, T., Takahashi, K.-I. & Yukioka, M. (1992) Heat shock proteins in human and mouse embryonic cells after exposure to heat shock or teratogenic agents. *Teratog. Carcinog. Mutag.*, **11**, 235–244

Hood, R.D. & Bishop, S.L. (1972) Teratogenic effects of sodium arsenate in mice. *Arch. environ. Health*, **24**, 62–65

Hood, R.D. & Harrison, W.P. (1982) Effects of prenatal arsenite exposure in the hamster. *Bull. environ. Contam. Toxicol.*, **29**, 671–678

Hood, R.D. & Vedel-Macrander, G.C. (1984) Evaluation of the effect of BAL (2,3-dimercaptopropanol) on arsenite-induced teratogenesis in mice. *Toxicol. appl. Pharmacol.*, **73**, 1–7

Hood, R.D., Harrison, W.P. & Vedel, G.C. (1982) Evaluation of arsenic metabolites for prenatal effects in the hamster. *Bull. environ. Contam. Toxicol.*, **29**, 679–687

Hopenhayn-Rich, C., Smith, A.H. & Goeden, H.M. (1993) Human studies do not support the methylation threshold hypothesis for the toxicity of inorganic arsenic. *Environ. Res.*, **60**, 161–177

Hopenhayn-Rich, C., Biggs, M.L., Fuchs, A., Bergoglio, R., Tello, E.E., Nicolli, H. & Smith, A.H. (1996a) Bladder cancer mortality associated with arsenic in drinking water in Argentina. *Epidemiology*, **7**, 117–124

Hopenhayn-Rich, C., Biggs, M.L., Kalman, D.A., Moore, L.E. & Smith, A.H. (1996b) Arsenic methylation patterns before and after changing from high to lower concentrations of arsenic in drinking water. *Environ. Health Perspect.*, **104**, 1200–1207

Hopenhayn-Rich, C., Biggs, M.L., Smith, A.H., Kalman, D.A. & Moore, L.E. (1996c) Methylation study of a population environmentally exposed to arsenic in drinking water. *Environ. Health Perspect.*, **104**, 620–628

Hopenhayn-Rich, C., Biggs, M.L. & Smith, A.H. (1998) Lung and kidney cancer mortality associated with arsenic in drinking water in Córdoba, Argentina. *Int. J. Epidemiol.*, **27**, 561–569

Hopenhayn-Rich, C., Browning, S., Hertz-Picciotto, I., Ferreccio, C., Peralta, C. & Gibb, H. (2000) Chronic arsenic exposure and risk of infant mortality in two areas of Chile. *Environ. Health Perspect.*, **108**, 667–673

Hotta, N. (1989) [Clinical aspects of chronic arsenic poisoning due to environmental and occupational pollution in and around a small refining spot.] *Nippon Taishitsugaku Zasshi* [Jpn. J. Const. Med.], **53**, 49–70 (in Japanese)

Hsueh, Y.-M., Cheng, G.-S., Wu, M.-M., Yu, H.-S., Kuo, T.-L. & Chen, C.-J. (1995) Multiple risk factors associated with arsenic-induced skin cancer: Effects of chronic liver disease and malnutritional status. *Br. J. Cancer*, **71**, 109–114

Hsueh, Y.-M., Chiou, H.-Y., Huang, Y.-L., Wu, W.-L., Huang, C.-C., Yang, M.-H., Lue, L.-C., Chen, G.-S. & Chen, C.-J. (1997) Serum β-carotene level, arsenic methylation capability, and incidence of skin cancer. *Cancer Epidemiol. Biomarkers Prev.*, **6**, 589–596

Hsueh, Y.-M., Huang, Y.-L., Huang, C.-C., Wu, W.-L., Chen, H.-M., Yang, M.-H., Lue, L.-C. & Chen, C.-J. (1998) Urinary levels of inorganic and organic arsenic metabolites among residents in an arseniasis-hyperendemic area in Taiwan. *J. Toxicol. environ. Health*, **A54**, 431–444

Huang, R.Y., Lee, T.C. & Jan, K.Y. (1986) Cycloheximide suppresses the enhancing effect of sodium arsenite on the clastogenicity of ethyl methanesulphonate. *Mutagenesis*, **1**, 467–470

Hughes, M.F. (2002) Arsenic toxicity and potential mechanisms of action. *Toxicol. Lett.*, **133**, 1–16

Hughes, M.F. & Kenyon, E.M. (1998) Dose-dependent effects on the disposition of monomethyl-arsonic acid and dimethylarsinic acid in the mouse after intravenous administration. *J. Toxicol. environ. Health*, **A53**, 95–112

Hughes, M.F., Mitchell C.T., Edwards, B.C. & Rahman, M.S. (1995) In vitro percutaneous absorption of dimethylarsinic acid in mice. *J. Toxicol. environ. Health*, **45**, 279–290

Hughes, M.F., Del Razo, L.M. & Kenyon, E.M. (2000) Dose-dependent effects on tissue distribution and metabolism of dimethylarsinic acid in the mouse after intravenous administration. *Toxicology*, **143**, 155–166

Hutchinson, J. (1888) On some examples of arsenic-keratosis of the skin and of arsenic-cancer. *Trans. pathol. Soc. London*, **39**, 352–393

IARC (1980) *IARC Monographs on the Evaluation of the Carcinogenic Risk of Chemicals to Humans*, Vol. 23, *Some Metals and Metallic Compounds*, Lyon, IARC*Press*, pp. 39–143

IARC (1987) *IARC Monographs on the Evaluation of Carcinogenic Risks to Humans*, Suppl. 7, *Overall Evaluations of Carcinogenicity: An Updating of* IARC Monographs *Volumes 1 to 42*, Lyon, IARC*Press*, pp. 100–106

Infante-Rivard, C., Olson, E., Jacques, L. & Ayotte, P. (2001) Drinking water contaminants and childhood leukemia. *Epidemiology*, **12**, 13–19

Ishinishi, N., Tomita, M. & Hisanaga, A. (1980) Study on chronic toxicity of arsenic trioxide in rats with special reference to the liver damages. *Fukuoka Acta med.*, **71**, 27–40

Ishinishi, N., Yamamoto, A., Hisanga, A. & Imamasu, T. (1983) Tumorigenicity of arsenic trioxide to the lung in Syrian golden hamsters by intermittent instillations. *Cancer Lett.*, **21**, 141–147

Jaafar, R., Omar, I., Jidon, A.J., Wan-Khamizar, B.W.K., Siti-Aishah, B.M.A. & Sharifah-Noor-Akmal, S.H. (1993) Skin cancer caused by chronic arsenical poisoning — A report of three cases. *Med. J. Malaysia*, **48**, 86–92

Jäättelä, M. (1999) Escaping cell death: Survival proteins in cancer. *Exp. Cell Res.*, **248**, 30–43

Jackson, R. & Grainge, J.W. (1975) Arsenic and cancer. *Can. med. Assoc. J.*, **113**, 396–399

Jamett, A., Santander, M., Peña, L., Muñoz, L. & Gras, N. (1992) Arsenic levels in hair samples of inhabitants of the secind region of Chile. In: *Proceedings of an International Seminar on Arsenic in the Environment and its Incidence on Health*, Santiago, Universidad de Chile, pp. 87–90

Jha, A.N., Noditi, M., Nilsson, R. & Natarajan, A.T. (1992) Genotoxic effects of sodium arsenite on human cells. *Mutat. Res.*, **284**, 215–221

Jiang, G., Lu, X., Gong, Z., Cullen, W.R. & Le, X.C. (2003) Trivalent arsenic species: Analysis, stability, and interaction with a protein. In: Chappell, W.R., Abernathy, C.O. & Calderon, R.L., eds, *Arsenic Exposure and Health Effects*, Elsevier Science, 51–67

Johnston, D., Oppermann, H., Jackson, J. & Levinson, W. (1980) Induction of four proteins in chick embryo cells by sodium arsenite. *J. biol. Chem.*, **255**, 6975–6980

Kaltreider, R.C., Pesce, C.A., Ihnat, M.A., Lariviere, J.P. & Hamilton, J.W. (1999) Differential effects of arsenic(III) and chromium(VI) on nuclear transcription factor binding. *Mol. Carcinog.*, **25**, 219–229

Karagas, M.R., Morris, J.S., Weiss, J.E., Spate, V., Baskett, C. & Greenberg, E.R. (1996) Toenail samples as an indicator of drinking water arsenic exposure. *Cancer Epidemiol. Biomarkers Prev.*, **5**, 849–852

Karagas, M.R., Tosteson, T.D., Blum, J., Morris, J.S., Baron, J.A. & Klaue, B. (1998) Design of an epidemiologic study of drinking water arsenic exposure and skin and bladder cancer risk in a U.S. population. *Environ. Health Perspect.*, **106** (Suppl. 4), 1047–1050

Karagas, M.R., Tosteson, T.D., Blum, J., Klaue, B., Weiss, J.E., Stannard, V., Spate, V. & Morris, J.S. (2000) Measurement of low levels of arsenic exposure: A comparison of water and toenail concentrations. *Am. J. Epidemiol.*, **152**, 84–90

Karagas, M.R., Le, C.X., Morris, S., Blum, J., Lu, X., Spate, V., Carey, M., Stannard, V., Klaue, B. & Tosteson, T.D. (2001a) Markers of low level arsenic exposure for evaluating human cancer risks in a US population. *Int. J. occup. Med. Environ. Health*, **14**, 171–175

Karagas, M.R., Stukel, T.A., Morris, J.S., Tosteson, T.D., Weiss, J.E., Spencer, S.K. & Greenberg, E.R. (2001b) Skin cancer risk in relation to toenail arsenic concentrations in a US population-based case-control study. *Am. J. Epidemiol.*, **153**, 559–565

Karagas, M.R., Stukel, T.A. & Tosteson, T.D. (2002) Assessment of cancer risk and environmental levels of arsenic in New Hampshire. *Int. J. Hyg. environ. Health*, **205**, 85–94

Kasai, H. & Nishimura, S. (1984) Hydroxylation of deoxyguanosine at the C-8 position by ascorbic acid and other reducing agents. *Nucleic Acids Res.*, **12**, 2137–2145

Kasai, H., Yamaizumi, Z., Berger, M. & Cadet, J. (1992) Photosensitized formation of 7,8-dihydro-8-oxo-2'-deoxyguanosine (8-hydroxy-2'-deoxyguanosine) in DNA by riboflavin: A non singlet oxygen mediated reaction. *J. Am. chem. Soc.*, **114**, 9692–9694

Kashiwada, E., Kuroda, K. & Endo, G. (1998) Aneuploidy induced by dimethylarsinic acid in mouse bone marrow cells. *Mutat. Res.*, **413**, 33–38

Kato, K., Yamanaka, K., Nakano, M., Hasegawa, A., Oku, N. & Okada, S. (1997) Cell-nuclear accumulation of 72-kDa stress protein induced by dimethylated arsenics. *Biol. pharm. Bull.*, **20**, 364–369

Kato, K., Yamanaka, K., Hasegawa, A. & Okada, S. (1999) Dimethylarsinic acid exposure causes accumulation of Hsp72 in cell nuclei and suppresses apoptosis in human alveolar cultured (L-132) cells. *Biol. pharm. Bull.*, **22**, 1185–1188

Kato, K., Yamanaka, K., Nakano, M., Hasegawa, A. & Okada, S. (2000) 72-kDA stress protein (Hsp72) induced by administration of dimethylarsinic acid to mice accumulates in alveolar flat cells of lung, a target organ for arsenic carcinogenesis. *Biol. pharm. Bull.*, **23**, 1212–1215

Katsnelson, B.A., Neizvestinova, Y.M. & Blokhin, V.A. (1986) [Stomach carcinogenesis induction by chronic treatment with arsenic.] *Vopr. Onkol.*, **32**, 68–73 (in Russian)

Kenyon, E.M. & Hughes, M.F. (2001) A concise review of the toxicity and carcinogenicity of dimethylarsinic acid. *Toxicology*, **160**, 227–236

Keyse, S.M. & Tyrrell, R.M. (1989) Heme oxygenase is the major 32-kDa stress protein induced in human skin fibroblasts by UVA radiation, hydrogen peroxide, and sodium arsenite. *Proc. natl Acad. Sci. USA*, **86**, 99–103

Kilburn, K.H. (1997) Neurobehavioral impairment from long-term residential arsenic exposure. In: Abernathy, C.O., Calderon, R.L. & Chappell, W.R., eds, *Arsenic Exposure and Health Effects*, London, Chapman & Hall, pp. 159–175

Kingsley, G.R. & Schaffert, R.R. (1951) Microdetermination of arsenic and its application to biological material. *Anal. Chem.*, **23**, 914–919

Kinniburgh, D.G. & Smedley, P.L., eds (2001) *Arsenic Contamination of Groundwater in Bangladesh* (British Geological Survey Technical Report WC/00/19), Keyworth, British Geological Survey

Kitchin, K.T. (2001) Recent advances in arsenic carcinogenesis: Modes of action, animal model systems, and methylated arsenic metabolites. *Toxicol. appl. Pharmacol.*, **172**, 249–261

Kitchin, K.T. & Ahmad, S. (2003) Oxidative stress as a possible mode of action for arsenic carcinogenesis. *Toxicol. Lett.*, **137**, 3–13

Klaassen, C.D. (1974) Biliary excretion of arsenic in rats, rabbits, and dogs. *Toxicol. appl. Pharmacol.*, **29**, 447–457

Klaassen, C.D. & Cagen, S.Z. (1981) Metallothionein as a trap for reactive organic intermediates. *Adv. exp. Med. Biol.*, **136**, 633–646

Koch, I., Feldmann, J., Wang, L., Andrewes, P., Reimer, K.J. & Cullen, W.R. (1999) Arsenic in the Meager Creek hot springs environment, British Columbia, Canada. *Sci. total Environ.*, **236**, 101–117

Koljonen, T. (1992) *Geochemical Atlas of Finland*, Espoo, Geological Survey of Finland

Kondo, H., Ishiguro, Y., Ohno, K., Nagase, M., Toba, M. & Takagi, M. (1999) Naturally occurring arsenic in the groundwaters in the southern region of Fukuoka Prefecture, Japan. *Water Res.*, **33**, 1967–1972

Korte, N.E. & Fernando, Q. (1991) A review of arsenic (III) in groundwater. *Crit. Rev. environ. Control*, **21**, 1–39

Kreppel, H., Bauman, J.W., Liu, J., McKim, J.M., Jr & Klaassen, C.D. (1993) Induction of metallothionein by arsenicals in mice. *Fundam. appl. Toxicol.*, **20**, 184–189

Kuo, T.L. (1968) Arsenic content of artesian well water in endemic area of chronic arsenic poisoning. *Rep. Inst. Pathol. natl Taiwan Univ.*, **20**, 7–13

Kurokawa, M., Ogata, K., Idemori, M., Tsumori, S., Miyaguni, H., Inoue, S. & Hotta, N. (2001) Investigation of skin manifestations of arsenicism due to intake of arsenic-contaminated groundwater in residents of Samta, Jessore, Bangladesh. *Arch. Dermatol.*, **137**, 102–103

Kurttio, P., Komulainen, H., Hakala, E., Kahelin, H. & Pekkanen, J. (1998) Urinary excretion of arsenic species after exposure to arsenic present in drinking water. *Arch. environ. Contam. Toxicol.*, **34**, 297–305

Kurttio, P., Pukkala, E., Kahelin, H., Auvinen, A. & Pekkanen, J. (1999) Arsenic concentrations in well water and risk of bladder and kidney cancer in Finland. *Environ. Health Perspect.*, **107**, 705–710

Lacayo, M.L., Cruz, A., Calero, S., Lacayo, J. & Fomsgaard, I. (1992) Total arsenic in water, fish and sediments from Lake Xolotlán, Managua, Nicaragua. *Bull. environ. Contam. Toxicol.*, **49**, 463–470

Lahermo, P., Alfthan, G. & Wang, D. (1998) Selenium and arsenic in the environment in Finland. *J. environ. Pathol. Toxicol. Oncol.*, **17**, 205–216

Lai, M.-S., Hsueh, Y.-M., Chen, C.-J., Shyu, M.-P., Chen, S.-Y., Kuo, T.-L., Wu, M.-M. & Tai, T.-Y. (1994) Ingested inorganic arsenic and prevalence of diabetes mellitus. *Am. J. Epidemiol.*, **139**, 484–492

Lalor, G., Rattray, R., Simpson, P. & Vutchkov, M.K. (1999) Geochemistry of an arsenic anomaly in St. Elizabeth, Jamaica. *Environ. Geochem. Health*, **21**, 3–11

Lamble, K.J. & Hill, S.J. (1996) Arsenic speciation in biological samples by on-line high-performance liquid chromatography–microwave digestion–hydride generation–atomic absorption spectrometry. *Anal. chim. Acta*, **344**, 261–270

Langner, H.W., Jackson, C.R., McDermott, T.R. & Inskeep, W.P. (2001) Rapid oxidation of arsenite in a hot spring ecosystem, Yellowstone National Park. *Environ. Sci. Technol.*, **35**, 3302–3309

Lantzsch, H. & Gebel, T. (1997) Genotoxicity of selected metal compounds in the SOS chromotest. *Mutat. Res.*, **389**, 191–197

Larochette, N., Decaudin, D., Jacotot, E., Brenner, C., Marzo, I., Susin, S.A., Zamzami, N. , Xie, Z., Reed, J. & Kroemer, G. (1999) Arsenite induces apoptosis via a direct effect on the mitochondrial permeability transition pore. *Exp. Cell Res.*, **249**, 413–421

Larsen, N.A., Nielsen, B., Pakkenberg, H., Christoffersen, P., Damogarrd, E. & Heydorn, K. (1974) The concentrations of arsenic, manganese and selenium in organs from normal and uremic persons determined by activation analysis. *Ugeskr. Laeger*, **136**, 2586–2590

Le, X.C. (2002) Arsenic speciation in the environment and humans. In: Frankenberger, W.T., Jr, ed., *Environmental Chemistry of Arsenic*, New York, Marcel Dekker, pp. 95–116

Le, X.C. & Ma, M. (1997) Speciation of arsenic compounds by using ion-pair chromatography with atomic spectrometry and mass spectrometry detection. *J. Chromatogr.*, **A764**, 55–64

Lee, T.-C., Oshimura, M. & Barrett, J.C. (1985a) Comparison of arsenic-induced cell transformation, cytotoxicity, mutation and cytogenetic effects in Syrian hamster embryo cells in culture. *Carcinogenesis*, **6**, 1421–1426

Lee, T.-C., Huang, R.-Y. & Jan, K.-Y. (1985b) Sodium arsenite enhances the cytotoxicity, clastogenicity, and 6-thioguanine-resistant mutagenicity of ultraviolet light in Chinese hamster ovary cells. *Mutat. Res.*, **148**, 83–89

Lee, T.-C., Wang-Wuu, S., Huang, R.Y., Lee, K.C.C. & Jan, K.Y. (1986a) Differential effects of pre- and posttreatment of sodium arsenite on the genotoxicity of methyl methanesulfonate in Chinese hamster ovary cells. *Cancer Res.*, **46**, 1854–1857

Lee, T.-C., Lee, K.C., Tzeng, Y.J., Huang, R.Y. & Jan, K.Y. (1986b) Sodium arsenite potentiates the clastogenicity and mutagenicity of DNA crosslinking agents. *Environ. Mutag.*, **8**, 119–128

Lee, T.-C., Tanaka, N., Lamb, P.W., Gilmer, T.M. & Barrett, J.C. (1988) Induction of gene amplification by arsenic. *Science*, **241**, 79–81

Le, X.-C., Cullen, W.R. & Reimer, K.J. (1994a) Effect of cysteine on the speciation of arsenic by using hydride generation atomic absorption spectrometry. *Anal. chim. Acta*, **285**, 277–285

Le, X.-C., Cullen, W.R. & Reimer, K.J. (1994b) Human urinary arsenic excretion after one-time ingestion of seaweed, crab, and shrimp. *Clin. Chem.*, **40**, 617–624

Le, X.C., Cullen, W.R. & Reimer, K.J. (1994c) Speciation of arsenic compounds in some marine organisms. *Environ. Sci. Technol.*, **28**, 1598–1604

Le, X.C., Li, X.-F., Lai, V., Ma, M., Yalcin, S. & Feldmann, J. (1998) Simultaneous speciation of selenium and arsenic using elevated temperature liquid chromatography separation with inductively coupled plasma mass spectrometry detection. *Spectrochim. Acta*, **B53**, 899–909

Le, X.C., Yalcin, S. & Ma, M. (2000a) Speciation of submicrogram per liter levels of arsenic in water: On-site species separation integrated with sample collection. *Environ. Sci. Technol.*, **34**, 2342–2347

Le, X.C., Ma, M., Cullen, W.R., Aposhian, H.V., Lu, X. & Zheng, B. (2000b) Determination of monomethylarsonous acid, key arsenic methylation intermediate, in human urine. *Environ. Health Perspect.*, **108**, 1015–1018

Lee-Chen, S.F., Yu, C.T. & Jan, K.Y. (1992) Effect of arsenite on the DNA repair of UV-irradiated Chinese hamster ovary cells. *Mutagenesis*, **7**, 51–55

Lerda, D. (1994) Sister-chromatid exchange (SCE) among individuals chronically exposed to arsenic in drinking water. *Mutat. Res.*, **312**, 111–120

Lerda, D.E. & Prosperi, C.H. (1996) Water mutagenicity and toxicology in Rio Tercero (Cordoba, Argentina). *Water Res.*, **30**, 819–824

Lewis, D.R., Southwick, J.W., Ouellet-Hellstrom, R., Rench, J. & Calderon, R.L. (1999) Drinking water arsenic in Utah: A cohort mortality study. *Environ. Health Perspect.*, **107**, 359–365

Li, J.-H. & Rossman, T.G. (1989) Inhibition of DNA ligase activity by arsenite: A possible mechanism of its comutagenesis. *Mol. Toxicol.*, **2**, 1–9

Li, W., Wanibuchi, H., Salim, E.I., Yamamoto, S., Yoshida, K., Endo, G. & Fukushima, S. (1998) Promotion of NCI-Black-Reiter male rat bladder carcinogenesis by dimethylarsinic acid an organic arsenic compound. *Cancer Lett.*, **134**, 29–36

Liebscher, K. & Smith, H. (1968) Essential and nonessential trace elements. A method of determining whether an element is essential or nonessential in human tissue. *Arch. environ. Health*, **17**, 881–890

Lin, T.-H., Huang, Y.-L. & Wang, M.-Y. (1998) Arsenic species in drinking water, hair, fingernails, and urine of patients with blackfoot disease. *J. Toxicol. environ. Health*, **A53**, 85–93

Lin S., Shi, Q., Nix, F.B., Styblo, M., Beck, M.A., Herbin-Davis, K.M., Hall, L.L., Simeonsson, J.B. & Thomas, D.J. (2002) A novel *S*-adenosyl-L-methionine:arsenic(III) methyltransferase from rat liver cytosol. *J. biol. Chem.*, **277**, 10795–10803

Lindgren, A., Vahter, M. & Dencker, L. (1982) Autoradiographic studies on the distribution of arsenic in mice and hamster administered [74]As-arsenite or -arsenate. *Acta pharmacol. toxicol.*, **51**, 253–265

Lindgren, A., Danielsson, B.R.G., Dencker, L. & Vahter, M. (1984) Embryotoxicity of arsenite and arsenate: Distribution in pregnant mice and monkeys and effects on embryonic cells *in vitro*. *Acta pharmacol. toxicol.*, **54**, 311–320

Liou, S.-H., Lung, J.-C., Chen, Y.-H., Yang, T., Hsieh, L.-L., Chen, C.-J. & Wu, T.-N. (1999) Increased chromosome-type chromosome aberration frequencies as biomarkers of cancer risk in a blackfoot endemic area. *Cancer Res.*, **59**, 1481–1484

Liu, J., Liu, Y., Goyer, R.A., Achanzar, W. & Waalkes, M.P. (2000) Metallothionein-I/II null mice are more sensitive than wild-type mice to the hepatotoxic and nephrotoxic effects of chronic oral or injected inorganic arsenicals. *Toxicol. Sci.*, **55**, 460–467

Liu, J., Kadiiska, M.B., Liu, Y., Lu, T., Qu, W. & Waalkes, M.P. (2001) Stress-related gene expression in mice treated with inorganic arsenicals. *Toxicol. Sci.*, **61**, 314–320

Liu, S.X., Athar, M., Lippai, I., Waldren, C. & Hei, T.K. (2001) Induction of oxyradicals by arsenic: Implication for mechanism of genotoxicity. *Proc. natl Acad. Sci. USA*, **98**, 1643–1648

Lo, M.C. (1975) *Report on the Investigation of Arsenic Content of Well Water in the Province of Taiwan*, Nan-Tour, Taiwan Provincial Institute of Sanitary Department

Londesborough, S., Mattusch, J. & Wennrich, R. (1999) Separation of organic and inorganic arsenic species by HPLC–ICP–MS. *Fresenius J. anal. Chem.*, **363**, 577–581

Lopez, S., Miyashita, Y. & Simons, S.S., Jr (1990) Structurally based, selective interaction of arsenite with steroid receptors. *J. biol. Chem.*, **265**, 16039–16042

Lu, F.J., Shih, S.R., Liu, T.M. & Shown, S.H. (1990) The effect of fluorescent humic substances existing in the well water of blackfoot disease endemic areas in Taiwan on prothrombin time and activated partial thromboplastin time in vitro. *Thromb. Res.*, **57**, 747–753

Luo, Z.D., Zhang, Y.M., Ma, L., Zhang, G.Y., He, X., Wilson, R., Byrd, D.M., Griffiths, J.G., Lai, S., He, L., Grumski, K. & Lamm, S.H. (1997) Chronic arsenicism and cancer in Inner Mongolia — Consequences of well-water arsenic levels greater than 50 µg/L. In: Abernathy, C.O., Calderon,

R.L. & Chappell, W.R., eds, *Arsenic Exposure and Health Effects*, London, Chapman & Hall, pp. 55–68

Lynn, S., Gurr, J.-R., Lai, H.-T. & Jan, K.-Y. (2000) NADH oxidase activation is involved in arsenite-induced oxidative DNA damage in human vascular smooth muscle cells. *Circ. Res.*, **86**, 514–519

Ma, L., Luo, Z., Zhang, Y., Zhang, G., Tai, S., Liang, S., Ren, S. & Zhang, M. (1996) [Current status of endemic arsenic poisoning in Inner Mongolia.] *Chinese J. public Health*, **15** (Suppl. 3), 15–39 (in Chinese)

Ma, H.Z., Xia, Y.J., Wu, K.G., Sun, T.Z. & Mumford, J.L. (1999) Human exposure to arsenic and health effects in Bayingnormen, Inner Mongolia. In: Chappell, W.R., Abernathy, C.O. & Calderon, R.L., eds, *Arsenic Exposure and Health Effects*, Amsterdam, Elsevier Science, pp. 127–131

Ma, L.Q., Komar, K.M., Tu, C., Zhang, W., Cai, Y. & Kennelley, E.D. (2001) A fern that hyper-accumulates arsenic. *Nature*, **409**, 579

Mahieu, P., Buchet, J.P., Roels, H.A. & Lauwerys, R. (1981) The metabolism of arsenic in humans acutely intoxicated by As_2O_3. Its significance for the duration of BAL therapy. *Clin. Toxicol.*, **18**, 1067–1075

Maitani, T., Saito, N., Abe, M., Uchiyama, S. & Saito, Y. (1987) Chemical form-dependent induction of hepatic zinc-thionein by arsenic administration and effect of co-administered selenium in mice. *Toxicol. Lett.*, **39**, 63–70

Maiti, S. & Chatterjee, A.K. (2001) Effects on levels of glutathione and some related enzymes in tissues after an acute arsenic exposure in rats and their relationship to dietary protein deficiency. *Arch. Toxicol.*, **75**, 531–537

Mäki-Paakkanen, J., Kurttio, P., Paldy, A. & Pekkanen, J. (1998) Association between the clastogenic effect in peripheral lymphocytes and human exposure to arsenic through drinking water. *Environ. mol. Mutag.*, **32**, 301–313

Mallick, S. & Rajagopal, N.R. (1995) The mischief of oxygen on groundwater. In: *Post Conference Report: Experts' Opinions, Recommendations and Future Planning for Groundwater Problem of West Bengal*, Calcutta, School of Environmental Studies, Jadavpur University, pp. 71–73

Mandal, B.K. & Suzuki, K.T. (2002) Arsenic round the world: A review. *Talanta*, **58**, 201–235

Mandal, B.K., Roy Chowdhury, T., Samanta, G., Basu, G.K., Chowdhury, P.P., Chanda, C.R., Lodh, D., Karan, N.K., Dhar, R.K., Tamili, D.K., Das, D., Saha, K.C & Chakraborti, D. (1996) Arsenic in groundwater in seven districts of West Bengal, India — The biggest arsenic calamity in the world. *Current Sci.*, **70**, 976–986

Mandal, B.K., Roy Chowdhury, T., Samanta, G., Basu, G.K., Chowdhury, P.P., Chanda, C.R., Lodh, D., Karan, N.K., Dhar, R.K., Tamili, D.K., Das, D., Saha, K.C. & Chakraborti, D. (1997) Chronic arsenic toxicity in West Bengal (Letter). *Current Sci.*, **72**, 114–117

Mandal, B.K., Chowdhury, T.R., Samanta, G., Mukherjee, D.P., Chanda, C.R., Saha, K.C. & Chakraborti, D. (1998) Impact of safe water for drinking and cooking on five arsenic-affected families for 2 years in West Bengal, India. *Sci. total Environ.*, **218**, 185–201

Mandal, B.K., Biswas, B.K., Dhar, R.K., Roy Chowdhury, T., Samanta, G., Basu, G.K., Chanda, C.R., Saha, K.C., Chakraborti, D., Kabir, S. & Roy, S. (1999) Groundwater arsenic contamination and sufferings of people in West Bengal, India and Bangladesh. Status report up to March, 1998. In: Sarkar, B., eds, *Metals and Genetics*, New York, Kluwer Academic/Plenum Publishers, pp. 41–65

Mandal, B.K., Ogra, Y. & Suzuki, K.T. (2001) Identification of dimethylarsinous and monomethyl-arsonous acids in human urine of the arsenic-affected areas in West Bengal, India. *Chem. Res. Toxicol.*, **14**, 371–378

Mandal, B.K., Ogra, Y. & Suzuki, K.T. (2003) Speciation of arsenic in human nail and hair from arsenic-affected area by HPLC-inductively coupled argon plasma mass spectrometry. *Toxicol. appl. Pharmacol.*, **189**, 73–83

Marafante, E., Bertolero, F., Edel, J., Pietra, R. & Sabbioni, E. (1982) Intracellular interaction and biotransformation of arsenite in rats and rabbits. *Sci. total Environ.*, **24**, 27–39

Marafante, E., Vahter, M. & Envall, J. (1985) The role of the methylation in the detoxication of arsenate in the rabbit. *Chem.-biol. Interact.*, **56**, 225–238

Marafante, E., Vahter, M., Norin, H., Envall, J., Sandström, M., Christakopoulos, A. & Ryhage, R. (1987) Biotransformation of dimethylarsinic acid in mouse, hamster and man. *J. appl. Toxicol.*, **7**, 111–117

Mass, M.J. & Wang, L. (1997) Arsenic alters cytosine methylation patterns of the promoter of the tumor suppressor gene *p53* in human lung cells: A model for a mechanism of carcinogenesis. *Mutat. Res.*, **386**, 263–277

Mass, M.J., Tennant, A., Roop, B.C., Cullen, W.R., Styblo, M., Thomas, D.J. & Kligerman, A.D. (2001) Methylated trivalent arsenic species are genotoxic. *Chem. Res. Toxicol.*, **14**, 355–361

Matisoff, G., Khourey, C.J., Hall, J.F., Varnes, A.W. & Strain, W.H. (1982) The nature and source of arsenic in Northeastern Ohio ground water. *Groundwater*, **20**, 446–456

Matschullat, J. (2000) Arsenic in the geosphere — A review. *Sci. total Environ.*, **249**, 297–312

Matschullat, J., Borba, R.P., Deschamps, E., Figueireo, B.R., Gabrio, T. & Schwenk, M. (2000) Human and environmental contamination in the Iron Quadrangle, Brazil. *Appl. Geochem.*, **15**, 193–202

Matsui, M., Nishigori, C., Toyokuni, S., Takada, J., Akaboshi, M., Ishikawa, M., Imamura, S. & Miyachi, Y. (1999) The role of oxidative DNA damage in human arsenic carcinogenesis: Detection of 8-hydroxy-2′-deoxyguanosine in arsenic-related Bowen's disease. *J. invest. Dermatol.*, **113**, 26–31

McKenzie, E.J., Brown, K.L., Cady, S.L. & Campbell, K.A. (2001) Trace metal chemistry and silicification of microorganisms in geothermal sinter, Taupo Volcanic Zone, New Zealand. *Geothermics*, **30**, 483–502

McLaren, S.J. & Kim, N.D. (1995) Evidence for a seasonal fluctuation of arsenic in New Zealand's longest river and the effect of treatment on concentrations in drinking water. *Environ. Pollut.*, **90**, 67–73

Mealey J., Brownell, G.L. & Sweet, W.H. (1959) Radioarsenic in plasma, urine, normal tissue, and intracranial neoplasms. *Arch. Neurol. Psychiatr.*, **81**, 310–320

Meíja, J., Carrizales, L., Rodriguez, V.M., Jiménez-Capdeville, M.E. & Díaz-Barriga, F. (1999) [A method for health risk assessment of mining sites.] *Salud publica Méx.*, **41** (Suppl. 2), S132–S140

Meranger, J.C., Subramanian, K.S. & McCurdy, R.F. (1984) Arsenic in Nova Scotian groundwater. *Sci. total Environ.*, **39**, 49–55

Milton, A.H. & Rahman, M. (1999) Environmental pollution and skin involvement pattern of chronic arsenicosis in Bangladesh. *J. occup. Health*, **41**, 207–208

Milton, A.H., Hasan, Z., Rahman, A. & Rahman, M. (2001) Chronic arsenic poisoning and respiratory effects in Bangladesh. *J. occup. Health*, **43**, 136–140

Ministerio de Salud de la República de Chile (1986) *Primera Jornada Sobre Arsenicismo Laboral y Ambiental II Region* (Departamento de Programas del Ambiente Informe Técnico), Santiago, Servicio de Salud Antofagasta

Mirkes, P.E. & Cornel, L. (1992) A comparison of sodium arsenite-hyperthermia-induced stress responses and abnormal development in cultured postimplantation rat embryos. *Teratology*, **46**, 251–259

Mirkes, P.E., Doggett, B. & Cornel, L. (1994) Induction of a heat shock response (HSP 72) in rat embryos exposed to selected chemical teratogens. *Teratology*, **49**, 135–142

Moore, L.E., Warner, M.L., Smith, A.H., Kalman, D. & Smith, M.T. (1996) Use of the fluorescent micronucleus assay to detect the genotoxic effects of radiation and arsenic exposure in exfoliated human epithelial cells. *Environ. mol. Mutag.*, **27**, 176–184

Moore, L.E., Smith, A.H., Hopenhayn-Rich, C., Biggs, M.L., Kalman, D.A. & Smith, M.T. (1997a) Micronuclei in exfoliate bladder cells among individuals chronically exposed to arsenic in drinking water. *Cancer Epidemiol. Biomarkers Prev.*, **6**, 31–36

Moore, L.E., Smith, A.H., Hopenhayn-Rich, C., Biggs, M.L., Kalman, D.A. & Smith, M.T. (1997b) Decrease in bladder cell micronucleus prevalence after intervention to lower the concentration of arsenic in drinking water. *Cancer Epidemiol. Biomarkers Prev.*, **6**, 1051–1056

Moore, M.M., Harrington-Brock, K. & Doerr, C.L. (1997) Relative genotoxic potency of arsenic and its methylated metabolites. *Mutat. Res.*, **386**, 279–290

Morikawa, T., Wanibuchi, H., Morimura, K., Ogawa, M. & Fukushima, S. (2000) Promotion of skin carcinogenesis by dimethylarsinic acid in *keratin (K6)/ODC* transgenic mice. *Jpn. J. Cancer Res.*, **91**, 579–581

Morris, J.S., Schmid, M., Newman, S., Scheuer, P.J. & Sherlock, S. (1974) Arsenic and noncirrhotic portal hypertension. *Gastroenterology*, **66** , 86–94

Morrissey, R.E. & Mottet, N.K. (1983) Arsenic-induced exencephaly in the mouse and associated lesions occurring during neurulation. *Teratology*, **28**, 399–411

Morton, W., Starr, G., Pohl, D., Stoner, J., Wagner, S. & Weswig, P. (1976) Skin cancer and water arsenic in Lane County, Oregon. *Cancer*, **37**, 2523–2532

Mukherjee, S.C., Rahman, M.M., Chowdhury, U.K., Sengupta, M.K., Lodh, D., Chanda, C.R., Saha, K.C. & Chakraborti, D. (2003) Neuropathy in arsenic toxicity from groundwater arsenic contamination in West Bengal, India. *J. environ. Sci. Health*, **A38**, 165–183

Muller, W.U., Streffer, C. & Fisher-Lahdo, C. (1986) Toxicity of sodium arsenite in mouse embryos in vitro and its influence on radiation risk. *Arch. Toxicol.*, **59**, 172–175

Nakadaira, H., Endoh, K., Katagiri, M. & Yamamoto, M. (2002) Elevated mortality from lung cancer associated with arsenic exposure for a limited duration. *J. occup. environ. Med.*, **44**, 291–299

Nakamuro, K. & Sayato, Y. (1981) Comparative studies of chromosomal aberration induced by trivalent and pentavalent arsenic. *Mutat. Res.*, **88**, 73–80

Nakano, M., Yamanaka, K., Hasegawa, A., Sawamura, R. & Okada, S. (1992) Preferential increase of heterochromatin in venular endothelium of lung in mice after administration of dimethylarsinic acid, a major metabolite of inorganic arsenics. *Carcinogenesis*, **13**, 391–393

National Health and Medical Research Council and Agriculture and Resource Management Council of Australia and New Zealand (1996) *Australian Drinking Water Guidelines Summary*, Canberra

National Research Council (1999) *Arsenic in Drinking Water*, Washington DC, National Academy Press, pp. 27–82

National Research Council (2001) *Arsenic in Drinking Water. 2001 Update*, Washington DC, National Academy of Sciences, National Academy Press

Nemec, M.D., Holson, J.F., Farr, C.H. & Hood, R.D. (1998) Developmental toxicity assessment of arsenic acid in mice and rabbits. *Reprod. Toxicol.*, **12**, 647–658

Nesnow, S., Roop, B.C., Lambert, G., Kadiiska, M., Mason, R.P., Cullen, W.R. & Mass, M.J. (2002) DNA damage induced by methylated trivalent arsenicals is mediated by reactive oxygen species. *Chem. Res. Toxicol.*, **15**, 1627–1634

Neubauer, O. (1947) Arsenical cancer: A review. *Br. J. Cancer*, **i**, 192–251

Neumann, E. & Schwank, R. (1960) Multiple malignant and benign epidermal and dermal tumours following arsenic. *Acta derm.-venereol.*, **40**, 400–409

Nevens, F., Fevery, J., Van Steenbergen, W., Sciot, R., Desmet, V. & De Groote, J. (1990) Arsenic and non-cirrhotic portal hypertension. A report of eight cases. *J. Hepatol.*, **11**, 80–85

Ng, J.C., Johnson, D., Imray, P., Chiswell, B. & Moore, M.R. (1998) Speciation of arsenic metabolites in the urine of occupational workers and experimental rats using an optimised hydride cold- trapping method. *Analyst*, **123**, 929–933

Ng, J.C., Seawright, A.A., Qi, L., Garnett, C.M., Chiswell, B. & Moore, M.R. (1999) Tumours in mice induced by exposure to sodium arsenate in drinking water. In: Chappell, W.R., Abernathy, C.O. & Calderon, R.L., eds, *Arsenic Exposure and Health Effects*, Amsterdam, Elsevier Science, pp. 217–223

Ng, J.C., Qi, L., Wang, J.-P., Xiao, X., Shahin, M., Moore, M.R. & Prakash, A.S. (2001) Mutations in C57Bl/6J and metallothionein knock-out mice induced by chronic exposure to sodium arsenate in drinking-water. In: Chappell, W.R., Abernathy, C.O. & Calderon, R.L., eds, *Arsenic Exposure and Health Effects IV*, Amsterdam, Elsevier Science, pp. 231–242

Nichols, T.A., Morris, J.S., Mason, M.M., Spate, V.L., Baskett, C.K., Cheng, T.P., Tharp, C.J., Scott, J.A., Horsman, T.L., Colbert, J.W., Rawson, A.E., Karagas, M.R. & Stannard, V. (1998) The study of human nails as an intake monitor for arsenic using neutron activation analysis. *J. radioanal. nucl. Chem.*, **236**, 51–56

Nickson, R., McArthur, J., Burgess, W., Ahmed, K.M., Ravenscroft, P. & Rahman, M. (1998) Arsenic poisoning of Bangladesh groundwater. *Nature*, **395**, 338

Nickson, R.T., McArthur, J.M., Ravenscroft, P., Burgess, W.G. & Ahmed, K.M. (2000) Mechanism of arsenic release to groundwater, Bangladesh and West Bengal. *Appl. Geochem.*, **15**, 403–413

Nimick, D.A., Moore, J.N., Dalby, C.E. & Savka, M.W. (1998) The fate of geothermal arsenic in the Madison and Missouri Rivers, Montana and Wyoming. *Water Resource Res.*, **34**, 3051–3067

Nishikawa, T., Wanibuchi, H., Ogawa, M., Kinoshita, A., Morimura, K., Hiroi, T., Funae, Y., Kishida, H., Nakae, D. & Fukushima, S. (2002) Promoting effects of monomethylarsonic acid, dimethylarsinic acid and trimethylarsine oxide on induction of rat liver preneoplastic glutathione *S*-transferase placental form positive foci: A possible reactive oxygen species mechanism. *Int. J. Cancer*, **100**, 136–139

Noda, Y., Suzuki, T., Kohara, A., Hasegawa, A., Yotsuyanagi, T., Hayashi, M., Sofuni, T., Yamanaka, K. & Okada, S. (2002) In vivo genotoxicity evaluation of dimethylarsinic acid in Muta™ Mouse. *Mutat. Res.*, **513**, 205–212

Nordenson, I. & Beckman, L. (1982) Occupational and environmental risks in and around a smelter in northern Sweden. VII. Reanalysis and follow-up of chromosomal aberrations in workers exposed to arsenic. *Hereditas*, **96**, 175–181

Nordenson, I., Sweins, A. & Beckman, L. (1981) Chromosome aberrations in cultured human lymphocytes exposed to trivalent and pentavalent arsenic. *Scand. J. Work Environ. Health*, **7**, 277–281

Nordstrom, D.K. (2002) Worldwide occurences of arsenic in ground water. *Science*, **296**, 2143–2145

O'Neil, M.J., ed. (2001) *The Merck Index*, 13th Ed., Whitehouse Station, NJ, Merck & Co., pp. 135–138

Okoji, R.S., Yu, R.C., Maronpot, R.R. & Froines, J.R. (2002) Sodium arsenite administration via drinking water increases genome-wide and Ha-ras DNA hypomethylation in methyl-deficient C57BL/6J mice. *Carcinogenesis*, **23**, 777–785

Ostrosky-Wegman, P., Gonsebatt, M.E., Montero, R., Vega, L., Barba, H., Espinosa, J., Palao, A., Cortinas, C., García-Vargas, G., Del Razo, L.M. & Cebrián, M. (1991) Lymphocyte proliferation kinetics and genotoxic findings in a pilot study on individuals chronically exposed to arsenic in Mexico. *Mutat. Res.*, **250**, 477–482

Oya-Ohta, Y., Kaise, T. & Ochi, T. (1996) Induction of chromosomal aberrations in cultured human fibroblasts by inorganic and organic arsenic compounds and the different roles of glutathione in such induction. *Mutat. Res.*, **357**, 123–129

Page, G.W. (1981) Comparison of groundwater and surface water for patterns and levels of contamination by toxic substances. *Environ. Sci. Technol.*, **15**, 1475–1481

Pan, T.C., Lin, T.H., Tseng, C.L., Yang, M.H. & Huang, C.W. (1993) Trace elements in hair of Blackfoot disease. *Biol. trace Elem. Res.*, **39**, 117–128

Pande, S.P., Deshpande, L.S. & Kaul, S.N. (2001) Laboratory and field assessment of arsenic testing field kits in Bangladesh and West Bengal, India. *Environ. Monit. Assess.*, **68**, 1–18

Pandey, P.K., Khare, R.N., Sharma, R., Sar, S.K., Pandey, M. & Binayake, P. (1999) Arsenicosis and deteriorating groundwater quality: Unfolding crisis in central-east Indian region. *Current Sci.*, **77**, 686–693

Pandey, P.K., Yadav, S., Nair, S. & Bhui, A. (2002) Arsenic contamination of the environment. A new perspective from central-east India. *Environ. int.*, **28**, 235–245

Pazirandeh, A., Brati, A.H. & Marageh, M.G. (1998) Determination of arsenic in hair using neutron activation. *Appl. Radiat. Isot.*, **49**, 753–759

Peña, L., Jamett, A., Santander, M., Gras, N. & Muñoz, L. (1992) [Concentration of trace elements in children in northern Chile.] *Rev. med. Chile*, **120**, 20–24 (in Spanish)

Penedo, M. & Zigarán, A. (2002) *Hidroarsenicismo en la Provincia de Córdoba. Actualización del Mapa de Riesgo e Incidencia*, Córdoba, CEPROCOR y Ministerio de Salud y Seguridad Social de la Provincia de Córdoba

Peng, B., Sharma, R., Mass, M.J. & Kligerman, A.D. (2002) Induction of genotoxic damage is not correlated with the ability to methylate arsenite in vitro in the leukocytes of four mammalian species. *Environ. mol. Mutag.*, **39**, 323–332

Pershagen, G. & Björklund, N.E. (1985) On the pulmonary tumorigenicity of arsenic trisulfide and calcium arsenate in hamsters. *Cancer Lett.*, **27**, 99–104

Pershagen, G., Nordberg, G. & Björklund, N.E. (1984) Carcinomas of the respiratory tract in hamsters given arsenic trioxide and/or benzo[a]pyrene by pulmonary route. *Environ. Res.*, **34**, 227–241

Peters, S.C., Blum, J.D., Klaue, B. & Karagas, M.R. (1999) Arsenic occurrence in New Hampshire drinking water. *Environ. Sci. Technol.*, **33**, 1328–1333

Petrick, J.S., Jagadish, B., Mash, E.A. & Aposhian, H.V. (2001) Monomethylarsonous acid (MMA[III]) and arsenite: LD_{50} in hamsters and in vitro inhibition of pyruvate dehydrogenase. *Chem. Res. Toxicol.*, **14**, 651–656

Pfeifer, H.-R. & Zobrist, J. (2002) Arsenic in drinking water — Also a problem in Switzerland? *EAWAG News*, **53**, 15–17

Pillai, A., Sunita, G. & Gupta, V.K. (2000) A new system for the spectrophotometric determination of arsenic in environmental and biological samples. *Anal. chim. Acta*, **408**, 111–115

Pomroy, C., Charbonneau, S.M., McCullough, R.S. & Tam, G.K.H. (1980) Human retention studies with [74]As. *Toxicol. appl. Pharmacol.*, **53**, 550–556

Pongratz, R. (1998) Arsenic speciation in environmental samples of contaminated soils. *Sci. total Environ.*, **224**, 133–141

Quamruzzaman, Q., Roy, S., Rahman, M., Mia, S. & Arif, A.I. (1999) Rapid action programme: Emergency arsenic mitigation programme in two hundred villages in Bangladesh. In: Chappell, W., Abernathy, C.O. & Calderon, R.L., eds, *Arsenic Exposure and Health Effects*, Oxford, Elsevier Science, pp. 363–366

Queirolo, F. Stegen, S., Mondaca, J., Cortés, R., Rojas, R., Contreras, C., Munoz, L., Schwuger, M.J. & Ostapczuk, P. (2000a) Total arsenic, lead, cadmium, copper, and zinc in some salt rivers in the northern Andes of Antofagasta, Chile. *Sci. total Environ.*, **255**, 85–95

Queirolo, F., Stegen, S., Restovic, M., Paz, M., Ostapczuk, P., Schwuger, M.J. & Muñoz, L. (2000b) Total arsenic, lead and cadmium levels in vegetables cultivated at the Andean villages of northern Chile. *Sci. total Environ.*, **255**, 75–84

Radabaugh, T.R. & Aposhian, H.V. (2000) Enzymatic reduction of arsenic compounds in mammalian systems: Reduction of arsenate to arsenite by human liver arsenate reductase. *Chem. Res. Toxicol.*, **13**, 26–30

Rahman, M. (2002) Arsenic and hypertension in Bangladesh. *Bull. World Health Org.*, **80**, 173

Rahman, M. & Axelson, O. (2001) Arsenic ingestion and health effects in Bangladesh: Epidemiological observations. In: Chappell, W.R., Abernathy, C.O. & Calderon, R.L., eds, *Arsenic Exposure and Health Effects IV*, Amsterdam, Elsevier, pp. 193–199

Rahman, M.S., Hall, L.L. & Hughes, M.F. (1994) *In vitro* percutaneous absorption of sodium arsenate in B6C3F$_1$ mice. *Toxicol. in Vitro*, **8**, 441–448

Rahman, M., Tondel, M., Ahmad, S.K. & Axelson, O. (1998) Diabetes mellitus associated with arsenic exposure in Bangladesh. *Am. J. Epidemiol.*, **148**, 198–203

Rahman, M., Tondel, M., Chowdhury, I.A. & Axelson, O. (1999a) Relations between exposure to arsenic, skin lesions, and glucosuria. *Occup. environ. Med.*, **56**, 277–281

Rahman, M., Tondel, M., Ahmad, S.A., Chowdhury, I.A., Faruquee, M.H. & Axelson, O. (1999b) Hypertension and arsenic exposure in Bangladesh. *Hypertension*, **33**, 74–78

Rahman, M.M., Chowdhury, U.K., Mukherjee, S.C., Mondal, B.K., Paul, K., Lodh, D., Biswas, B.K., Chanda, C.R., Basu, G.K., Saha, K.C., Roy, S., Das, R., Palit, S.K., Quamruzzaman, Q. & Chakraborti, D. (2001) Chronic arsenic toxicity in Bangladesh and West Bengal, India — A review and commentary. *Clin. Toxicol.*, **39**, 683–700

Rahman, M.M., Mukherjee, D., Sengupta, M.K., Chowdhury, U.K., Lodh, D., Chanda, C.R., Roy, S., Selim, M., Quamruzzaman, Q., Milton, A.H., Shahidullah, S.M., Rahman, M.T. & Chakraborti, D. (2002) Effectiveness and reliability of arsenic field kits: Are the million dollar screening projects effective or not? *Environ. Sci. Technol.*, **36**, 5385–5394

Rahman, M.M., Mandal, B.K., Roy Chowdhury, T., Sengupta, M.K., Chowdhury, U.K., Lodh, D., Chanda, C.R., Basu, G.K., Mukherjee, S.C., Saha, K.C. & Chakraborti, D. (2003) Arsenic groundwater contamination and sufferings of people in North 24-Parganas, one of the nine arsenic affected districts of West Bengal, India: The seven years study report. *J. environ. Sci. Health*, **A38**, 25–59

Ramírez, P., Eastmond, D.A., Laclette, J.P. & Ostrosky-Wegman, P. (1997) Disruption of micro-tubule assembly and spindle formation as a mechanism for the induction of aneuploid cells by sodium arsenite and vanadium pentoxide. *Mutat. Res.*, **386**, 291–298

Rasmussen, R.E. & Menzel, D.B. (1997) Variation in arsenic-induced sister chromatid exchange in human lymphocytes and lymphoblastoid cell lines. *Mutat. Res.*, **386**, 299–306

Rice, C.M., Ahcroft, W.A., Batten, D.J., Boyce, A.J., Caulfield, J.B.D., Fallick, A.E., Hole, M.J., Jones, E., Pearson, M.J., Rogers, G., Saxton, J.M., Stuart, F.M., Trewin, N.H. & Turner, G. (1995) A Devonian auriferous hot-spring, system, Rhynie, Scotland. *J. geol. Soc.*, **152**, 229–250

Rivara, M.I., Cebrián, M., Corey, G., Hernández, M. & Romieu, I. (1997) Cancer risk in an arsenic-contaminated area of Chile. *Toxicol. ind. Health*, **13**, 321–338

Rogers, E.H., Chernoff, N. & Kavlock, R.J. (1981) The teratogenic potential of cacodylic acid in the rat and mouse. *Drug chem. Toxicol.*, **4**, 49–61

Romo-Kröger, C.M. & Llona, F. (1993) A case of atmospheric contamination at the slopes of the Los Andes mountain range. *Atmos. Environ.*, **27A**, 401–404

Romo-Kröger, C.M., Morales, J.R., Dinator, M.I. & Llona, F. (1994) Heavy metals in the atmosphere coming from a copper smelter in Chile. *Atmos. Environ.*, **28**, 705–711

Rosenberg, H.G. (1974) Systemic arterial disease and chronic arsenicism in infants. *Arch. Pathol.*, **97**, 360–365

Rossman, T.G., Stone, D., Molina, M. & Troll, W. (1980) Absence of arsenite mutagenicity in E. coli and Chinese hamster cells. *Environ. Mutag.*, **2**, 371–379

Rossman, T.G., Uddin, A.N., Burns, F.J. & Bosland, M.C. (2001) Arsenite is a cocarcinogen with solar ultraviolet radiation for mouse skin: An animal model for arsenic carcinogenesis. *Toxicol. appl. Pharmacol.*, **176**, 64–71

Roth, F. (1957) After-effects of chronic arsenism in Moselle wine-makers. *Dtsch. med. Wochenschr.*, **82**, 211–217

Roussel, R.R. & Barchowsky, A. (2000) Arsenic inhibits NF-κB-mediated gene transcription by blocking IκB kinase activity and IκBα phosphorylation and degradation. *Arch. Biochem. Biophys.*, **377**, 204–212

RoyChoudhury, A., Das, T., Sharma, A. & Talukder, G. (1996) Dietary garlic extract in modifying clastogenic effects of inorganic arsenic in mice: Two-generation studies. *Mutat. Res.*, **359**, 165–170

Roychowdhury, T., Uchino, T., Tokunaga, H. & Ando, M. (2002) Survey of arsenic in food compo-sites from an arsenic-affected area of West Bengal, India. *Food chem. Toxicol.*, **40**, 1611–1621

Roychowdhury, T., Tokunaga, H. & Ando, M. (2003) Survey of arsenic and other heavy metals in food composites and drinking water and estimation of dietary intake by the villagers from an arsenic-affected area of West Bengal, India. *Sci. total Environ.*, **308**, 15–35

Rudnay, P. & Börzsönyi, M. (1981) [The tumorigenic effect of treatment with arsenic trioxide.] *Magyar Onkol.*, **25**, 73–77 (in Hungarian)

Rutishauser, U., Acheson, A., Hall, A.K., Mann, D.M. & Sunshine, J. (1988) The neural cell adhesion molecule (NCAM) as a regulator of cell–cell interactions. *Science*, **240**, 53–57

Ryabukhin, Y.S. (1978) *Activation Analysis of Hair as an Indicator of Contamination of Man by Environmental Trace Element Pollutants* (IAEA Report 50), Vienna, International Atomic Energy Agency

Saad, A. & Hassanien, M.A. (2001) Assessment of arsenic level in the hair of the nonoccupational Egyptian population: Pilot study. *Environ. int.*, **27**, 471–478

Sabbioni, E., Fischbach, M., Pozzi, G., Pietra, R., Gallorini, M. & Piette, J.L. (1991) Cellular retention, toxicity and carcinogenic potential of seafood arsenic. I. Lack of cytotoxicity and transforming activity of arsenobetaine in the BALB/3T3 cell line. *Carcinogenesis*, **12**, 1287–1291

Saha, K.C. (1984) Melanokeratosis from arsenic contaminated tubewell water. *Indian J. Dermatol.*, **29**, 37–46

Saha, K.C. (1995) Chronic arsenical dermatoses from tube-well water in West Bengal during 1983–87. *Indian J. Dermatol.*, **40**, 1–12

Saha, K.C. (2001) Cutaneous malignancy in arsenicosis. *Br. J. Dermatol.*, **145**, 185

Saha, K.C. & Chakraborti, D. (2001) Seventeen years experience of arsenicosis in West Bengal, India. In: Chappell, W.R., Abernathy, C.O. & Calderon, R.L., eds, *Arsenic Exposure and Health Effects IV*, New York, Elsevier, pp. 387–395

Saha, K.C. & Poddar, D. (1986) Further studies on chronic arsenical dermatosis. *Indian J. Dermatol.*, **31**, 29–33

Saleha Banu, B., Danadevi, K., Jamil, K., Ahuja, Y.R., Visweswara, R.K. & Ishaq, M. (2001) In vivo genotoxic effect of arsenic trioxide in mice using comet assay. *Toxicology*, **162**, 171–177

Salim, E.I., Wanibuchi, H., Morimura, K., Wei, M., Mitsuhashi, M., Yoshida, K., Endo, G. & Fukushima, S. (2003) Carcinogenicity of dimethylarsinic acid in p53$^{+/-}$ heterozygous knockout and wild type C57BL/6J mice. *Carcinogenesis*, **24**, 335–342

Sampayo-Reyes, A., Zakharyan, R.A., Healy, S.M. & Aposhian, H.V. (2000) Monomethylarsonic acid reductase and monomethylarsonous acid in hamster tissue. *Chem. Res. Toxicol.*, **13**, 1181–1186

Sancha, A.M. (1997) *Determinación de Línea Base de Arsénico ambiental* (Proyecto FONDEF 2 – 24), Santiago, Facultad de Ciencias Físicas y Matemáticas de la Universidad de Chile

Sandoval, H. & Venturino, H. (1987) [Environmental contamination with arsenic in Chile.] *Cuad. méd. Soc.*, **28**, 30–37 (in Spanish)

Santolaya, B.R., Slazar, L., Santolaya, C.R., Sandoval, M. & Alfaro, R. (1995) *Arsénico Impacto Sobre el Hombre y su Entorno. II Region de Chile (Antofagasta)* (Programa Ambiente 02 (1992–1993)), Chuquicamata, Centro de Investigaciones Ecobiológicas y Médicas de Altura, División Chuqicamata Codelco Chile

Santra, A., Das Gupta, J., De, B.K., Roy, B. & Guha Mazumder, D.N. (1999) Hepatic manifestations in chronic arsenic toxicity. *Indian J. Gastroenterol.*, **18**, 152–155

Santra, A., Maiti, A., Das, S., Lahiri, S., Charkaborty, S.K. & Guha Mazumder, D.N. (2000) Hepatic damage caused by chronic arsenic toxicity in experimental animals. *Clin. Toxicol.*, **38**, 395–405

Sarin, S.K., Sharma, G., Banerjee, S., Kathayat, R. & Malhotra, V. (1999) Hepatic fibrogenesis using chronic arsenic ingestion: Studies in a murine model. *Indian J. exp. Biol.*, **37**, 147–151

de Sastre, M.S.R., Varillas, A. & Kirschbaum, P. (1992) Arsenic content in water in the northwest area of Argentina. In: *International Seminar Proceedings (91–99) on Arsenic in the Environment and its Incidence on Health*, Santiago, Universidad de Chile, pp. 123–130

Sato, M. & Bremner, I. (1993) Oxygen free radicals and metallothionein. *Free Radic. Biol. Med.*, **14**, 325–337

Schaumlöffel, N. & Gebel, T. (1998) Heterogeneity of the DNA damage provoked by antimony and arsenic. *Mutagenesis*, **13**, 281–286

Schiller, C.M., Fowler, B.A. & Woods, J.S. (1977) Effects of arsenic on pyruvate dehydrogenase activation. *Environ. Health Perspect.*, **19**, 205–207

Scott, N., Hatlelid, K.M., MacKenzie, N.E. & Carter, D.E. (1993) Reactions of arsenic(III) and arsenic(V) species with glutathione. *Chem. Res. Toxicol.*, **6**, 102–106

Sekhar, K.C., Chary, N.S., Kamala, C.T., Rao, V.R., Balaram, V. & Anjaneyulu, Y. (2003) Risk assessment and pathway study of arsenic in industrially contaminated sites of Hyderabad: A case study. *Environ. int.*, **29**, 601–611

Shibata, Y. & Morita, M. (1989) Speciation of arsenic by reversed-phase high performance liquid chromatography–inductively coupled plasma mass spectrometry. *Anal. Sci.*, **5**, 107–109

Shibutani, S., Takeshita, M. & Grollman, A.P. (1991) Insertion of specific bases during DNA synthesis past the oxidation-damaged base 8-oxodG. *Nature*, **349**, 431–434

Shiobara, Y., Ogra, Y. & Suzuki, K.T. (2001) Animal species differences in the uptake of dimethyl-arsinous acid (DMA^{III}) by rat blood cells. *Chem. Res. Toxicol.*, **14**, 1446–1452

Shirachi, D.Y., Johansen, M.G., McGowan, J.P. & Tu, S.-H. (1983) Tumorigenic effect of sodium arsenite in rat kidney. *Proc. West. pharmacol. Soc.*, **26**, 413–415

Shotyk, W., Cheburkin, A.K., Appleby, P.G., Fankhauser, A. & Kramers, J.D. (1996) Two thousand years of atmospheric arsenic, antimony, and lead deposition recorded in an ombrotrophic peat bog profile, Jura Mountains, Switzerland. *Earth planet. Sci. Lett.*, **145**, E1–E7

Shraim, A., Chiswell, B. & Olszowy, H. (1999) Speciation of arsenic by hydride generation–atomic absorption spectrometry (HG–AAS) in hydrochloric reaction medium. *Talanta*, **50**, 1109–1127

Shraim, A., Chiswell, B. & Olszowy, H. (2000) Use of perchloric acid as a reaction medium for speciation of arsenic by hydride generation–atomic absorption spectrometry. *Analyst*, **125**, 949–953

Shraim, A., Hirano, S. & Yamauchi, H. (2001) Extraction and speciation of arsenic in hair using HPLC-ICPMS. *Anal. Sci.*, **17** (Suppl.), i1729–i1732

Shrestha, R.R., Maskey, A. & Khadka, P.K. (2002) Groundwater arsenic contamination in Nepal: A new challenge for water supply sector. In: *Fifth International Conference on Arsenic Exposure and Health — Speakers' Abstracts*, Denver, CO, University of Colorado

Silva, R.A., Muñoz, S.E., Perez, C.A. & Eynard, A.R. (2000) Effects of dietary fat on benz-a-pyrene-induced forestomach tumorigenesis in mice chronically exposed to arsenic. *Exp. Toxicol. Pathol.*, **52**, 11–16

Simeonova, P.P. & Luster, M.I. (2000) Mechanisms of arsenic carcinogenicity: Genetic or epigenetic mechanisms? *J. environ. Pathol. Toxicol. Oncol.*, **19**, 281–286

Simeonova, P.P., Wang, S., Toriuma, W., Kommineni, V., Matheson, J., Unimye, N., Kayama, F., Harki, D., Vallyathan, V. & Luster, M.I. (2000) Arsenic mediates cell proliferation and gene expression in the bladder epithelium: Association with activating protein-1 transactivation. *Cancer Res.*, **60**, 3445–3453

Siripitayakunkit, U., Visudhiphan, P., Pradipasen, M. & Vorapongsathron, T. (1999) Association between chronic arsenic exposure and children's intelligence in Thailand. In: Chappell, W.R., Abernathy, C.O. & Calderon, R.L., eds, *Arsenic Exposure and Health Effects*, Amsterdam, Elsevier Science, pp. 141–149

SM 3113 (1999) *Standard Methods for the Examination of Water and Wastewater. Metals by Electrothermal Atomic Absorption Spectrometry*, American Water Works Association

SM 3114 (1999) *Standard Methods for the Examination of Water and Wastewater. Arsenic and Selenium Hydride by Hydride Generation/Atomic Absorption Spectrometry*, American Water Works Association

SM 3120 (1999) *Standard Methods for the Examination of Water and Wastewater. Metals by Plasma Emission Spectrometry*, American Water Works Association

Smedley, P.L. (1996) Arsenic in rural groundwater in Ghana. *J. Afr. Earth Sci.*, **22**, 459–470

Smedley, P.L. & Kinniburgh, D.G. (2002) A review of the source, behaviour and distribution of arsenic in natural waters. *Appl. Geochem.*, **17**, 517–568

Smedley, P.L., Nicolli, H.B., Macdonald, D.M.J., Barros, A.J. & Tullio, J.O. (2002) Hydrochemistry of arsenic and other inorganic constituents in groundwaters from La Pampa, Argentina. *Appl. Geochem.*, **17**, 259–284

Smith, A.H., Hopenhayn-Rich, C., Bates, M.N., Goeden, H.M., Hertz-Picciotto, I., Duggan, H.M., Wood, R., Kosnett, M.J. & Smith, M.T. (1992) Cancer risks from arsenic in drinking water. *Environ. Health Perspect.*, **97**, 259–267

Smith, A.H., Goycolea, M., Haque, R. & Biggs, M.L. (1998) Marked increase in bladder and lung cancer mortality in a region of northern Chile due to arsenic in drinking water. *Am. J. Epidemiol.*, **147**, 660–669

Smith, A.H., Lingas, E.O. & Rahman, M. (2000a) Contamination of drinking water by arsenic in Bangladesh: A public health emergency. *Bull. World Health Org.*, **78**, 1093–1103

Smith, A.H., Arroyo, A.P., Guha Mazumder, D.N., Kosnett, M.J., Hernandez, A.L., Beeris, M., Smith, M.M. & Moore, L.E. (2000b) Arsenic-induced skin lesions among the Atacameño people in northern Chile despite good nutrition and centuries of exposure. *Environ. Health Perspect.*, **108**, 617–620

Smith, A.H., Lopipero, P.A., Bates, M.N. & Steinmaus, C.M. (2002) Arsenic epidemiology and drinking water standards. *Science*, **296**, 2145–2146

Sommers, S.G. & McManus, R.G. (1953) Multiple arsenical cancers of skin and internal organs. *Cancer*, **6**, 347–359

Southwick, J.W., Western, A.E. & Beck, M.M. (1983) An epidemiological study of arsenic in drinking water in Millard County, Utah. In: Lederer, W. & Fensterheim, R., eds, *Arsenic: Industrial, Biomedical, Environmental Perspectives*, New York, Van Nostrand Reinhold, pp. 210–225

Sriwana, T., van Bergen, M.J., Sumarti, S., de Hoog, J.C.M., van Os, B.J.H., Wahyuningsih, R. & Dam, M.A.C. (1998) Volcanogenic pollution by acid water discharges along Ciwidey River, West Java (Indonesia). *J. geochem. Explor.*, **62**, 161–182

Steering Committee Arsenic Investigation Project (1991) *National Drinking Water Mission Submission Project on Arsenic Pollution in Groundwater in West Bengal* (Final Report), Kalkota, Public Health and Environment Department, Government of West Bengal

Sternowsky, H.-J., Moser, B. & Szadkowsky, D. (2002) Arsenic in breast milk during the first 3 months of lactation. *Int. J. Hyg. environ. Health*, **205**, 405–409

Stocker, J., Balluch, D., Gsell, M., Harms, H., Feliciano, J., Daunert, S., Malik, K.A. & van der Meer, J.R. (2003) Development of a set of simple bacterial biosensors for quantitative and rapid measurements of arsenite and arsenate in potable water. *Environ. Sci. Technol.*, **37**, 4743–4750

Stoica, A., Pentecost, E. & Martin, M.B. (2000) Effects of arsenite on estrogen receptor-α expression and activity in MCF-7 breast cancer cells. *Endocrinology*, **141**, 3595–3602

Styblo, M., Yamauchi, H. & Thomas, D.J. (1995) Comparative *in vitro* methylation of trivalent and pentavalent arsenicals. *Toxicol. appl. Pharmacol.*, **135**, 172–178

Styblo, M., Delnomdedieu, M. & Thomas, D.J. (1996) Mono- and dimethylation of arsenic in rat liver cytosol in vitro. *Chem.-biol. Interact.*, **99**, 147–164

Styblo, M., Del Razo, L.M., LeCluyse, E.L., Hamilton, G.A., Wang, C., Cullen, W.R. & Thomas, D.J. (1999) Metabolism of arsenic in primary cultures of human and rat hepatocytes. *Chem. Res. Toxicol.*, **12**, 560–565

Styblo, M., Del Razo, L.M., Vega, L., Germolec, D.R., LeCluyse, E.L., Hamilton, G.A., Reed, W., Wang, C., Cullen, W.R. & Thomas, D.J. (2000) Comparative toxicity of trivalent and penta-valent inorganic and methylated arsenicals in rat and human cells. *Arch. Toxicol.*, **74**, 289–299

Sun, G., Pi, J., Li, B., Guo, X., Yamauchi, H. & Yoshida, T. (2001) Progresses on researches of endemic arsenism in China: Population at risk, intervention actions, and related scientific issues. In: Chappell, W.R., Abernathy, C.O. & Calderon, R.L., eds, *Arsenic Exposure and Health Effects IV*, Amsterdam, Elsevier, pp. 79–85

Swart, D.J. & Simeonsson, J.B. (1999) Development of an electrothermal atomization laser-excited atomic fluorescence spectrometry procedure for direct measurements of arsenic in diluted serum. *Anal. Chem.*, **71**, 4951–4955

Szuler, I.M., Williams, C.N., Hindmarsh, J.T. & Park-Dincsoy, H. (1979) Massive variceal hemorrhage secondary to presinusoidal portal hypertension due to arsenic poisoning. *Can. med. Assoc. J.*, **120**, 168–171

Tabacova, S., Hunter E.S., III & Gladen, B.C. (1996) Developmental toxicity of inorganic arsenic in whole embryo culture: Oxidation state, dose, time, and gestational age dependence. *Toxicol. appl. Pharmacol.*, **138**, 298–307

Taiwan Environmental Protection Agency (2000) [*Standards for Drinking Water Quality*], Taipei (in Chinese)

Takatsu, A. & Uchiumi, A. (1998) Abnormal arsenic accumulation by fish living in a naturally aci-dified lake. *Analyst*, **123**, 73–75

Takeuchi, X. (1979) Embryotoxicity of arsenic acid: Light and electron microscopy of its effect on neurolation-stage rat embryo. *J. toxicol. Sci.*, **4**, 405–416

Tam, G.K.H., Charbonneau, S.M., Bryce, F., Pomroy, C. & Sandi, E. (1979) Metabolism of inorganic arsenic ([74]As) in humans following oral ingestion. *Toxicol. appl. Pharmacol.*, **50**, 319–322

Tamaki, S. & Frankenberger, W.T. (1992) Environmental biochemistry of arsenic. *Rev. environ. Contam. Toxicol.*, **124**, 79–110

Tandukar, N., Bhattacharya, P. & Mukherjee, A.B. (2001) Preliminary assessment of arsenic conta-mination in groundwater in Nepal. In: *Proceedings of the International Conference on Arsenic in the Asia-Pacific Region: Managing Arsenic for our Future, Adelaide, Australia, November 2001*, Collingwood, CSIRO-Land and Water, p. 103

Tay, C.H. (1974) Cutaneous manifestations of arsenic poisoning due to certain Chinese herbal medicine. *Austr. J. Dermatol.*, **15**, 121–131

Thomas, D.J., Styblo, M. & Lin, S. (2001) The cellular metabolism and systemic toxicity of arsenic. *Toxicol. appl. Pharmacol.*, **176**, 127–144

Thompson, T.S., Le, M.D., Kasick, A.R. & Macaulay, T.J. (1999) Arsenic in well water supplies in Saskatchewan. *Bull. environ. Contam. Toxicol.*, **63**, 478–483

Thornalley, P.J. & Vašák, M. (1985) Possible role for metallothionein in protection against radiation-induced oxidative stress. Kinetics and mechanism of its reaction with superoxide and hydroxyl radicals. *Biochim. biophys. Acta*, **827**, 36–44

Tian, D., Ma, H., Feng, Z., Xia, Y., Le, X.C., Ni, Z., Allen, J., Collins, B., Schreinemachers, D. & Mumford, J.L. (2001) Analyses of micronuclei in exfoliated epithelial cells from individuals chronically exposed to arsenic via drinking water in Inner Mongolia, China. *J. Toxicol. environ. Health*, **A64**, 473–484

Tice, R.R., Yager, J.W., Andrews, P. & Crecelius, E. (1997) Effect of hepatic methyl donor status on urinary excretion and DNA damage in B6C3F1 mice treated with sodium arsenite. *Mutat. Res.*, **386**, 315–334

Tinwell, H., Stephens, S.C. & Ashby, J. (1991) Arsenite as the probable active species in the human carcinogenicity of arsenic: Mouse micronucleus assays on Na and K arsenite, orpiment, and Fowler's solution. *Environ. Health Perspect.*, **95**, 205–210

Tondel, M., Rahman, M., Magnuson, A., Chowdhury, I.A., Faruquee, M.H. & Ahmad, S.A. (1999) The relationship of arsenic levels in drinking water and the prevalence rate of skin lesions in Bangladesh. *Environ. Health Perspect.*, **107**, 727–729

Tsai, S.-M., Wang, T.-N. & Ko, Y.-C. (1999) Mortality for certain diseases in areas with high levels of arsenic in drinking water. *Arch. environ. Health*, **54**, 186–193

Tseng, W.-P. (1977) Effects and dose–response relationships of skin cancer and blackfoot disease with arsenic. *Environ. Health Perspect.*, **19**, 109–119

Tseng, W.P., Chu, H.M., How, S.W., Fong, J.M., Lin, C.S. & Yeh, S. (1968) Prevalence of skin cancer in an endemic area of chronic arsenicism in Taiwan. *J. natl Cancer Inst.*, **40**, 453–463

Tseng, C.-H., Chong, C.-K., Chen, C.-J. & Tai, T.-Y. (1996) Dose–response relationship between peripheral vascular disease and ingested inorganic arsenic among residents in blackfoot disease endemic villages in Taiwan. *Atherosclerosis*, **120**, 125–133

Tseng, C.-H., Tai, T.-Y., Chong, C.-K., Tseng, C.-P., Lai, M.-S., Lin, B.J,. Chiou, H.-Y., Hsueh, Y.-M., Hsu, K.-H. & Chen, C.-J. (2000) Long-term arsenic exposure and incidence of non-insulin-dependent diabetes mellitus: A cohort study in arseniasis-hyperendemic villages in Taiwan. *Environ. Health Perspect.*, **108**, 847–851

Tsuda, T., Nagira, T., Yamamoto, M., Kurumatani, N., Hotta, N., Harada, M. & Aoyama, H. (1989) Malignant neoplasms among residents who drank well water contaminated by arsenic from a King's Yellow factory. *Sangyo Ika Daigaku Zasshi*, **11** (Suppl.), 289–301

Tsuda, T., Babazono, A., Yamamoto, E., Kurumatani, N., Mino, Y., Ogawa, T., Kishi, Y. & Aoyama, H. (1995) Ingested arsenic and internal cancer: A historical cohort study followed for 33 years. *Am. J. Epidemiol.*, **141**, 198–209

Turpeinen, R., Pantsar-Kallio, M., Häggblom, M. & Kairesalo, T. (1999) Influence of microbes on the mobilization, toxicity and biomethylation of arsenic in soil. *Sci. total Environ.*, **236**, 173–180

Turpeinen, R., Pantsar-Kallio, M. & Kairesalo, T. (2002) Role of microbes in controlling the speciation of arsenic and production of arsines in contaminated soils. *Sci. total Environ.*, **285**, 133–145

UNESCAP-UNICEF-WHO (2001) *Meeting on Geology and Health: Solving the Arsenic Crisis in the Asia-Pacific Region, 2–4 May 2001, Bangkok*

UNIDO (2001) *Study Report under the Project Concerted Action on Elimination/Reduction of Arsenic in Groundwater, West Bengal, India* (Project NC/IND/99/967), Vols I and II, New Delhi, United Nations Industrial Development Organization

US Environmental Protection Agency (1994a) *Method 200.7. Methods for the Determination of Metals in Environmental Samples*, Washington DC

US Environmental Protection Agency (1994b) *Method 200.8. Methods for the Determination of Metals in Environmental Samples*, Washington DC

US Environmental Protection Agency (1996a) *Method 1638. Determination of Trace Elements in Ambient Waters by Inductively Coupled Plasma–Mass Spectrometry*, Washington DC

US Environmental Protection Agency (1996b) *Method 7063. Arsenic in Aqueous Samples and Extracts by Anodic Stripping Boltammetry (AVS)*, Washington DC

US Environmental Protection Agency (1996c) *Method 1632. Inorganic Arsenic in Water by Hydride Generation Quartz Furnace Atomic Absorption*, Washington DC

US Environmental Protection Agency (2001) National primary drinking water regulations; arsenic and clarifications to compliance and new source contaminants monitoring; final rule. *Fed. Regist.*, **66**, 6976–7066

US Environmental Protection Agency (2002) Drinking water regulations for arsenic and classifications to compliance and new source contaminants monitoring. In: *Implementation Guidance for Arsenic Rules*, Washington DC

US Environmental Protection Agency-Battelle (2002a) *The Environmental Technology Verification Program, EPA/ETV Joint Verification Statement, Arsenic Test Kit, Quick™*, Washington DC

US Environmental Protection Agency-Battelle (2002b) *The Environmental Technology Verification Program, EPA/ETV Joint Verification Statement, Arsenic Test Kit, AS 75*, Washington DC

Vahter, M. (1999a) Variation in human metabolism of arsenic. In: Chappell, W.R., Abernathy, C.O. & Calderon, R.L., eds, *Arsenic Exposure and Health Effects*, Oxford, Elsevier Science, pp. 267–279

Vahter, M. (1999b) Methylation of inorganic arsenic in different mammalian species and population groups. *Sci. Prog.*, **82**, 69–88

Vahter, M. (2002) Mechanism of arsenic biotransformation. *Toxicology*, **181–182**, 211–217

Vahter, M. & Concha, G. (2001) Role of metabolism in arsenic toxicity. *Pharmacol. Toxicol.*, **89**, 1–5

Vahter, M. & Marafante, E. (1987) Effects of low dietary intake of methionine, choline or proteins on the biotransformation of arsenite in the rabbit. *Toxicol. Lett.*, **37**, 41–46

Vahter, M. & Marafante, E. (1989) Intracellular distribution and chemical forms of arsenic in rabbits exposed to arsenate. *Biol. trace Elem. Res.*, **21**, 233–239

Vahter, M. & Norin, H. (1980) Metabolism of ^{74}As-labeled trivalent and pentavalent inorganic arsenic in mice. *Environ. Res.*, **21**, 446–457

Vahter, M., Marafante, E., Lindgren, A. & Dencker, L. (1982) Tissue distribution and subcellular binding of arsenic in marmoset monkeys after injection of ^{74}As-arsenite. *Arch. Toxicol.*, **51**, 65–77

Vahter, M., Marafante, E. & Dencker, L. (1984) Tissue distribution and retention of ^{74}As-dimethylarsinic acid in mice and rats. *Arch. environ. Contam. Toxicol.*, **13**, 259–264

Vahter, M., Couch, R., Nermell, B. & Nilsson, R. (1995a) Lack of methylation of inorganic arsenic in the chimpanzee. *Toxicol. appl. Pharmacol.*, **133**, 262–268

Vahter, M., Concha, G., Nermell, B., Nilsson, R., Dulout, F. & Natarajan, A.T. (1995b) A unique metabolism of inorganic arsenic in native Andean women. *Eur. J. Pharmacol.*, **293**, 455–462

Van Duuren, B.L., Melchionne, S., Seidman, I. & Pereira, M.A. (1986) Chronic bioassays of chlorinated humic acids in B6C3F1 mice. *Environ. Health Perspect.*, **69**, 109–117

Venturino, H. (1987) [Determination of urinary arsenic concentration in different regions of Chile. Epidemiological study] *Cuad. méd. Soc.*, **28**, 38–40 (in Spanish)

Vig, B.K., Figueroa, M.L., Cornforth, M.N. & Jenkins, S.H. (1984) Chromosome studies in human subjects chronically exposed to arsenic in drinking water. *Am. J. ind. Med.*, **6**, 325–338

Vogel, A.E. (1954) Special tests for small amounts of arsenic. In: Vogel, A.E., ed., *A Text-book of Macro and Semimicro Qualitative Inorganic Analysis*, London, Longmans, pp. 242–247

Waalkes, M.P., Ward, J.M., Liu, J. & Diwan, B.A. (2003) Transplacental carcinogenicity of inorganic arsenic in the drinking water: Induction of hepatic, ovarian, pulmonary and adrenal tumors in mice. *Toxicol. appl. Pharmacol.*, **186**, 7–17

Wagner, S.L., Maliner, J.S., Morton, W.E. & Braman, R.S. (1979) Skin cancer and arsenical intoxication from well water. *Arch. Dermatol.*, **115**, 1205–1207

Wang, L. (1996) Current status of arsenic-related health hazards in Xinjiang. *Chinese J. public Health*, **15** (Suppl. 3), 53–58

Wang, C. & Lazarides, E. (1984) Arsenite-induced changes in methylation of the 70,000 Dalton heat shock proteins in chicken embryo fibroblasts. *Biochem. biophys. Res. Commun.*, **119**, 735–743

Wang, T.-S., Kuo, C.-F., Jan, K.-Y. & Huang, H. (1996) Arsenite induces apoptosis in Chinese hamster ovary cells by generation of reactive oxygen species. *J. cell. Physiol.*, **169**, 256–268

Wang, T.S., Shu, Y.F., Liu, Y.C., Jan, K.Y. & Huang, H. (1997) Glutathione peroxidase and catalase modulate the genotoxicity of arsenite. *Toxicology*, **121**, 229–237

Wang, C.-H., Jeng, J.-S., Yip, P.-K., Chen, C.-L., Hsu, L.-I., Hsueh, Y.-M., Chiou, H.-Y., Wu, M.-M. & Chen, C.-J. (2002) Biological gradient between long-term arsenic exposure and carotid atherosclerosis. *Circulation*, **105**, 1804–1809

Wanibuchi, H., Yamamoto, S., Chen, H., Yoshida, K., Endo, G., Hori, T. & Fukushima, S. (1996) Promoting effects of dimethylarsinic acid on *N*-butyl-*N*-(4-hydroxybutyl)nitrosamine-induced urinary bladder carcinogenesis in rats. *Carcinogenesis*, **17**, 2435–2439

Wanibuchi, H., Hori, T., Meenakshi, V., Ichihara, T., Yamamoto, S., Yano, Y., Otani, S., Nakae, D., Konishi, Y. & Fukushima, S. (1997) Promotion of rat hepatocarcinogenesis by dimethylarsinic acid: Association with elevated ornithine decarboxylase activity and formation of 8-hydroxy-deoxyguanosine in the liver. *Jpn. J. Cancer Res.*, **88**, 1149–1154

Warner, K.L. (2001) Arsenic in shallow and deep glacial drift aquifers — Illinois. In: *Proceedings of the 2001 USGS Workshop on Arsenic in the Environment, 21–22 February 2001, Denver, Colorado*, Denver, CO, USGS Arsenic Studies Group, US Geological Survey

Warner, M.L., Moore, L.E., Smith, M.T., Kalman, D.A., Fanning, E. & Smith, A.H. (1994) Increased micronuclei in exfoliated bladder cells of individuals who chronically ingest arsenic-contaminated water in Nevada. *Cancer Epidemiol. Biomarkers Prev.*, **3**, 583–590

Wei, M., Wanibuchi, H., Yamamoto, S., Li, W. & Fukushima, S. (1999) Urinary bladder carcinogenicity of dimethylarsinic acid in male F344 rats. *Carcinogenesis*, **20**, 1873–1876

Wei, M., Wanibuchi, H., Morimura, K., Iwai, S., Yoshida, K., Endo, G., Nakae, D. & Fukushima, S. (2002) Carcinogenicity of dimethylarsinic acid in male F344 rats and genetic alterations in induced urinary bladder tumors. *Carcinogenesis*, **23**, 1387–1397

Welch, A.H., Lico, M.S. & Hughes, J.L. (1988) Arsenic in ground water of the western United States. *Ground Water*, **26**, 333–347

Welch, A.H., Helsel, D.R., Focazio, M.J. & Watkins, S.A. (1999) Arsenic in ground water supplies of the United States. In: Chappell, W.R., Abernathy, C.O. & Calderon, R.L., eds., *Arsenic Exposure and Health Effects*, Oxford, Elsevier Science, pp. 9–17

Welch, A.H., Westjohn, D.B., Helsel, D.R. & Wanty, R.B. (2000) Arsenic in ground water of the United States: Occurrence and geochemistry. *Ground Water*, **38**, 589–604

Wester, R.C., Maibach, H.I., Sedik, L., Melendres, J. & Wade, M. (1993) In vivo and in vitro percutaneous absorption and skin decontamination of arsenic from water and soil. *Fundam. appl. Toxicol.*, **20**, 336–340

WHO (1981) *Arsenic* (Environmental Health Criteria 18), Geneva, International Programme on Chemical Safety

WHO (1998) *Guidelines for Drinking-water Quality*, 2nd Ed., Vol. 2, *Health Criteria and Other Supporting Information*, Geneva

WHO (2000) *Guidelines for Drinking-water Quality Training Pack* [http://www.who.int/water_sanitation_sanitation_health/Training_mat/GDWQtraining.htm]

WHO (2001) *Arsenic and Arsenic Compounds* (Environmental Health Criteria 224), 2nd Ed., Geneva, International Programme on Chemical Safety

Wijeweera, J.B., Gandolfi, A.J., Parrish, A. & Lantz, R.C. (2001) Sodium arsenite enhances AP-1 and NFκB DNA binding and induces stress protein expression in precision-cut rat lung slices. *Toxicol. Sci.*, **61**, 283–294

van Wijk, R., Welters, M., Souren, J.E.M., Ovelgonne, H. & Wiegant, F.A.C. (1993) Serum-stimulated cell cycle progression and stress protein synthesis in C3H10T1/2 fibroblasts treated with sodium arsenite. *J. cell. Physiol.*, **155**, 265–272

Wildfang, E., Zakharyan, R.A. & Aposhian, H.V. (1998) Enzymatic methylation of arsenic compounds. VI. Characterization of hamster liver arsenite and methylarsonic acid methyltransferase activities *in vitro*. *Toxicol. appl. Pharmacol.*, **152**, 366–375

Wildfang, E., Radabaugh, T.R. & Aposhian, H.V. (2001) Enzymatic methylation of arsenic compounds. IX. Liver arsenite methyltransferase and arsenate reductase activities in primates. *Toxicology*, **168**, 213–221

Willhite, C.C. (1981) Arsenic-induced axial skeletal (dysraphic) disorders. *Exp. mol. Pathol.*, **34**, 145–158

Williams, M., Fordyce, F., Paijitprapapon, A. & Charoenchaisri, P. (1996) Arsenic contamination in surface drainage and groundwater in part of the southeast Asian tin belt, Nakhon Si Thammarat Province, southern Thailand. *Environ. Geol.*, **27**, 16–33

Winski, S.L. & Carter, D.E. (1995) Interactions of rat red blood cell sulfhydryls with arsenate and arsenite. *J. Toxicol. environ. Health.*, **46**, 379–397

Winski, S.L. & Carter, D.E. (1998) Arsenate toxicity in human erythrocytes: Characterization of morphologic changes and determination of the mechanism of damage. *J. Toxicol. environ. Health*, **A53**, 345–355

Wlodarczyk, B., Bennett, G.D., Calvin, J.A., Craig, J.C. & Finnell, R.H. (1996) Arsenic-induced alterations in embryonic transcription factor gene expression: Implications for abnormal neural development. *Develop. Genet.*, **18**, 306–315

Wong, O., Whorton, M.D., Foliart, D.E. & Lowengart, R. (1992) An ecologic study of skin cancer and environmental arsenic exposure. *Int. Arch. occup. environ. Health*, **64**, 235–241

Woods, J.S. & Fowler, B.A. (1978) Altered regulation of mammalian hepatic heme biosynthesis and urinary porphyrin excretion during prolonged exposure to sodium arsenate. *Toxicol. appl. Pharmacol.*, **43**, 361–371

Woolson, E.A. (1977) Generation of alkylarsines from soil. *Weed Sci.*, **25**, 412–416

Wu, W. & Welsh, M.J. (1996) Expression of the 25-kDa heat-shock protein (HSP27) correlates with resistance to the toxicity of cadmium chloride, mercuric chloride, *cis*-platinum(II)-diammine dichloride, or sodium arsenite in mouse embryonic stem cells transfected with sense or antisense HSP27 cDNA. *Toxicol. appl. Pharmacol.*, **141**, 330–339

Wu, M.-M., Kuo, T.-L., Hwang, Y.-H. & Chen, C.-J. (1989) Dose–response relation between arsenic concentration in well water and mortality from cancers and vascular diseases. *Am. J. Epidemiol.*, **130**, 1123–1132

Wu, M.-M., Chiou, H.-Y., Wang, T.-W., Hsueh, Y.-M., Wang, I.-H., Chen, C.-J. & Lee, T.-C. (2001) Association of blood arsenic levels with increased reactive oxidants and decreased antioxidant capacity in human population of northeastern Taiwan. *Environ. Health Perspect.*, **109**, 1011–1017

Wyatt, C.J., Lopez Quiroga, V., Olivas Acosta, R.T. & Méndez, R.O. (1998a) Excretion of arsenic (As) in urine of children, 7–11 years, exposed to elevated levels of As in the city water supply in Hermosillo, Sonora, Mexico. *Environ. Res.*, **A78**, 19–24

Wyatt, C.J., Fimbres, C., Romo, L., Méndez, R.O. & Grijalva, M. (1998b) Incidence of heavy metal contamination in water supplies in northern Mexico. *Environ. Res.*, **76**, 114–119

Yager, J.W. & Wiencke, J.K. (1997) Inhibition of poly(ADP-ribose) polymerase by arsenite. *Mutat. Res.*, **386**, 345–351

Yalçin, S. & Le, X.C. (2001) Speciation of arsenic using solid phase extraction cartridges. *J. environ. Monit.*, **3**, 81–85

Yamamoto, A., Hisanaga, A. & Ishinishi, L.N. (1987) Tumorigenicity of inorganic arsenic compounds following intratracheal instillations to the lungs of hamsters. *Int. J. Cancer*, **40**, 220–223

Yamamoto, S., Konishi, Y., Matsuda, T., Murai, T., Shibata, M.-A., Matsui-Yuasa, I., Otani, S., Kuroda, K., Endo, G. & Fukushima, S. (1995) Cancer induction by an organic arsenic compound, dimethylarsinic acid (cacodylic acid), in F344/DuCrj rats after pretreatment with five carcinogens. *Cancer Res.*, **55**, 1271–1276

Yamanaka, K. & Okada, S. (1994) Induction of lung-specific DNA damage by metabolically methylated arsenics via the production of free radicals. *Environ. Health Perspect.*, **102** (Suppl. 3), 37–40

Yamanaka, K., Hasegawa, A., Sawamura, R. & Okada, S. (1989) Dimethylated arsenics induce DNA strand breaks in lung *via* the production of active oxygen in mice. *Biochem. biophys. Res. Commun.*, **165**, 43–50

Yamanaka, K., Hoshino, M., Okamoto, M., Sawamura, R., Hasegawa, A. & Okada, S. (1990) Induction of DNA damage by dimethylarsine, a metabolite of inorganic arsenics, is for the major part likely due to its peroxyl radical. *Biochem. biophys. Res. Commun.*, **168**, 58–64

Yamanaka, K., Hasegawa, A., Sawamura, R. & Okada, S. (1991) Cellular response to oxidative damage in lung induced by the administration of dimethylarsinic acid, a major metabolite of inorganic arsenics, in mice. *Toxicol. appl. Pharmacol.*, **108**, 205–213

Yamanaka, K., Ohtsubo, K., Hasegawa, A., Hayashi, H., Ohji, H., Kanisawa, M. & Okada, S. (1996) Exposure to dimethylarsinic acid, a main metabolite of inorganic arsenics, strongly promotes tumorigenesis initiated by 4-nitroquinoline 1-oxide in the lungs of mice. *Carcinogenesis*, **17**, 767–770

Yamanaka, K., Katsumata, K., Ikuma, K., Hasegawa, A., Nakano, M. & Okada, S. (2000) The role of orally administered dimethylarsinic acid, a main metabolite of inorganic arsenics, in the promotion and progression of UVB-induced skin tumorigenesis in hairless mice. *Cancer Lett.*, **152**, 79–85

Yamanaka, K., Mizoi, M., Kato, K., Hasegawa, A., Nakano, M. & Okada, S. (2001a) Oral administration of dimethylarsinic acid, a main metabolite of inorganic arsenic, in mice promotes skin tumorigenesis initiated by dimethylbenz(a)anthracene with or without ultraviolet B as a promoter. *Biol. pharm. Bull.*, **24**, 510–514

Yamanaka, K., Takabayashi, F., Mizoi, M., An, Y., Hasegawa, A. & Okada, S. (2001b) Oral exposure of dimethylarsinic acid, a main metabolite of inorganic arsenics, in mice leads to an increase in 8-oxo-2'-deoxyguanosine level, specifically in the target organs for arsenic carcinogenesis. *Biochem. biophys. Res. Commun.*, **287**, 66–70

Yamauchi, H. & Yamamura, Y. (1979) Urinary inorganic arsenic and methylarsenic excretion following arsenate-rich seaweed ingestion (author's transl). *Sangyo Igaku*, **21**, 47–54 (in Japanese)

Yamauchi, H. & Yamamura, Y. (1984) Metabolism and excretion of orally administered dimethylarsinic acid in the hamster. *Toxicol. appl. Pharmacol.*, **74**, 134–140

Yamauchi, H. & Yamamura, Y. (1985) Metabolism and excretion of orally administrated arsenic trioxide in the hamster. *Toxicology*, **34**, 113–121

Yamauchi, H., Yamato, N. & Yamamura, Y. (1988) Metabolism and excretion of orally and intraperitoneally administered methylarsonic acid in the hamster. *Bull. environ. Contam. Toxicol.*, **40**, 280–286

Yeh, S. (1963) Studies on endemic chronic arsenism in southwest coast of Taiwan. *Rep. Inst. Pathol. natl Taiwan Univ.*, **14**, 1–23 (in Chinese)

Yeh, S. (1973) Skin cancer in chronic arsenicism. *Hum. Pathol.*, **4**, 469–485

Yeh, S., How, S.W. & Lin, C.S. (1968) Arsenical cancer of skin. Histologic study with special reference to Bowen's disease. *Cancer*, **21**, 312–339

Yoshida, T., Shimamura, T. & Shigeta, S. (1986) Immunological effects of arsenic compounds on mouse spleen cells *in vitro*. *Tokai J. exp. clin. Med.*, **11**, 353–359

Yoshida, T., Shimamura, T. & Shigeta, S. (1987) Enhancement of the immune response *in vitro* by arsenic. *Int. J. Immunopharmacol.*, **9**, 411–415

Yoshida, K., Chen, H., Inoue, Y., Wanibuchi, H., Fukushima, S., Kuroda, K. & Endo, G. (1997) The urinary excretion of arsenic metabolites after a single oral administration of dimethylarsinic acid to rats. *Arch. environ. Contam. Toxicol.*, **32**, 416–421

Yu, R.C., Hsu, K.-H., Chen, C.-J. & Froines, J.R. (2000) Arsenic methylation capacity and skin cancer. *Cancer Epidemiol. Biomarkers Prev.*, **9**, 1259–1262

Zakharova, T., Tatano, F. & Menshikov, V. (2002) Health cancer risk assessment for arsenic exposure in potentially contaminated areas by fertilizer plants: A possible regulatory approach applied to a case study in Moscow Region, Russia. *Regul. Toxicol. Pharmacol.*, **36**, 22–33

Zakharyan, R.A. & Aposhian, H.V. (1999) Enzymatic reduction of arsenic compounds in mammalian systems: The rate-limiting enzyme of rabbit liver arsenic biotransformation is MMA^V reductase. *Chem. Res. Toxicol.*, **12**, 1278–1283

Zakharyan, R., Wu, Y., Bogdan, G.M. & Aposhian, H.V. (1995) Enzymatic methylation of arsenic compounds: Assay, partial purification, and properties of arsenite methyltransferase and monomethylarsonic acid methyltransferase of rabbit liver. *Chem. Res. Toxicol.*, **8**, 1029–1038

Zakharyan, R.A., Wildfang, E. & Aposhian, H.V. (1996) Enzymatic methylation of arsenic compounds. III. The marmoset and tamarin, but not the rhesus, monkeys are deficient in methyltransferases that methylate inorganic arsenic. *Toxicol. appl. Pharmacol.*, **140**, 77–84

Zakharyan, R.A., Sampayo-Reyes, A., Healy, S.M., Tsaprailis, G., Board, P.G., Liebler, D.C. & Aposhian, H.V. (2001) Human monomethylarsonic acid (MMA^V) reductase is a member of the glutathione-S-transferase superfamily. *Chem. Res. Toxicol.*, **14**, 1051–1057

Zaldívar, R. (1974) Arsenic contamination of drinking water and foodstuffs causing endemic chronic poisoning. *Beitr. Pathol.*, **151**, 384–400

Zaldívar, R. & Guillier, A. (1977) Environmental and clinical investigations on endemic chronic arsenic poisoning in infants and children. *Zbl. Bakt. Hyg. I. Abt. Orig. B*, **165**, 226–234

Zaldívar, R., Villar, I., Robinson, H., Wetterstrand, W.H., Ghai, G.L. & Brain, C. (1978) [Chronic arsenic poisoning. Epidemiological, clinical, pathological, toxicological and nutritional data.] *Rev. méd. Chile*, **106**, 1027–1030 (in Spanish)

Zaldívar, R., Prunés, L. & Ghai, G.L. (1981) Arsenic dose in patients with cutaneous carcinomata and hepatic haemangio-endothelioma after environmental and occupational exposure. *Arch. Toxicol.*, **47**, 145–154

Zhang, X., Cornelis, R., De Kimpe, J., Mees, L. & Lameire, N. (1997) Speciation of arsenic in serum, urine, and dialysate of patients on continuous ambulatory peritoneal dialysis. *Clin. Chem.*, **43**, 406–408

Zhang, X., Cornelis, R., De Kimpe, J., Mees, L. & Lameire, N. (1998a) Study of arsenic-protein binding in serum of patients of continuous ambulatory peritoneal dialysis. *Clin. Chem.*, **44**, 141–147

Zhang, X., Cornelis, R., Mees, L. Vanholder, R. & Lameire, N. (1998b) Chemical speciation of arsenic in serum of uraemic patients. *Analyst*, **123**, 13–17

Zhao, C.Q., Young, M.R., Diwan, B.A., Coogan, T.P. & Waalkes, M.P. (1997) Association of arsenic-induced malignant transformation with DNA hypomethylation and aberrant gene expression. *Proc. natl Acad. Sci. USA*, **94**. 10907–10912

Zhong, C.X. & Mass, M.J. (2001) Both hypomethylation and hypermethylation of DNA associated with arsenite exposure in cultures of human cells identified by methylation-sensitive arbitrarily-primed PCR. *Toxicol. Lett.*, **122**, 223–234

Zierler, S., Theodore, M., Cohen, A. & Rothman, K.J. (1988) Chemical quality of maternal drinking water and congenital heart disease. *Int. J. Epidemiol.*, **17**, 589–594

MONOGRAPHS ON CHLORAMINE, CHLORAL AND CHLORAL HYDRATE, DICHLOROACETIC ACID, TRICHLOROACETIC ACID AND 3-CHLORO-4-(DICHLOROMETHYL)-5-HYDROXY-2(5*H*)-FURANONE

INTRODUCTION TO THE MONOGRAPHS ON CHLORAMINE, CHLORAL AND CHLORAL HYDRATE, DICHLOROACETIC ACID, TRICHLOROACETIC ACID AND 3-CHLORO-4-(DICHLOROMETHYL)-5-HYDROXY-2(5H)-FURANONE

Studies of Cancer in Humans

No specific studies on chlorinated drinking-water are available in the literature. However, a number of studies that address the risk of several cancers associated with drinking-water also evaluate the risk linked to chlorinated compounds, and are described below.

1.   Cohort studies

Wilkins and Comstock (1981) (Tables 1 and 2) followed up 14 553 men and 16 227 women over 25 years of age, who were residents of Washington County, MD (USA), using data from a census conducted in 1963. Cancer incidence and mortality rates in 1963–75 were assessed in two subcohorts: people who were exposed to chlorinated surface water (average chloroform concentration, 107 µg/L) and those who used water from deep wells that had not been chlorinated. After adjusting for potential confounders, incidence rates were reported for several subsites of cancer. A slightly elevated but non-significant risk for urinary bladder cancer was observed for both men and women. There were no differences in risk for cancers of the kidney or liver between exposed and non-exposed cases, but a risk for liver cancer was suggested in women.

Koivusalo *et al.* (1997) (Tables 1 and 2) reported a historical cohort study of 621 431 persons in 56 towns in Finland with different water sources and practices of water treatment and thus different levels of exposure to chlorination by-products. The cohort was restricted to persons who, in 1970, were living in their town of birth. Information from 1970 onwards on occupation, migration, death, place of residence and water-pipe connection was obtained from Statistics Finland. Cancer incidence for 26 organ sites among these individuals in 1971–93 was ascertained by record linkage with the Finnish National Cancer Registry.

Table 1. Studies on the incidence of urinary bladder cancer associated with chlorinated drinking-water

Study	Population/end-point	No. of cases	Exposure	Risk estimate	Comments
Cohort studies					
Wilkins & Comstock (1981)	Residents of Washington County, MD, USA 30 780 persons (14 553 men, 16 227 women), ≥ 25 years of age, followed up 1963–75	52	Chlorinated surface water (average chloroform concentration, 107 µg/L) versus non-chlorinated deep wells	RR (95% CI) Men 1.8 (0.8–4.8) Women 1.6 (0.5–6.3)	Adjusted for differences between cohorts in age, marital status, education, smoking history, church attendance, housing, persons per room
Doyle et al. (1997)	Iowa Women's Health Study (USA): 41 836 women aged 55–69 years, followed up 1986–93	42	1108 municipal water supplies (1979 and 1986–87) Chloroform concentration in 1986–87 (µg/L) < limit of detection 1–2 3–13 14–287 p for trend	1.0 0.9 (0.4–2.0) 1.2 (0.6–2.7) 0.6 (0.3–1.6) 0.46	Adjusted for age, education, smoking, physical activity, fruit and vegetable intake, energy intake, body mass index and waist-to-hip ratio
Koivusalo et al. (1997)	Finland, incidence 1971–93 56 towns – 32% of country population	313 464 men 307 967 women	Estimates of mutagenic potency of drinking-water; 3000 net rev/L increase in average exposure to mutagenicity	RR (95% CI) Men 1.03 (0.8–1.3) Women 1.5 (1.01–2.2)	Record-linkage study Adjusted for age, time-period, urbanization and social status. Cancers of ureter and urethra included

Table 1 (contd)

Study	Population/end-point	No. of cases	Exposure	Risk estimate	Comments
Case–control studies					
Cantor *et al.* (1987)	10 areas in the USA: Atlanta, Connecticut, Detroit, Iowa, New Jersey, New Mexico, New Orleans, Seattle, San Francisco, Utah Incidence, 1-year period starting December 1997	2805 5258 population controls (men and women)	Duration of consumption of chlorinated surface drinking-water in subjects with tap-water consumption above median (1.44 L/day)	OR (95% CI) *Men* *Years* 0 1.0 1–19 1.1 (0.7–1.6) 20–39 1.1 (0.7–1.5) 40–59 1.2 (0.8–1.7) ≥ 60 1.2 (0.7–2.1) *p* for trend = 0.44 *Women* 0 1.0 1–19 1.8 (0.8–3.7) 20–39 1.5 (0.7–3.1) 40–59 2.2 (1.0–4.8) ≥ 60 3.2 (1.2–8.7) *p* for trend = 0.02	Adjusted for age, smoking habit, high-risk occupation, population size of usual residence and reporting centre
McGeehin *et al.* (1993)	Colorado, USA Incidence 1990–91	327 261 controls with other cancers excluding lung and colorectal cancer (men and women, all white)	Lifetime exposure to chlorinated water from individual histories of residence and water source	OR (95% CI) *Years* 0 1.0 1–10 0.7 (0.4–1.3) 11–20 1.4 (0.8–2.5) 21–30 1.5 (0.8–2.9) > 30 1.8 (1.1–2.9)	Adjusted for coffee consumption, smoking, tap-water intake, family history of bladder cancer, sex, medical history of bladder infection or kidney stone
Vena *et al.* (1993)	Western New York, USA Incidence, 1979–85	351 855 population controls (men, white)	Daily intake of tap-water	*Age < 65 years* *No. of cups* 0–5 1.00 6–7 1.3 (0.7–2.4) 8–9 1.6 (0.9–3.0) 10–39 2.6 (1.5–4.5) *p* for trend < 0.001	Adjusted for age, education, cigarette smoking (pack–years), and coffee, carotene and non-tap-water intake

Table 1 (contd)

Study	Population/end-point	No. of cases	Exposure	Risk estimate	Comments
Vena et al. (1993) (contd)				*Age ≥ 65 years*	
				0–5 1.00	
				6–7 1.3 (0.8–2.1)	
				8–9 1.4 (0.8–2.5)	
				10–39 3.0 (1.8–5.0)	
				p for trend < 0.001	
King & Marrett (1996)	Ontario, Canada Incidence, September 1992–May 1994	696 1545 population controls (men and women)	Years of consumption of chlorinated surface drinking-water	*Years* 1.0	Adjusted for age, sex, log pack–years of smoking, current smoking, education and calorie intake
				0–9 1.0	
				10–19 1.04 (0.7–1.5)	
				20–34 1.2 (0.9–1.5)	
				≥ 35 1.4 (1.1–1.8)	
			Trihalomethane–years	*Quartiles (µg/L–years)*	
				0–583 1.0	
				584–1505 1.2 (0.9–1.6)	
				1506–1956 1.08 (0.8–1.4)	
				1957–6425 1.4 (1.1–1.9)	
			Level of trihalomethanes in water source	*Level (µg/L)* 1.0	
				0–24 1.0	
				25–74 1.4 (1.0–2.0)	
				≥ 75 1.7 (1.1–2.5)	
				p for trend 0.006	
Freedman et al. (1997)	Washington County, MD, USA Incidence, 1975–92	294 2326 population controls	Duration of residence with municipal water source	OR (95% CI)	Adjusted for age, sex, smoking and urbanicity
				Men	
				Years *Cases*	
				None 54 1.0	
				1–10 63 1.1 (0.6–1.9)	
				11–20 41 1.1 (0.6–1.9)	
				21–30 31 1.3 (0.7–2.5)	
				31–40 11 1.5 (0.6–3.3)	
				> 40 9 2.2 (0.8–5.1)	

Table 1 (contd)

Study	Population/end-point	No. of cases	Exposure			Risk estimate	Comments
Freedman et al. (1997) (contd)						*Women*	
				Years	Cases	1.0	
				None	25	0.7 (0.3–1.7)	
				1–10	28	0.7 (0.3–1.8)	
				11–20	15	0.6 (0.2–1.6)	
				21–30	7	0.7 (0.2–2.2)	
				31–40	5	0.6 (0.2–2.2)	
				> 40	4		
Cantor et al. (1998)	Iowa, USA Incidence, 1986–89	1123 bladder cancers 1983 population controls (men and women)	Total lifetime trihalomethanes (THM) estimated from lifetime residential histories, water utility survey and measurements of water samples	*THM (g)*		*Men*	Adjusted for age, study period, education, high-risk occupation and cigarette smoking (6 strata)
				≤ 0.04		1.0	
				0.05–0.12		1.3 (1.0–1.7)	
				0.13–0.34		1.1 (0.8–1.5)	
				0.35–1.48		1.2 (0.9–1.6)	
				1.49–2.41		1.3 (0.8–2.0)	
				≥ 2.42		1.8 (1.2–2.7)	
						p for trend = 0.05	
						Women	
				≤ 0.04		1.0	
				0.05–0.12		1.2 (0.8–1.8)	
				0.13–0.34		0.9 (0.6–1.6)	
				0.35–1.48		1.0 (0.6–1.7)	
				1.49–2.41		0.9 (0.9–2.0)	
				≥ 2.42		0.6 (0.3–1.4)	
						p for trend = 0.54	

Table 1 (contd)

Study	Population/end-point	No. of cases	Exposure	Risk estimate	Comments
Koivusalo *et al.* (1998)	Finland Incidence, 1991–92	732 914 population controls (men and women)	Mutagenic potency of drinking-water estimated from historical exposure at past residence, past water source and historical data on water quality and treatment	*Men* 1.2 (0.9–1.7) *Women* 1.2 (0.7–2.0)	Adjusted for age, smoking, socioeconomic status
			3000-net rev/L increase in average exposure to mutagenicity among subjects with ≥ 30 years of exposure		
			Tertiles of exposure among subjects with ≥ 30 years of exposure (net rev/L):	*Men* (552 cases)	
			Non-exposed	1.0	
			Low (1–999)	1.2 (0.8–1.6)	
			Medium (1000–2499)	0.97 (0.7–1.4)	
			High (≥ 2500)	1.4 (0.9–2.0)	
				Women (180 cases)	
			Non-exposed	1.0	
			Low (1–999)	1.2 (0.7–2.0)	
			Medium (1000–2499)	1.3 (0.7–2.4)	
			High (≥ 2500)	1.2 (0.6–2.2)	

RR, relative risk; CI, confidence interval; OR, odds ratio

Table 2. Cohort studies of cancer at other sites and chlorinated drinking-water

Reference	Population/follow-up	Exposure	Site and no. of cases		Relative risk (95% CI)	Comments
Wilkins & Comstock (1981)	Residents of Washington County, MD, USA 30 780 persons (14 553 men, 16 227 women) ≥ 25 years of age followed up 1963–75	Chlorinated surface water (average chloroform concentration, 107 µg/L) versus non-chlorinated deep wells	Liver	12		Adjusted for differences between cohorts in age, marital status, education, smoking history, church attendance, housing, persons per room
			Men		0.7 (0.2–3.5)	
			Women		1.8 (0.6–6.8)	
			Kidney	18		
			Men		0.8 (0.3–2.7)	
			Women		1.01 (0.3–6.0)	
Doyle et al. (1997)	Iowa Women's Health Study (USA) 41 836 women aged 55–69 years, followed up 1986–93	1108 municipal water supplies (1979 and 1986–87) Chloroform concentration in 1986–87 (µg/L)				Adjusted for age, education, smoking, physical activity, fruit and vegetable intake, energy intake, BMI and waist-to-hip ratio
		< limit of detection	Kidney	30	1.0	
		1–2			0.5 (0.2–1.6)	
		3–13			1.2 (0.5–3.1)	
		14–287			0.9 (0.3–2.3)	p for trend = 0.82
			Colon	178		
		1–2			1.1 (0.7–1.7)	
		3–13			1.4 (0.9–2.2)	
		14–287			1.7 (1.1–2.5)	p for trend < 0.01
			Rectum and anus	78		
		1–2			0.8 (0.4–1.5)	
		3–13			0.8 (0.4–1.5)	
		14–287			1.1 (0.6–1.9)	p for trend = 0.89

Table 2 (contd)

Reference	Population/follow-up	Exposure	Site and no. of cases		Relative risk (95% CI)	Comments
Doyle *et al.* (1997) (contd)						
		1–2	Lung	143	1.2 (0.8–2.1)	
		3–13			1.8 (1.1–3.0)	
		14–287			1.6 (0.97–2.6)	*p* for trend = 0.025
		1–2	Melanoma	44	2.6 (0.99–6.6)	
		3–13			1.3 (0.4–4.0)	
		14–287			3.4 (1.3–8.6)	*p* for trend = 0.049
		1–2	All cancers	983	1.04 (0.9–1.3)	
		3–13			1.2 (1.03–1.5)	
		14–287			1.3 (1.1–1.5)	*p* for trend < 0.01
Koivusalo *et al.* (1997)	Finland 56 towns; 32% of Finnish population, 1971–93	Chlorinated/ unchlorinated water supplies; mutagenicity assessment	621 431 persons (307 967 women, 313 464 men) *Both sexes*			Adjusted for age, time period, urbanization and social status. Cancers of the ureter and urethra included in bladder cancer
			Colon		0.9 (0.8–1.04)	
			Rectum		1.04 (0.9–1.3)	
			Oesophagus		1.4 (0.9–2.1)	
			Pancreas		1.01 (0.8–1.2)	
			Kidney		1.03 (0.8–1.3)	
			Brain and nervous system		1.00 (0.9–1.2)	
			Non-Hodgkin lymphoma		1.2 (0.9–1.5)	
			Leukaemia		1.04 (0.9–1.3)	
			Women			
			Colon		0.95 (0.8–1.9)	
			Rectum		1.4 (1.03–1.9)	
			Oesophagus		1.9 (1.02–3.5)	
			Breast		1.1 (1.01–1.2)	
			Pancreas		1.1 (0.8–1.5)	
			Kidney		1.03 (0.7–1.4)	
			Brain and nervous system		1.08 (0.9–1.4)	
			Non-Hodgkin lymphoma		1.4 (0.98–1.98)	
			Leukaemia		1.08 (0.8–1.5)	

Table 2 (contd)

Reference	Population/follow-up	Exposure	Site and no. of cases	Relative risk (95% CI)	Comments
Koivusalo *et al.* (1997) (contd)			*Men*		
			Colon	0.8 (0.7–1.04)	
			Rectum	0.9 (0.7–1.09)	
			Oesophagus	0.9 (0.5–1.7)	
			Prostate	0.97 (0.8–1.1)	
			Pancreas	0.9 (0.7–1.2)	
			Kidney	1.04 (0.8–1.4)	
			Brain and nervous system	0.9 (0.7–1.2)	
			Non-Hodkgin lymphoma	1.03 (0.8–1.4)	
			Leukaemia	1.02 (0.8–1.3)	

CI, confidence interval

Historical exposure estimates used information on past residence, past water source and historical data on water quality and treatment during the years 1955 and 1970. Exposure was expressed as estimates of mutagenic potency of drinking-water through regression models developed by Vartiainen *et al.* (1988), which used indicators of water quality and water treatment in the past (1986–87) to estimate mutagenic potency with the Ames test using *Salmonella typhimurium* strain TA100. Vartiainen *et al.* (1988) showed a correlation of 0.63 between the level of mutagenicity (TA100) and levels of trihalomethanes in water, but did not report on 3-chloro-4-(dichloromethyl)-5-hydroxy-2-(5*H*)-furanone (MX). Kronberg and Vartiainen (1988), however, reported that between 15 and 57% (mean, 33%) of the muta-genicity in 23 Finnish tap-water samples could be explained by MX. They also reported a correlation of 0.89 between the level of mutagenicity (TA100) in water and concentration of MX. Koivusalo *et al.* (1997) evaluated 24 cancer sites separately for men and women. Rela-tive risks were calculated using a continuous variable for the mutagenicity level of a 3000-net revertants (rev)/L exposure. Significantly elevated risks were observed among women only for cancers of the urinary bladder, rectum and oesophagus. The relative risk for female breast cancer was 1.1 (95% CI, 1.01–1.22). No increased risk for colon cancer was shown in this study.

The Iowa Women's Health Study in the USA comprised a cohort of 41 836 women, aged 55–69 in 1986, followed up for cancer incidence through to 31 December 1993 (Doyle *et al.*, 1997) (Tables 1 and 2). The source of drinking-water for each cohort member was assessed from a postal survey in 1989. After excluding women who reported having drunk municipal or private well-water for less than 10 years previously, 28 237 women were included in the analysis. Data on 252 municipal water supplies in 1979 and 856 muni-cipal water systems in 1986–87 were used to assess exposure. All women who lived in the same community and reported drinking municipal water were assigned the same concen-tration of trihalomethanes, including chloroform. The concentrations of chloroform in 1986–87 were categorized as below the detection limit (reference) or 1–2, 3–13 and 14–287 µg/L. The incidences of a number of cancers were estimated in each category. Relative risks were adjusted for age, education, smoking status, pack–years of smoking, physical activity, all fruit and vegetable intake, total energy intake, body mass index and waist-to-hip ratio. In comparison with women who used groundwater sources, women drinking water from surface sources were at increased risk for colon cancer and all cancers combined, with a clear dose–response relationship. In addition, trends in risk were found for cancer of the lung (*p* for trend = 0.025) and melanoma (*p* for trend = 0.049).

2.   Case–control studies

2.1   *Cancers of the urinary bladder* (Table 1)

Cantor *et al.* (1987) interviewed 2805 patients with urinary bladder cancer aged 21–84 years at the time of diagnosis and 5258 population-based controls in a case–control study in 10 geographical areas in the USA. Controls under the age of 65 years were selected by random-digit dialling and controls over the age of 65 from rosters of the US Health Care Financing Agency. The cases were newly diagnosed and histologically confirmed during 1977–78. Cases and controls were frequency-matched on sex, age and geographical area. All subjects were administered a questionnaire at home by trained interviewers, which included questions on consumption of tap-water during a typical week 1 year before the interview. A lifetime history of residence and water sources was ascertained. A total of 1102 water utilities were visited, and utility personnel were interviewed; the water sources were then categorized for chlorination status (chlorination or no chlorination) during various periods. The residential histories of the subjects were linked with year-by-year data on water source and treatment. Logistic regression models with adjustment for sex, age, study area, smoking, high-risk occupation and population size were used with various indicators of water quality. Odds ratios for bladder cancer were elevated among those with both elevated intakes of drinking-water and long-term consumption of chlorinated surface water. Among subjects who had lived for 60 years or longer at residences supplied with chlorinated surface water, odds ratios for successive quintiles of tap-water ingestion were: 1.0 ($\leq$ 0.8 L/day, referent), 0.8, 1.1, 1.7 and 2.0 (p for trend = 0.014). [Confidence intervals about the point estimates were not available.]

A population-based study of 327 histologically verified cases of urinary bladder cancer from the State Cancer Registry matched to 261 controls with other cancers was conducted in Colorado (USA) during 1990–91 (McGeehin et al., 1993). After physician approval, telephone interviews were carried out to obtain individual histories of residence and water source from living patients and controls; the response rates were 78% and 75%, respectively. These data were linked to data from water utilities and records of the Colorado Department of Health. The total lifetime concentration of trihalomethanes was calculated for each subject as a time-weighted mean from data for each water system in Colorado in 1989. The mean lifetime concentration was 616 µg/L for cases and 422 µg/L for controls ($p < 0.001$). Adjustment for potential confounders resulted in odds ratios that increased with years of exposure to chlorinated water. More than 34 years of exposure to chlorinated water, contrasted with no such exposure, was associated with increased risks in both nonsmokers (odds ratio, 2.9; 95% CI, 1.2–7.4) and smokers (odds ratio, 2.1; 95% CI, 1.1–3.8).

Vena et al. (1993) reported a case–control study among male residents of three counties in western New York State (USA). Cases were diagnosed during 1979–85 and were confirmed histologically. Physician approval was obtained for 972 cases. Of these, 719 were eligible according to inclusion criteria and 351 cases of transitional-cell bladder cancer were interviewed. Controls (aged 35–90 years) were residents of the same counties, were matched to individual cases by sex, age and neighbourhood of residence and were interviewed during the period of case ascertainment. All 850 were included in the bladder cancer study [response rates not indicated]. Usual dietary habits including consumption of tap-water and non-tap-water were ascertained for the year prior to diagnosis (cases) or interview (controls). Total tap-water consumption also included beverages prepared with tap-water. More than 70% of subjects spent more than 90% of their lives using a public source of tap-water that historically had been treated with chlorine. The odds ratios for bladder cancer increased with increasing number of cups of tap-water consumed daily (p for trend < 0.001). There was an almost threefold increase for intake of more than 10 cups of tap-water daily compared with an intake of less than 5 cups per day. No excess risk was observed in subjects who had used the public water supply for more than 50 years compared with those who had used it for less than 50 years. [The Working Group noted that the non-exposed group included subjects with a lengthy exposure time to chlorinated drinking-water. This paper did not analyse associations with water source or level of trihalomethanes.]

King and Marrett (1996) reported a population-based case–control study of bladder cancer among residents of Ontario, Canada (excluding the north), aged 25–74 years. Cases diagnosed in the period 1 September 1992 to 1 May 1994 were identified from records of the Ontario Cancer Registry. After exclusions due to illness, death and lack of physician approval, 1262 patients were contacted out of 1694 eligible cases, and 927 cases completed a questionnaire (84%). Controls were selected by random-digit dialling. Of 2768 eligible controls, 2494 were contacted and 2118 completed the questionnaire (87%). The overall response rates were 73% for cases and 72% for controls. The period of exposure ascertainment extended from 1950 to 1990. Summer trihalomethane levels were estimated by modelling data from 1988–92 for 114 water-treatment plants. Predictors in the model were characteristics of the raw water (source, depth of intake pipe, temperature), pretreatment procedures (dose of chlorination, chloramination), treatments employed (coagulation, polyelectrolytes, activated carbon) and post-treatment procedures (dose of chlorination, dechlorination). Separate models were created for surface water with and without chlorination pretreatment and for groundwater sources. Information from subjects was collected by telephone, using a structured interview after they had received a mailed questionnaire. Items included demographic data, smoking history, usual diet, lifetime residence (with information on water source) and usual water consumption prior to diagnosis. Water exposures were estimated by linking residential histories with the relevant data on treatment plant by time and geographical area. After restricting the population for analysis to persons with 30 or more years of identified exposure, 696 cases of bladder cancer and 1545 controls remained. Analyses were adjusted for age, sex, smoking habit, education and calorie intake. Odds ratios increased up to 1.4 with increasing duration of use of a chlorinated surface water source. Results from another analysis, in which trihalomethane–years was used as the exposure variable, showed a similar increase in risk. In addition, among subjects with relatively homogeneous exposures for at least 30 years, a trend in risk with increasing trihalomethane levels was observed ($p = 0.006$).

Freedman et al. (1997) conducted a case–control study of bladder cancer among residents of Washington County, MD, USA. Cases were white residents reported to the local cancer registry between 1975 and 1992 ($n = 294$). All cases had been enumerated in a special county census conducted by the Johns Hopkins School of Public Health and Hygiene in 1975 ($n = 2326$). Controls, frequency-matched by age ($\pm$ 5 years) and sex, were randomly selected from the 1975 census population. Data collected in the 1975 census included age, sex, ethnicity, years of schooling, smoking status, residence, source of drinking-water and other variables. Duration of exposure to chlorinated surface water was derived from census information on the length of residence in a household supplied with municipal water at the time of the census. With minor exceptions, municipalities in Washington County, MD, had been supplied for at least 30 years with chlorinated surface water. Odds ratios were adjusted for age, sex, smoking history and a number of other potential confounding factors. The risk for bladder cancer was weakly associated with duration of exposure to municipal drinking-water.

Koivusalo *et al.* (1998) conducted a population-based case–control study in Finland of 732 cases of bladder cancer, 703 cases of kidney cancer and 914 controls. The cases were identified by the Finnish Cancer Registry from persons diagnosed in 1991–92. Controls, frequency-matched by age and sex to the case series, were randomly selected from the national population registry. The overall response rate (cases and controls) was 68% of those eligible. Information on several confounding factors (tobacco use, socioeconomic status, intake of coffee and other beverages, past urinary tract infections and urolithiasis, occupational history, weight, height) and water-source history was obtained through a questionnaire. Historical exposure estimates were made using the methods described above (Koivusalo *et al.*, 1997). Exposure estimates covered the period 1950–87, and only persons with information on exposure for at least 30 years were included in the analysis. Using a continuous variable for the estimated level of mutagenicity through exposure to water, odds ratios were calculated for a 3000-net rev/L increase in average exposure between 1950 and 1987 in logistic regression models adjusted for age, sex, socioeconomic status and smoking habit. Overall, after adjustment for confounding, a small excess risk for bladder cancer was observed for an increase of 3000 net rev/L in men and women. In other analyses, subjects were placed in categories of exposure (low, 1–999 net rev/L; medium, 1000–2499 net rev/L; high, ≥ 2500 net rev/L) and risk was compared with unexposed individuals, but the odds ratios remained non-significant.

Cantor *et al.* (1998) reported a case–control study of bladder cancer among residents of Iowa (USA) aged 40–85 years. Patients with histologically confirmed bladder cancer were identified through the State Health Registry of Iowa, supplemented by a rapid reporting system, during 1986–89. Controls were frequency-matched to cases by 5-year age group and sex. Controls under 65 years of age were randomly selected from driver's licence listings, whereas controls aged 65 and older were selected from rosters of the US Health Care Financing Agency. When a case was deceased or incapable of responding (approximately 10%), a close family member or friend was invited to participate. Patients, controls and proxies completed a self-administered questionnaire on demographic data, smoking history, occupational history, further indicators of lifestyle and medical conditions and the frequency of consumption as an adult inside and outside the home of beverages containing tap-water and other beverages. Lifetime residential histories were recorded, and the water source at each location was identified. Missing information important to the analysis of drinking-water exposures was completed by telephone interview. All 280 Iowa water utilities that supplied at least 1000 persons were contacted for historical information, and at each utility an interviewer collected one or two samples from the clear well where the water enters the distribution system or from nearby the system. The lifetime exposure to total trihalomethanes (g) and the lifetime average trihalomethane concentration (μg/L) were calculated for 1123 cases and 1983 controls. The logistic regression models included adjustment for age, study period, level of education, high-risk occupation and cigarette smoking. Odds ratios increased with increasing total lifetime dose of trihalomethanes for men but not women. Results for average lifetime dose of trihalomethanes followed similar patterns [data not shown in table].

2.2   Colorectal cancer (Table 3)

On the basis of their previous findings of an excess risk for mortality from colon cancer associated with exposure to chlorinated water supplies (Kanarek & Young, 1982), Young et al. (1987) conducted a case–control study of 347 incident cases of colon cancer and 639 cancer controls, excluding gastrointestinal and urinary tract cancers, identified in the Wisconsin Cancer Reporting System. A group of 611 population controls was selected from state driver's licence listings. White men and women in whom colon cancer had been diagnosed when they were aged 35–90 were considered to be eligible. The self-administered questionnaire gathered information on background variables, past water sources, water-drinking and bathing habits, home treatment of tap-water, and medical, occupational, social and lifestyle histories. Overall, 45% of the cases of colon cancer and 48% of controls participated in the study. After accounting for deaths, an overall response rate (all cases and controls) of 65% was achieved. The exposure of each study subject to trihalomethanes was estimated from an algorithm based on a survey of 81 Wisconsin water supplies, historical data from water facilities, the residential history of the subjects, data on individual water use and other information. Average estimated trihalomethane concentrations of 10–40 μg/L or > 40 μg/L at the place of residence in the years 1951, 1961, 1971 and 1981 were used as exposure estimates, and odds ratios were calculated for each of these years (reference category, < 10 μg/L trihalomethane). Using the population control series, no association of estimated exposure to trihalomethanes with risk for colon cancer was observed. Similar results were found when cancer controls were used. In addition, analyses were conducted using cumulative lifetime exposure to trihalomethanes over the past 30, the past 20 and the past 10 years. Regardless of the control group used in the analysis, results were consistent with the authors' conclusion that exposure to trihalomethanes is not associated with colon cancer in Wisconsin.

Hildesheim et al. (1998) reported results from a case–control study of colon and rectal cancer among residents of Iowa, USA, aged 40–85 years. Patients with histologically confirmed cancers of the colon and rectum were identified through the State Health Registry of Iowa during 1987; for the controls, the same mailed questionnaire and back-up telephone interview were used as those in the study of Cantor et al. (1998; see Section 2.1). The lifetime exposure to total trihalomethanes (g) and lifetime average trihalomethane concentrations (μg/L) were calculated for 560 colon cancer patients, 537 rectal cancer patients and 1983 controls from data on water samples and from interviews. About 15% of the patients

Table 3. Case–control studies of colorectal cancer and chlorinated drinking-water

Study	Population/end-point	No. of cases	Exposure	OR (95% CI)	Comments
Young et al. (1987)	Wisconsin, USA Incidence, 1951–81	347 colon cancer 639 other cancer controls 611 population controls Age, 35–90 years (both sexes)	Total trihalomethane concentration at place of residence (µg/L) in 1951	<10 1.0 10–40 1.2 (0.6–2.3) >40 0.98 (0.4–2.3)	OR for colon cancer adjusted for sex, age and population size of place of residence. General population controls
			Cumulative total trihalomethane exposure (mg) over lifetime	<100 1.0 100–300 1.1 (0.7–1.8) >300 0.7 (0.4–1.2)	
Hildesheim et al. (1998)	Iowa, USA Incidence, 1986–89	560 colon cancer 537 rectal cancer 1983 population controls Age, 40–85 years (both sexes)	Total lifetime trihalo-methane (g)	≤0.04 1.0 0.05–0.12 1.3 (1.0–1.6) 0.13–0.34 1.3 (0.9–1.8) 0.35–1.48 1.5 (1.1–2.1) 1.49–2.41 1.9 (1.2–3.0) ≥2.42 1.6 (1.0–2.6)	Rectal cancer Adjusted for age and sex p for trend = 0.08
			Lifetime average trihalo-methane concentration (µg/L)	≤0.7 1.0 0.8–2.2 1.05 (0.8–1.4) 2.3–8.0 1.2 (0.9–1.7) 8.1–32.5 1.2 (0.9–1.7) 32.6–46.3 1.7 (1.1–2.6) ≥46.4 1.7 (1.1–2.6)	Rectal cancer Adjusted for age and sex p for trend = 0.01
King et al. (2000)	Southern Ontario, Canada Incidence, 1992–94	767 colon cancer 661 rectal cancer 1545 population controls Age, 30–74 years (both sexes)	Consumption of chlori-nated drinking-water (years)	*Men* 0–9 1.0 10–19 1.7 (1.1–2.7) 20–34 1.3 (0.96–1.9) ≥35 1.5 (1.1–2.1) *Women* 0–9 1.0 10–19 0.6 (0.3–0.9) 20–34 0.9 (0.6–1.2) ≥35 0.7 (0.5–1.1)	Colon cancer Adjusted for sex, age, education, body mass index and intake of energy, cholesterol, calcium, alcohol and coffee

Table 3 (contd)

Study	Population/end-point	No. of cases	Exposure	OR (95% CI)	Comments
King *et al.* (2000) (contd)			Trihalomethane level (µg/L)	*Men*	
			0–24	1.0	
			25–74	1.5 (0.99–2.4)	
			≥ 75	1.9 (1.2–3.1)	
				p for trend = 0.005	
				Women	
			0–24	1.0	
			25–74	0.5 (0.3–0.8)	
			≥ 75	0.9 (0.5–1.7)	
				p for trend = 0.211	
			Exposure to trihalo-methanes ≥ 75 µg/L (years)	*Men*	
			0–9	1.0	
			10–19	1.1 (0.9–1.5)	
			20–34	1.5 (0.99–2.3)	
			≥ 35	2.1 (1.2–3.7)	
				Women	
			0–9	1.0	
			10–19	0.9 (0.7–1.3)	
			20–34	0.9 (0.5–1.6)	
			≥ 35	1.2 (0.6–2.4)	

OR, odds ratio; CI, confidence interval

were interviewed by proxy. The logistic regression models included adjustment for sex, age, study period, education, high-risk occupation and cigarette smoking. There was a suggestion of a trend of increasing risk for rectal cancer with lifetime concentration of trihalomethanes and for average lifetime dose of trihalomethanes. No such trend was observed for colon cancer.

King *et al.* (2000) reported a population-based case–control study of colon and rectal cancer among residents of Ontario, Canada, aged 30–74 years. Of 1722 cases of colon and 1530 cases of rectal cancer identified by the Ontario Cancer Registry from 1 September 1992 to 1 May 1994, physician consent to contact patients was obtained for 1338 cases of colon and 1169 cases of rectal cancer. Questionnaires were completed by 991 and 875 cases of colon and rectal cancer, respectively. Analyses were conducted on 767 cases of colon cancer, 661 cases of rectal cancer, and 1545 controls with information on exposure for at least 30 of the 40 years prior to diagnosis (cases) or completion of the questionnaire (controls). Selection and numbers of controls, interviews and exposure assessments were conducted as in King and Marrett (1996; described in Section 2.1). Risk for colon cancer was elevated among men exposed to water with high levels of trihalomethanes, but not among women. Among men with homogeneous exposures for at least 30 years, adjusted odds ratios were also elevated, with a clear exposure–response relationship. This effect was not observed in women. When the exposure metric was years of exposure to ≥ 75 µg/L trihalomethane, the odds ratios for colon cancer were again elevated in men only. There was no association of risk for rectal cancer in either sex with number of years of exposure to water containing elevated levels of trihalomethanes.

2.3   *Other cancer sites* (Table 4)

(*a*)   *Kidney cancer*

Koivusalo *et al.* (1998) conducted a population-based case–control study in Finland of 732 cases of bladder cancer, 703 cases of kidney cancer and 914 controls (described in detail in Section 2.1). An exposure-related excess risk was observed among men only for a 3000 net rev/L increase in average exposure to chlorination by-products. No significant risk was observed when cases were placed in tertiles of exposure, although a weak association was suggested.

(*b*)   *Brain cancer*

Cantor *et al.* (1999) reported a case–control study of brain cancer among residents of Iowa (USA), aged 40–85 years. Patients with histologically confirmed glioma were identified through the State Health Registry of Iowa during 1984–87; for the controls, the same mailed questionnaire and back-up telephone interview were used as those in the study of Cantor *et al.* (1998; see Section 2.1). The lifetime exposure to total trihalomethanes (g) and the lifetime average trihalomethane concentrations (µg/L) were calculated for 291 cases and 1983 controls from data on water samples and from interviews. Elevated risks were observed among men, but not women, with duration of exposure to

Table 4. Case–control studies of cancer at other sites and chlorinated drinking-water

Cancer site	Study	Population/end-point	No. of cases	Exposure	OR (95% CI)	Comments
Kidney	Koivusalo et al. (1998)	Finland Incidence 1991–92	703 (386 men, 317 women) and 914 population controls (621 men, 293 women)	Mutagenicity assessment; 3000 net rev/L increase Tertiles of exposure (net rev/L): Non-exposed, Low (1–999), Medium (1000–2499), High (≥2500) Non-exposed, Low, Medium, High	≥30 years of estimable exposure Both sexes 1.3 (1.0–1.7) Women 1.1 (0.7–1.7) Men 1.5 (1.1–2.1) Women 1.0 0.9 (0.6–1.5) 1.3 (0.8–2.1) 1.1 (0.7–1.9) Men 1.0 1.2 (0.8–1.7) 1.3 (0.8–1.8) 1.6 (1.0–2.4)	Calculated for all those with at least 30 years of known exposure. Adjusted for age, smoking, socioeconomic status and sex
Brain	Cantor et al. (1999)	Residents of Iowa, USA, aged 40–85 years. Incidence 1984–87	291 glioma (155 men, 136 women) and 1983 population controls (1308 men, 675 women)	Chlorinated surface water. Water utilities surveyed, measurements of trihalomethanes, personal questionnaire for past exposure	Both sexes (years of exposure to ≥75 µg/L) 0 1.0 1–19 1.1 (0.8–1.6) 20–39 1.6 (1.0–2.6) ≥40 1.3 (0.8–2.3) p for trend = 0.1 Women 0 1.0 1–19 1.0 (0.6–1.6) 20–39 1.6 (0.8–3.0) ≥40 0.7 (0.3–1.6) p for trend = 0.4	Adjusted for sex, age, farming occupation and population size; 74.4% of cases had proxy respondents. Cases and controls with ≥70% of lifetime with known source selected. Excluded population better educated and more urban

Table 4 (contd)

Cancer site	Study	Population/end-point	No. of cases	Exposure	OR (95% CI)	Comments
Brain (contd)	Cantor et al. (1999) (contd)				*Men*	
					0 1.0	
					1–19 1.3 (0.8–2.1)	
					20–39 1.7 (0.9–3.3)	
					≥ 40 2.5 (1.2–5.0)	
					p for trend = 0.04	
				Lifetime average trihalo-methane concentration (μg/L)	*Both sexes*	
				≤ 0.7	1.0	
				0.8–2.2	0.9 (0.6–1.3)	
				2.3–32.5	0.9 (0.6–1.4)	
				≥ 32.6	1.1 (0.7–1.8)	
					p for trend = 0.3	
					Women	
				≤ 0.7	1.0	
				0.8–2.2	0.9 (0.5–1.5)	
				2.3–32.5	0.8 (0.5–1.5)	
				≥ 32.6	0.9 (0.4–1.8)	
					p for trend = 0.9	
					Men	
				≤ 0.7	1.0	
				0.8–2.2	0.9 (0.6–1.6)	
				2.3–32.5	1.0 (0.6–1.8)	
				≥ 32.6	1.4 (0.7–2.9)	
					p for trend = 0.04	

Table 4 (contd)

Cancer site	Study	Population/end-point	No. of cases	Exposure	OR (95% CI)	Comments
Pancreas	Ijsselmuiden et al. (1992)	Washington County, MD, USA. Incidence 1975–89	101 (47 men, 54 women) 206 population controls (96 men, 110 women) All white	Chlorinated drinking-water, as of 1975 census. Non-municipal (chlorinated) Municipal (chlorinated)	1.0 2.2 (1.2–3.95)	Adjusted for age and current cigarette smoking. Non-municipal but chlorinated water used as baseline for odds ratios
	Kukkula & Löfroth (1997)	Turku area, Finland. Study base: 220 000 persons Incidence 1989–91	183 (71 men, 112 women); 360 matched controls	Residence in an area supplied by chlorinated drinking-water until 1981	*Exposure (years)* 0 0.33 (0.2–0.7) 1 0.54 (0.3–1.2) 5 0.66 (0.3–1.3) 10 0.53 (0.3–1.07) 15 0.32 (0.1–0.8) 20 0.20 (0.04–0.9)	No adjustment for confounders. OR calculated from exposure data of the discordant case–control set. Total trihalomethanes often > 200 µg/L at end of distribution system
Acute lymphocytic leukaemia	Infante-Rivard et al. (2001)	Quebec, Canada. Incidence 1980–93	491 aged 0–9 years and 491 population controls (boys and girls)	Trihalomethanes, metals (As, Cd, Cr, Pb, Zn) and nitrates in drinking-water. Municipality–exposure matrix based on historical data.	Water chlorinated *Prenatal* Part of the time 1.6 (0.7–3.7) Always 0.8 (0.5–1.2) Cumulative exposure (total trihalomethanes) > 95th percentile 0.8 (0.4–1.8) 25th–75th percentile 1.1 (0.8–1.7) >75% percentile 1.2 (0.7–1.8)	Adjusted for maternal age and level of schooling Baseline: never Baseline: ≤ 95th percentile Baseline ≤ 24th percentile

Table 4 (contd)

Cancer site	Study	Population/ end-point	No. of cases	Exposure	OR (95% CI)	Comments
Acute lymphocytic leukaemia (contd)	Infante-Rivard et al. (2001) (contd)			Water chlorinated	*Postnatal*	Baseline: never
				Part of the time	1.4 (0.7–2.5)	
				Always	0.9 (0.6–1.3)	
				Cumulative exposure (total trihalomethanes)		
				> 95th percentile	1.5 (0.8–3.0)	Baseline: ≤ 95th percentile
				25th–75th percentile	1.1 (0.8–1.6)	Baseline: ≤ 24th percentile
				> 75% percentile	0.9 (0.6–1.4)	
	Infante-Rivard et al. (2002)	Québec, Canada. Incidence 1980–83 Case-only study	161 cases from earlier study (2001)	*GSTT1*-null Total trihalomethanes > 95th percentile		Postnatal exposure
				Average	9.1 (1.4–57.8)	
				Cumulative	2.5 (0.6–10.5)	
				CYP2E1*5 Total trihalomethanes ≥ 75th percentile		
				Average	4.1 (0.8–21.5)	
				Cumulative	5.96 (0.7–53.8)	

OR, odds ratio; CI, confidence interval

chlorinated surface waters with levels of trihalomethanes of about 75 µg/L. For lifetime average exposure to trihalomethanes, the odds ratio among men increased to 1.4 (95% CI, 0.7–2.9) for levels > 32.6 µg/L. Among women, there was no association of risk with average trihalomethane level.

(c) Pancreatic cancer

A population-based case–control study on pancreatic cancer was conducted in the Turku area, Finland (Kukkula & Löfroth, 1997). All 183 cases diagnosed during 1989–91 from a study base of approximately 220 000 persons were included in the study. Two random controls were selected for each case. Source of drinking-water (chlorinated or non-chlorinated) was identified for each subject for all residences in the last 20 years prior to diagnosis. Exposure to chlorinated drinking-water was not associated with risk for pancreatic cancer, with odds ratios ranging from 0.2 to 0.7 depending on the length of exposure. [The Working Group noted that the study did not provide information on individual water-drinking habits or on potential confounding factors, and that the exposure time window of 20 years prior to diagnosis is short.]

Ijsselmuiden *et al.* (1992) conducted a case–control study of pancreatic cancer among residents of Washington County, MD (USA). Cases were residents reported at the local cancer registry from 1975 to 1989. All cases had been enumerated in a special county census conducted by the Johns Hopkins School of Public Health and Hygiene in 1975. Information on pancreatic cancer was obtained from hospital records and death certificates. Controls were randomly selected from the 1975 census population. The total study population comprised 101 white cases and 206 white controls (participation rates above 95% for both groups). Data collected in the 1975 census included age, sex, race, years of schooling, smoking, residence, source of drinking-water and other variables. A small validation study indicated that the population in this county was relatively stable and that there were few changes in water sources between 1963 and 1975. Chlorinated municipal drinking-water was used as a source of drinking-water by 79% of cases and 63% of controls, yielding an odds ratio of 2.2 (95% CI, 1.2–3.95) adjusted for age and smoking, relative to users of private well-water. [The Working Group noted that the information collected in the census on residence and source of drinking-water was cross-sectional.]

(d) Childhood leukaemia

A population-based case–control study of 491 cases of acute lymphoblastic leukaemia cases aged 0–9 years in the Province of Québec, Canada, was reported by Infante-Rivard *et al.* (2001). Cases were diagnosed in 1980–93 and controls (*n* = 491) were selected from a population-based registry of children, individually matched to cases on age (within 24 months), sex and region of residence at the calendar date of diagnosis. Information was collected by telephone interview of parents on the child's residential history, including information on source of drinking-water. Data on drinking-water quality, with information on individual trihalomethane levels, were collected from municipal water supplies and from data held by the Ministry of Environment. Exposure indices were developed after merging residential histories with historical data on water quality. Odds ratios were estimated by conditional logistic regression with adjustment for maternal age and level of schooling. During the prenatal period, as well as the postnatal period, no substantial increase in risk was observed; however, a small increase was shown during the postnatal period for cumulative exposure to trihalomethanes (above the 95th percentile).

Subsequently, Infante-Rivard *et al.* (2002) conducted a case–case analysis among a subset of 161 childhood cases of acute lymphocytic leukaemia from their earlier study (Infante-Rivard *et al.*, 2001) for whom information on exposure was available as well as information on genotyping for polymorphisms in the *GSTT1* and *CYP2E1* genes that are involved in the metabolism of trihalomethanes. Risk was assessed for gene–exposure interaction. For *GSTT1*, 25 cases had the null genotype and 136 carried the normal allele. Twelve cases carried at least one *CYP2E1*5* allele and 125 were normal for the *CYP2E1* genotype. The risk for acute lymphoblastic leukaemia associated with average exposure to total trihalomethanes was elevated among children homozygous for *GSTT1* deletion (null genotype). An increase in risk was also suggested among children bearing the *CYP2E1*-null genotype. The authors noted in conclusion that this preliminary study shows suggestive but imprecise results which should be repeated.

References

Cantor, K.P., Hoover, R., Hartge, P., Mason, T.J., Silverman, D.T., Altman, R., Austin, D.F., Child, M.A., Key, C.R., Marrett, L.D., Myers, M.H., Narayana, A.S., Levin, L.I., Sullivan, J.W., Swanson, G.M., Thomas, D.B. & West, D.W. (1987) Bladder cancer, drinking water source, and tap water consumption: A case–control study. *J. natl Cancer Inst.*, **79**, 1269–1279

Cantor, K.P., Lynch, C.F., Hildesheim, M.E., Dosemeci, M., Lubin, J., Alavanja, M. & Craun, G. (1998) Drinking water source and chlorination byproducts. I. Risk of bladder cancer. *Epidemiology*, **9**, 21–28

Cantor, K.P., Lynch, C.F., Hildesheim, M.E., Dosemeci, M., Lubin, J., Alavanja, M. & Craun, G. (1999) Drinking water source and chlorination byproducts in Iowa. III. Risk of brain cancer. *Am. J. Epidemiol.*, **150**, 552–560

Doyle, T.J., Zheng, W., Cerhan, J.R., Hong, C.-P., Sellers, T.A., Kushi, L.H. & Folsom, A.R. (1997) The association of drinking water source and chlorination by-products with cancer incidence among postmenopausal women in Iowa: A prospective cohort study. *Am. J. public Health*, **87**, 1168–1176

Freedman, D.M., Cantor, K.P., Lee, N.L., Chen, L.-S., Lei, H.-H., Ruhl, C.E. & Wang, S.S. (1997) Bladder cancer and drinking water: A population-based case–control study in Washington County, Maryland (United States). *Cancer Causes Control*, **8**, 738–744

Hildesheim, M.E., Cantor, K.P., Lynch, C.F., Dosemeci, M., Lubin, J., Alavanja, M. & Craun, G. (1998) Drinking water source and chlorination byproducts. II. Risk of colon and rectal cancers. *Epidemiology*, **9**, 29–35

Ijsselmuiden, C.B., Gaydos, C., Feighner, B., Novakoski, W.L., Serwadda, D., Caris, L.H., Vlahov, D. & Comstock, G.W. (1992) Cancer of the pancreas and drinking water: A population-based case–control study in Washington County, Maryland. *Am. J. Epidemiol.*, **136**, 836–842

Infante-Rivard, C., Olson, E., Jacques, L. & Ayotte, P. (2001) Drinking water contaminants and childhood leukemia. *Epidemiology*, **12**, 13–19

Infante-Rivard, C., Amre, D. & Sinnett, D. (2002) GSTT1 and CYP2E1 polymorphisms and trihalomethanes in drinking water: Effect on childhood leukemia. *Environ. Health Perspect.*, **110**, 591–593

Kanarek, M.S. & Young, T.B. (1982) Drinking water treatment and risk of cancer death in Wisconsin. *Environ. Health Perspect.*, **46**, 179–186

King, W.D. & Marrett, L.D. (1996) Case–control study of bladder cancer and chlorination by-products in treated water (Ontario, Canada). *Cancer Causes Control*, **7**, 596–604

King, W.D., Marrett, L.D. & Woolcott, C.G. (2000) Case–control study of colon and rectal cancers and chlorination by-products in treated water. *Cancer Epidemiol. Biomarkers Prev.*, **9**, 813–818

Koivusalo, M., Pukkala, E., Vartiainen, T., Jaakkola, J.J.K. & Hakulinen, T. (1997) Drinking water chlorination and cancer — A historical cohort study in Finland. *Cancer Causes Control*, **8**, 192–200

Koivusalo, M., Hakulinen, T., Vartiainen, T., Pukkala, E., Jaakkola, J.J.K. & Tuomisto, J. (1998) Drinking water mutagenicity and urinary tract cancers: A population-based case–control study in Finland. *Am. J. Epidemiol.*, **148**, 704–712

Kronberg, L. & Vartiainen, T. (1988) Ames mutagenicity and concentration of the strong mutagen 3-chloro-4-(dichloromethyl)-5-hydroxy-2(5*H*)-furanone and of its geometric isomer E-2-chloro-3-(dichloromethyl)-4-oxo-butenoic acid in chlorine-treated tap waters. *Mutat. Res.*, **206**, 177–182

Kukkula, M. & Löfroth, G. (1997) Chlorinated drinking water and pancreatic cancer. *Eur. J. public Health*, **7**, 297–301

McGeehin, M.A., Reif, J.S., Becher, J.C. & Mangione, E.J. (1993) Case–control study of bladder cancer and water disinfection methods in Colorado. *Am. J. Epidemiol.*, **138**, 492–501

Vartiainen, T., Liimatainen, A., Kauranen, P. & Hiisvirta, L. (1988) Relations between drinking water mutagenicity and water quality parameters. *Chemosphere*, **17**, 189–202

Vena, J.E., Graham, S., Freudenheim, J., Marshall, J., Zielezny, M., Swanson, M. & Sufrin, G. (1993) Drinking water, fluid intake, and bladder cancer in western New York. *Arch. environ. Health*, **48**, 191–198

Wilkins, J.R., III & Comstock, G.W. (1981) Source of drinking water at home and site-specific cancer incidence in Washington County, Maryland. *Am. J. Epidemiol.*, **114**, 178–190

Young, T.B., Wolf, D.A. & Kanarek, M.S. (1987) Case–control study of colon cancer and drinking water trihalomethanes in Wisconsin. *Int. J. Epidemiol.*, **16**, 190–197

CHLORAMINE

1. Exposure Data

1.1 Chemical and physical data

1.1.1 *Nomenclature*

Chem. Abstr. Serv. Reg. No.: 10599-90-3
Chem. Abstr. Name: Chloramide
IUPAC Systematic Name: Chloramide
Synonyms: Chloroamine; monochloramine; monochloroamine; monochloroammonia

1.1.2 *Structural and molecular formulae and relative molecular mass*

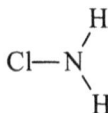

$$Cl—N \begin{matrix} H \\ \diagup \\ \diagdown \\ H \end{matrix}$$

ClH_2N Relative molecular mass: 51.47

1.1.3 *Chemical and physical properties of the pure substance* (Ura & Sakata, 1986)

(*a*) *Description*: Colourless liquid with a strong pungent odour
(*b*) *Melting-point*: –66 °C
(*c*) *Stability*: Stable in aqueous solution, but unstable and explosive in pure form

1.1.4 *Technical products and impurities*

Monochloramine is not known to be a commercial product but is generated *in situ* as needed.

1.1.5 *Analysis*

Chloramines have been determined in water by amperometric titration. Free chlorine is titrated at a pH of 6.5–7.5, a range at which chloramines react slowly. Chloramines are then titrated in the presence of the correct amount of potassium iodide in the pH range 3.5–4.5. The tendency of monochloramine to react more readily with iodide than dichloramine provides a means for estimation of monochloramine content (American Public Health Association/American Water Works Association/Water Environment Federation, 1999).

In another method, *N,N*-diethyl-*para*-phenylenediamine is used as an indicator in the titrimetric procedure with ferrous ammonium sulfate. In the absence of an iodide ion, free chlorine reacts instantly with the *N,N*-diethyl-*para*-phenylenediamine indicator to produce a red colour. Subsequent addition of a small amount of iodide ion (as potassium iodide) catalytically causes monochloramine to colour. For accurate results, careful pH and temperature control is essential. A simplified colorimetric version of this procedure is commonly used for monitoring levels of chloramine in municipal water systems (American Public Health Association/American Water Works Association/Water Environment Federation, 1999).

1.2   Production and use

1.2.1   *Production*

Chloramines are inorganic nitrogen compounds that contain one or more chlorine atoms attached to a nitrogen atom. A familiar example is monochloramine, the existence of which has been known since the beginning of the nineteenth century. Monochloramine is prepared in pH-controlled reactions by the action of hypochlorous acid or chlorine on ammonia (Ura & Sakata, 1986; Wojtowicz, 1993).

Free chlorine reacts readily with ammonia to form chloramines: monochloramine, dichloramine and trichloramine (nitrogen trichloride). The presence and concentrations of these chloramines depend chiefly on pH, temperature, initial chlorine-to-nitrogen ratio, absolute chlorine demand and reaction time. Both free chlorine and chloramines may be present simultaneously. Chloramines may be formed in water supplies during the treatment of raw waters containing ammonia or by the addition of ammonia or ammonium salts. Typical water treatment practices are designed to produce mainly monochloramine. Chloramines are also formed in swimming pools from the chlorination of ammonia or other nitrogen-containing contaminants (American Public Health Association/American Water Works Association/Water Environment Federation, 1999).

The chemistry of chloramines is fairly well understood and can be summarized in its simplest form by three reversible reactions involving ammonia (NH_3), hypochlorous acid ($HOCl$), monochloramine (NH_2Cl), dichloramine ($NHCl_2$) and trichloramine (NCl_3):

$$NH_3 + HOCl \rightleftharpoons NH_2Cl + H_2O \quad (1)$$

$$NH_2Cl + HOCl \rightleftharpoons NHCl_2 + H_2O \quad (2)$$

$$NHCl_2 + HOCl \rightleftharpoons NCl_3 + H_2O \quad (3)$$

Monochloramine is generally the dominant species produced during drinking-water disin-fection. The production of dichloramine is favoured as the chlorine-to-nitrogen ratio increases and the pH decreases. Hydrolysis reactions (the reverse reactions in equations 1–3 above) are also of considerable interest because they generate hypochlorous acid, which may be important in the formation of the chlorine-substituted organic compounds, hypobromous acid–hypobromite and bromamines. Also, the hydrolysis reaction with monochloramine liberates ammonia. At equilibrium, no more than several per cent of monochloramine is hydrolysed to free chlorine (Diehl *et al.*, 2000).

In addition to hydrolysis and conversion to trichloramine (equation 3), dichloramine decomposes by several other reactions, two of which are base-catalysed. The decomposi-tion of dichloramine accelerates as pH, the concentration of bases (e.g. alkalinity), or both increase. Therefore, dichloramine is much less stable than monochloramine under most conditions of practical interest; however, its greater instability does not necessarily mean that dichloramine is of little significance in the formation of disinfection by-products (Diehl *et al.*, 2000).

Monochloramine decomposes in a stepwise fashion, being converted first to dichlor-amine; the subsequent decomposition of dichloramine is primarily responsible for loss of total residual chlorine. The two major pathways for the decomposition of monochloramine are its hydrolysis and subsequent reaction with free chlorine (equations 1 and 2 above) and general acid catalysis:

$$NH_2Cl + NH_2Cl + H^+ \rightleftharpoons NHCl_2 + NH_3 + H^+ \quad (4)$$

The rate of general acid catalysis (equation 4) is a function of the concentration of proton donors and increases as their concentration increases and pH decreases. It has been noted that the carbonate system, via carbonic acid and bicarbonate, may significantly increase the rate of acid-catalysed decomposition at concentrations and pH values typical of many drinking-water systems. Therefore, the decomposition rate of both monochloramine and dichloramine may increase as alkalinity increases (Diehl *et al.*, 2000).

Data on the production of monochloramine for water treatment are not available.

1.2.2   *Use*

Chloramines, including monochloramine, are used in synthetic reactions (e.g. as an oxidizing agent for trisubstituted phosphines) and as bleaching agents, disinfectants and bactericides because they function as chlorinating agents and oxidants (undergoing hydrolysis to varying degrees to form hypochlorous acid). Chloramine is widely used in the primary and secondary disinfection of drinking-water (chloramination, in-situ NH_2Cl formation, is increasingly being used to disinfect public water supplies in order to reduce the formation of trihalomethanes). The use of chloramine in water treatment is increasing in the USA to meet stricter standards for disinfection by-products. In addition, chloramine is generally thought to produce a more stable residual than free chlorine and thus to provide more protection against bacterial regrowth in the distribution system (Ura & Sakata, 1986; Wojtowicz, 1993; Zhang & DiGiano, 2002).

A recent survey of water-treatment plants that serve more than 10 000 people in the USA showed an increase in the percentage of those that use chloramine disinfection, from 20% in 1989 to 29% in 1998 (Oldenburg *et al.*, 2002). In addition, the Environmental Protection Agency (1998) in the USA has predicted that an additional 25% will change to chloramine disinfection to comply with stricter standards for disinfection by-products.

Monochloramine has been used in the synthesis of a wide range of other chemicals. The chemistry of monochloramine involves chlorination, amination, addition, condensation, redox, acid-base and decomposition reactions. When monochloramine is added to ketones, it forms vicinal chloroamines and when condensed with aldehydes gives *N*-chloroimines. In the presence of excess base, monochloramine decomposes to ammonia and nitrogen. The reaction of equimolar amounts of monochloramine and caustic with excess ammonia is the basis of the industrial-scale production of hydrazine and, in 1990, it was estimated that 55 000 tonnes of monochloramine were consumed worldwide in the production of hydrazine (Wojtowicz, 1993). Monochloramine is also used in organic synthesis for preparation of amines and substituted hydrazines (Wojtowicz, 1993).

1.3   Occurrence

1.3.1   *Natural occurrence*

Chloramines are not known to occur as natural products. *N*-Chloramines, including monochloramines, are formed *in vivo* secondary to the formation of hypochlorous acid in phagocytes and neutrophils (Carr *et al.*, 2001).

1.3.2   *Occupational exposure*

In studies of occupational exposures around swimming pools, Hery *et al.* (1995) measured concentrations of chloramines (expressed as trichloramine) ranging from values below the limit of detection (around 0.05 mg/m^3) to 1.94 mg/m^3 in the air of swimming pools, depending on the characteristics of the pool, water temperature, water-surface stirring and air filtration. Massin *et al.* (1998) measured the levels of trichloramine in the air of indoor swimming pools and found average concentrations of 0.24 mg/m^3 (standard deviation [SD], 0.17) in public pools and 0.67 mg/m^3 (SD, 0.37) in leisure-centre pools. Two public pools had concentrations as high as 0.60 mg/m^3 and two leisure-centre pools had concentrations as low as 0.14 mg/m^3. Thickett *et al.* (2002) measured concentrations of trichloramine in air varying from 0.1 to 0.57 mg/m^3 in indoor swimming pools, where they found cases of occupational asthma among swimming instructors and lifeguards.

Highly irritant and odour-intensive chloramine compounds are also responsible for the typical 'indoor swimming pool smell'. Due to their different volatilities, pool water releases dichloramine about three times faster and trichloramine about 300 times faster than monochloramine (Holzwarth *et al.*, 1984). Dichloramine imparts a chlorinous odour to water, while monochloramine does not. Trichloramine has a strong unpleasant odour at concentrations in water as low as 0.02 mg/L (Kirk & Othmer, 1993).

Hery *et al.* (1998) measured total concentrations of chloramines ranging from 0.4 to 16 mg/m^3 in ambient air samples and from 0.2 to 5 mg/m^3 in personal air samples in a green salad-processing plant. In personal air samples, concentrations of soluble chlorine (hypochlorite, mono- and dichloramine) ranged from < 0.1 to 3.7 mg/m^3, while concentrations of trichloramine ranged from < 0.1 to 2.3 mg/m^3. Area samples in the plant contained 0.1–10.9 mg/m^3 soluble chlorine and 0.1–5.9 mg/m^3 trichloramine.

1.3.3   *Air*

See Section 1.3.2, Occupational exposure.

1.3.4   *Water*

Chloramines, mainly monochloramine, are used to provide a disinfection residual in drinking-water distribution systems where it is difficult to maintain free chlorine residual or where the formation of disinfection by-products is a problem (Vikesland *et al.*, 2001). Levels up to 4 mg/L are typically added and decrease with length of residence to around 0.6 mg/L (Zhang & DiGiano, 2002). The decay of chloramines in water distribution systems is dependent on, for example, pH, temperature, length of residence, assimilable organic carbon, and nitrite and bromide content (Vikesland *et al.*, 2001; Zhang & DiGiano, 2002).

Knowledge of the occurrence of organic chloramines in swimming-pool water is very limited, as there are currently no suitable analytical methods for their determination and characterization. Organic chloramines are formed when urine is present in the pools. The distribution of total nitrogen among relevant nitrogen compounds in swimming pools has been calculated. Although more than 80% of the total nitrogen content in urine is present in the form of urea and the ammonia content in urine, at approximately 5%, is comparatively low, swimming-pool water exhibits considerable concentrations of ammonia, which are evidently formed secondarily by degradation of urea following chemical reactions with chlorine. Ammonia reacts rapidly with hypochlorous acid to form monochloramine, dichloramine and trichloramine, whereas nitrogen-containing organic compounds, such as amino acids, may react with hypochlorite to form organic chloramines (Taras, 1953; Isaac & Morris, 1980; WHO, 2000).

1.4   Regulations and guidelines

The WHO (1998) has established a drinking-water guideline for monochloramine of 3 mg/L. The WHO also notes that insufficient data on health effects are available for di- and trichloramine to establish a numerical guideline.

Australia and New Zealand have also established a guideline of 3 mg/L for mono-chloramine (National Health and Medical Research Council and Agriculture and Resource Management Council of Australia and New Zealand, 1996). In Canada, the maximum allowable concentration for total chloramines has been established at 3.0 mg/L (Health Canada, 2003).

The Environmental Protection Agency (1998) in the USA established a maximum residual disinfectant level for chloramine of 4.0 mg/L (as Cl_2). This level can be exceeded under emergency conditions such as a major line break or a natural disaster.

2. Studies of Cancer in Humans

See Introduction to the monographs on chloramine, chloral and chloral hydrate, di-chloroacetic acid, trichloroacetic acid and 3-chloro-4-(dichloromethyl)-5-hydroxy-2(5*H*)-furanone.

3. Studies of Cancer in Experimental Animals

3.1   Oral administration

3.1.1   *Mouse*

Groups of 70 male and 70 female $B6C3F_1$ mice, 5–6 weeks of age, were given chlora-minated, deionized, charcoal-filtered drinking-water[1] that contained 50, 100 or 200 ppm [mg/L] monochloramine expressed as available atomic chlorine or deionized, charcoal-filtered drinking-water (controls). The dose levels of chloramine were based on a previous study that indicated that 200 ppm was the maximum concentration that mice will drink. Interim sacrifices of 9–10 mice each were carried out at 14 or 15 and 66 weeks, leaving 50–51 mice of each sex that were killed after 103–104 weeks of exposure. At 2 years, survival of mice that received chloraminated drinking-water was not significantly different from that of controls. Within the first week of chloramine treatment, a dose-related decrease in drinking-water consumption was observed in both males and females, which continued throughout the study. There was also a dose-related decrease in the body weights of males and females administered chloramine, which occurred earlier and to a greater extent in females than in males. The chloraminated water decreased the body weight of female mice by approximately 10% (50 ppm), 15% (100 ppm) and 30% (200 ppm). No neoplastic or

---

[1] The chloramine formulations were prepared from buffered sodium hypochlorite (pH 9) stock solutions to which dilute ammonium hydroxide was added. The sodium hypochlorite stock solution was prepared by bubbling chlorine gas into charcoal-filtered, deionized water and was raised to pH 9 with sodium bicarbonate/carbonate.

non-neoplastic lesions associated with the consumption of chloraminated drinking-water was observed in male or female mice (National Toxicology Program, 1992).

3.1.2   *Rat*

Groups of 70 male and 70 female Fischer 344/N rats, 5–6 weeks of age, were given chloraminated, deionized, charcoal-filtered drinking-water[1] that contained 50, 100 or 200 ppm [mg/mL] monochloramine expressed as available atomic chlorine or deionized, charcoal-filtered drinking-water (controls). The dose levels of chloramine were based on a previous study that indicated that 200 ppm was the maximum concentration that rats will drink. Interim sacrifices of 9–10 rats each were carried out at 15 and 66 weeks, leaving 50–51 rats of each sex that were killed after 103–104 weeks of exposure. Survival of male and female rats that received chloraminated drinking-water was not significantly different from that of controls except for the low-dose male group in which survival was greater. Within the first week of chloramine administration, a dose-related decrease in drinking-water consumption by males and females was observed, which remained reduced throughout the study. Chloraminated water did not affect the body weight of female and male rats until weeks 97 and 101, at which time a 10% reduction relative to controls occurred in the high-dose female and male rats, respectively. The incidence of mononuclear-cell leukaemia was higher in female rats that received the high dose of chloramine than in controls: 8/50 (16%), 11/50 (22%), 15/50 (30%) and 16/50 (32%) for the 0-, 50-, 100- and 200-ppm chloramine-treated groups, respectively. When adjusted for intercurrent mortality (Kaplan-Meier estimated tumour incidence), the rate of leukaemia was 20.8, 29.0, 39.9 and 41.4% for the 0-, 50-, 100- and 200-ppm chloramine-treated groups, respectively. The results were analysed by the life table test (Haseman, 1984) for an overall dose–response trend, resulting in a *p*-value of 0.02. This was followed by pairwise comparison between the controls and the treated groups resulting in *p*-values of 0.280, 0.077 and 0.036 for the 50-, 100- and 200-ppm chloramine-treated groups, respectively. The authors reported that the incidence of leukaemia in the concurrent control group was lower than that in untreated historical controls (170/680 [25%], with a range of 14–36%). There was a marginal increase in splenic histiocytic lymphoid hyperplasia in females (controls, 3/50; 50 ppm, 4/50; 100 ppm, 2/50; and 200 ppm, 6/50). Since this apparent increase lacked a dose–response relationship and was only marginally significant, it was not considered by the authors to be related to the chloraminated water. There were no other neoplastic or non-neoplastic lesions associated with the consumption of chloraminated drinking-water. The report concluded that there was no evidence for the carcinogenic activity of chloraminated drinking-water in male rats and equivocal evi-

---

[1] The chloramine formulations were prepared from buffered sodium hypochlorite (pH 9) stock solutions to which dilute ammonium hydroxide was added. The sodium hypochlorite stock solution was prepared by bubbling chlorine gas into charcoal-filtered, deionized water and was raised to pH 9 with sodium bicarbonate/ carbonate.

dence in female rats based on an increased incidence of mononuclear-cell leukaemia (National Toxicology Program, 1992).

3.2   Administration with known carcinogens or modifying factors

Rat: *Helicobacteria pylori* has been associated with an increased risk for stomach cancer that could result from the interaction of ammonia produced by its urease activity and hypochlorous acid produced by neutrophils. Ammonia and hypochlorous acid interact to form chloramine. Hence, administering ammonia and hypochlorous acid to rats in order to form chloramine has been used as a model for the ulcerogenic and carcinogenic activity of the bacterium (Iishi *et al.*, 1997; Iseki *et al.*, 1998; Narahara *et al.*, 2001; Kato *et al.*, 2002).

Early studies demonstrated that the combination of 20% ammonium acetate in the diet and 30 mM sodium hypochlorite in the drinking-water enhanced the incidence of *N*-methyl-*N'*-nitro-*N*-nitrosoguanidine (MNNG)-induced stomach cancer, which was increased from 10/20 (50%) to 17/20 (85%) by the combination. The number of stomach cancers per tumour-bearing rat was not altered: 1.8 ± 0.4 versus 1.7 ± 0.2 (Iishi *et al.*, 1997). In another report, the combination increased the incidence of rats with stomach cancer from 12/19 (63%) to 19/19 (100%), while again not altering the number of stomach cancers per tumour-bearing rat: 1.3 ± 0.2 versus 1.9 ± 0.2 (Iseki *et al.*, 1998). Neither 20% ammonium acetate nor 30 mM sodium hypochlorite administered alone affected the incidence of MNNG-induced stomach cancer (Iishi *et al.*, 1997; Iseki *et al.*, 1998).

Groups of 20 male Wistar rats [age unspecified] received 25 µg/mL MNNG in the drinking-water for 25 weeks and, at week 26, received 20% ammonium acetate in the diet and 30 mM sodium hypochlorite in the drinking-water. Some groups of rats also received subcutaneous injections of 25 or 50 mg/kg bw ambroxol, a clinical mucoregulatory agent with antioxidant activity, every other day, to determine its ability to prevent the enhancement of stomach cancer by chloramine. All surviving rats were killed at week 52 and were evaluated for stomach tumours as were rats that survived to at least week 50. The incidence of MNNG-induced stomach adenocarcinomas was significantly increased ($p < 0.01$) from 7/20 (35%) in animals not administered chloramine to 17/20 (85%) in animals administered chloramine. Ambroxol reduced the incidence of stomach cancer in rats treated with MNNG plus chloramine to 6/19 (32%) and 5/20 (25%) for the 25- and 50-mg/kg doses, respectively. Neither dose level of ambroxol affected the incidence of stomach cancer induced by MNNG. In rats that did not receive chloramine as chloraminated water, the incidence of MNNG-induced stomach cancer was 8/20 (40%) and 9/20 (45%) for the low- and high-dose ambroxol-treated groups, respectively (Narahara *et al.*, 2001).

4. Other Data Relevant to an Evaluation of Carcinogenicity and its Mechanisms

4.1 Absorption, distribution, metabolism and excretion

4.1.1 *Humans*

No studies on the absorption, distribution, metabolism and excretion of monochloramine in humans have been reported.

4.1.2 *Experimental systems*

Male Sprague-Dawley rats (220–240 g) were given 1.1 mg per animal [^{36}Cl]-labelled chloramine [NH_2 ^{36}Cl] orally as 3 mL of a solution containing 370 mg/L chloramine. The peak plasma concentration of ^{36}Cl (10.3 μg/L) was reached 8 h after dosing, and the absorption and elimination half-lives were 2.5 h and 38.8 h, respectively. The distribution of radioactivity was highest in plasma and lowest in fat. Approximately 25% and 2% of the administered dose of radiolabelled chloramine was excreted in the urine and faeces, respectively, during 120 h after treatment. Only 0.35% of the administered dose of monochloramine was present in plasma as [^{36}Cl]chloride 120 h after treatment. No evidence for enzymatic intervention in the metabolism of monochloramine was presented (Abdel-Rahman *et al.*, 1983).

The effect of 15 μM monochloramine as hepatic function was investigated in isolated perfused male Sprague-Dawley rat liver. The uptake of monochloramine averaged 98%. Approximately 0.7% of the amount taken up by the liver was reduced by glutathione (GSH) and appeared in the bile in the form of GSH disulfide (Bilzer & Lauterburg, 1991).

Monochloramine inactivated bovine liver catalase *in vitro*, either by reacting with reduced nicotinamide adenine dinucleotide phosphate (NADPH) or by hydrolysis to hypochlorous acid, which may ligate with the haeme iron of catalase, or both (Mashino & Fridovich, 1988). Subsequent studies showed that the reaction of monochloramine with NADH yields a pyridine chlorohydrin (Prütz *et al.*, 2001).

Monochloramine inhibited purified rat liver enzymes, N^{10}-formyl tetrahydrofolate dehydrogenase (0.56–3.35 μM) and formaldehyde dehydrogenase (2.7–101 μM), *in vitro* (Minami *et al.*, 1993).

4.2   Toxic effects

4.2.1   *Humans*

    (*a*)   *Inhalation*

Chloramines are volatile and, as a result, an important route of human exposure is through inhalation. Numerous case reports have documented the effects of inhalation exposures in households as a result of mixing bleach and liquid ammonia for cleaning. A more limited set of information relates to the inhalation of chloramine in occupational settings. Few studies have investigated the potential effects of inhaling chloramine while bathing. The following is a summary of representative studies of these exposure conditions.

Instances of inhalation of chloramines in the home have been reported as case studies, which give little opportunity for accurate assessment of exposure. However, they do provide a qualitative description of the symptoms and damage that ensued.

Tanen *et al.* (1999) documented acute lung injury that progressed to severe pneumonitis, caused by the use of combined hypochlorite and ammonia for cleaning in an occupational setting. A tracheostomy was necessary, but the patient recovered within 7 days.

Hery *et al.* (1998) reported an analysis of concentrations of chloramine and complaints of irritation of the eye and upper respiratory tract of workers in a green-salad processing plant. The irritant agents were chloramines that resulted from the reaction of hypochlorite with nitrogen compounds in the sap proteins that were released when the vegetables were cut. The study documented the relationship between concentrations of chloramine in the processing water and those in the air at different workstations. The air measurements included nitrogen trichloride (trichloroamine) as a separate entity and clearly showed that the exposure was not entirely composed of monochloramine. Concentrations of chlorine (5–60 mg/L) in combination with amino-containing chemicals resulted in levels of total soluble chlorine compounds in the air of 0.1–0.8 mg/m^3 chlorine and 0.2–0.9 mg/m^3 nitrogen trichloride. This exposure assessment was linked to a second study in which irritation phenomena were studied among 334 swimming-pool lifeguards (Massin *et al.*, 1998). Exposure of these individuals was primarily by inhalation, and eye, nose and throat irritation showed a dose–response relationship with exposures to nitrogen trichloride. There was little evidence that chronic respiratory symptoms increased with increasing exposure (e.g. chronic bronchitis). The irritant effects became apparent at nominal air concentrations of 0.5 mg/m^3 nitrogen trichloride. The extent to which nitrogen trichloride would be formed in the air from use of chloraminated water in confined spaces (e.g. showers) in the home has not been studied.

A recent account of case studies of two lifeguards and one swimming instructor reported that two subjects had decreased forced expiratory volume in response to a challenge of nitrogen trichloride at 0.5 mg/m^3 and the third showed a positive response in the workplace (Thickett *et al.*, 2002). These authors concluded that nitrogen trichloride was a cause of occupational asthma.

(b)   *Ingestion*

A few studies have specifically evaluated the potential effects of ingested chloramine. Lubbers and Bianchine (1984) and Lubbers *et al.* (1984) studied acute and repeated doses of monochloramine in comparison with other drinking-water disinfectants. In a rising-dose tolerance study (Lubbers & Bianchine, 1984), groups of 10 adult male volunteers drank two 500-mL portions of water containing 0.01, 1.0, 8, 18 or 24 mg/L chloramine, or hypochlorite, sodium chlorite, sodium chlorate or chlorine dioxide or served as controls. A battery of measures, including blood and urine biochemistry, blood cell counts and morphology, and physical examination were conducted following each dose. No statistically significant changes were noted in the chloramine-treated group. In the sub-sequent repeated-dose study (Lubbers *et al.*, 1984), 10 volunteers were required to drink a 500-mL portion of water containing 5 ppm [5 mg/L] chloramine daily for 12 weeks followed by an 8-week wash-out period. The volunteers were required to consume the entire 500-mL portion within 15 min. The other treatment groups involved comparisons with other drinking-water treatments as described in the rising-dose study. The same set of biochemical, clinical and haematological tests and physical examinations were con-ducted. The only statistically significant change in the chloramine-treated group was an increase in triiodothyronine (T_3) uptake. However, this change remained within the normal range and was considered to be clinically irrelevant.

Wones *et al.* (1993) reported the results of a study focused on measures of lipid meta-bolism and thyroid function that involved three groups of 16 adult male volunteers who received 1.5 L distilled water or 2 or 15 ppm [2 or 15 mg/L] chloramine daily for 4 weeks, followed by a 4-week acclimatization period during which all subjects were provided distilled drinking-water and a standardized diet that was relatively high in total fat, satu-rated fat and cholesterol. The diet was maintained whereas the water provided was changed in the subsequent 4-week experimental period. The only statistically significant change was a 12% increase in plasma apolipoprotein B in men consuming 15 ppm chlor-amine. The only parameter that tended to parallel this trend was an increase in plasma tri-glycerides, but this was not statistically significant ($p = 0.07$).

(c)   *Haemodialysis patients*

Two studies are representative of broad findings in patients undergoing haemodialysis treatment when the municipal drinking-water that was used in the preparation of dialysis fluid contained appreciable concentrations of chloramine. Fluck *et al.* (1999) reported that chloramine-induced haemolysis was responsible for erythropoeitin-resistant anaemia that developed in a haemodialysis unit in the United Kingdom. An increase in mean methaemo-globinaemia of 23% and a fall in mean haptoglobin of 21% was noted in these patients. Following installation of columns that contained activated carbon to remove the chlor-amine, these effects were reversed. The concentrations of chloramine prior to installation of the columns were 0.25–0.3 ppm [mg/L] but fell to < 0.1 ppm [mg/L] following installa-tion of carbon column filtration.

There is some evidence of varying sensitivity to the haemolytic effects of chloramine among haemodialysis patients. Weinstein *et al.* (2000) studied differences between 24 responders and nine non-responders to chloramine in dialysis fluid. The initial water analysis indicated a concentration of 0.19 mg/L chloramine in the water supply. An inverse correlation between concentrations of serum GSH and haemoglobin was observed in these patients.

4.2.2   *Experimental systems*

(*a*)   *Inhalation*

No study of exposure to chloramine via inhalation has been reported. One study in mice of sensory irritation responses to nitrogen trichloride found that the airborne concentration of nitrogen trichloride that resulted in a 50% decrease in respiratory rate was 2.5 ppm [12.3 µg/mL] (Gagnaire *et al.*, 1994). The maximal response was observed within 10 min.

(*b*)   *Oral*

Monochloramine was administered in the drinking-water at concentrations of 0, 1, 10 and 100 mg/L to male Sprague-Dawley rats (Abdel-Rahman *et al.*, 1984) and haematological parameters were examined after 3 and 10 months of treatment. Only four animals per group were sampled at each time-point. A statistically significant decrease in haematocrit was observed after 3 months of treatment with 10 and 100 mg/L. While a similar trend was apparent at 10 months, the results were not statistically significant.

Daniel *et al.* (1990) compared the subchronic toxicity of chloramine with that of chlorine and chlorine dioxide in Sprague-Dawley rats. Chloramine was administered in the drinking-water at concentrations of 0, 25, 50, 100 and 200 mg/L for 90 days to groups of five males or females. Parameters of haematology and clinical chemistry were measured after 90 days. Tissues were sampled and examined for gross lesions and histopathology. At the 200-mg/L dose, weight gain was decreased and the relative kidney weights were increased by chloramine in both male and female rats. Male animals had significantly decreased haematocrits at 100 mg/L and decreased red blood cell counts at both 100 and 200 mg/L. No treatment-related pathology was noted.

The effects of monochloramine were compared with those of chlorine in 10 male and female B6C3F$_1$ mice given 0, 12.5, 25, 50, 100 and 200 mg/L of the two compounds in the drinking-water (Daniel *et al.*, 1991). Body weight gain was significantly decreased at 100 and 200 mg/L in both male and female mice. Red blood cell counts, haemoglobin and haematocrit levels in male mice were significantly increased at 200 mg/L. Only red blood cell counts were elevated in female mice at this dose. Relative kidney weights were increased at 200 mg/L in males and females, but not at lower doses of monochloramine. No treatment-related pathology was observed.

Bercz *et al.* (1982) studied the ability of chloramine to induce haematological effects and alterations in clinical chemistry measures including serum thyroxine (T$_4$) concentrations in adult African green monkeys (five males and seven females). Chloramine was administered in the drinking-water at a concentration of 100 mg/L for a period of 6 weeks (total body dose, 10 mg/kg bw per day). No significant effects were observed.

An immunotoxicological evaluation was made in groups of 12 male Sprague-Dawley rats treated with sodium hypochlorite or 9, 19 or 38 ppm [mg/L] chloramine in the drinking-water (Exon *et al.*, 1987). No remarkable findings were noted with respect to relative spleen and thymus weights, delayed hypersensitivity reactions, natural killer cell activity or the phagocytic activity of macrophages. A substantial and dose-related increase in prostaglandin E_2 activity was observed in adherent resident peritoneal cells from animals treated with 19 and 38 ppm. However, no corresponding alterations in immune function were found.

A study of plasma cholesterol and thyroid hormone levels was conducted in white Carneau pigeons treated with various disinfectants (Revis *et al.*, 1986). Increases in plasma levels of low-density lipoprotein cholesterol were observed with 2 and 15 ppm [mg/L] hypochlorite (chlorine pH 8.5), and 15 ppm [mg/L] chloramine and chlorine dioxide when administered with a high-cholesterol diet. These changes appeared to be correlated with an increase in the size of atherosclerotic plaques and a very large decrease in T_4 and T_3 concentrations in serum. It must be noted that these results are substantially different from those observed in primates (Bercz *et al.*, 1982), in an independent study in pigeons (Penn *et al.*, 1990), in rats (Poon *et al.*, 1997) and in humans (Wones *et al.*, 1993) receiving comparable treatments with a variety of disinfectants, including chloramine.

Poon *et al.* (1997) specifically examined the effects of 200 ppm [mg/L] (equivalent to 21.6 mg/kg per day) monochloramine in drinking-water in male Sprague-Dawley rats treated for 13 weeks. The study included a group of control rats that were given the same volume of water as that consumed by the chloramine-treated animals. The animals were monitored for changes in levels of T_4, hepatic drug metabolizing enzymes, haematological parameters and measures of immune response. The authors concluded that the minor changes seen in these parameters were largely associated with decreased water and food consumption.

The role of monochloramine and related organic chloramines formed *in situ* has been studied extensively with respect to the development of stomach lesions. This interest arises from investigations of mechanisms by which *H. pylori* might induce gastritis, peptic ulcers and stomach cancer (Suzuki *et al.*, 1992; Dekigai *et al.*, 1995; Xia & Talley, 2001). This organism produces a high concentration of ammonia in the stomach of infected patients, which is postulated to interact with neutrophil-derived hypochlorous acid to produce chloramine.

Monochloroamine exerts biological effects in isolated systems at concentrations that would commonly be seen in drinking-water. At concentrations of 1–10 µM [0.5–5 mg/L], monochloramine inhibited the growth of the normal rat gastric mucosa cell line, RGM-L (Naito *et al.*, 1997). This growth inhibition was in part attributed to evidence of increased apoptosis (characterized by a fraction of subdiploid cells and nuclear fragmentation) with concomitant accumulation of cells in the G2/M phase of the cell cycle.

The application of chloramine to the serosal side of the colonic mucosa of rats at concentrations of 50 μM increased mucosal secretion of prostaglandin E_2 (Tamai *et al.*, 1991). The 50% maximum effective for monochloramine was 3.2 μM [0.165 mg/L] compared with 6.5 μM [0.34 mg/L] for hypochlorous acid and 6.5 μM [0.22 mg/L] for hydrogen peroxide.

Apoptosis induced by etoposide was inhibited in human Jurkat T cells (acute T-cell leukaemia cell line) by 70 μM [3.6 mg/L] monochloramine (Than *et al.*, 2001). This effect was associated with an increase in the number of cells arrested in the G_0/G_1 phase and a decrease in the number of cells in the S phase. Cells that survived had an increased incidence of aneuploidy, probably attributable to treatment with etoposide.

Monochloramine, at a concentration range of 30–50 μM [1.5–2.6 mg/L], inhibited the respiratory burst that is induced in neutrophils by phorbol esters. These concentrations did not affect cell viability, but decreased protein kinase C activity, which fell to zero with 70 μM [3.6 mg/L] chloramine (Ogino *et al.*, 1997).

Experiments that examined the ability of monochloramine solutions to damage the gastric mucosa used higher concentrations than those encountered in drinking-water. Umeda *et al.* (1999) gave 5 mL/kg [30.9 mg/kg] of a 120-mM [6.18 g/L] concentration of chloramine to rats to induce gastric mucosal damage. However, the aim of the study was to demonstrate that Lafutidine, a histamine H_2-receptor antagonist, substantially decreased the size of the lesions produced. The authors indicated that concentrations of 60 mM [3.09 g/L] were needed to induce lesions.

(c) Dermal

Solutions of up to 1000 mg/L monochloramine produced no evidence of hyperplasia when applied to the skin of Sencar mice for 10 min per day for 4 days (Robinson *et al.*, 1986). This was in sharp contrast to results obtained with hypochlorous acid, which more than doubled skin thickness in parallel experiments at concentrations as low as 300 mg/L.

4.3 Reproductive and developmental effects

4.3.1 Humans

No data were available to the Working Group.

4.3.2 Experimental systems

Carlton *et al.* (1986) examined the effects of chloramine on reproductive parameters of Long-Evans rats. Twelve males were treated by gavage with either 0, 2.5, 5 or 10 mg/kg bw monochloramine for 56 days prior to mating and throughout the 10-day mating period; 24 females were treated with the same doses for 14 days before mating and throughout mating, gestation and lactation, and 21 days postpartum. No clinical signs of toxicity, haematological changes or changes in reproductive parameters were found. No treatment-related histopathology was observed in the reproductive tracts of either male or female rats.

4.4 Genetic and related effects

4.4.1 *Humans*

No data were available to the Working Group.

4.4.2 *Experimental systems* (see Table 1 for details and references)

All the results given in Section 4.4.2 correspond to single studies.

After exposure of *Bacillus subtilis* cells or *B. subtilis* DNA to monochloramine, single-strand breaks as well as loss of DNA-transforming activity were observed. Monochloramine did not induce a mutagenic effect in *Salmonella typhimurium* strain TA100 and was a weak mutagen in the reversion of trpC to trp⁺ in *B. subtilis*. [The Working Group noted that the procedure followed by these authors was less sensitive than the standard Maron and Ames plating method.]

The water obtained after treatment of a non-mutagenic raw water with monochloramine showed mutagenic activity to *S. typhimurium* in the presence and in the absence of metabolic activation (Cheh *et al.*, 1980). [The Working Group noted that the mutagenic activity is probably not due to monochloramine but to by-products of the reaction of monochloramine with organic carbon contained in the raw water.] Treatment of fulvic acid solutions or of drinking-water with monochloramine induced the appearance of significant mutagenicity in *S. typhimurium* strains TA100 and TA98 (Cozzie *et al.*, 1993; DeMarini *et al.*, 1995) [The Working Group noted that the mutagenic activity observed was not attributable to monochloramine itself but to by-products of the reaction of monochloramine with organic compounds.]

Monochloramine caused double-strand DNA breakage in plasmid pUC18, and chlorogenic acid prevented this effect. *In vitro*, it induced DNA double-strand breaks and chromatin condensation in rabbit gastric mucosal cells and human stomach carcinoma KATO III cells, and DNA fragmentation in rabbit gastric mucosal cells and in human stomach adenocarcinoma MKN45 cells. The precursors of monochloramine, ammonia and hypochlorous acid did not induce DNA double-strand breaks or chromatin condensation in either rabbit gastric mucosal or KATO III cells (Suzuki *et al.*, 1997).

In a single in-vivo study, monochloramine did not induce micronucleus formation, chromosomal aberrations or aneuploidy in the bone marrow of CD-1 mice or sperm abnormalities in B6C3F₁ mice. Monochloramine induced the formation of micronuclei in erythrocytes of newt larvae *in vivo*.

Table 1. Genetic and related effects of chloramine

Test system	Result[a] Without exogenous metabolic system	With exogenous metabolic activation	Dose[b] (LED or HID)	Reference
DNA strand breaks, *Bacillus subtilis* culture	+	NT	112 µM [5.7]	Shih & Lederberg (1976a)
Salmonella typhimurium TA100, reverse mutation	−	NT	60 µM [3]	Thomas et al. (1987)
Bacillus subtilis, reverse mutation	+	NT	56 µM [2.9]	Shih & Lederberg (1976b)
DNA double-strand breaks, plasmid pUC18 DNA *in vitro*	+	NT	3 mM [154.5]	Shibata et al. (1999)
DNA double-strand breaks, rabbit gastric mucosal cells *in vitro*	+	NT	0.1 mM [5.15]	Suzuki et al. (1997)
DNA fragmentation, rabbit gastric mucosal cells *in vitro*	+	NT	0.1 mM [5.15]	Suzuki et al. (1997)
DNA double-strand breaks, human gastric carcinoma KATO III cells *in vitro*	+	NT	0.1 mM [5.15]	Suzuki et al. (1997)
Chromatin condensation, rabbit gastric mucosal cells *in vitro*	+	NT	0.1 mM [5.15]	Suzuki et al. (1997)
DNA fragmentation, human gastric adenocarcinoma MKN45 cells *in vitro*	+	NT	0.001 mM [0.05]	Suzuki et al. (1998)
DNA fragmentation, human gastric carcinoma KATO III cells *in vitro*	−	NT	0.001 mM [0.05]	Suzuki et al. (1998)
Chromatin condensation, human gastric carcinoma KATO III cells *in vitro*	+	NT	0.1 mM [5.15]	Suzuki et al. (1997)
Micronucleus formation, CD-1 mouse bone-marrow erythrocytes *in vivo*	−		200[c] po × 5	Meier et al. (1985)
Micronucleus formation, *Pleurodeles waltl* newt larvae peripheral erythrocytes *in vivo*	+		0.15[d]	Gauthier et al. (1989)
Chromosomal aberrations, CD-1 mouse bone-marrow cells *in vivo*	−		200[c] po × 5	Meier et al. (1985)
Aneuploidy, bone-marrow cells of CD-1 mouse *in vivo*	−		200[c] po × 5	Meier et al. (1985)
Sperm morphology, B6C3F1 mice *in vivo*	−		200[c] po × 5	Meier et al. (1985)

[a] +, positive; −, negative; NT, not tested
[b] LED, lowest effective dose; HID, highest ineffective dose; in-vitro tests, µg/mL; in-vivo tests, mg/kg bw; po, orally
[c] Amount of chloramine (in µg) administered daily to each animal by gavage for 5 days
[d] Larvae reared in chloramine-containing water

5.   Summary of Data Reported and Evaluation

5.1   Exposure data

Chloramine, formed by the reaction of ammonia with chlorine, is increasingly being used in the disinfection of drinking-water. Monochloramine, dichloramine and trichloramine are in equilibrium, with monochloramine predominating. Exposure to milligram-per-litre levels occurs through ingestion of chloraminated water. Chloramines are also formed in swimming pools from the reaction of chlorine with nitrogen-containing contaminants, and trichloramine has been measured in swimming-pool air. Chloramine generated *in situ* is also used as an intermediate in the production of hydrazines, organic amines and other industrial chemicals.

5.2   Human carcinogenicity data

Several studies were identified that analysed risk with respect to one or more measures of exposure to complex mixtures of disinfection by-products that are found in most chlorinated and chloraminated drinking-water. No data specifically on chloramine were available to the Working Group.

5.3   Animal carcinogenicity data

Chloraminated drinking-water (predominantly in the form of monochloramine) was tested for carcinogenicity by oral administration in female and male mice and rats without demonstrating clear evidence of carcinogenic activity. In carcinogen-initiated rats, chloramine generated by ammonium acetate (in feed) and sodium hypochlorite (in drinking-water) promoted stomach cancer.

5.4   Other relevant data

^{36}Cl-Labelled chloramine is readily absorbed after oral administration to rats. About 25% of the administered radioactivity is excreted in the urine over 120 h.

Nitrogen trichloride (trichloroamine) is volatilized from food-processing water disinfected with chloramine and from swimming-pool waters disinfected with chlorine, and reacts with ammonia in water to form chloramine. Upon inhalation, it produces lung irritation and may be a cause of occupational asthma. Ingestion of monochloramine produced no clinical abnormalities in male volunteers at concentrations as high as 15 mg/L. No reproductive or developmental effects have been associated with monochloramine.

Chloramine induced single-strand breaks and loss of DNA-transforming activity and was a weak mutagen in *Bacillus subtilis*. It was not mutagenic to *Salmonella typhimurium*. *In vitro*, chloramine caused double-strand DNA breakage in plasmid pUC18, and DNA fragmentation and DNA double-strand breaks as well as chromatin condensation in rabbit gastric mucosal cells and human stomach cancer cells. Monochloramine did not induce micronuclei, chromosomal aberration, aneuploidy or sperm abnormality in mice *in vivo*, but induced the formation of micronuclei in erythrocytes of newt larvae *in vivo*.

5.5   Evaluation

There is *inadequate evidence* in humans for the carcinogenicity of chloramine.

There is *inadequate evidence* in experimental animals for the carcinogenicity of monochloramine.

Overall evaluation

Chloramine is *not classifiable as to its carcinogenicity to humans (Group 3)*.

6.   References

Abdel-Rahman, M.S., Waldron, D.M. & Bull, R.J. (1983) A comparative kinetics study of mono-chloramine and hypochlorous acid in rat. *J. appl. Toxicol.*, **3**, 175–179

Abdel-Rahman, M.S., Suh, D.H. & Bull, R.J. (1984) Toxicity of monochloramine in rat: An alternative drinking water disinfectant. *J. Toxicol. environ Health*, **13**, 825–834

American Public Health Association/American Water Works Association/Water Environment Federation (1999) Method 4500-Cl chlorine (residual). In: *Standard Methods for the Examination of Water and Wastewater*, 20th Ed., Washington, DC

Bercz, J.P., Jones, L., Garner, L., Murray, D., Ludwig, D.A. & Boston, J. (1982) Subchronic toxicity of chlorine dioxide and related compounds in drinking water in the nonhuman primate. *Environ. Health Perspect.*, **46**, 47–55

Bilzer, M., & Lauterburg, B.H. (1991) Effects of hypochlorous acid and chloramines on vascular resistance, cell integrity, and biliary glutathione disulfide in the perfused rat liver: Modulation by glutathione. *J. Hepatol.*, **13**, 84–89

Carlton, B.D., Barlett, P., Basaran, A., Colling, K., Osis, I. & Smith, M.K. (1986) Reproductive effects of alternative disinfectants. *Environ. Health Perspect.*, **69**, 237–241

Carr, A.C., Hawkins, C.L., Thomas, C.R., Stocker, R. & Frei, B. (2001) Relative reactivities of *N*-chloramines and hypochlorous acid with human plasma constituents. *Free Radic. Biol. Med.*, **30**, 526–536

Cheh, A.M., Skochdopole, J., Koski, P. & Cole, L. (1980) Nonvolatile mutagens in drinking water: Production by chlorination and destruction by sulfite. *Science*, **207**, 90–92

Cozzie, D.A., Kanniganti, R., Charles, M.J., Johnson, J.D. & Ball, L.M. (1993) Formation and characterization of bacterial mutagens from reaction of the alternative disinfectant mono-chloramine with model aqueous solutions of fulvic acid. *Environ. mol. Mutag.*, **21**, 237–246

Daniel, F.B., Condie, L.W., Robinson, M., Stober, J.A., York, R.G., Olson, G.R. & Wang, S.-R. (1990) Comparative subchronic toxicity studies of three disinfectants. *J. Am. Water Works Assoc.*, **82**, 61–69

Daniel, F.B., Ringhand, H.P., Robinson, M., Stober, J.A., Olson, G.R. & Page, N.P. (1991) Comparative subchronic toxicity of chlorine and monochloramine in the B6C3F1 mouse. *J. Am. Water Works Assoc.*, **83**, 68–75

Dekigai, H., Murakami, M. & Kita, T. (1995) Mechanism of *Helicobacter pylori*-associated gastric mucosal injury. *Dig. Dis. Sci.*, **40**, 1332–1339

DeMarini, D.M., Abu-Shakra, A., Felton, C.F., Patterson, K.S. & Shelton, M.L. (1995) Mutation spectra in Salmonella of chlorinated, chloraminated, or ozonated drinking water extracts: Comparison to MX. *Environ. mol. Mutag.*, **26**, 270–285

Diehl, A.C., Speitel, G.E., Jr, Symons, J.M., Krasner, S.W., Hwang, C.J. & Barrett, S.E. (2000) DBP formation during chloramination. *J. Am. Water Works Assoc.*, **92**, 76–90

Environmental Protection Agency (1998) National primary drinking water regulations; disinfectants and disinfection byproducts; final rule. *Fed. Regist.*, **63**, 69390–69476

Exon, J.H., Koller, L.D., O'Reilly, C.A. & Bercz, J.P. (1987) Immunotoxicologic evaluation of chlorine-based drinking water disinfectants, sodium hypochlorite and monochloramine. *Toxicology*, **44**, 257–269

Fluck, S., McKane, W., Cairns, T., Fairchild, V., Lawrence, A., Lee, J., Murray, D., Polpitiye, M., Palmer, A. & Taube, D. (1999) Chloramine-induced haemolysis presenting as erythropoietin resistance. *Nephrol. Dial. Transplant.*, **14**, 1687–1691

Gagnaire, F., Azim, S., Bonnet, P., Hecht, G. & Hery, M. (1994) Comparison of the sensory irritation response in mice to chlorine and nitrogen trichloride. *J. appl. Toxicol.*, **14**, 405–409

Gauthier, L., Levi, Y. & Jaylet, A. (1989) Evaluation of the clastogenicity of water treated with sodium hypochlorite or monochloramine using a micronucleus test in newt larvae (*Pleurodeles waltl*). *Mutagenesis*, **4**, 170–173

Haseman, J.K. (1984) Statistical issues in the design, analysis and interpretation of animal carcinogenicity studies. *Environ. Health Perspect.*, **58**, 385–392

Health Canada (2003) *Summary of Guidelines for Canadian Drinking Water Quality*, Ottawa

Hery, M., Hecht, G., Gerger, J.M., Gendre, J.C., Hubert, G. & Rebuffaud, I. (1995). Exposure to chloramines in the atmosphere of indoor swimming pools. *Ann. occup. Hyg.*, **39**, 427–439

Hery, M., Gerber, J.M., Hecht, G., Subra, I., Possoz, C., Aubert, S., Dieudonne, M. & Andre, J.C. (1998) Exposure to chloramines in a green salad processing plant. *Ann. occup. Hyg.*, **42**, 437–451

Holzwarth, G., Balmer, R.G. & Soni, L. (1984) The fate of chlorine and chloramines in cooling towers. *Water Res.*, **18**, 1421–1427

Iishi, H., Tatsuta, M., Baba, M., Mikuni, T., Yamamoto, R., Iseki, K., Yano, H., Uehara, H. & Nakaizumi, A. (1997) Enhancement by monochloramine of the development of gastric cancers in rats: A possible mechanism of *Helicobacter pylori*-associated gastric carcinogenesis. *J. Gastroenterol.*, **32**, 435–441

Isaac, R.A. & Morris, J.C. (1980) Rates of transfer of active chlorine between nitrogenous substrates. In: Jolley, R.L., ed., *Water Chlorination*, Vol. 3, Ann Arbor, MI, Ann Arbor Science, pp. 183–191

Iseki, K., Tatsuta, M., Iishi, H., Baba, M., Mikuni, T., Hirasawa, R., Yano, H., Uehara, H. & Nakaizumi, A. (1998) Attenuation by methionine of monochloramine-enhanced gastric carcinogenesis induced by N-methyl-N'-nitro-N-nitrosoguanidine in Wistar rats. *Int. J. Cancer*, **76**, 73–76

Kato, S., Umeda, M., Takeeda, M., Kanatsu, K. & Takeuchi, K. (2002) Effect of taurine on ulcerogenic response and impaired ulcer healing induced by monochloramine in rat stomachs. *Aliment. Pharmacol. Ther.*, **16** (Suppl. 2), 35–43

Kirk, R.E. & Othmer, D.F., eds (1993) *Encyclopedia of Chemical Technology*, 4th Ed., Vol. 5, New York, John Wiley & Sons, p. 916

Lubbers, J.R. & Bianchine, J.R. (1984) Effects of the acute rising dose administration of chlorine dioxide, chlorate and chlorite to normal healthy adult male volunteers. *J. environ. Pathol. Toxicol. Oncol.*, **5**, 215–228

Lubbers, J.R., Chauhan, S., Miller, J.K. & Bianchine, J.R. (1984) The effects of chronic administration of chlorine dioxide, chlorite and chlorate to normal healthy adult male volunteers. *J. environ. Pathol. Toxicol. Oncol.*, **5**, 229–238

Mashino, T. & Fridovich, I. (1988) NADPH mediates the inactivation of bovine liver catalase by monochloroamine. *Arch. Biochem. Biophys.*, **265**, 279–285

Massin, N., Bohadana, A.B., Wild, P., Héry, M., Toamain, J.P. & Hubert, G. (1998) Respiratory symptoms and bronchial responsiveness in lifeguards exposed to nitrogen trichloride in indoor swimming pools. *Occup. environ. Med.*, **55**, 258–263

Meier, J.R., Bull, R.J., Stober, J.A. & Cimino, M.C. (1985) Evaluation of chemicals used for drinking water disinfection for production of chromosomal damage and sperm-head abnormalities in mice. *Environ. Mutag.*, **7**, 201–211

Minami, M., Inagaki, H., Katsumata, M., Miyake, K. & Tomoda, A. (1993) Inhibitory action of chloramine on formate-metabolizing system. Studies suggested by an unusual case record. *Biochem. Pharmacol.*, **45**, 1059–1064

Naito, Y., Yoshikawa, T., Fujii, T., Boku, Y., Yagi, N., Dao, S., Yoshida, N., Kondo, M., Matsui, H., Ohtani-Fujita, N. & Sakai, T. (1997) Monochloramine-induced cell growth inhibition and apoptosis in a rat gastric mucosal cell line. *J. clin. Gastroenterol.*, **25** (Suppl. 1), S179–S185

Narahara, H., Tatsuta, M., Iishi, H., Baba, M., Mikuni, T., Uedo, N., Sakai, N. & Yano, H. (2001) Attenuation by ambroxol of monochloramine-enhanced gastric carcinogenesis: A possible prevention against *Helicobacter pylori*-associated gastric carcinogenesis. *Cancer Lett.*, **168**, 117–124

National Health and Medical Research Council and Agriculture and Resource Management Council of Australia and New Zealand (1996) *Australian Drinking Water Guidelines Summary*, Canberra

National Toxicology Program (1992) *Toxicology and Carcinogenesis Studies of Chlorinated Water (CAS Nos. 7782-50-5 and 7681-52-9) and Chloraminated Water (CAS No. 10599-90-3) (Deionized and Charcoal-Filtered) in F344/N Rats and B6C3F1 Mice (Drinking Water Studies)* (Tech. Rep. Ser. No. 392; NIH Publ. No. 92-2847), Research Triangle Park, NC

Ogino, T., Kobuchi, H., Sen, C.K., Roy, S., Packer, L. & Maguire, J.J. (1997) Monochloramine inhibits phorbol ester-inducible neutrophil respiratory burst activation and T cell interleukin-2 receptor expression by inhibiting inducible protein kinase C activity. *J. biol. Chem.*, **272**, 26247–26252

Oldenburg, P.S., Regan, J.M., Harringston, G.W. & Noguera, D.R. (2002) Kinetics of *Nitrosomonas europaea* inactivation by chloramine. *J. Am. Water Works Assoc.*, **94**, 100–110

Penn, A., Lu, M.-X. & Parkes, J.L. (1990) Ingestion of chlorinated water has no effect upon indicators of cardiovascular disease in pigeons. *Toxicology*, **63**, 301–313

Poon, R., Lecavalier, P., Tryphonas, H., Bondy, G., Chen, M., Chu, I., Yagminas, A., Valli, V.E., D'Amour, M. & Thomas, B. (1997) Effects of subchronic exposure of monochloramine in drinking water on male rats. *Regul. Toxicol. Pharmacol.*, **25**, 166–175

Prütz, W.A., Kissner, R. & Koppenol, W.H. (2001) Oxidation of NADH by chloramines and chloramides and its activation by iodide and by tertiary amines. *Arch. Biochem. Biophys.*, **393**, 297–307

Revis, N.W., McCauley, P., Bull, R. & Holdsworth, G. (1986) Relationship of drinking water disinfectants to plasma cholesterol and thyroid hormone levels in experimental studies. *Proc. natl Acad. Sci. USA*, **83**, 1485–1489

Robinson, M., Bull, R.J., Schamer, M. & Long, R.E. (1986) Epidermal hyperplasia in mouse skin following treatment with alternative drinking water disinfectants. *Environ. Health Perspect.*, **69**, 293–300

Shibata, H., Sakamoto, Y., Oka, M. & Kono, Y. (1999) Natural antioxidant, chlorogenic acid, protects against DNA breakage caused by monochloramine. *Biosci. Biotechnol. Biochem.*, **63**, 1295–1297

Shih, K.L. & Lederberg, J. (1976a) Effects of chloramine on *Bacillus subtilis* deoxyribonucleic acid. *J. Bacteriol.*, **125**, 934–945

Shih, K.L. & Lederberg, J. (1976b) Chloramine mutagenesis in *Bacillus subtilis*. *Science*, **192**, 1141–1143

Suzuki, M., Miura, S., Suematsu, M., Fukumura, D., Kurose, I., Suzuki, H., Kai, A., Kudoh, Y., Ohashi, M. & Tsuchiya, M. (1992) *Helicobacter pylori*-associated ammonia production enhances neutrophil-dependent gastric mucosal cell injury. *Am. J. Physiol.*, **263**, G719–G725

Suzuki, H., Mori, M., Suzuki, M., Sakurai, K., Miura, S. & Ishii, H. (1997) Extensive DNA damage induced by monochloramine in gastric cells. *Cancer Lett.*, **115**, 243–248

Suzuki, H., Seto, K., Mori, M., Suzuki, M., Miura, S. & Ishii, H. (1998) Monochloramine induced DNA fragmentation in gastric cell line MKN45. *Am. J. Physiol.*, **275**, G712–G716

Tamai, H., Kachur, J.F., Baron, D.A., Grisham, M.B. & Gaginella, T.S. (1991) Monochloramine, a neutrophil-derived oxidant, stimulates rat colonic secretion. *J. Pharmacol. exp. Ther.*, **257**, 887–894

Tanen, D.A., Graeme, K.A. & Raschke, R. (1999) Severe lung injury after exposure to chloramine gas from household cleaners. *New Engl. J. Med.*, **341**, 848–849

Taras, M.J. (1953) Effect of free residual chlorination on nitrogen compounds in water. *J. Am. Water Works Assoc.*, **45**, 4761

Than, T.A., Ogino, T., Omori, M. & Okada, S. (2001) Monochloramine inhibits etoposide-induced apoptosis with an increase in DNA aberration. *Free Radic. Biol. Med.*, **30**, 932–940

Thickett, K.M., McCoach, J.S., Gerber, J.M., Sadhra, S. & Burge, P.S. (2002) Occupational asthma caused by chloramines in indoor swimming-pool air. *Eur. respir. J.*, **19**, 827–832

Thomas, E.L., Jefferson, M.M., Bennett, J.J. & Learn, D.B. (1987) Mutagenic activity of chloramines. *Mutat. Res.*, **188**, 35–43

Umeda, M., Fujita, A., Nishiwaki, H. & Takeuchi, K. (1999) Monochloramine and gastric lesions. Effect of lafutidine, a novel histamine H_2-receptor antagonist, on monochloramine-induced gastric lesions in rats: Role of capsaicin-sensitive sensory neurons. *J. Gastroenterol. Hepatol.*, **14**, 859–865

Ura, Y. & Sakata, G. (1986) Chloroamines. In: Gerhartz, W., Yamamoto, Y.S., Campbell, F.T., Pfefferkorn, R. & Rounsaville, J.F., eds, *Ullmann's Encyclopedia of Industrial Chemistry*, 5th Ed., Vol. A6, New York, VCH Publishers, pp. 553–558

Vikesland, P.J., Ozekin, K. & Valentine, R.L. (2001) Monochloramine decay in model and distribution system waters. *Water Res.*, **35**, 1766–1776

Weinstein, T., Chagnac, A., Korzets, A., Boaz, M., Ori, Y., Herman, M., Malachi, T. & Gafter, U. (2000) Haemolysis in haemodialysis patients: Evidence for impaired defence mechanisms against oxidative stress. *Nephrol. Dial. Transplant.*, **15**, 883–887

WHO (1998) *Guidelines for Drinking-water Quality*, 2nd Ed., Vol. 2, *Health Criteria and Other Supporting Information and Addendum to Vol. 2*, Geneva

WHO (2000) Chemical hazards. In: *Guidelines for Safe Recreational-water Environments*, Vol. 2, *Swimming Pools, Spas and Similar Recreational-water Environment, Final Draft for Consultation, August 2000*, Geneva, pp. 4-1–4-28

Wojtowicz, J.A. (1993) Chloramines and bromamines. In: Kroschwitz, J.I. & Howe-Grant, M. eds, *Kirk-Othmer Encyclopedia of Chemical Technology*, 4th Ed., Vol. 5, New York, John Wiley & Sons, pp. 911–932

Wones, R.G., Deck, C.C., Stadler, B., Roark, S., Hogg, E. & Frohman, L.A. (1993) Effects of drinking water monochloramine on lipid and thyroid metabolism in healthy men. *Environ. Health Perspect.*, **99**, 369–374

Xia, H.H.-X. & Talley, N.J. (2001) Apoptosis in gastric epithelium induced by *Helicobacter pylori* infection: Implications in gastric carcinogenesis. *Am. J. Gastroenterol.*, **96**, 16–26

Zhang, W. & DiGiano, F.A. (2002) Comparison of bacterial regrowth in distribution systems using free chlorine and chloramine: A statistical study of causative factors. *Water Res.*, **36**, 1469–1482

CHLORAL AND CHLORAL HYDRATE

These substances were considered by a previous Working Group, in February 1995 (IARC, 1995). Since that time, new data have become available and these have been incorporated into the monograph and taken into consideration in the present evaluation.

1. Exposure Data

1.1 Chemical and physical data

1.1.1 Nomenclature

Chloral

Chem. Abstr. Serv. Reg. No.: 75-87-6
Chem. Abstr. Name: Trichloroacetaldehyde
IUPAC Systematic Name: Chloral
Synonyms: Anhydrous chloral; 2,2,2-trichloroacetaldehyde; trichloroethanal; 2,2,2-tri-chloroethanal

Chloral hydrate

Chem. Abstr. Serv. Reg. No.: 302-17-0
Deleted CAS Number: 109128-19-0
Chem. Abstr. Name: 2,2,2-Trichloro-1,1-ethanediol
IUPAC Systematic Name: Chloral hydrate
Synonyms: Chloral monohydrate; trichloroacetaldehyde hydrate; trichloroacetaldehyde monohydrate; 1,1,1-trichloro-2,2-dihydroxyethane

1.1.2   *Structural and molecular formulae and relative molecular mass*

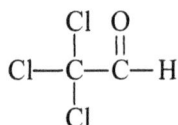

$$\begin{array}{ccc} & Cl & O \\ & | & \| \\ Cl- & C-C & -H \\ & | \\ & Cl \end{array}$$

C₂HCl₃O   Chloral   Relative molecular mass: 147.39

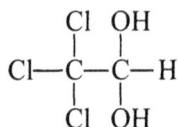

$$\begin{array}{ccc} & Cl & OH \\ & | & | \\ Cl- & C-C & -H \\ & | & | \\ & Cl & OH \end{array}$$

C₂H₃Cl₃O₂   Chloral hydrate   Relative molecular mass: 165.40

1.1.3   *Chemical and physical properties of the pure substances*

Chloral

   (*a*)   *Description*: Liquid (Lide, 2000)
   (*b*)   *Boiling-point*: 97.8 °C (Lide, 2000)
   (*c*)   *Melting-point*: –57.5 °C (Lide, 2000)
   (*d*)   *Density*: 1.512 at 20 °C/4 °C (Lide, 2000)
   (*e*)   *Spectroscopy data*: Infrared (prism [4626, 4426]), ultraviolet [5-3], nuclear
           magnetic resonance [8241] and mass [814] spectral data have been reported
           (Weast & Astle, 1985; Sadtler Research Laboratories, 1991)
   (*f*)   *Solubility*: Very soluble in water in which it is converted to chloral hydrate;
           soluble in diethyl ether and ethanol (Lide, 2000)
   (*g*)   *Volatility*: Vapour pressure, 6.66 kPa at 25 °C (Lide, 2000)
   (*h*)   *Stability*: Polymerizes under the influence of light and in presence of sulfuric
           acid to form a white solid trimer called metachloral (Budavari, 1998)
   (*i*)   *Conversion factor*: mg/m³ = 6.03 × ppm[a]

Chloral hydrate

   (*a*)   *Description*: Large, monoclinic plates with an aromatic, penetrating and slightly·
           bitter odour and a slightly bitter, caustic taste (O'Neil, 2001)
   (*b*)   *Boiling-point*: 98 °C, decomposes into chloral and water (O'Neil, 2001)
   (*c*)   *Melting-point*: 57 °C (Lide, 2000)
   (*d*)   *Density*: 1.9081 at 20 °C/4 °C (Lide, 2000)

---

[a] Calculated from: mg/m³ = (molecular weight/24.45) × ppm, assuming normal temperature (25 °C) and pressure (760 mm Hg)

(e) *Spectroscopy data*: Infrared (prism [5423]), nuclear magnetic resonance [10362] and mass [1054] spectral data have been reported (Weast & Astle, 1985; Sadtler Research Laboratories, 1991)

(f) *Solubility*: Freely soluble in water; freely soluble in acetone and methyl ethyl ketone; moderately or sparingly soluble in benzene, carbon tetrachloride, petroleum ether, toluene and turpentine (O'Neil, 2001)

(g) *Volatility*: Vapour pressure, 4.7 kPa at 20 °C; slowly evaporates on exposure to air (Jira *et al.*, 1986; O'Neil, 2001)

(h) *Octanol/water partition coefficient (P)*: Log P, 0.99 (Hansch *et al.*, 1995)

(i) *Conversion factor*: $mg/m^3 = 6.76 \times ppm$[a]

1.1.4 *Technical products and impurities*

Trade names for chloral include Grasex and Sporotal 100; trade names for chloral hydrate include Ansopal, Aquachloral, Chloradorm, Chloraldurat, Chloralix, Dormel, Elixnocte, Escre, Hydral, Lanchloral, Lorinal, Medianox, Nervifene, Noctec, Novochlorhydrate, Nycton, Phaldrone, Rectules, Somnos, Suppojuvent Sedante, Tosyl, Trawotox and Welldorm (Royal Pharmaceutical Society of Great Britain, 2002).

Technical-grade chloral ranges in purity from 94 to 99 wt%, with water being the main impurity. Other impurities sometimes present include chloroform, hydrogen chloride, dichloroacetaldehyde and phosgene (Jira *et al.*, 1986).

US Pharmacopeia specifies that USP-grade chloral hydrate must contain not less than 99.5% $C_2H_3Cl_3O_2$ (Pharmacopeial Convention, 1990). Chloral hydrate is available as a 500-mg capsule, as a 325-, 500- and 650-mg suppository and as a 250- and 500-mg/mL syrup (Medical Economics Co., 1996).

1.1.5 *Analysis*

A common analytical method to determine chloral is to treat it for 2 min with 1 M sodium hydroxide, which cleaves chloral into chloroform and sodium formate; the excess alkali is then titrated with acid. Alternatively, chloral is treated with quinaldine ethyl iodide to form a blue cyanine dye, the quantity of which is measured spectrophotometrically. Gas chromatography (GC) can be used for quantitative analysis of chloral and its hydrate, which breaks down to chloral on vaporization (Jira *et al.*, 1986).

Chloral hydrate has been determined in water using liquid–liquid extraction and GC with electron capture detection (GC–ECD). This method has been applied to drinking-water, water at intermediate stages of treatment and raw source water, and had a detection limit of 0.005 μg/L (Environmental Protection Agency, 1995; American Public Health Association/American Water Works Association/Water Environment Federation, 1999).

---

[a] Calculated from: mg/m^3 = (molecular weight/24.45) × ppm, assuming normal temperature (25 °C) and pressure (760 mm Hg)

In a national survey of chlorinated disinfection by-products in Canadian drinking-water conducted in 1993, the minimum quantifiable limit for this method was 0.1 μg/L (Health Canada, 1995).

An analytical survey of 16 drinking-water sources in various areas of Australia was conducted to determine the occurrence of disinfection by-products, including chloral hydrate, using a method based on a US Environmental Protection Agency standard method for the determination of chlorination disinfection by-products and chlorinated solvents in drinking-water by liquid–liquid extraction and GC–ECD (Simpson & Hayes, 1998).

The Association of Official Analytical Chemists has reported a spectrophotometric method for the determination of chloral hydrate in drugs, based on the reaction of quinaldine ethyl iodide with chloral hydrate to produce a stable blue cyanine dye with an absorption maximum at about 605 nm (Helrich, 1990).

Mishchikhin and Felitsyn (1988) described a method for the GC determination of chloral hydrate in biological materials using four columns with different packings. The elution of the compounds was monitored with two flame ionization detectors, and the limit of detection was approximately 0.01 mg/sample.

Koppen *et al.* (1988) described the determination of trichloroethylene metabolites, including chloral hydrate, in rat liver homogenate. The method was based on selective thermal conversion of chloral hydrate into chloroform, which is determined using head-space-GC with ECD.

Liquid chromatographic methods have been developed for the determination of chloral hydrate. Optimal reversed-phase separations were achieved after derivatization of chloral hydrate with 1,2-benzenedithiol and ultraviolet detection at 220 nm. Alternatively, chloral hydrate can be reacted with sodium hydroxide to form sodium formate which is then analysed by anion-exchange liquid chromatography, with suppressed conductivity detection. Detection limits as low as 0.2 μg/L were reported (Bruzzoniti *et al.*, 2001).

1.2   Production and use

1.2.1   *Production*

Chloral was first synthesized by J. von Liebig in 1832 by chlorinating ethanol; it was introduced as a hypnotic agent by Liebreich in 1869. Chloral is produced commercially by the chlorination of acetaldehyde or ethanol in hydrochloric acid, during which anti-mony trichloride may be used as a catalyst. Chloral hydrate is distilled from the reaction mixture and is then mixed with concentrated sulfuric acid, the heavier acid layer is drawn off, and chloral is fractionally distilled. Chloral hydrate is produced by adding water again to chloral (Jira *et al.*, 1986; O'Neil, 2001).

Available information indicates that chloral (anhydrous) is produced by 14 companies in China, seven companies in India and one company each in Brazil, France, Japan, Mexico, Russia and the USA. Available information also indicates that chloral hydrate is produced by four companies in China, three companies in Germany, two companies in

Japan and one company each in Mexico, Russia and Spain (Chemical Information Services, 2002a).

Available information indicates that chloral (anhydrous) is formulated into pharmaceutical products by one company in the United Kingdom, and that chloral hydrate is formulated into pharmaceutical products by three companies each in Spain and the USA, two companies each in Belgium, France, Hungary, Switzerland and the United Kingdom and one company each in Argentina, Australia, Canada, Indonesia, Moldova and Singapore (Chemical Information Services, 2002b).

1.2.2 Use

Chloral was the first hypnotic drug and is a precursor in the commercial synthesis of the insecticide DDT, which was introduced in 1941. Chloral was an important chemical in the 1960s, but its importance has declined steadily since then because the use of DDT and other chlorinated insecticides has been restricted in many countries. Much smaller amounts are used to make other insecticides (methoxychlor, naled, trichlorfon and dichlorvos), a herbicide (trichloroacetic acid) and hypnotic drugs (chloral hydrate, chloral betaine, α-chloralose and triclofos sodium). Chloral hydrate also possesses anticonvulsant and muscle relaxant properties (Jira *et al.*, 1986; Williams & Holladay, 1995).

As a hypnotic, chloral hydrate is principally used for the short-term (2-week) treatment of insomnia. It is used post-operatively to allay anxiety and to induce sedation and/or sleep. It is also used post-operatively as an adjunct to opiates and other analgesics to control pain. It is effective in reducing anxiety associated with the withdrawal of alcohol and other drugs such as opiates and barbiturates. It also has been used to produce sleep prior to electroencephalogram evaluations. In the USA, it has been widely used for sedation of children before diagnostic, dental or medical procedures. Following oral administration, chloral hydrate is converted rapidly to trichloroethanol, which is largely responsible for its hypnotic action; for oral use, it is sometimes given in a flavoured syrup. Externally, chloral hydrate has a rubefacient action and has been used as a counter-irritant. It is administered by mouth as a liquid or as gelatin capsules. It has also been dissolved in a bland fixed oil and given by enema or as suppositories (Gennaro, 2000; Royal Pharmaceutical Society of Great Britain, 2002).

The usual sedative dose of chloral hydrate for adults is 250 mg three times daily after meals. The usual hypnotic dose for adults is 500 mg–1 g 15–30 min before bedtime. When chloral hydrate is administered in the management of alcohol withdrawal symptoms, the usual dose is 500 mg–1 g repeated at 6-h intervals if needed. Generally, single doses or daily dosage for adults should not exceed 2 g. The sedative dose of chloral hydrate for children is 8 mg/kg bw or 250 mg/m^2 of body surface area three times a day, with a maximum dose of 500 mg three times a day. The hypnotic dose for children is 50 mg/kg bw or 1.5 g/m^2 of body surface area, with a maximum single dose of 1 g. As a premedication before electroencephalogram evaluation, children have been given chloral hydrate at a dose of 20–25 mg/kg bw (Medical Economics Co., 1996).

1.3   Occurrence

1.3.1   *Natural occurrence*

Chloral is not known to occur as a natural product.

1.3.2   *Occupational exposure*

The National Occupational Exposure Survey conducted between 1981 and 1983 indicated that 11 278 employees in the USA were potentially exposed to chloral (National Institute for Occupational Safety and Health, 1994). The estimate is based on a survey of companies and did not involve measurements of actual exposures.

Chloral has been detected in the work environment during spraying and casting of polyurethane foam (Boitsov *et al.*, 1970). It has also been identified as an autoxidation product of trichloroethylene during extraction of vegetable oil (McKinney *et al.*, 1955). It has been identified in the output of etching chambers in semiconductor processing (Ohlson, 1986).

1.3.3   *Air*

No data were available to the Working Group.

1.3.4   *Water*

Chloral is formed and rapidly converted to chloral hydrate during aqueous chlorination of humic substances and amino acids (Miller & Uden, 1983; Sato *et al.*, 1985; Trehy *et al.*, 1986; Italia & Uden, 1988). It may therefore occur (as chloral hydrate) in drinking-water as a result of chlorine-based disinfection of raw waters containing natural organic substances (see IARC, 1991). The concentrations of chloral hydrate measured in water are summarized in Table 1. Concentrations were higher in the summer than in the winter (LeBel *et al.*, 1997; Williams *et al.*, 1997). Kim *et al.* (2002) measured concentrations of chloral hydrate ranging from 0.19 to 30 µg/L in a model swimming pool system.

Chloral has also been detected in the spent chlorination liquor from the bleaching of sulfite pulp after oxygen treatment, at concentrations of < 0.1–0.5 g/ton of pulp (Carlberg *et al.*, 1986). It has been found in trace amounts after photocatalytic degradation of trichloroethylene in water (Glaze *et al.*, 1993).

1.3.5   *Other*

Chloral is an intermediate metabolite of trichloroethylene in humans, and chloral hydrate has been found in the plasma of humans following anaesthesia with trichloroethylene (Cole *et al.*, 1975; Davidson & Beliles, 1991).

Table 1. Concentrations of chloral (as chloral hydrate) in water

Water type (location)	Concentration (μg/L)	Reference
Treatment plant (Canada)	< 0.1–15.1	Williams *et al.* (1997)
Treatment plant and distribution system (Canada)	< 0.1–23.4	LeBel *et al.* (1997)
Drinking-water (USA)[a]	7.2–18.2	Uden & Miller (1983)
Drinking-water (USA)[a]	0.01–5.0	Environmental Protection Agency (1988)
Drinking-water (USA)[a]	1.7–3.0	Krasner *et al.* (1989)
Drinking-water (USA)[a]	0.14–6.7	Koch & Krasner (1989)
Drinking-water (USA)[a]	6.3–28	Jacangelo *et al.* (1989)
Drinking-water and distribution system (USA)[a]	< 0.5–92	Blank *et al.* (2002)
Distribution system (Canada)	< 0.1–22.5	Williams *et al.* (1997)
Distribution system (Australia)	0.2–19	Simpson & Hayes (1998)
Swimming pool (Germany)	265	Baudisch *et al.* (1997)
Swimming pool (Germany)	0.3–67.5	Mannschott *et al.* (1995)

[a] Samples taken in water leaving the treatment plant

1.4   Regulations and guidelines

The WHO (1998) has established a provisional guideline of 10 μg/L for chloral and chloral hydrate in drinking-water. A provisional guideline is established when there is some evidence of a potential health hazard but where available data on health effects are limited, or where an uncertainty factor greater than 1000 has been used in the derivation of the tolerable daily intake.

The drinking-water guideline for chloral and chloral hydrate in Australia and New Zealand is 20 μg/L (National Health and Medical Research Council and Agriculture and Resource Management Council of Australia and New Zealand, 1996). This guideline also notes that the minimization of the concentration of all chlorination by-products is encouraged by reducing the amount of naturally occurring organic material in the source water, reducing the amount of chlorine added or using an alternative disinfectant, without compromising disinfection.

The European Union (European Commission, 1998) and Canada (Health Canada, 2003) have not set a guideline value but also encourage the reduction of concentrations of total disinfection by-products. The Environmental Protection Agency (1998) controls the formation of unregulated disinfection by-products in the USA, which would include chloral and chloral hydrate, with regulatory requirements for the reduction of the precursors, in this case, total organic carbon. The Stage 1 Disinfectants/Disinfection By-Product Rule mandates a reduction in the percentage of total organic carbon from source water and to finished water on the basis of total organic carbon in source water and its alkalinity. 'Enhanced' coagulation or 'enhanced' softening are mandated as a treatment technique for this reduction, unless the levels of total organic carbon or disinfection by-products in source water are low.

2. Studies of Cancer in Humans

See Introduction to the monographs on chloramine, chloral and chloral hydrate, di-chloroacetic acid, trichloroacetic acid and 3-chloro-4-(dichloromethyl)-5-hydroxy-2(5H)-furanone.

3. Studies of Cancer in Experimental Animals

3.1   Oral administration

3.1.1   *Mouse*

Groups of male C57BL × C3HF$_1$ mice [B6C3F$_1$ mice], 15 days of age, were fasted for 3 h and then administered a single intragastric dose of 5 (25 animals) or 10 (20 animals) μg/g bw crystalline chloral hydrate (USP grade) dissolved in distilled water. A group of 35 controls was administered distilled water only. Groups of 6–10 mice used for cell proli-feration studies were killed 24 h after receiving chloral hydrate. The remainder were weaned at 4 weeks of age, and animals were killed when moribund or at intervals up to 92 weeks after treatment. The student t-test was used for evaluating significance between the treatment and control groups. [Data on survival and body weight were not presented.] Liver tumours [identified as nodules] were found in mice killed between 48 and 92 weeks following treatment, were classified as hyperplastic, adenomatous or trabecular and were grouped for statistical analysis. [The authors did not account for all animals.] The inci-dences were: control, 2/19 (10.5%; two trabecular); 5 μg/kg, 3/9 (33.3%; one hyper-plastic, one adenomatous and one trabecular); and 10 μg/kg, 6/8 (75%; three adenomatous and three trabecular; $p < 0.05$) (Rijhsinghani et al., 1986).

A group of 40 male B6C3F$_1$ mice, 28 days of age, was exposed to a target concentration of 1 g/L chloral hydrate (measured dose, 0.95 ± 0.14 g/L; 95% pure with no identifiable impurities) in distilled water for 104 weeks. Five mice were examined at 30 and another five at 60 weeks. A group of 33 control animals was given distilled water alone. Statistical methods included a one-factor analysis of variance (performed as two-tailed tests) for conti-nuous variables (e.g. body and organ weights, water consumption). Data on survival and tumour counts were analysed using the log rank test and tumour prevalence was analysed by Fisher's exact test. Statistical significance was set conservatively ($p \leq 0.03$). Marginal significance was set at $0.03 < p < 0.05$. Time-weighted water consumption was 175.5 ± 18.6

for animals administered chloral hydrate versus 197.0 ± 16.4 mL/kg bw per day for those in the control group, which yielded a mean daily dose of 166 mg/kg bw per day. The slight increase in the mortality observed in the chloral hydrate-treated group was not statistically different; tissues were examined at various intervals in five animals from the control and treated groups. No tumours were found in either control or chloral hydrate-treated animals at 30 weeks. At 60 weeks, no tumours were observed in the control group (0/5); hepato-cellular carcinomas (2/5 [40%]) were found in chloral hydrate-treated animals. At 104 weeks, treatment with chloral hydrate marginally increased the prevalence (7/24 [29%] versus 1/20 [5%]) and multiplicity (0.29 versus 0.10 per animal) of hepatocellular adenomas and significantly increased the prevalence (11/24 [46%] versus 2/20 [10%]; $p < 0.03$) and multiplicity (0.75 versus 0.25 per animal) of hepatocellular carcinoma in comparison with controls. The values for animals bearing both adenomas and carcinomas were 17/24 (71%) versus 3/20 (15%) ($p \leq 0.01$) and 1.04 versus 0.25 per animal ($p \leq 0.01$) in comparison with controls (Daniel *et al.*, 1992a).

Groups of 72 male B6C3F$_1$ mice, 28–30 days of age, were exposed to chloral hydrate (> 99% pure) dissolved in deionized water for 104 weeks. Target concentrations were 0.05, 0.5 and 1.0 g/L, and the measured concentrations of chloral hydrate were 0.12 ± 0.02, 0.58 ± 0.04 and 1.28 ± 0.20 g/L. Water consumption, time-weighted over 104 weeks, was 111.7 (control), 112.8 (0.12 g/L), 112.1 (0.58 g/L) and 114.5 (1.28 g/L) mL/kg bw per day, which yielded mean daily doses of 0, 13.5, 65.0 and 146.6 mg/kg bw per day, respectively, time-weighted over the duration of the study. Continuous variables were analysed using a one-way analysis of variance. Parametric and non-parametric tests were used where appro-priate. Tumour prevalence (number of animals with a lesion per number of animals examined) was analysed using Fisher's exact and Fisher-Irwin (for trends with dose) tests. Tumour multiplicity (number of lesions per animal) was analysed using log-rank tests and a log-rank monotone trend test. Using one-sided tests, $p \leq 0.05$ was considered to be signi-ficant and $0.05 < p < 0.1$ was considered to be of marginal significance. No alterations in body weight were seen in any of the chloral hydrate-treated groups compared with controls. Six animals from each group were killed at 26, 52 and 78 weeks, and all remaining animals were killed at 104 weeks. Unscheduled deaths were: vehicle control, 16/72; 13.5 mg/kg per day, 9/72; 65.0 mg/kg per day, 19/72; and 146.6 mg/kg per day, 28/72. The increased number of unscheduled deaths in the high-dose group was not significant when compared with survival in the control group (log-rank test). The total number of animals examined for pathology over the course of the study was: 65/72, 69/72, 62/72 and 60/72 for the control, low, mid and high doses, respectively. Tumour prevalence and multiplicity were calculated in animals that survived longer than 78 weeks; the numbers of animals in the analysis were 42 controls, 46 low-dose, 39 mid-dose and 32 high-dose. The prevalence and multiplicity of hepatocellular adenomas were 21.4% and 0.21 ± 0.06 for controls, 43.5% and 0.65 ± 0.12 ($p \leq 0.05$) for the low dose, 51.3% and 0.95 ± 0.18 ($p \leq 0.05$) for the mid dose and 50.0% and 0.72 ± 0.15 ($p \leq 0.05$) for the high dose. The values for hepatocellular carcinoma were 54.8% and 0.74 ± 0.12 for controls, 54.3% and 0.72 ± 0.11 for the low dose, 59.0% and 1.03 ± 0.19 for the mid dose and 84.4% and 1.31 ± 0.17 ($p \leq 0.05$) for the high dose. The prevalence and multiplicity values for animals bearing either adenoma or carcinoma were 64.5% and 0.95 ± 0.13 for controls, 78.3% and 1.37 ± 0.16 ($p \leq 0.05$, multiplicity only) for the low dose, 79.5% and 1.97 ± 0.25 ($p \leq 0.05$) for the mid dose and 90.6% and 2.03 ± 0.24 ($p \leq 0.05$) for the high dose. One adenoma occurred in the mid-dose group at 26 weeks (George *et al.*, 2000).

Groups of 48 female B6C3F$_1$ mice, 28 days of age, received 0 (control), 25, 50 and 100 mg/kg bw chloral hydrate (purity ~99.5%) in distilled water by gavage for 104 weeks. No differences in survival or mean body weights were observed among the treated and control animals. The incidence of hyperplasia of the pituitary gland pars distalis was 3/45 (6.7%) control, 6/44 (13.6%) low-dose, 4/47 (8.5%) mid-dose and 0/41 (0%) high-dose animals. The incidence of pars distalis adenoma was 0/45 control, 2/44 (5%) low-dose, 0/47 mid-dose and 5/41 (12%) high-dose mice ($p = 0.0237$). The historical incidence of control groups was 15/308 (4.9%) with a range of 0–6%. The incidences of malignant lymphoma were: 9/48 (19%) control, 7/48 (15%) low-dose, 8/48 (17%) mid-dose and 15/48 (31%) high-dose mice. The historical incidence for malignant lymphoma in control groups was 92/374 (24.6%) with a range of 21–43%. In a second experiment, groups of 48 female mice, 28 days of age, received 0 or 100 mg/kg bw chloral hydrate by gavage on 5 days per week. Eight mice from each group were killed at 3, 6 or 12 months, after which treatment was discontinued for the remaining animals for the duration of the study (104 weeks). Survival was similar among all treatment groups. The mean body weights of animals receiving 100 mg/kg bw chloral hydrate for 3 and 6 months tended to be greater than those of vehicle controls between 52 and 104 weeks. Incidences of pars distalis adenoma were 0/45 controls, 3/36 (8%) at 3 months, 1/36 (3%) at 6 months, 1/33 (3%) at 12 months and 5/41 (12%) at 24 months ($p = 0.0237$). In a third experiment, groups of 48 female mice, 28 days of age, received a single dose of 0, 10, 25 or 50 mg/kg bw chloral hydrate by gavage and were held for 104 weeks. No differences were observed in survival, body weight gains or neoplasia among the control and chloral hydrate-treated groups. In a fourth experiment, groups of 48 male and 48 female mice, 15 days of age, received a single oral dose of 0, 10, 25 or 50 mg/kg bw chloral hydrate by gavage and were held for 104 weeks. No differences were observed in survival, body weight or neoplasia among the control and chloral hydrate-treated groups. The authors concluded that under the conditions of the 2-year gavage study, there was equivocal evidence of carcinogenic activity for chloral hydrate based on an increased incidence of pituitary gland par distalis adenoma. No increased carcinogenicity was found for single-dose administration of chloral hydrate to male or female mice. No increase in hepatocarcinogenicity was observed under any dose regimen (National Toxicology Program, 2002a).

Groups of 120 male B6C3F$_1$ mice, 6 weeks of age, received chloral hydrate (99% pure) in distilled water at doses of 0, 25, 50 or 100 mg/kg bw on 5 days per week for 104–105 weeks. Animals in each dose group were divided into two dietary groups: *ad libitum* and dietary controlled. Twelve animals per group were killed for interim evaluation at 15 months. Survival of chloral hydrate-treated mice was similar to that of vehicle control animals. Mean body weights of chloral hydrate-treated animals in each dietary group did not differ from those of the controls. Liver weights of all treated groups were greater than those of the controls, but not statistically significantly so. In the groups fed *ad libitum*, the incidence of combined hepatocellular adenomas and carcinomas was significantly greater in low-dose animals than in controls: control, 16/48 (33%); low-dose, 25/48 (52%; $p = 0.0437$); mid-dose, 23/47 (49%); and high-dose, 22/48 (46%). The incidences of hepatocellular carcinoma and of combined neoplasms occurred with a positive dose trend: control, 11/48 (23%); low-dose, 11/48 (23%); mid-dose, 14/48 (29%); and high-dose, 18/48 (38%; $p = 0.0450$); and the incidence of hepatocellular carcinomas in the high-dose dietary-controlled mice was significantly increased: control, 2/48 (4%); low-dose, 5/48 (10%); mid-dose, 4/48 (8%); and high-dose, 8/48 (17%; $p = 0.0422$) (National Toxicology Program, 2002b).

3.1.2 *Rat*

Groups of 50 male and 50 female Sprague-Dawley rats, 25–29 days of age [age at the start of treatment not specified], were administered chloral hydrate [purity not specified] dissolved in drinking-water [not specified] at concentrations that had been found to give doses of 15, 45 and 135 mg/kg bw per day for 124 weeks (males) and 128 weeks (females). Two control groups received drinking-water only. Concentrations of chloral hydrate were adjusted to the mean drinking-water intake of the previous week for each group to maintain a constant dose level relative to the body weights of the animals. None of the treatments with chloral hydrate had a significant effect on survival (Fisher's exact *t*-test) or on body weight gain, or feed and water consumption (Dunnett's multiple *t*-test). No chloral hydrate-related tumours (Peto test) or other lesions were observed at any organ site [complete histopathology performed] in the animals treated for 124 and 128 weeks. [The authors stated that chloral hydrate was ineffective in increasing neoplasia in rats but the data were not provided.] (Leuschner & Beuscher, 1998).

Groups of 78 male Fischer 344/N rats, 28–30 days of age, were administered chloral hydrate (> 99% pure) dissolved in deionized water at target concentrations of 0, 0.05, 0.5 and 2.0 g/L. Parametric and non-parametric tests were used where appropriate. Tumour prevalence (number of animals with a lesion per number of animals examined) was analysed using Fisher's exact and Fisher-Irwin (for trends with dose) tests. Tumour multiplicity (number of lesions per animal) was analysed using log-rank tests and a log-rank monotone trend test. Using one-sided tests, $p \leq 0.05$ was considered to be significant and $0.05 < p < 0.1$ was considered to be of marginal significance. The measured concentrations for the low-dose and mid-dose animals were the same as those in the experiment with mice. The measured concentrations of chloral hydrate were 0.12 ± 0.02, 0.58 ± 0.04 and 2.51 ± 0.13 g/L, respectively. Water consumption, time-weighted over 104 weeks, was 61.7 (control), 61.9 (0.12 g/L), 64.5 (0.58 g/L) and 64.8 (2.51 g/L) mL/kg bw per day, which yielded mean daily doses of 0, 7.4, 37.4, and 162.6 mg/kg bw per day time-weighted over the duration of the study, respectively. No alterations in body weight were seen in any of the chloral hydrate-treated groups compared with controls. Six animals from each group were killed at 13, 26, 52 and 78 weeks, and all remaining animals were killed at 104 weeks. Unscheduled deaths were: vehicle control, 18/78; 7.4 mg/kg, 14/78; 37.4 mg/kg, 16/78; and 162.6 mg/kg, 18/78. The total number of animals examined for pathology over the course of the study was 71/78, 70/78, 73/78 and 74/78 for the control, low-dose, mid-dose and high-dose groups, respectively. Prevalence and multiplicity were calculated in animals that survived longer than 78 weeks. The number of animals examined were 42 control, 44 low-dose, 44 mid-dose and 42 high-dose. The prevalence and multiplicity of hepatocellular adenomas, respectively, were 0% and 0 for controls, 7.1% and 0.07 ± 0.04 per animal for the low dose, 2.3% and 0.02 ± 0.02 per animal for the mid dose, and 4.5% and 0.05 ± 0.03 per animal for the high dose. The values for hepatocellular carcinoma were 2.4% and 0.02 ± 0.02 for controls, 7.1% and 0.07 ± 0.04 for the low dose, 0 for the mid dose and 2.3% and 0.02 ± 0.02 for the high dose. The values for animals bearing either adenoma or carcinoma were 2.4% and 0.02 ± 0.02 for controls, 14.3% and 0.14 ± 0.05 for the low dose, 2.3% and 0.02 ± 0.02 for the mid dose and 6.8% and 0.07 ± 0.04 for the high dose (George et al., 2000).

3.2   Intraperitoneal injection

3.2.1   Mouse

Groups of 22–24 male and 21–23 female neonatal B6C3F$_1$ mice were administered intraperitoneal injections of chloral hydrate [purity not specified] in fractionated doses at 8 and 15 days of age. A total dose of 1000 nmol was delivered in 30 μL dimethyl sulfoxide (DMSO) as one third and two thirds of the dose, and a total dose of 2000 nmol was delivered as three sevenths and four sevenths of the dose in 30 μL DMSO, respectively. As a positive control, 4-aminobiphenyl was used at total doses of 500 and 1000 nmol, respectively. The animals receiving 2000 nmol chloral hydrate were killed at 12 months of age while those receiving 1000 nmol were killed at 20 months of age. At 12 months, liver neoplasms (adenomas or carcinomas) were found in 5/24 (21% and 1.0 tumour per animal) male mice that received 2000 nmol chloral hydrate compared with 24/24 (100% and > 5.7 mean number of tumours per animal; $p < 0.0004$) mice treated with 4-aminobiphenyl. No neoplasms were observed in the DMSO controls. In male mice treated neonatally with 1000 nmol and kept for 20 months, 10/23 (43%) developed combined liver neoplasms and 1.4 tumours per animal compared with 7/23 (30%) and 1.4 tumours per animal for DMSO controls and 22/22 (100%; $p < 0.0004$) and 3.5 tumours per animal for the positive controls. No liver tumours were found in female mice that received chloral hydrate or DMSO alone. The incidence of liver adenomas in female mice treated with 500 nmol 4-aminobiphenyl was 9/23 (39%; $p < 0.006$), with 1.1 tumours per animal. The tumour incidence in male mice administered 2000 nmol chloral hydrate was one animal away from statistical significance; higher doses of chloral hydrate (2500 and 5000 nmol; one third and two thirds of the total dose injected intraperitoneally on days 8 and 15 of age, respectively) were evaluated in both male and female mice examined at 12 months of age;

1/24 male and 0/24 female mice had a liver adenoma in the low-dose treatment group. Liver adenomas were found in 2/22 males and 1/24 females in the high-dose group. One male developed a lung adenoma [no values were given either for the positive or vehicle control groups] (Von Tungeln *et al.*, 2002).

4. Other Data Relevant to an Evaluation of Carcinogenicity and its Mechanisms

4.1   Absorption, distribution, metabolism and excretion

4.1.1   *Humans*

The metabolic pathways of chloral hydrate and its metabolites in humans are depicted in Figure 1.

The disposition of chloral hydrate has been studied in critically ill neonates and children (Mayers *et al.*, 1991). Three groups of subjects (group 1, preterm infants; group 2, full-term infants; and group 3, toddler–child patients) were given a single oral dose of 50 mg/kg bw chloral hydrate. In contrast to data reported for adult humans (IARC, 1995), chloral hydrate was detected in plasma for several hours after administration. The half-lives for chloral hydrate were 1.0, 3.0, and 9.7 h and those for trichloroethanol were 39.8, 27.8 and 9.7 h in

Figure 1. Metabolic pathways of chloral hydrate and its metabolites

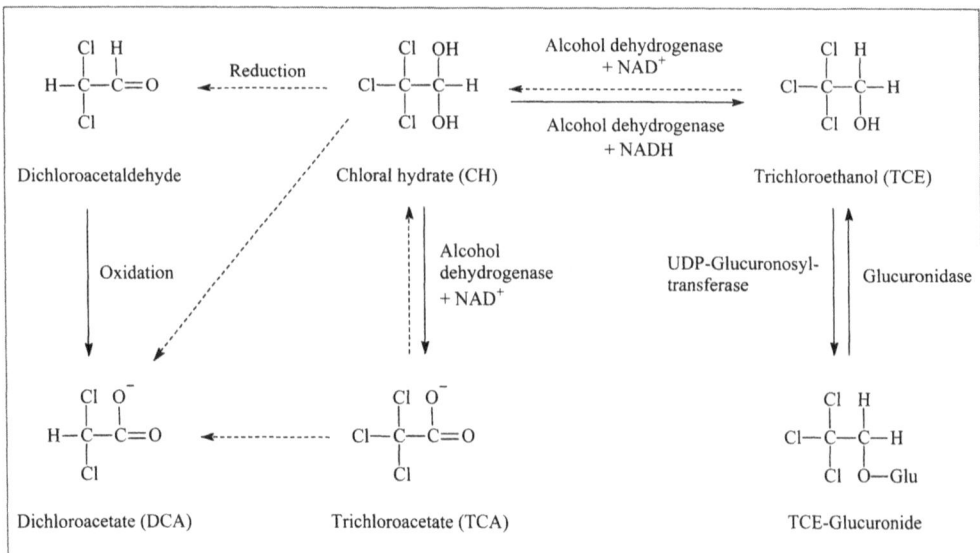

Dotted lines are proposed pathways and solid lines are established pathways.
NAD$^+$, nicotinamide adenine dinucleotide; NADH, reduced nicotinamide adenine dinucleotide

groups 1, 2 and 3, respectively. The plasma concentration of trichloroacetic acid did not decline even 6 days after treatment with chloral hydrate.

The kinetics and metabolism of chloral hydrate have been studied in children aged 3 months to 18 years diagnosed with congenital lactic acidosis (Henderson *et al.*, 1997). The children received chloral hydrate, dichloroacetic acid or both, and chloral hydrate, trichloroacetic acid, dichloroacetic acid and trichloroethanol were measured in blood. In one child given 50 mg/kg bw chloral hydrate, the plasma half-life of this compound was 9.7 h. Dichloroacetic acid was also detected in this child, with a half-life of 8.0 h, compared with a half-life of 0.5–1.5 h in children given 25 mg/kg bw dichloroacetic acid. In a child given 12.5 mg/kg bw [1,2-^{13}C]dichloroacetic acid orally 15 min before administration of 50 mg/kg bw chloral hydrate, the half-lives of dichloroacetic acid and trichloroethanol were 0.34 h and 9.7 h, respectively. In a child given 50 mg/kg bw chloral hydrate 20 h before or immediately after administration of 12.5 mg/kg bw [1,2-^{13}C]dichloroacetic acid, the half-lives of dichloroacetic acid and trichloroethanol were 3.18 h and 7.2 h, respectively. The authors note that the half-life of dichloroacetic acid was greater when it was formed as a metabolite of chloral hydrate than when given directly and that the half-life of dichloroacetic acid was prolonged in a child given two doses of chloral hydrate. These data indicate that chloral hydrate or a metabolite of chloral hydrate, or both, inhibit the metabolism of dichloroacetic acid. [The Working Group noted that reported blood concentrations of dichloroacetic acid after administration of chloral hydrate may be confounded by the observation that trichloroacetic acid, a metabolite of chloral hydrate, may be converted to dichloroacetic acid in freshly drawn blood (Ketcha *et al.*, 1996).]

The bioavailability and pharmacokinetics of chloral hydrate have been studied in human volunteers (Zimmermann *et al.*, 1998). Eighteen healthy male subjects (aged 20–31 years) were given 250 or 500 mg chloral hydrate either in immediate-release or enteric-coated modified-release capsules or as a solution. Because of the extensive first-pass metabolism of chloral hydrate, the bioavailability of trichloroethanol was used as a surrogate for its absorption. The bioavailability of chloral hydrate given in capsules amounted to 94.8–101.6% of that given as a solution. The terminal half-lives for the elimination of trichloroethanol and trichloroacetic acid from plasma were 9.3–10.2 and 89–94 h, respectively.

1-Trichloromethyl-2,3,4,9-tetrahydro-1*H*-β-carboline (also called 1-trichloromethyl-1,2,3,4-tetrahydro-β-carboline), a dopaminergic neurotoxin formed by the endogenous reaction of chloral hydrate and tryptamine, was identified at concentrations of up to 20 ng/mL in the blood of five humans (65–82 years old) suffering from Parkinson's disease, who were given 750–3000 mg chloral hydrate orally over 2–7 days (Bringmann *et al.*, 1999).

4.1.2   *Experimental systems*

The pharmacokinetics and metabolism of chloral hydrate have been investigated in male B6C3F$_1$ mice (Abbas *et al.*, 1996). In mice given 11.2, 112 and 336 mg/kg bw chloral hydrate intravenously, the clearances from the systemic circulation were 36, 20 and 7.6 L/h/kg and the half-lives were 0.09, 0.32, and 0.40 h, respectively, indicating saturable kinetics. Only 0.1–0.2% of the administered doses was eliminated unchanged in the urine. Trichloroethanol and trichloroethyl glucuronide were formed rapidly after treatment with chloral hydrate, with elimination half-lives of 0.26, 0.34, and 0.36 h for trichloroethanol and 0.21, 0.36, and 0.72 h for trichloroethyl glucuronide after the low, mid and high doses, respectively. Only a small amount of trichloroethanol was excreted in the urine (0.63–1.7% of the initial dose of chloral hydrate), but the excretion of trichloroethyl glucuronide amounted to 52–72% of the administered dose. Trichloroacetic acid was formed as a metabolite of chloral hydrate and was excreted in urine in amounts of 10–35% of the dose of the parent compound. Dichloroacetic acid was found in blood and liver, but not in the urine.

The effect of enterohepatic circulation on the pharmacokinetics of chloral hydrate was studied in male Fischer 344 rats (Merdink et al., 1999). Cannulae were introduced into the bile duct and jugular vein of the rats, and chloral hydrate was infused into the jugular vein at doses of 12, 48 or 192 mg/kg bw. The elimination of chloral hydrate from the blood showed biphasic kinetics in both bile duct-cannulated and control rats; the half-lives for the first and second phases of elimination were estimated to be 0.09 and 0.75 h, respectively. There was no difference in the elimination half-lives of chloral hydrate from the plasma or its whole-body clearance between control and bile-duct-cannulated rats. Less than 1% of the total body clearance of chloral hydrate was attributed to renal or biliary clearance, and the remainder was attributed to its metabolism to trichloroacetic acid and trichloroethanol. The metabolism of chloral hydrate to trichloroacetic acid predominated at low doses (12 and 48 mg/kg bw), whereas metabolism to trichloroethanol predominated at higher doses (192 mg/kg bw).

The extrahepatic metabolism of chloral hydrate was studied in male and female mongrel dogs with a hepatic bypass and control animals that were given 25 mg/kg bw intravenously (Hobara et al., 1987). The plasma and urine concentrations of unchanged chloral hydrate were higher in dogs with a hepatic bypass than in control animals (non-bypass dogs), as were serum concentrations of trichloroethanol. Trichloroethanol concentrations in urine were higher in bypass animals compared with controls except at 30 and 60 min after administration of chloral hydrate. Concentrations of trichloroethyl glucuronide and trichloroacetic acid in serum and urine were higher in controls than in bypass dogs at all sampling times. The authors concluded that there was significant extrahepatic conversion of chloral hydrate to trichloroethanol, but that formation of trichloroethyl glucuronide and trichloroacetic acid in extrahepatic tissues was lower than in animals with intact hepatic circulation.

The intestinal absorption of chloral hydrate was studied in male and female mongrel dogs fitted with jejunal, ileal and colonic loops that were perfused with a 264 µg/mL solution in phosphate buffer (pH 7.0) (Hobara et al., 1988). The absorbed fraction of chloral hydrate was about 50% in the jejunum and ileum, and about 40% in the colon.

The metabolism of chloral hydrate was investigated in male and female $B6C3F_1$ mice and Fischer 344 rats given one dose or 12 doses (over a 16-day period) of 50 or 200 mg/kg bw by gavage (Beland *et al.*, 1998; Beland, 1999). Trichloroacetic acid was the major metabolite detected in the plasma of both rats and mice. Plasma concentrations of trichloroethanol were higher in female mice and rats than in males; in contrast, plasma concentrations of trichloroethyl glucuronide were higher in male than in female mice. The half-lives of trichloroethanol and trichloroethyl glucuronide were greater in mice than in rats. Pharmacokinetic analysis showed that the elimination half-life of chloral hydrate from plasma was similar in both rats and mice. In mice, but not in rats, the rate of elimination of trichloroacetic acid was greater in animals given 12 doses than in animals given one dose.

The metabolism of chloral hydrate to dichloroacetic acid was studied in male $B6C3F_1$ mice given 50 mg/kg bw intravenously, but dichloroacetic acid was not detected in the blood (limit of detection, 0.2 µg/mL) (Merdink *et al.*, 1998).

The kinetics of the dismutation of chloral hydrate by guinea-pig pulmonary carbonyl reductase has been studied (Hara *et al.*, 1991). The enzyme irreversibly converted choral hydrate into trichloroacetic acid and trichloroethanol in the presence of the reduced or oxidized cofactors, of which nicotinamide adenine dinucleotide (phosphate) $(NADP^+)$ gave a higher reaction rate than did reduced $NADP^+$, and the concentration ratios of the two products to chloral hydrate utilized were 1:1. In the $NADP^+$-linked reaction trichloroacetic acid was the predominant product and its amount was compatible with that of trichloroethanol plus reduced $NADP^+$ produced, whereas in the reduced $NADP^+$-linked reaction equal amounts of trichloroacetic acid and trichloroethanol were formed and the cofactor was little oxidized. The steady-state kinetic measurements in the $NADP^+$-linked chloral hydrate oxidation were consistent with an ordered Bi-Bi mechanism which is the same as that for the secondary alcohol oxidation by the enzyme.

A comparison of the biotransformation of chloral hydrate to trichloroacetic acid and trichloroethanol in rat, mouse and human liver and blood is available (Lipscomb *et al.*, 1996). In hepatic $700 \times g$ supernatant fractions (containing microsomes and mitochondria) from male $B6C3F_1$ mice, male Fischer 344 rats and humans, nicotinamide adenine dinucleotide (NAD^+) supported the oxidation of chloral hydrate to trichloroacetic acid and reduced nicotinamide adenine dinucleotide (NADH) supported the reduction of chloral hydrate to trichloroethanol. Kinetic analysis showed that the K_m for the reduction of chloral hydrate to trichloroethanol was at least 10-fold lower than that for the formation of trichloroacetic acid in all three species. Studies with lysed whole blood showed that chloral hydrate is biotransformed to both trichloroacetic acid and trichloroethanol, but that the latter metabolite predominates. Formation of trichloroacetic acid in blood was greater in humans and mice than in rats, and more trichloroethanol was formed in rodent blood than in human blood.

The biotransformation of chloral hydrate was studied in hepatic microsomal fractions isolated from control and pyrazole-treated male $B6C3F_1$ mice (Ni *et al.*, 1994, 1996; Beland, 1999). Incubation of chloral hydrate in the presence of the free radical trapping agent, *N-tert*-butyl-α-phenylnitrone, with microsomes from control and pyrazole-treated mice [oxygen concentration not stated] and analysis by electron spin resonance spectroscopy showed the formation of a carbon-centered free radical with a spectrum similar to that of carbon tetrachloride and trichloroacetic acid. Incubation of [1-^{14}C]chloral hydrate with horseradish peroxidase and prostaglandin H synthase led to the formation of [^{14}C]carbon dioxide (3.0 and 5.2%, respectively), indicating the formation of one-electron intermediates (Beland, 1999).

Trichloroethanol binds to bovine serum albumin with a K_D of 3.3 mmol/L, inducing a conformational change in this protein (Solt & Johansson, 2002).

4.1.3   *Comparison of humans and animals*

The metabolic fate of chloral hydrate is qualitatively similar in humans and animals: trichloroacetic acid, dichloroacetic acid and trichloroethanol have been detected in blood, and trichloroethyl glucuronide is the major urinary metabolite. In humans, the half-life of chloral hydrate ranged from ~1 to ~10 h (Mayers *et al.*, 1991; Zimmermann *et al.*, 1998), whereas in mice and rats, the half-life of chloral hydrate was < 1 h (Abbas *et al.*, 1996; Merdink *et al.*, 1999).

4.2   **Toxic effects**

4.2.1   *Humans*

Chloral hydrate is used clinically as a sedative or hypnotic.

The lethal dose of chloral hydrate in humans is about 10 g; however, a fatal outcome was reported after ingestion of 4 g, and recovery has been seen after a dose of 30 g. The toxic effects that have been described after overdoses of chloral hydrate include irritation of the mucous membranes in the alimentary tract, depression of respiration and induction of cardiac arrhythmia. Habitual use of chloral hydrate is reported to cause unspecified hepatic and renal damage (Goodman Gilman *et al.*, 1991, 1996).

4.2.2   *Experimental systems*

The oral 50% lethal dose (LD_{50}) of chloral hydrate in rats was 480 mg/kg bw (Goldenthal, 1971); LD_{50s} in mice were reported to be 1442 mg/kg bw in males and 1265 mg/kg bw in females (Sanders *et al.*, 1982). The immediate cause of death after administration of lethal doses of chloral hydrate appeared to be inhibition of respiration.

The subchronic toxicity of chloral hydrate has been studied in CD1 mice and Sprague-Dawley rats. Administration of chloral hydrate to mice by gavage at daily doses of 14.4 and 144 mg/kg bw for 14 consecutive days resulted in an increase in relative liver weight and a decrease in spleen size. No other changes were seen. Administration of chloral hydrate to mice in drinking-water for 90 days at concentrations of 0.07 and 0.7 mg/mL resulted in dose-related hepatomegaly in males only and significant changes in hepatic microsomal enzymes in both males and females, indicative of hepatic toxicity (Sanders *et al.*, 1982).

After chloral hydrate was administered for 90 days in drinking-water to male and female Sprague-Dawley rats at a concentration of 0.3, 0.6, 1.2 or 2.4 mg/mL, only male rats receiving the highest dose showed significant decreases in food and water consumption and weight gain. Males also had an apparent increase in the incidence of focal hepatocellular necrosis and increased activities of serum enzymes. No liver damage was seen in female rats (Daniel *et al.*, 1992b).

Exposure of female CD1 mice to 100 ppm [603 µg/L] chloral for 6 h induced deep anaesthesia, which was fully reversible on cessation of exposure. Vacuolation of lung Clara cells, alveolar necrosis, desquamation of the bronchiolar epithelium and alveolar oedema were observed. Cytochrome P450 enzyme activity was reduced, although the activities of ethoxycoumarin *O*-diethylase and glutathione *S*-transferase were unaffected (Odum *et al.*, 1992).

Male Sprague-Dawley rats were administered chloral hydrate in the drinking-water for 7 days at concentrations of 0.13, 1.35 or 13.5 mg/L (Poon *et al.*, 2000). No changes were observed in body or organ weights. In the high-dose group, a threefold increase in palmitoyl-coenzyme A oxidase, a marker of peroxisome proliferation was observed. Hepatic, but not serum, cholesterol and triglycerides were decreased and glutathione *S*-transferases and glutathione levels were elevated at this dose. A significant increase in levels of serum trichloroacetic acid was also observed in the high-dose group; the authors concluded that the hepatic effects were probably attributable to this metabolite.

The effect of dietary restriction on the carcinogenicity of chloral hydrate was studied in male B6C3F$_1$ mice. In mice fed diets *ad libitum*, the only histopathological effect was an increase in the incidence of glomerulosclerosis with 25 and 100 mg/kg chloral hydrate. There appeared to be a small increase in the incidence of renal tubule degeneration at the high dose in the group fed controlled diets. Although a significant increase in hepatocellular carcinomas was observed at 100 mg/kg per day in the dietary-controlled groups (see Section 3), there was no associated increase in non-neoplastic pathology in the liver. However, there was evidence of peroxisome proliferation, as indicated by a significant induction of cytochrome P450 4A and lauric acid ω-hydroxylase activity, only at doses of 100 mg/kg chloral hydrate and above in dietary-controlled animals. Significant increases in palmitoyl coenzyme A fatty acid hydroxylase activity were observed in the dietary-controlled and calorically restricted animals, but only at doses of ≥ 250 mg/kg. The carcinogenesis bioassays in all three dietary groups were conducted at doses ≤ 100 mg/kg (National Toxicology Program, 2002b).

Incubation of chloral hydrate with male B6C3F$_1$ mouse liver microsomes resulted in increased amounts of lipid peroxidation products (malondialdehyde and formaldehyde). This effect was inhibited by α-tocopherol or menadione (Ni *et al.*, 1994).

4.3   Reproductive and prenatal effects

4.3.1   *Humans*

Little information is available on the possible adverse effects of chloral hydrate on human pregnancy. Chloral hydrate is known to cross the human placenta at term (Bernstine *et al.*, 1954), but its use during relatively few pregnancies did not cause a detectable increase in abnormal outcomes (Heinonen *et al.*, 1977). Some data suggest that prolonged administration of sedative doses of chloral hydrate to newborns increases the likelihood of hyperbilirubinaemia (Lambert *et al.*, 1990).

Low levels of chloral hydrate have been found in breast milk. Although breastfeeding infants may be sedated by chloral hydrate in breast milk, the highest concentration measured in the milk (about 15 µg/mL) was considerably lower than that which would be measured in blood at a clinically active dose (100 µg/mL) (Bernstine *et al.*, 1956; Wilson, 1981).

4.3.2   *Experimental systems*

In-vitro evidence indicates that chloral hydrate induces a meiotic delay in maturing mouse oocytes (Eichenlaub-Ritter & Betzendahl, 1995; Eichenlaub-Ritter *et al.*, 1996). This effect on oocyte maturation apparently dominates the aneugenic effects. The concentrations required for meiotic delay (10 µg/mL) were two to three orders of magnitude below those required to affect tubulin polymerization (Brunner *et al.*, 1991; Wallin & Hartley-Asp, 1993). Meiotic arrest occurred at 125 µg/mL.

Chloral hydrate or a metabolite reached the testis and produced spermatid micronuclei of mice given single intraperitoneal injections of 41, 83 or 165 mg/kg bw and evaluated at 49 days following treatment (Allen *et al.*, 1994). The spermatogonial sperm-cell phase was significantly affected at all three doses. There was some evidence of increased frequency of micronucleated spermatids when treatment was in the preleptotene stage with the mid dose; the response was not dose-dependent. The leptotene–zygotene and the diakinesis-metaphase I phases appeared to be insensitive to chloral hydrate.

The parameters of reproduction were evaluated in male Fischer 344 rats administered 0.5 and 2 g/L chloral hydrate in the drinking-water for 52 weeks beginning at 28 days of age. At the high dose, corresponding to 188 mg/kg bw, chloral hydrate significantly decreased both the percentage of motil and progressively motil sperm, and shifted the straight-line velocity distribution of sperm to a lower modal velocity range (Klinefelter *et al.*, 1995). Administration of doses of chloral hydrate that are somewhat higher than the human therapeutic dose to pregnant mice (21.3 and 204.8 mg/kg per day in drinking-water during the gestational period) did not increase the incidence of gross external malformations in the offspring and did not impair normal development of pups (Kallman *et al.*, 1984).

4.4   Genetic and related effects

4.4.1   *Humans*

No data were available to the Working Group.

4.4.2   *Experimental systems* (see Table 2 for details and references)

The results obtained with chloral hydrate in a collaborative European Union project on aneuploidy have been summarized (Adler, 1993; Natarajan, 1993; Parry, 1993), and the genetic and related effects of chloral hydrate have been reviewed (Leuschner & Beuscher, 1998; Beland, 1999; Moore & Harrington-Brock, 2000).

(a)   DNA binding

In single studies in mice *in vivo*, chloral hydrate increased the level of malondialdehyde-derived adduct and 8-oxo-2′-deoxyguanosine in liver DNA (Von Tungeln *et al.*, 2002) but radioactively labelled chloral hydrate did not bind to liver DNA (Keller & Heck, 1988). Malondialdehyde adducts were formed when chloral hydrate was incubated with calf thymus DNA and microsomes from male B6C3F$_1$ mouse liver *in vitro* (Ni *et al.*, 1995).

(b)   Mutation and allied effects

Chloral hydrate did not induce mutation in most strains of *Salmonella typhimurium*, but did in four of six studies with *S. typhimurium* TA100 and in two studies with *S. typhimurium* TA104. The latter response was inhibited by the free-radical scavengers α-tocopherol and menadione (Ni *et al.*, 1994; Beland, 1999). It did not induce reverse mutation in *Saccharomyces cerevisiae* D$_7$ in one study.

Chloral hydrate did not induce mitotic crossing over in *Aspergillus nidulans* in the absence of metabolic activation. In *S. cerevisiae*, weak induction of mitotic gene conversion after metabolic activation and of meiotic recombination in the absence of metabolic activation was seen, whereas no intrachromosomal recombination was observed.

Chloral hydrate clearly induced aneuploidy in various fungi in the absence of metabolic activation. The results of a single study in Chinese spring wheat were negative with respect to induction of chromosome loss and gain.

In *Drosophila melanogaster*, chloral hydrate induced somatic mutations in a wing-spot test, but was either negative or inconclusive in sex-linked lethal mutation induction experiments.

In single studies, it was reported that chloral hydrate did not produce DNA–protein associations in rat liver nuclei or DNA single-strand breaks or alkaline-labile sites in rat primary hepatocytes *in vitro*.

In a single in-vitro study, a weak positive response was observed in the gene mutation test on a mouse lymphoma cell line.

In single in-vitro studies, a dose-dependent induction of sister chromatid exchanges was seen in chloral hydrate-treated Chinese hamster ovary cells.

Table 2. Genetic and related effects of chloral hydrate

Test system	Result[a] Without exogenous metabolic system	Result[a] With exogenous metabolic system	Dose[b] (LED or HID)	Reference
SOS chromotest, *Escherichia coli* PQ37	–	–	10 000	Giller *et al.* (1995)
Salmonella typhimurium TA100, TA1535, TA98, reverse mutation	–	–	10 000	Waskell (1978)
Salmonella typhimurium TA100, TA1537, TA1538, TA98, reverse mutation	+	+	1000	Haworth *et al.* (1983)
Salmonella typhimurium TA100, reverse mutation	–	–	5000 µg/plate	Leuschner & Leuschner (1991)
Salmonella typhimurium TA100, reverse mutation	+	+	2000 µg/plate	Ni *et al.* (1994)
Salmonella typhimurium TA100, reverse mutation, liquid medium	–	+	300	Giller *et al.* (1995)
Salmonella typhimurium TA100, reverse mutation	+	+	1000 µg/plate	Beland (1999)
Salmonella typhimurium TA104, reverse mutation	+	+	1000 µg/plate	Ni *et al.* (1994)
Salmonella typhimurium TA104, reverse mutation	+	+	1000	Beland (1999)
Salmonella typhimurium TA1535, reverse mutation	–	–	1850	Leuschner & Leuschner (1991)
Salmonella typhimurium TA1535, TA1537, reverse mutation	–	–	6667	Haworth *et al.* (1983)
Salmonella typhimurium TA1535, reverse mutation	–	–	10 000	Beland (1999)
Salmonella typhimurium TA98, reverse mutation	–	–	7500	Haworth *et al.* (1983)
Salmonella typhimurium TA98, reverse mutation	?	–	10 000 µg/plate	Beland (1999)
Saccharomyces cerevisiae D7, reverse mutation	–	–	3300	Bronzetti *et al.* (1984)
Saccharomyces cerevisiae RSY6, intrachromosomal recombination	–	NT	16.5	Howlett & Schiestl (2000)
Aspergillus nidulans, diploid strain 35 × 17, mitotic crossing-over	–	NT	1650	Crebelli *et al.* (1985)
Aspergillus nidulans, diploid strain 30, mitotic crossing-over	–	NT	6600	Käfer (1986)
Aspergillus nidulans, diploid strain NH, mitotic crossing-over	–	NT	1000	Kappas (1989)
Aspergillus nidulans, diploid strain P1, mitotic crossing-over	–	NT	990	Crebelli *et al.* (1991)

Table 2 (contd)

Test system	Result[a]		Dose[b] (LED or HID)	Reference
	Without exogenous metabolic system	With exogenous metabolic system		
Saccharomyces cerevisiae D7, mitotic gene conversion	–	(+)	2500	Bronzetti *et al.* (1984)
Aspergillus nidulans, diploid strain 35 × 17, haploids and nondisjunctional diploids	+	NT	825	Crebelli *et al.* (1985)
Aspergillus nidulans, diploid strain 30 conidia, aneuploidy	+	NT	825	Käfer (1986)
Aspergillus nidulans, haploid conidia, aneuploidy and polyploidy	+	NT	1650	Käfer (1986)
Aspergillus nidulans, diploid strain NH, nondisjunctional mitotic segregants	+	NT	450	Kappas (1989)
Aspergillus nidulans, diploid strain P1, nondisjunctional diploids and haploids	+	NT	660	Crebelli *et al.* (1991)
Aspergillus nidulans, haploid strain 35, hyperploidy	+	NT	2640	Crebelli *et al.* (1991)
Saccharomyces cerevisiae, meiotic recombination	?	NT	3300	Sora & Agostini Carbone (1987)
Saccharomyces cerevisiae, disomy in meiosis	+	NT	2500	Sora & Agostini Carbone (1987)
Saccharomyces cerevisiae, diploids in meiosis	+	NT	3300	Sora & Agostini Carbone (1987)
Saccharomyces cerevisiae D61.M, mitotic chromosomal malsegregation	+	NT	1000	Albertini (1990)
Saccharomyces cerevisiae diploid strain D6, monosomy	+	NT	1000	Parry *et al.* (1990)
Seedlings of hexaploid Chinese spring wheat, Neatby's strain, chromosomal loss and gain	–	NT	5000	Sandhu *et al.* (1991)
Haemanthus katherinae endosperm, apolar mitosis, *in vitro*	+	NT	200	Molè-Bajer (1969)
Drosophila melanogaster, somatic mutation wing spot test	+		825	Zordan *et al.* (1994)
Drosophila melanogaster, induction of sex-linked lethal mutation	?		37.2 feed	Beland (1999)
Drosophila melanogaster, induction of sex-linked lethal mutation	–		67.5 inj	Beland (1999)

Table 2 (contd)

Test system	Result[a]		Dose[b] (LED or HID)	Reference
	Without exogenous metabolic system	With exogenous metabolic system		
DNA–protein cross-links, rat liver nuclei *in vitro*	–	NT	41 250	Keller & Heck (1988)
DNA single-strand breaks (alkaline unwinding), rat primary hepatocytes *in vitro*	–	NT	1650	Chang *et al.* (1992)
Gene mutation, mouse lymphoma L5178Y/TK[+/–]–3.7.2C cell line *in vitro*	(+)		1000	Harrington-Brock *et al.* (1998)
Sister chromatid exchange, Chinese hamster ovary cells *in vitro*	+	+	100	Beland (1999)
Micronucleus formation (kinetochore-positive), Chinese hamster C1-1 cells *in vitro*	+	NT	165	Degrassi & Tanzarella (1988)
Micronucleus formation (kinetochore-negative), Chinese hamster C1-1 cells *in vitro*	–	NT	250	Degrassi & Tanzarella (1988)
Micronucleus formation (kinetochore-positive), Chinese hamster LUC2 cells *in vitro*	+	NT	400	Parry *et al.* (1990)
Micronucleus formation (kinetochore-positive), Chinese hamster LUC2 cells *in vitro*	+	NT	400	Lynch & Parry (1993)
Micronucleus formation, Chinese hamster V79 cells *in vitro*	+	NT	316	Seelbach *et al.* (1993)
Micronucleus formation, mouse lymphoma L5178Y/TK[+/–]–3.7.2C cell line *in vitro*	–	NT	1300	Harrington-Brock *et al.* (1998)
Micronucleus formation, mouse lymphoma L5178Y/TK[+/–]–3.7.2C cell line *in vitro*	+	NT	500	Nesslany & Marzin (1999)
Chromosomal aberrations, Chinese hamster CHED cells *in vitro*	+	NT	20	Furnus *et al.* (1990)
Chromosomal aberrations, Chinese hamster ovary cells *in vitro*	+	+	1000	Beland (1999)
Chromosomal aberrations, mouse lymphoma L5178Y/TK[+/–]–3.7.2C cell line *in vitro*	(+)	NT	1250	Harrington-Brock *et al.* (1998)

Table 2 (contd)

Test system	Result[a]		Dose[b] (LED or HID)	Reference
	Without exogenous metabolic system	With exogenous metabolic system		
Multipolar mitotic spindles, Chinese hamster DON.Wg3H cells *in vitro*	+	NT	500	Parry *et al.* (1990)
Multipolar mitotic spindles, Chinese hamster DON.Wg3H cells *in vitro*	+	NT	50	Warr *et al.* (1993)
Aneuploidy, Chinese hamster CHED cells *in vitro*	+[c]	NT	10	Furnus *et al.* (1990)
Aneuploidy, primary Chinese hamster embryonic cells *in vitro*	+[c]	NT	250	Natarajan *et al.* (1993)
Aneuploidy (hypoploidy), Chinese hamster LUC2 p4 cells *in vitro*	+	NT	250	Warr *et al.* (1993)
Aneuploidy, mouse lymphoma L5178Y/TK[+/−]–3.7.2C cell line *in vitro*	–	NT	1300	Harrington-Brock *et al.* (1998)
Tetraploidy and endoreduplication, Chinese hamster LUC2 p4 cells *in vitro*	+	NT	500	Warr *et al.* (1993)
Lacking mitotic spindle, Chinese hamster DON.Wg3H cells *in vitro*	+	NT	250	Parry *et al.* (1990)
Lacking mitotic spindle, metaphase defects, Chinese hamster LUC1 cells *in vitro*	+	NT	50	Parry *et al.* (1990)
Chromosomal dislocation from mitotic spindle, Chinese hamster DON.Wg3H cells *in vitro*	+	NT	500	Parry *et al.* (1990); Warr *et al.* (1993)
Bivalent chromosomes in meiosis I, MF1 mouse oocytes *in vitro* (16 h treatment)	+	NT	10	Eichenlaub-Ritter *et al.* (1996)
Aberrant meiosis I spindle (fusiform poles), meiotic arrest, hypoploidy, MF1 mouse oocytes *in vitro* (16 h treatment)	+	NT	125	Eichenlaub-Ritter *et al.* (1996)
Inhibition of spindle elongation, PtK2 rat kangaroo kidney epithelial cells *in vitro*	+	NT	1000	Lee *et al.* (1987)
Inhibition of chromosome-to-pole movement, PtK2 rat kangaroo kidney epithelial cells *in vitro*	–	NT	1000	Lee *et al.* (1987)
Breakdown of mitotic microtubuli, PtK2 rat kangaroo kidney epithelial cells *in vitro*	+	NT	1000	Lee *et al.* (1987)

Table 2 (contd)

Test system	Result[a]		Dose[b] (LED or HID)	Reference
	Without exogenous metabolic system	With exogenous metabolic system		
Porcine brain tubulin assembly inhibition *in vitro*	+	NT	9900	Brunner *et al.* (1991)
Porcine brain tubulin disassembly inhibition *in vitro*	+	NT	40	Brunner *et al.* (1991)
Bovine brain tubulin assembly inhibition *in vitro*	(+)	NT	165	Wallin & Hartley-Asp (1993)
Centriole migration block (kinetochore ultrastructure change), Chinese hamster cells (clone 237) *in vitro*	+	NT	1000	Alov & Lyubskii (1974)
Cell transformation, Syrian hamster embryo cells (24 h treatment)	+	NT	350	Gibson *et al.* (1995)
Cell transformation, Syrian hamster embryo cells (7 day treatment)	+	NT	1	Gibson *et al.* (1995)
Cell transformation, Syrian hamster dermal cell line 21NSR (24 h treatment)	+	NT	50	Parry *et al.* (1996)
Inhibition of intercellular communication, B6C3F1 mouse and Fischer 344 rat hepatocytes *in vitro*	–	NT	83	Klaunig *et al.* (1989)
Inhibition of intercellular communication, Sprague-Dawley rat liver Clone 9 cell *in vitro*	+	NT	165	Benane *et al.* (1996)
DNA single-strand breaks (alkaline unwinding), human lymphoblastoid CCRF-CEM cells *in vitro*	–	NT	1650	Chang *et al.* (1992)
Gene mutation, *tk* locus, human lymphoblastoid H2E1 V2[d] cell line *in vitro*	+	NT	1000	Beland (1999)
Gene mutation, *hprt* locus, human lymphoblastoid H2E1 V2[d] cell line	+	NT	1000	Beland (1999)
Sister chromatid exchange, human lymphocytes *in vitro*	(+)	NT	54	Gu *et al.* (1981)
Micronucleus formation, isolated human lymphocytes *in vitro*	–	T	1500	Vian *et al.* (1995)
Micronucleus formation, human lymphocytes in whole blood *in vitro*	+	NT	100	Migliore & Nieri (1991)
Micronucleus formation, human lymphocytes *in vitro*	(+)	NT	100	Ferguson *et al.* (1993)

Table 2 (contd)

Test system	Result[a] Without exogenous metabolic system	Result[a] With exogenous metabolic system	Dose[b] (LED or HID)	Reference
Micronucleus formation, human lymphocytes in vitro	+	-	100	Van Hummelen & Kirsch-Volders (1992)
Micronucleus formation, human lymphoblastoid AHH-1[d] cell line in vitro	+	NT	100	Parry et al. (1996)
Micronucleus formation, human lymphoblastoid MCL-5[d] cell line in vitro	-	NT	500	Parry et al. (1996)
Micronucleus formation (kinetochore-positive), human diploid LEO fibroblasts in vitro	+	NT	120	Bonatti et al. (1992)
Aneuploidy (double Y induction), human lymphocytes in vitro	+	NT	250	Vagnarelli et al. (1990)
Aneuploidy (hyperdiploidy and hypodiploidy), human lymphocytes in vitro	+	NT	50	Sbrana et al. (1993)
Polyploidy, human lymphocytes in vitro	+	NT	137	Sbrana et al. (1993)
C-Mitosis, human lymphocytes in vitro	+	NT	75	Sbrana et al. (1993)
Host-mediated assay, Saccharomyces cerevisiae D7 recovered from CD1 mouse lungs, mitotic gene conversion, in vivo	(+)		500 po × 1	Bronzetti et al. (1984)
DNA single-strand breaks (alkaline unwinding), male Sprague-Dawley rat liver in vivo	+		300 po × 1	Nelson & Bull (1988)
DNA single-strand breaks (alkaline unwinding), male Fischer 344 rat liver in vivo	-		1650 po × 1	Chang et al. (1992)
DNA single-strand breaks (alkaline unwinding), male B6C3F$_1$ mouse liver in vivo	+		100 po × 1	Nelson & Bull (1988)
DNA single-strand breaks (alkaline unwinding), male B6C3F$_1$ mouse liver in vivo	-		825 po × 1	Chang et al. (1992)
Micronucleus formation, male and female NMRI mice, bone-marrow erythrocytes in vivo	-		500 ip × 1	Leuschner & Leuschner (1991)

Table 2 (contd)

Test system	Result[a] Without exogenous metabolic system	Result[a] With exogenous metabolic system	Dose[b] (LED or HID)	Reference
Micronucleus formation, BALB/c mouse spermatids *in vivo* (preleptotene spermatocytes treated)	–		83 ip × 1	Russo & Levis (1992a)
Micronucleus formation, male BALB/c mouse bone-marrow erythrocytes and early spermatids (diakinesis/metaphase I and metaphase II stages treated) *in vivo*	+		83 ip × 1	Russo & Levis (1992b)
Micronucleus formation (kinetochore-positive and -negative), male BALB/c mouse bone-marrow erythrocytes *in vivo*	+		200 ip × 1	Russo *et al.* (1992)
Micronucleus formation, male (C57Bl/Cne × C3H/Cne)F₁ mouse bone-marrow erythrocytes *in vivo*	–		400 ip × 1	Leopardi *et al.* (1993)
Micronucleus formation, C57B1 mouse spermatids *in vivo* (spermatogonial stem cells and preleptotene spermatocytes treated)	+		41 ip × 1	Allen *et al.* (1994)
Micronucleus formation, male Swiss CD-1 mouse bone-marrow erythrocytes *in vivo*	+		200 ip × 1	Marrazini *et al.* (1994)
Micronucleus formation, B6C3F₁ mouse spermatids after spermatogonial stem-cell treatment *in vivo*	+		165 ip × 1	Nutley *et al.* (1996)
Micronucleus formation, B6C3F₁ mouse spermatids after meiotic cell treatment *in vivo*	–		413 ip × 1	Nutley *et al.* (1996)
Micronucleus formation, male (102/E1×C3H/E1)F₁, BALB/c mouse peripheral-blood erythrocytes *in vivo*	–		200 ip × 1	Grawé *et al.* (1997)
Micronucleus formation, male B6C3F₁ mouse bone-marrow erythrocytes *in vivo*	+		500 ip × 3	Beland (1999)
Micronucleus formation, *Pleurodeles waltl newt larvae* peripheral erythrocytes *in vivo*	+		200[e]	Giller *et al.* (1995)

Table 2 (contd)

Test system	Result[a] Without exogenous metabolic system	Result[a] With exogenous metabolic system	Dose[b] (LED or HID)	Reference
Chromosomal aberrations, male and female (102/E1 × C3H/E1)F₁ mouse bone-marrow cells in vivo	–		600 ip × 1	Xu & Adler (1990)
Chromosomal aberrations, male and female Sprague-Dawley rat bone-marrow cells in vivo	–		1000 po × 1	Leuschner & Leuschner (1991)
Chromosomal aberrations, BALB/c mouse spermatogonia treated, spermatogonia observed in vivo	–		83 ip × 1	Russo & Levis (1992b)
Chromosomal aberrations, (C57B1/Cne × C3H/Cne)F₁ mouse secondary spermatocytes (staminal gonia–pachytene treated), in vivo	+		82.7 ip × 1	Russo et al. (1984)
Chromosomal aberrations (translocations, breaks and fragments), (C57B1/Cne × C3H/Cne)F₁ mouse primary and secondary spermatocytes (from differentiating spermatogonia–pachytene stages treated), in vivo	–		413 ip × 1	Liang & Pacchierotti (1988)
Chromosomal aberrations, male Swiss CD-1 mouse bone-marrow erythrocytes in vivo	–		400 ip × 1	Marrazini et al. (1994)
Chromosomal aberrations, ICR mouse oocytes in vivo	–		600 ip × 1	Mailhes et al. (1993)
Gonosomal and autosomal univalents (C57B1/Cne × C3H/Cne)F₁ mouse primary spermatocytes (from differentiating spermatogonia–pachytene stages treated)	–	NT	413 ip × 1	Liang & Pacchierotti (1988)
Aneuploidy (C57B1/Cne × C3H/Cne)F₁ mouse secondary spermatocytes (from differentiating spermatogonia–pachytene stages treated) in vivo	+		82.7 ip × 1	Russo et al. (1984)
Aneuploidy (C57B1/Cne × C3H/Cne)F₁ mouse secondary spermatocytes in vivo (from differentiating spermatogonia–pachytene stages treated) in vivo	(+)		165 ip × 1	Liang & Pacchierotti (1988)
Aneuploidy (hypoploidy and hyperploidy), ICR mouse oocytes in vivo	–[f]		200 ip × 1	Mailhes et al. (1988)

Table 2 (contd)

Test system	Result[a]		Dose[b] (LED or HID)	Reference
	Without exogenous metabolic system	With exogenous metabolic system		
Polyploidy, male and female (102/E1 × C3H/E1)F$_1$ mouse bone-marrow cells *in vivo*	−		600 ip × 1	Xu & Adler (1990)
Aneuploidy (102/E1 × C3H/E1)F$_1$ mouse secondary spermatocytes *in vivo*	+		200 ip × 1	Miller & Adler (1992)
Aneuploidy, male (C57Bl/Cne × C3H/Cne)F$_1$ mouse bone marrow *in vivo*.	+[c]		400 ip × 1	Leopardi *et al.* (1993)
Aneuploidy (C57Bl/Cne × C3H/Cne)F$_1$ mouse secondary spermatocytes *in vivo*	−		400 ip × 1	Leopardi *et al.* (1993)
Aneuploidy (hypoploidy and hyperploidy), ICR mouse oocytes *in vivo*	−[f]		600 ip × 1	Mailhes *et al.* (1993)
Hyperploidy, male Swiss CD-1 mouse bone-marrow erythrocytes *in vivo*	+		200 ip x 1	Marrazini *et al.* (1994)
Trichloroethanol				
λ Prophage induction, *Escherichia coli* WP2	−	−	650 000[g]	DeMarini *et al.* (1994)
Salmonella typhimurium TA100, TA98, reverse mutation	−	−	7500 µg/plate	Waskell (1978)
Salmonella typhimurium TA100, reverse mutation	−	−	0.5 µg/cm^3 vapour[j]	DeMarini *et al.* (1994)
Salmonella typhimurium TA104, reverse mutation	−	+	2500 µg/plate	Beland (1999)
Inhibition of intercellular communication, male Sprague-Dawley rat liver Clone 9 cell *in vitro*	+	NT	150	Benane *et al.* (1996)

[a] +, considered to be positive; (+), considered to be weakly positive in an inadequate study; −, considered to be negative; ?, considered to be inconclusive (variable responses in several experiments within an inadequate study); NT, not tested

[b] LED, lowest effective dose; HID, highest ineffective dose; in-vitro tests, µg/mL; in-vivo tests, mg/kg bw; feed, feeding; inj, injection; po, orally; ip, intraperitoneally

[c] Negative for induction of polyploidy

[d] AHH-1: native strain expressing CYP1A1; MCL-5 and H2E1 V2: engineered strains expressing other CYP

[e] Larvae reared in chloral hydrate-containing water

[f] Slight induction of hypoploid cells may have been due to technical artefacts.

[g] Estimated from graph in paper

Chloral hydrate increased the frequency of micronuclei in Chinese hamster cell lines. The micronuclei produced probably contained whole chromosomes and not chromosome fragments, as the micronuclei could all be labelled with antikinetochore antibodies (Parry *et al.*, 1990; Lynch & Parry, 1993). When performed on mouse lymphoma cell line, the micronucleus test was negative in one study and positive in another.

Chloral hydrate also produced chromosomal aberrations *in vitro* in Chinese hamster cells and in mouse lymphoma cell line L5178Y/TK$^{+/-}$ -3.7.2C. It produced chromosomal dislocation from the mitotic spindle in Chinese hamster Don.Wg.3H cells *in vitro*.

Aneuploidy was induced *in vitro* by chloral hydrate in Chinese hamster cells but not in mouse lymphoma cell line L5178Y/TK$^{+/-}$-3.7.2C. In a study on kangaroo rat kidney epithelial cells, chloral hydrate inhibited spindle elongation and broke down mitotic microtubuli, although it did not inhibit pole-to-pole movement of chromosomes. It produced multipolar spindles and a total lack of mitotic spindles in Chinese hamster Don.Wg.3H cells. In a study on cultured mouse oocytes, chloral hydrate induced bivalent chromosomes, aberrant spindle, hypoploidy and meiotic arrest.

Cell transformation was observed in Syrian hamster embryo and dermal cells exposed to chloral hydrate *in vitro*.

Chloral hydrate induced DNA single-strand breaks and *TK* and *HPRT* loci gene mutations in human lymphoblastoid cells and weakly induced sister chromatid exchange in cultured human lymphocytes. It did not inhibit cell-to-cell communication in mouse or rat hepatocytes *in vitro*, but did in a study on Sprague-Dawley rat liver clone 9 cells.

In human diploid fibroblasts *in vitro*, chloral hydrate induced micronuclei that contained kinetochores. Micronuclei were induced in studies with human whole blood cultures but not in one study with isolated lymphocytes. The differences seen in the micronucleus test were attributed to differences between whole blood and purified lymphocyte cultures (Vian *et al.*, 1995), but this hypothesis has not been tested. Moreover, micronuclei were induced *in vitro* in human lymphoblastoid cell line AHH-1 (native strain expressing CYP1A1), but not in human lymphoblastoid cell line MCL-5 (engineered strain expressing other CYPs).

In human lymphocytes *in vitro*, it induced aneuploidy, C-mitosis and polyploidy.

Chloral hydrate increased the rate of mitotic gene conversion in a host-mediated assay with *S. cerevisiae* D$_7$ recovered from CD$_1$ mouse lungs.

One study showed induction of single-strand breaks in liver DNA of both rats and mice treated with chloral hydrate *in vivo*; another study in both species found no such effect.

In vivo, micronuclei were induced in mouse bone-marrow erythrocytes in four of six studies after treatment with chloral hydrate; in one of these studies, the use of antikinetochore antibodies suggested induction of micronuclei containing both whole chromosomes and fragments. Micronuclei were not produced in peripheral erythrocytes of two strains of male mice treated *in vivo*. Chloral hydrate induced micronuclei in the spermatids of mice treated *in vivo* in three of five studies. The cell-cycle stage chosen for treatment is crucial, as micronuclei are induced after spermatogonial treatment of stem cells and not meiotic cells; the premeiotic S-phase, preleptotene, was concluded to be sensitive only to clastogenic agents. In one of the studies that showed an effect, only kinetochore-negative micronuclei were induced, but kinetochore-negative micronuclei were also produced by another established aneuploidogen, vincristine sulfate. The finding may therefore suggest not induction of fragments harbouring micronuclei but an inability of the antibody to label kinetochores in the micronuclei, which may lose their normal kinetochore protein composition and structure (Allen *et al.*, 1994).

In a single study, chloral hydrate induced micronuclei in erythrocytes of newt larvae.

In vivo, the frequency of chromosomal aberrations in spermatogonia, primary and secondary spermatocytes and oocytes was not increased in single studies after treatment with chloral hydrate. Chloral hydrate induced chromosomal aberrations in mouse secondary spermatocytes in one study after treatment *in vivo*, but failed to induce chromosomal aberration in mouse and rat bone marrow treated *in vivo*.

In single studies, chloral hydrate induced aneuploidy and hyperploidy in the bone marrow of mice treated *in vivo*. It increased the rate of aneuploidy in mouse secondary spermatocytes in three of four studies, but failed to show increased frequency of numerical or structural chromosome changes in mouse oocytes. It did not produce polyploidy in oocytes and in bone marrow or gonosomal or autosomal univalents in primary spermatocytes of mice treated *in vivo*.

Genetic effects of trichloroethanol

Trichloroethanol, a reduction product of chloral hydrate, induced formation of stable DNA adducts when incubated with calf thymus DNA and microsomes from male B6C3F$_1$ mouse liver *in vitro* (Ni *et al.*, 1995), but caused mutation in *S. typhimurium* strain TA104 in the presence of metabolic system but not in strain TA100 or TA98. It failed to induce λ prophage in *E. coli*. It inhibited gap-junctional intercellular communication in rat hepatocytes *in vitro*.

4.5   Mechanistic considerations

Chloral hydrate is genotoxic *in vivo* and *in vitro*, and its induction of liver tumours in mice appears to be associated with peroxisome proliferation.

5.   Summary of Data Reported and Evaluation

5.1   Exposure data

Chloral is a chlorinated aldehyde that found extensive use, beginning in the 1940s, as a precursor in the production of the insecticide DDT and, to a lesser extent, of other insecticides and pharmaceuticals. This use of chloral has declined steadily since the 1960s, especially in those countries where the use of DDT has been restricted. Chloral is readily converted to chloral hydrate in the presence of water. Chloral hydrate is used as a sedative before medical procedures and to reduce anxiety related to withdrawal from drugs. Wider exposure to chloral hydrate occurs at microgram-per-litre levels in drinking-water and swimming pools as a result of chlorination.

5.2   Human carcinogenicity data

Several studies were identified that analysed risk with respect to one or more measures of exposure to complex mixtures of disinfection by-products that are found in most chlorinated and chloraminated drinking-water. No data specifically on chloral or chloral hydrate were available to the Working Group.

5.3   Animal carcinogenicity data

Following administration in drinking-water, chloral hydrate increased the incidence of hepatocellular neoplasms in male mice in two studies. An increased trend with dose for hepatocellular carcinoma in one study with dietary-restricted male mice was reported. Chloral hydrate was not active in male or female rats in two studies. Chloral hydrate increased the incidence of adenomas in the pars distalis of the pituitary gland in female mice only at the highest dose in one gavage study, but did not induce tumours of the pars distalis in male mice.

5.4   Other relevant data

Chloral hydrate is a sedative and hypnotic with some clinical uses. These late effects are also evident in animal studies.

Chloral hydrate is metabolized to trichloroacetic acid, trichloroethanol (which is converted to trichloroethyl glucuronide) and dichloroacetic acid in humans and in rodents.

There is limited evidence from a single study that chloral hydrate affects sperm, but no evidence of actual reproductive or developmental toxicity has been shown.

Chloral hydrate is a well-established aneuploidogenic agent that also has some mutagenic activity. *In vivo*, it clearly induced aneuploidy and micronuclei in mammals, whereas chromosomal aberrations were not found in most studies. Conflicting results were obtained with regard to the induction of DNA damage in chloral hydrate-treated mammals. In human cells *in vitro*, chloral hydrate induced aneuploidy, micronuclei and gene mutations. Sister chromatid exchange and DNA strand-break studies yielded inconclusive and negative results, respectively. Chloral hydrate clearly induced micronuclei in Chinese hamster cells, whereas findings in mouse lymphoma cells were conflicting. It failed to induce DNA damage, but caused weak mutagenicity and a clear induction of aneuploidy, chromosomal aberrations and sister chromatid exchange in rodent cells *in vitro*. It induced the formation of micronuclei in erythrocytes of newt larvae *in vivo*. In fungi, chloral hydrate clearly induced aneuploidy, while mitotic recombination and gene conversion assays were inconclusive. Induction of somatic mutation (but not sex-linked mutation) by chloral hydrate was demonstrated in insects. In bacteria, the compound induced base-pair substitution mutations. When incubated with calf thymus DNA, chloral hydrate induced the formation of malonaldehyde-related DNA adducts.

5.5   Evaluation

There is *inadequate evidence* in humans for the carcinogenicity of chloral and chloral hydrate.

There is *limited evidence* in experimental animals for the carcinogenicity of chloral hydrate.

Overall evaluation

Chloral hydrate is *not classifiable as to its carcinogenicity to humans (Group 3)*.

6.   References

Abbas, R.R., Seckel, C.S., Kidney, J.K. & Fisher, J.W. (1996) Pharmacokinetic analysis of chloral hydrate and its metabolism in B6C3F1 mice. *Drug. Metab. Dispos.*, **24**, 1340–1346

Adler, I.-D. (1993) Synopsis of the in vivo results obtained with the 10 known or suspected aneugens tested in the CEC collaborative study. *Mutat. Res.*, **287**, 131–137

Albertini, S. (1990) Analysis of nine known or suspected spindle poisons for mitotic chromosome malsegregation using *Saccharomyces cerevisiae* D61.M. *Mutagenesis*, **5**, 453–459

Allen, J.W., Collins, B.W. & Evansky, P.A. (1994) Spermatid micronucleus analyses of trichloro-ethylene and chloral hydrate effects in mice. *Mutat. Res.*, **323**, 81–88

Alov, I.A. & Lyubskii, S.L. (1974) Experimental study of the functional morphology of the kine-tochore in mitosis. *Byull. éksp. Biol. Med.*, **78**, 91–94

American Public Health Association/American Water Works Association/Water Environment Federation (1999) *Standard Methods for the Examination of Water and Wastewater*, 20th Ed., Washington, DC [CD-ROM]

Baudisch, C., Pansch, G., Prösch, J. & Puchert, W. (1997) [*Determination of Volatile Halogenated Hydrocarbons in Chlorinated Swimming Pool Water. Research Report*], Aussnstelle Schwerin, Landeshygieneinstitut Mecklenburg-Vorpommern (in German) [cited in WHO (2000) *Guidelines for Safe Recreational-Water Environments*, Vol. 2, *Swimming Pools, Spas and Similar Recreational-Water Environments*, Geneva]

Beland, F.A. (1999) NTP technical report on the toxicity and metabolism studies of chloral hydrate (CAS No. 302-17-0). Administered by gavage to F344/N rats and B6C3F$_1$ mice. *Toxic. Rep. Ser.*, **59**, 1–66, A1–E7

Beland, F.A., Schmitt, T.C., Fullerton, N.F. & Young, J.F. (1998) Metabolism of chloral hydrate in mice and rats after single and multiple doses. *J. Toxicol. environ. Health*, **A54**, 209–226

Benane, S.G., Blackman, C.F. & House, D.E. (1996) Effect of perchloroethylene and its meta-bolites on intercellular communication in clone 9 rat liver cells. *J. Toxicol. environ. Health*, **48**, 427–437

Bernstine, J.B., Meyer, A.E. & Hayman, H. (1954) Maternal and foetal blood estimation following the administration of chloral hydrate during labour. *J. Obstet. Gynaecol.*, **61**, 683–685

Bernstine, J.B., Meyer, A.E. & Bernstine, R.L. (1956) Maternal blood and breast milk estimation following the administration of chloral hydrate during the puerperium. *J. Obstet. Gynaecol.*, **63**, 228–231

Blank, V., Shukairy, H.M. & McLain, J. (2002) Unregulated organic DBPs in ICR finished water and distribution systems. In: McGuire, M.J., McLain, J. & Obdensky, A., eds, *Information Collection Rule Analysis*, Denver, CO, American Water Works Research Foundation

Boitsov, A.N., Rotenberg, Y.S. & Mulenkova, V.G. (1970) [On the toxicological evaluation of chloral in the process of its liberation during filling and pouring of foam polyurethanes.] *Gig. Tr. prof. Zabol.*, **14**, 26–29 (in Russian)

Bonatti, S., Cavalieri, Z., Viaggi, S. & Abbondandolo, A. (1992) The analysis of 10 potential spindle poisons for their ability to induce CREST-positive micronuclei in human diploid fibro-blasts. *Mutagenesis*, **7**, 111–114

Bringmann, G., God, R., Fähr, S., Feineis, D., Fornadi, K. & Fornadi, F. (1999) Identification of the dopaminergic neurotoxin 1-trichloromethyl-1,2,3,4-tetrahydro-β-carboline in human blood after intake of the hypnotic chloral hydrate. *Anal. Biochem.*, **270**, 167–175

Bronzetti, G., Galli, A., Corsi, C., Cundari, E., Del Carratore, R., Nieri, R. & Paolini, M. (1984) Genetic and biochemical investigation on chloral hydrate in vitro and in vivo. *Mutat. Res.*, **141**, 19–22

Brunner, M., Albertini, S. & Würgler, F.E. (1991) Effects of 10 known or suspected spindle poisons in the *in vitro* porcine brain tubulin assembly assay. *Mutagenesis*, **6**, 65–70

Bruzzoniti, M.C., Mentasti, E., Sarzanini, C. & Tarasco, E. (2001) Liquid chromatographic methods for chloral hydrate determination. *J. Chromatogr.*, **A920**, 283–289

Budavari, S., ed. (1998) *The Merck Index*, 12th Ed., Version 12:2, Whitehouse Station, NJ, Merck & Co. [CD-ROM]

Carlberg, G.E., Drangsholt, H. & Gjøs, N. (1986) Identification of chlorinated compounds in the spent chlorination liquor from differently treated sulphite pulps with special emphasis on mutagenic compounds. *Sci. total Environ.*, **48**, 157–167

Chang, L.W., Daniel, F.B. & DeAngelo, A.B. (1992) Analysis of DNA strand breaks induced in rodent liver in vivo, hepatocytes in primary culture, and a human cell line by chlorinated acetic acids and chlorinated acetaldehydes. *Environ. mol. Mutag.*, **20**, 277–288

Chemical Information Services (2002a) *Directory of World Chemical Producers (Version 2002.1)*, Dallas, TX

Chemical Information Services (2002b) *Worldwide Bulk Drug Users Directory (Version 2002)*, Dallas, TX [http://www.chemical.info.com]

Cole, W.J., Mitchell, R.G. & Salamonsen, R.F. (1975) Isolation, characterization and quantitation of chloral hydrate as a transient metabolite of trichloroethylene in man using electron capture gas chromatography and mass fragmentography. *J. Pharm. Pharmacol.*, **27**, 167–171

Crebelli, R., Conti, G., Conti, L. & Carere, A. (1985) Mutagenicity of trichloroethylene, trichloro-ethanol and chloral hydrate in *Aspergillus nidulans*. *Mutat. Res.*, **155**, 105–111

Crebelli, R., Conti, G., Conti, L. & Carere, A. (1991) *In vitro* studies with nine known or suspected spindle poisons: Results in tests for chromosome malsegregation in *Aspergillus nidulans*. *Mutagenesis*, **6**, 131–136

Daniel, F.B., DeAngelo, A.B., Stober, J.A., Olson, G.R. & Page, N.P. (1992a) Hepatocarcinogenicity of chloral hydrate, 2-chloroacetaldehyde, and dichloroacetic acid in the male B6C3F1 mouse. *Fundam. appl. Toxicol.*, **19**, 159–168

Daniel, F.B., Robinson, M., Stober, J.A., Page, N.P. & Olson, G.R. (1992b) Ninety-day toxicity study of chloral hydrate in the Sprague-Dawley rat. *Drug chem. Toxicol.*, **15**, 217–232

Davidson, I.W.F. & Beliles, R.P. (1991) Consideration on the target organ toxicity of trichloroethylene in terms of metabolite toxicity and pharmacokinetics. *Drug Metab. Rev.*, **23**, 493–599

Degrassi, F. & Tanzarella, C. (1988) Immunofluorescent staining of kinetochores in micronuclei: A new assay for the detection of aneuploidy. *Mutat. Res.*, **203**, 339–345

DeMarini, D.M., Perry, E. & Shelton, M.L. (1994) Dichloroacetic acid and related compounds: Induction of prophage in *E. coli* and mutagenicity and mutation spectra in Salmonella TA100. *Mutagenesis*, **9**, 429–437

Eichenlaub-Ritter, U. & Betzendahl, I. (1995) Chloral hydrate induced spindle aberrations, metaphase I arrest and aneuploidy in mouse oocytes. *Mutagenesis*, **10**, 477–486

Eichenlaub-Ritter, U., Baart, E., Yin, H. & Betzendahl, I. (1996) Mechanisms of spontaneous and chemically-induced aneuploidy in mammalian oogenesis: Basis of sex-specific differences in response to aneugens and the necessity for further tests. *Mutat. Res.*, **372**, 279–294

Environmental Protection Agency (1988) *Health and Environmental Effects Document for Chloral*, Washington, DC, Office of Solid Waste and Emergency Response, Office of Health and Environmental Assessment, 65 pp.

Environmental Protection Agency (1995) *Method 551.1. Determination of chlorination Disinfection Byproducts, Chlorinated Solvents, and Halogenated Pesticides/Herbicides in Drinking Water by Liquid–liquid Extraction and Gas Chromatography with Electron Capture Detection* (Revision 1.0), Cincinnati, OH, National Exposure Research Laboratory, Office of Research and Development

Environmental Protection Agency (1998) National primary drinking water regulations; disinfectants and disinfection byproducts; final rule. *Fed. Regist.*, December 16

European Commission (1998) *Council Directive 98/83/EC of 3 November 1998 on the Quality of Water Intended for Human Consumption*, Luxemburg, Office for Official Publications of the European Communities

Ferguson, L.R., Morcombe, P. & Triggs, C.N. (1993) The size of cytokinesis-blocked micronuclei in human peripheral blood lymphocytes as a measure of aneuploidy induction by set A compounds in the EEC trial. *Mutat. Res.*, **287**, 101–112

Furnus, C.C., Ulrich, M.A., Terreros, M.C. & Dulout, F.N. (1990) The induction of aneuploidy in cultured Chinese hamster cells by propionaldehyde and chloral hydrate. *Mutagenesis*, **5**, 323–326

Gennaro, A.R., ed. (2000) *Remington: The Science and Practice of Pharmacy*, 20th Ed., Baltimore, MD, Lippincott Williams & Wilkins, pp. 1417–1418

George, M.H., Moore, T., Kilburn, S., Olson, G.R. & DeAngelo, A.B. (2000) Carcinogenicity of chloral hydrate administered in drinking water to the male F344/N rat and male B6C3F1 mouse. *Toxicol. Pathol.*, **28**, 610–618

Gibson, D.P., Aardema, M.J., Kerckaert, G.A., Carr, G.J., Brauninger, R.M. & LeBoeuf, R.A. (1995) Detection of aneuploidy-inducing carcinogens in the Syrian hamster embryo (SHE) cell transformation assay. *Mutat. Res.*, **343**, 7–24

Giller, S., Le Curieux, F., Gauthier, L., Erb, F. & Marzin, D. (1995) Genotoxicity assay of chloral hydrate and chloropicrine. *Mutat. Res.*, **348**, 147–152

Glaze, W.H., Kenneke, J.F. & Ferry, L.J. (1993) Chlorinated byproducts from the TiO$_2$-mediated photodegradation of trichloroethylene and tetrachloroethylene in water. *Environ. Sci. Technol.*, **27**, 177–184

Goldenthal, E.I. (1971) A compilation of LD50 values in newborn and adult animals. *Toxicol. appl. Pharmacol.*, **18**, 185–207

Goodman Gilman, A., Rall, T.W., Nies, A.S. & Taylor, P., eds (1991) *The Pharmacological Basis of Therapeutics*, New York, Pergamon Press, pp. 357–364

Goodman Gilman, A., Hardman, J.G., Limbird, L.E., Molinoff, P.B. & Ruddon, R.W., eds (1996) *The Pharmacological Basis of Therapeutics*, 9th Ed., New York, McGraw-Hill, p. 381

Grawé, J., Nüsse, M. & Adler, I.-D. (1997) Quantitative and qualitative studies of micronucleus induction in mouse erythrocytes using flow cytometry. I. Measurement of micronucleus induction in peripheral blood polychromatic erythrocytes by chemicals with known and suspected genotoxicity. *Mutagenesis*, **12**, 1–8

Gu, Z.W., Sele, B., Jalbert, P., Vincent, M., Vincent, F., Marka, C., Chmara, D. & Faure, J. (1981) [Induction of sister chromatid exchange by trichloroethylene and its metabolites.] *Toxicol. Eur. Res.*, **3**, 63–67 (in French)

Hansch, C., Leo, A. & Hoekman, D. (1995) *Exploring QSAR: Hydrophobic, Electronic, and Steric Constants*, Washington, DC, American Chemical Society, p. 4

Hara, A., Yamamoto, H., Deyashiki, Y., Nakayama, T., Oritani, H. & Sawada, H. (1991) Aldehyde dismutation catalyzed by pulmonary carbonyl reductase: Kinetic studies of chloral hydrate metabolism to trichloroacetic acid and trichloroethanol. *Biochim. biophys. Acta*, **1075**, 61–67

Harrington-Brock, K., Doerr, C.L. & Moore, M.M. (1998) Mutagenicity of three disinfection by-products: Di- and trichloroacetic acid and chloral hydrate in L5178Y/TK$^{+/-}$ -3.7.2C mouse lymphoma cells. *Mutat. Res.*, **413**, 265–276

Haworth, S., Lawlor, T., Mortelmans, K., Speck, W. & Zeiger, E. (1983) Salmonella mutagenicity test results for 250 chemicals. *Environ. Mutag.*, **Suppl. 1**, 3–142

Health Canada (1995) *A National Survey of Chlorinated Disinfection By-products in Canadian Drinking Water* (95-EHD-197), Ottawa, Ontario, Minister of Supply and Services Canada

Health Canada (2003) *Summary of Guidelines for Canadian Drinking Water Quality*, Ottawa

Heinonen, O.P., Slone, D. & Shapiro, S. (1977) *Birth Defects and Drugs in Pregnancy*, Littleton, MA, PSG Publishing Co.

Helrich, K., ed. (1990) *Official Methods of Analysis of the Association of Official Analytical Chemists*, 15th Ed., Vol. 1, Arlington, VA, Association of Official Analytical Chemists, p. 562

Henderson, G.N., Yan, Z., James, M.O., Davydova, N. & Stacpoole, P.W. (1997) Kinetics and metabolism of chloral hydrate in children: Identification of dichloroacetate as a metabolite. *Biochem. biophys. Res. Commun.*, **235**, 695–698

Hobara, T., Kobayashi, H., Kawamoto, T., Iwamoto, S. & Sakai, T. (1987) Extrahepatic metabolism of chloral hydrate, trichloroethanol and trichloroacetic acid in dogs. *Pharmacol. Toxicol.*, **61**, 58–62

Hobara, T., Kobayashi, H., Kawamoto, T., Iwamoto, S. & Sakai, T. (1988) Intestinal absorption of chloral hydrate, free trichloroethanol and trichloroacetic acid in dogs. *Pharmacol. Toxicol.*, **62**, 250–258

Howlett, N.G. & Schiestl, R.H. (2000) Simultaneous measurement of the frequencies of intrachromo-somal recombination and chromosome gain using the yeast DEL assay. *Mutat. Res.*, **454**, 53–62

IARC (1991) *IARC Monographs on the Evaluation of the Carcinogenic Risks to Humans*, Vol. 52, *Chlorinated Drinking-water; Chlorination By-products; Some Other Halogenated Compounds; Cobalt and Cobalt Compounds*, Lyon, pp. 55–141

IARC (1995) *IARC Monographs on the Evaluation of Carcinogenic Risks to Humans*, Vol. 63, *Dry Cleaning, Some Chlorinate Solvents and Other Industrial Chemicals*, Lyon, pp. 245–269

Italia, M.P. & Uden P.C. (1988) Multiple element emission spectral detection gas chromatographic profiles of halogenated products from chlorination of humic acid and drinking water. *J. Chromatogr.*, **438**, 35–43

Jacangelo, J.G., Patania, N.L., Reagan, K.M., Aieta, E.M., Krasner, S.W. & McGuire, M.J. (1989) Ozonation: Assessing its role in the formation and control of disinfection by-products. *J. Am. Water Works Assoc.*, **81**, 74–84

Jira, R., Kopp, E. & McKusick, B.C. (1986) Chloroacetaldehydes. In: Gerhartz, W., Yamamoto, Y.S., Campbell, F.T., Pfefferkorn, R. & Rounsaville, J.F., eds., *Ullmann's Encyclopedia of Industrial Chemistry*, 5th rev. Ed., Volume A6, New York, VCH Publishers, pp. 527–536

Käfer, E. (1986) Tests which distinguish induced crossing-over and aneuploidy from secondary segregation in Aspergillus treated with chloral hydrate and γ-rays. *Mutat. Res.*, **164**, 145–166

Kallman, M.J., Kaempf, G.L. & Balster, R.L. (1984) Behavioral toxicity of chloral in mice: An approach to evaluation. *Neurobehav. Toxicol. Teratol.*, **6**, 137–146

Kappas, A. (1989) On the mechanisms of induced aneuploidy in *Aspergillus nidulans* and validation of tests for genomic mutations. *Prog. clin. Biol. Res.*, **318**, 377–384

Keller, D.A. & Heck, H.d'A. (1988) Mechanistic studies on chloral toxicity: Relationship to trichloroethylene carcinogenesis. *Toxicol. Lett.*, **42**, 183–191

Ketcha, M.M., Stevens, D.K., Warren, D.A., Bishop, C.T. & Brashear, W.T. (1996) Conversion of trichloroacetic acid to dichloroacetic acid in biological samples. *J. anal. Toxicol.*, **20**, 236–241

Kim, H., Shim, J. & Lee, S. (2002) Formation of disinfection by-products in chlorinated swimming pool water. *Chemosphere*, **46**, 123–130

Klaunig, J.E., Ruch, R.J. & Lin, E.L.C. (1989) Effects of trichloroethylene and its metabolites on rodent hepatocyte intercellular communication. *Toxicol. appl. Pharmacol.*, **99**, 454–465

Klinefelter, G.R., Suarez, J.D., Roberts, N.L. & DeAngelo, A.B. (1995) Preliminary screening for the potential of drinking water disinfection byproducts to alter male reproduction. *Reprod. Toxicol.*, **9**, 571–578

Koch, B. & Krasner, S.W. (1989) Occurrence of disinfection by-products in a distribution system. In: *Proceedings of the AWWA, Annual Conference*, Part 2, Chicago, IL, American Water Works Association, pp. 1203–1230

Kojima, T., Suanhara, T. & Kuroda, H. (1990) [Determination of chloroacetic acids in tap water.] *Kagawa-ken Eisei Kenkyushoho*, (17), 122–128 [Analytical Abstract: 54(4):H124] (in Japanese)

Koppen, B., Dalgaard, L. & Christensen, J.M. (1988) Determination of trichloroethylene metabolites in rat liver homogenate using headspace gas chromatography. *J. Chromatogr.*, **442**, 325–332

Krasner, S.W., McGuire, M.J., Jacangelo, J.G., Patania, N.L., Reagan, K.M. & Aieta, E.M. (1989) The occurrence of disinfection by-products in US drinking water. *J. Am. Water Works Assoc.*, **81**, 41–53

Lambert, G.H., Muraskas, J., Anderson, C.L. & Myers, T.F. (1990) Direct hyperbilirubinemia associated with chloral hydrate administration in the newborn. *Pediatrics*, **86**, 277–281

LeBel, G.L., Benoit, F.M. & Williams, D.T. (1997) A one-year survey of halogenated disinfection by-products in the distribution system of treatment plants using three different disinfection processes. *Chemosphere*, **34**, 2301–2317

Lee, G.M., Diguiseppi, J., Gawdi, G.M. & Herman, B. (1987) Chloral hydrate disrupts mitosis by increasing intracellular free calcium. *J. Cell Sci.*, **88**, 603–612

Leopardi, P., Zijno, A., Bassani, B. & Pacchierotti, F. (1993) In vivo studies on chemically induced aneuploidy in mouse somatic and germinal cells. *Mutat. Res.*, **287**, 119–130

Leuschner, J. & Beuscher, N. (1998) Studies on the mutagenic and carcinogenic potential of chloral hydrate. *Arzneim.-Forsch./Drug Res.*, **48**, 961–968

Leuschner, J. & Leuschner, F. (1991) Evaluation of the mutagenicity of chloral hydrate in vitro and in vivo. *Arzneim.-Forsch./Drug Res.*, **41**, 1101–1103

Liang, J.C. & Pacchierotti, F. (1988) Cytogenetic investigation of chemically-induced aneuploidy in mouse spermatocytes. *Mutat. Res.*, **201**, 325–335

Lide (2000) Properties of Organic Compounds on CD-ROM, Version 6, Chapman & Hall/CRC, London

Lipscomb, J.C., Mahle, D.A., Brashear, W.T. & Garrett, C.M. (1996) A species comparison of chloral hydrate metabolism in blood and liver. *Biochem. biophys. Res. Commun.*, **227**, 340–350

Lynch, A.M. & Parry, J.M. (1993) The cytochalasin-B micronucleus/kinetochore assay *in vitro*: Studies with 10 suspected aneugens. *Mutat. Res.*, **287**, 71–86

Mailhes, J.B., Preston, R.J., Yuan, Z.P. & Payne, H.S. (1988) Analysis of mouse metaphase II oocytes as an assay for chemically induced aneuploidy. *Mutat. Res.*, **198**, 145–152

Mailhes, J.B., Aardema, M.J. & Marchetti, F. (1993) Investigation of aneuploidy induction in mouse oocytes following exposure to vinblastine-sulfate, pyrimethamine, diethylstilbestrol diphosphate, or chloral hydrate. *Environ. mol. Mutag.*, **22**, 107–114

Mannschott, P., Erdinger, L. & Sonntag, H.-G. (1995) [Halogenated organic compounds in swimming pool water.] *Zbl. Hyg.*, **197**, 516–533 (in German)

Marrazzini, A., Betti, C., Bernacchi, F., Barrai, I. & Barale, R. (1994) Micronucleus test and metaphase analyses in mice exposed to known and suspected spindle poisons. *Mutagenesis*, **9**, 505–515

Mayers, D.J., Hindmarsh, K.W., Sankaran, K., Gorecki, D.K.J. & Kasian, G.F. (1991) Chloral hydrate disposition following single-dose administration to critically ill neonates and children. *Dev. Pharmacol. Ther.*, **16**, 71–77

McKinney, L.L., Uhing, E.H., White, J.L. & Picken, J.C., Jr (1955) Vegetable oil extraction. Autoxidation products of trichloroethylene. *Agric. Food Chem.*, **3**, 413–419

Medical Economics Co. (1996) *PDR®: Generics*, 2nd Ed., Montvale, NJ, Medical Economics Data Production Co., pp. 584–586

Merdink, J.L., Gonzalez-Leon, A., Bull, R.J. & Schultz, I.R. (1998) The extent of dichloroacetate formation from trichloroethylene, chloral hydrate, trichloroacetate, and trichloroethanol in B6C3F1 mice. *Toxicol. Sci.*, **45**, 33–41

Merdink, J.L., Stenner, R.D., Stevens, D.K., Parker, J.C. & Bull, R.J. (1999) Effect of entero-hepatic circulation on the pharmacokinetics of chloral hydrate and its metabolites in F344 rats. *J. Toxicol. environ. Health*, **A56**, 357–368

Migliore, L. & Nieri, M. (1991) Evaluation of twelve potential aneuploidogenic chemicals by the *in vitro* human lymphocyte micronucleus assay. *Toxicol. in Vitro*, **5**, 325–336

Miller, B.M. & Adler, I.-D. (1992) Aneuploidy induction in mouse spermatocytes. *Mutagenesis*, **7**, 69–76

Miller, J.W. & Uden, P.C. (1983) Characterization of nonvolatile aqueous chlorination products of humic substances. *Environ. Sci. Technol.*, **17**, 150–157

Mishchikhin, V.A. & Felitsin, F.P. (1988) [Gas-chromatographic detection of chloroform, carbon tetrachloride, dichloroethane, trichloroethylene and chloralhydrate in biological material.] *Sud. med. Ekspert.*, **31**, 30–33 (in Russian)

Molè-Bajer, J. (1969) Fine structural studies of apolar mitosis. *Chromosoma*, **26**, 427–448

Moore, M.M. & Harrington-Brock, K. (2000) Mutagenicity of trichloroethylene and its meta-bolites: Implications for the risk assessment of trichloroethylene. *Environ. Health Perspect.*, **108**, Suppl. 2, 215–223

Natarajan, A.T. (1993) An overview of the results of testing of known or suspected aneugens using mammalian cells *in vitro*. *Mutat. Res.*, **287**, 113–118

Natarajan, A.T., Duivenvoorden, W.C.M., Meijers, M. & Zwanenburg, T.S.B. (1993) Induction of mitotic aneuploidy using Chinese hamster primary embryonic cells. Test results of 10 chemicals. *Mutat. Res.*, **287**, 47–56

National Health and Medical Research Council and Agriculture and Resource Management Council of Australia and New Zealand (1996) *Australian Drinking Water Guidelines Summary*, Canberra

National Institute for Occupational Safety and Health (1994) *National Occupational Exposure Survey (1981–1983)*, Cincinnati, OH

National Toxicology Program (2002a) *Toxicology and Carcinogenesis Studies of Chloral Hydrate* (NTP Technical Report 502), Research Triangle Park, NC

National Toxicology Program (2002b) *Toxicology and Carcinogenesis Studies of Chloral Hydrate* (ad libitum *and dietary controlled*) (NTP Technical Report 503), Research Triangle Park, NC

Nelson, M.A. & Bull, R.J. (1988) Induction of strand breaks in DNA by trichloroethylene and metabolites in rat and mouse liver *in vivo*. *Toxicol. appl. Pharmacol.*, **94**, 45–54

Nesslany, F. & Marzin, D. (1999) A micromethod for the *in vitro* micronucleus assay. *Mutagenesis*, **14**, 403–410

Ni, Y.-C., Wong, T.-Y., Kadlubar, F.F. & Fu, P.P. (1994) Hepatic metabolism of chloral hydrate to free radical(s) and induction of lipid peroxidation. *Biochem. biophys. Res. Commun.*, **204**, 937–943

Ni, Y.-C., Kadlubar, F.F. & Fu, P.P. (1995) Formation of malondialdehyde-modified 2′-deoxyguanosinyl adduct from metabolism of chloral hydrate by mouse liver microsomes. *Biochem. biophys. Res. Commun.*, 216, 1110–1117

Ni, Y.-C., Wong, T.-Y., Lloyd, R.V., Heinze, T.M., Shelton, S., Casciano, D., Kadlubar, F.F. & Fu, P.P. (1996) Mouse liver microsomal metabolism of chloral hydrate, trichloroacetic acid, and trichloroethanol leading to induction of lipid peroxidation *via* a free radical mechanism. *Drug Metab. Dispos.*, **24**, 81–90

Nutley, E.V., Tcheong, A.C., Allen, J.W., Collins, B.W., Ma, M., Lowe, X.R., Bishop, J.B., Moore, D.H., II & Wyrobek, A.J. (1996) Micronuclei induced in round spermatids of mice after stem-cell treatment with chloral hydrate: Evaluations with centromeric DNA probes and kineto-chore antibodies. *Environ. mol. Mutag.*, **28**, 80–89

Odum, J., Foster, J.R. & Green, T. (1992) A mechanism for the development of Clara cell lesions in the mouse lung after exposure to trichloroethylene. *Chem.-biol. Interact.*, **83**, 135–153

Ohlson, J. (1986) Dry etch chemical safety. *Solid State Technol.*, **July**, 69–73

O'Neil, M.J., ed. (2001) *The Merck Index*, 13th Ed., Whitehouse Station, NJ, Merck & Co, p. 355

Parry, J.M. (1993) An evaluation of the use of in vitro tubulin polymerisation, fungal and wheat assays to detect the activity of potential chemical aneugens. *Mutat. Res.*, **287**, 23–28

Parry, J.M., Parry, E.M., Warr, T., Lynch, A. & James, S. (1990) The detection of aneugens using yeasts and cultured mammalian cells. In: Mendelsohn, M.L. & Albertini, R.J., eds, *Mutation and the Environment. Part B, Metabolism, Testing Methods, and Chromosomes*, New York, Wiley-Liss, pp. 247–266

Parry, J.M., Parry, E.M., Bourner, R., Doherty, A., Ellard, S., O'Donovan, J., Hoebee, B., de Stoppelaar, J.M., Mohn, G.R., Önfelt, A., Renglin, A., Schultz, N., Söderpalm-Berndes, C., Jensen, K.G., Kirsch-Volders, M., Elhajouji, A., Van Hummelen, P., Degrassi, F., Antoccia, A., Cimini, D., Izzo, M., Tanzarella, C., Adler, I.-D., Kliesch, U., Schriever-Schwemmer, G., Gasser, P., Crebelli, R., Carere, A., Andreoli, C., Benigni, R., Leopardi, P., Marcon, F., Zinjo, Z., Natarajan, A.T., Boei, J.J.W.A., Kappas, A., Voutsinas, G., Zarani, F.E., Patrinelli, A., Pachierotti, F., Tiveron, C. & Hess, P. (1996) The detection and evaluation of aneugenic chemicals. *Mutat. Res.*, **353**, 11–46

Pharmacopeial Convention (1990) *The United States Pharmacopeia, 22nd rev./The National Formulary*, 17th rev., Rockville, MD, pp. 269–270

Poon, R., Nadeau, B. & Chu, I. (2000) Biochemical effects of chloral hydrate on male rats following 7-day drinking water exposure. *J. appl. Toxicol.*, **20**, 455–461

Rijhsinghani, K.S., Swerdlow, M.A., Ghose, T., Abrahams, C. & Rao, K.V.N. (1986) Induction of neoplastic lesions in the livers of $C_{57}BL \times C_3HF_1$ mice by chloral hydrate. *Cancer Detect. Prev.*, **9**, 279–288

Royal Pharmaceutical Society of Great Britain (2002) *Martindale, The Complete Drug Reference*, 33rd Ed., London, The Pharmaceutical Press [MicroMedex online]

Russo, A. & Levis, A.G. (1992a) Detection of aneuploidy in male germ cells of mice by means of a meiotic micronucleus assay. *Mutat. Res.*, **281**, 187–191

Russo, A. & Levis, A.G. (1992b) Further evidence for the aneuploidogenic properties of chelating agents: Induction of micronuclei in mouse male germ cells by EDTA. *Environ. mol. Mutag.*, **19**, 125–131

Russo, A., Pacchierotti, F. & Metalli, P. (1984) Nondisjunction induced in mouse spermatogenesis by chloral hydrate, a metabolite of trichloroethylene. *Environ. Mutag.*, **6**, 695–703

Russo, A., Stocco, A. & Majone, F. (1992) Identification of kinetochore-containing (CREST⁺) micronuclei in mouse bone marrow erythrocytes. *Mutagenesis*, **7**, 195–197

Sadtler Research Laboratories (1991) *Sadtler Standard Spectra, 1981–1991 Supplementary Index*, Philadelphia, PA

Sanders, V.M., Kauffmann, B.M., White, K.L., Jr, Douglas, K.A., Barnes, D.W., Sain, L.E., Bradshaw, T.J., Borzelleca, J.F. & Munson, A.E. (1982) Toxicology of chloral hydrate in the mouse. *Environ. Health Perspect.*, **44**, 137–146

Sandhu, S.S., Dhesi, J.S., Gill, B.S. & Svendsgaard, D. (1991) Evaluation of 10 chemicals for aneuploidy induction in the hexaploid wheat assay. *Mutagenesis*, **6**, 369–373

Sato, T., Mukaida, M., Ose, Y., Nagase, H. & Ishikawa, T. (1985) Chlorinated products from structural compounds of soil humic substances. *Sci. total Environ.*, **43**, 127–140

Sbrana, I., Di Sibio, A., Lomi, A. & Scarcelli, V. (1993) C-Mitosis and numerical chromosome aberration analyses in human lymphocytes: 10 known or suspected spindle poisons. *Mutat. Res.*, **287**, 57–70

Seelbach, A., Fissler, B. & Madle, S. (1993) Further evaluation of a modified micronucleus assay with V79 cells for detection of aneugenic effects. *Mutat. Res.*, **303**, 163–169

Simpson, K.L. & Hayes, K.P. (1998) Drinking water disinfection by-products: An Australian perspective. *Water Res.*, **32**, 1522–1528

Solt, K. & Johansson, J.S. (2002) Binding of the active metabolite of chloral hydrate, 2,2,2-trichloroethanol, to serum albumin demonstrated using tryptophan fluorescence quenching. *Pharmacology*, **64**, 152–159

Sora, S. & Agostini Carbone, M.L. (1987) Chloral hydrate, methylmercury hydroxide and ethidium bromide affect chromosomal segregation during meiosis of *Saccharomyces cerevisiae*. *Mutat. Res.*, **190**, 13–17

Trehy, M.L., Yost, R.A. & Miles, C.J. (1986) Chlorination byproducts of amino acids in natural waters. *Environ. Sci. Technol.*, **20**, 1117–1122

Uden, P.C. & Miller, J.W. (1983) Chlorinated acids and chloral in drinking water. *J. Am. Water Works Assoc.*, **75**, 524–527

Vagnarelli, P., De Sario, A. & De Carli, L. (1990) Aneuploidy induced by chloral hydrate detected in human lymphocytes with the Y97 probe. *Mutagenesis*, **5**, 591–592

Van Hummelen, P. & Kirsch-Volders, M. (1992) Analysis of eight known or suspected aneugens by the *in vitro* human lymphocyte micronucleus test. *Mutagenesis*, **7**, 447–455

Vian, L., Van Hummelen, P., Bichet, N., Gouy, D. & Kirsch-Volders, M. (1995) Evaluation of hydroquinone and chloral hydrate on the in vitro micronucleus test on isolated lymphocytes. *Mutat. Res.*, **334**, 1–7

Von Tungeln, L.S., Yi, P., Bucci, T.J., Samokyszyn, V.M., Chou, M.W., Kadlubar, F.F. & Fu, P.P. (2002) Tumorigenicity of chloral hydrate, trichloroacetic acid, trichloroethanol, malondialdehyde, 4-hydroxy-2-nonenal, crotonaldehyde, and acrolein in the $B6C3F_1$ neonatal mouse. *Cancer Lett.*, **185**, 13–19

Wallin, M. & Hartley-Asp, B. (1993) Effects of potential aneuploidy inducing agents on microtubule assembly *in vitro*. *Mutat. Res.*, **287**, 17–22

Warr, T.J., Parry, E.M. & Parry, J.M. (1993) A comparison of two in vitro mammalian cell cytogenetic assays for the detection of mitotic aneuploidy using 10 known or suspected aneugens. *Mutat. Res.*, **287**, 29–46

Waskell, L. (1978) A study of the mutagenicity of anesthetics and their metabolites. *Mutat. Res.*, **57**, 141–153

Weast, R.C. & Astle, M.J. (1985) *CRC Handbook of Data on Organic Compounds*, Volumes I and II, Boca Raton, Florida, CRC Press

WHO (1998) *Guidelines for Drinking-water Quality*, 2nd Ed., Vol. 2, *Health Criteria and Other Supporting Information*, Geneva (addendum to Vol. 2, pp. 281–283)

Williams, M. & Holladay, M.W. (1995) Hypnotics, sedatives, anticonvulsants, and anxiolytics. In: Kroschwitz, J.I. & Howe-Grant, M., eds., *Kirk-Othmer Encyclopedia of Chemical Technology*, 4th Ed., Vol. 13, New York, John Wiley & Sons, pp. 1082–1100

Williams, D.T., LeBel, G.L. & Benoit, F.M. (1997) Disinfection by-products in Canadian drinking water. *Chemosphere*, **34**, 299–316

Wilson, J.T. (1981) *Drugs in Breast Milk*, Lancaster, MTP

Xu, W. & Adler, I.-D. (1990) Clastogenic effects of known and suspect spindle poisons studied by chromosome analysis in mouse bone marrow cells. *Mutagenesis*, **5**, 371–374

Zimmermann, T., Wehling, M. & Schulz, H.-U. (1998) [The relative bioavailability and pharmacokinetics of chloral hydrate and its metabolites.] *Arzneim.-Forsch./Drug Res.*, **48**, 5–12 (in German)

Zordan, M., Osti, M., Pesce, M. & Costa, R. (1994) Chloral hydrate is recombinogenic in the wing spot test in *Drosophila melanogaster. Mutat. Res.*, **322**, 111–116

DICHLOROACETIC ACID

This substance was considered by a previous Working Group in February 1995 (IARC, 1995). Since that time, new data have become available, and these have been incorporated into the monograph and taken into consideration in the present evaluation.

1. Exposure Data

1.1   Chemical and physical data

1.1.1   *Nomenclature*

Chem. Abstr. Serv. Reg. No.: 79-43-6
Deleted CAS Reg. No.: 42428-47-7
Chem. Abstr. Name: Dichloroacetic acid
IUPAC Systematic Name: Dichloroacetic acid
Synonyms: DCA; DCA (acid); dichloracetic acid; dichlorethanoic acid; dichloroethanoic acid

1.1.2   *Structural and molecular formulae and relative molecular mass*

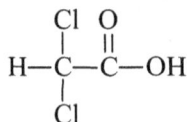

$$\underset{\underset{Cl}{|}}{\overset{\overset{Cl}{|}}{H-C-\overset{\overset{O}{\|}}{C}-OH}}$$

$C_2H_2Cl_2O_2$ Relative molecular mass: 128.94

1.1.3   *Chemical and physical properties of the pure substance*

(*a*)   *Description*: Colourless, highly corrosive liquid (Koenig *et al.*, 1986)
(*b*)   *Boiling-point*: 192 °C (Koenig *et al.*, 1986)
(*c*)   *Melting-point*: 13.5 °C (freezing-point) (Koenig *et al.*, 1986)
(*d*)   *Density*: 1.5634 at 20 °C/4 °C (Morris & Bost, 1991)

(e)   *Spectroscopy data*: Infrared (prism [2806]), nuclear magnetic resonance [166] and mass spectral data have been reported (Weast & Astle, 1985)

(f)   *Solubility*: Miscible with water; soluble in organic solvents such as alcohols, ketones, hydrocarbons and chlorinated hydrocarbons (Koenig *et al.*, 1986)

(g)   *Volatility*: Vapour pressure, 0.19 kPa at 20 °C (Koenig *et al.*, 1986)

(h)   *Stability*: Dissociation constant (K_a), 5.14×10^{-2} (Morris & Bost, 1991)

(i)   *Octanol/water partition coefficient (P)*: log P, 0.92 (Hansch *et al.*, 1995)

(j)   *Conversion factor*: $mg/m^3 = 5.27 \times ppm$[a]

1.1.4   Technical products and impurities

Dichloroacetic acid is commercially available as a technical-grade liquid with the following typical specifications: purity, 98.0% min.; monochloroacetic acid, 0.2% max.; trichloroacetic acid, 1.0% max.; and water, 0.3% max. (Clariant Corp., 2001; Clariant GmbH, 2002).

1.1.5   Analysis

Dichloroacetic acid has been determined in water using liquid–liquid extraction, conversion to its methyl ester and gas chromatography (GC) with electron capture detection. This method has been applied to drinking-water, groundwater, water at intermediate stages of treatment and raw source water, with a detection limit of 0.24 µg/L (Environmental Protection Agency, 1995; American Public Health Association/American Water Works Association/Water Environment Federation, 1999).

A similar method was used in 1993 in a national survey of chlorinated disinfection by-products in Canadian drinking-water. Methyl esters were analysed by GC–mass spectrometry with selected ion monitoring. The minimum quantifiable limit for this method was 0.01 µg/L (Health Canada, 1995; Williams *et al.*, 1997).

These methods were modified for an analytical survey of 16 drinking-water sources in Australia (Simpson & Hayes, 1998) and in a survey of treated water from 35 Finnish waterworks during different seasons (Nissinen *et al.*, 2002).

1.2   Production and use

Dichloroacetic acid was reported to be first synthesized in 1864 by the further chlorination of monochloroacetic acid with chlorine (Beilstein Online, 2002). Haloacetic acids were first detected in 1983 as disinfection by-products in chlorinated drinking-waters, 9 years after the discovery of trihalomethanes in chlorinated waters (Nissinen *et al.*, 2002).

---

[a] Calculated from: mg/m^3 = (molecular weight/24.45) × ppm, assuming normal temperature (25 °C) and pressure (760 mm Hg)

1.2.1 *Production*

The most common production method for dichloroacetic acid is the hydrolysis of dichloroacetyl chloride, which is produced by the oxidation of trichloroethylene. It can also be obtained by hydrolysis of pentachloroethane with 88–99% sulfuric acid or by oxidation of 1,1-dichloroacetone with nitric acid and air. In addition, dichloroacetic acid can be produced by catalytic dechlorination of trichloroacetic acid or ethyl trichloroacetate with hydrogen over a palladium catalyst (Koenig *et al.*, 1986; Morris & Bost, 1991).

Available information indicates that dichloroacetic acid is produced by two companies in the USA and one company each in China, Japan and Mexico (Chemical Information Services, 2002).

Available information indicates that dichloroacetic acid is formulated into pharmaceutical products by one company each in New Zealand and Turkey (Chemical Information Services, 2002).

1.2.2 *Use*

Dichloroacetic acid, particularly in the form of its esters, is an intermediate in organic synthesis, used in the production of glyoxylic acid, dialkoxy and diaroxy acids, and sulfonamides and in the preparation of iron chelates in the agricultural sector. It is also used as an analytical reagent in fibre manufacture (polyethylene terephthalate) and as a medicinal disinfectant (substitute for formalin) (Koenig *et al.*, 1986; Morris & Bost, 1991; Clariant GmbH, 2002).

Dichloroacetic acid is used in medical practice as a cauterizing agent. It rapidly penetrates and cauterizes the skin and keratins. Its cauterizing ability compares with that of electrocautery or freezing. It is used on calluses, hard and soft corns, xanthoma palpebrarum, seborrhoeic keratoses, in-grown nails, cysts and benign erosion of the cervix (Gennaro, 2000).

1.3 Occurrence

1.3.1 *Natural occurrence*

Dichloroacetic acid is not known to occur as a natural product.

1.3.2 *Occupational exposure*

The National Occupational Exposure Survey conducted between 1981 and 1983 indicated that 1592 employees in the USA were potentially exposed to dichloroacetic acid in 39 facilities (National Institute for Occupational Safety and Health, 1994). The estimate was based on a survey of companies and did not involve measurements of actual exposures.

1.3.3   *Air*

No data were available to the Working Group.

1.3.4   *Water*

Dichloroacetic acid is produced as a by-product during chlorination of water containing humic substances (Christman *et al.*, 1983; Miller & Uden, 1983; Legube *et al.*, 1985; Reckhow & Singer, 1990; Reckhow *et al.*, 1990). Consequently, it may occur in drinking-water after chlorine-based disinfection of raw waters containing natural organic substances (Hargesheimer & Satchwill, 1989; see IARC, 1991a) and in swimming pools (Stottmeister & Naglitsch, 1996; Kim & Weisel, 1998). The concentrations of dichloroacetic acid measured in various water sources are summarized in Table 1.

Geist *et al.* (1991) measured concentrations of dichloroacetic acid ranging from < 3 to 522 μg/L in surface water downstream from a paper mill in Austria, and Mohamed *et al.* (1989) measured concentrations ranging from 14 to 18 μg/L in effluent from a kraft pulp mill in Malaysia.

Levels of dichloroacetic acid tend to decline with length of time in the distribution system (Chen & Weisel, 1998), and concentrations tend to be higher in warmer seasons (LeBel *et al.*, 1997; Chen & Weisel, 1998). It has been identified as a major chlorinated by-product of the photocatalytic degradation of tetrachloroethylene in water and as a minor by-product of the degradation of trichloroethylene (Glaze *et al.*, 1993).

Dichloroacetic acid has also been detected in the Great Lakes, Canada (Scott *et al.*, 2002) and in fog samples (0.12–5.0 μg/L) at ecological research sites in north-eastern Bavaria, Germany (Römpp *et al.*, 2001). Clemens and Schöler (1992a) measured 1.35 μg/L dichloroacetic acid in rainwater in Germany. Precipitation samples in Canada contained dichloroacetic acid concentrations ranging from < 0.0004 to 2.4 μg/L and concentrations in Canadian lakes varied from < 0.0001 to 4.7 μg/L (Scott *et al.*, 2000). Dichloroacetic acid has a relatively short persistence (~ 4 days) in pond waters (Ellis *et al.*, 2001)

1.3.5   *Other*

Dichloroacetic acid has been reported as a biotransformation product of methoxyflurane (Mazze & Cousins, 1974) and dichlorvos (see WHO, 1989; IARC, 1991b), and it may occur in the tissues and fluids of animals treated with dichlorvos for helminthic infections (Schultz *et al.*, 1971).

Kim and Weisel (1998) measured the amount of dichloroacetic acid excreted by humans after having swum in a chlorinated pool, and this ranged from 25 to 960 ng per one urine void. The background excretion rate varied from 109 to 253 ng per one urine void. Weisel *et al.* (1999) measured an average excretion rate of 1.04 ng/min dichloroacetic acid in subjects exposed to low levels (1.76 μg/L) and 1.47 ng/min dichloroacetic acid in subjects exposed to high levels (32.7 μg/L) of dichloroacetic acid in water.

Table 1. Concentrations of dichloracetic acid in water

Water type (location)	Concentration range (µg/L)	Reference
Treatment plant and distribution system (Canada)	1.8–53.2	LeBel et al. (1997)
Treatment plant (Canada)	0.3–163.3	Williams et al. (1997)
Treatment plant and distribution system (Poland)	13.34–25.55	Dojlido et al. (1999)
Treatment plant and distribution system (South Korea)	0.8–5.2[a]	Shin et al. (1999)
Drinking-water (USA)[b]	63.1–133	Uden & Miller (1983)
Drinking-water (USA)[b]	5.0–7.3	Krasner et al. (1989)
Drinking-water (USA)[b]	4.7–23	Jacangelo et al. (1989)
Drinking-water (USA)[b]	8–79	Reckhow & Singer (1990)
Drinking-water (Spain)	ND–2.0[b]	Cancho et al. (1999)
Drinking-water and distribution system (USA)[b]	1–100	Obolensky et al. (2003)
Distribution system (Canada)	0.2–120.1	Williams et al. (1997)
Distribution system (USA)	1.7–14	Chen & Weisel (1998)
Distribution system (Australia)	1–46	Simpson & Hayes (1998)
Drinking-water (USA)	7.5–21	Lopez-Avila et al. (1999)
Drinking-water (USA)	0.33–110	Weisel et al. (1999)
Drinking-water (Canada)	8.1–12.7	Scott et al. (2000)
Drinking-water (Finland)	< 2.2–42	Nissinen et al. (2002)
Drinking-water (Spain)	0.2–21.5	Villanueva et al. (2003)
Swimming pool (Germany)	indoors: 0.2–10.6 open air: 83.5–181.0[c]	Clemens & Schöler (1992b)
Swimming pools (Germany)		Stottmeister & Naglitsch
Indoor	1.5–192	(1996)
Hydrotherapy	1.8–27	
Outdoor	6.2–562	
Swimming pool (USA)	52–647	Kim & Weisel (1998)

[a] Based on the assumption that dichloroacetic acid makes up approximately 40% of haloacetic acids
[b] Samples taken from water leaving the treatment plant
[c] The authors suggest that the higher levels found in open-air swimming pools may be due to greater input of organic matter.
ND, not detected

Dichloroacetic acid has been detected in spruce needles from the Black Forest in Germany and the Montafon region in Austria, both of which are considered to be relatively unpolluted areas, in the range of 10 to several hundred micrograms per kilogram (Frank et al., 1989).

1.4   Regulations and guidelines

The WHO (1998) has established a provisional guideline of 50 μg/L for dichloroacetic acid in drinking-water. A provisional guideline is established when there is some evidence of a potential health hazard but where available data on health effects are limited, or where an uncertainty factor greater than 1000 has been used in the derivation of the tolerable daily intake.

In Australia and New Zealand, the drinking-water guideline for dichloroacetic acid is 100 μg/L (National Health and Medical Research Council and Agriculture and Resource Management Council of Australia and New Zealand, 1996). This guideline also notes that minimizing the concentration of all chlorination by-products is encouraged by reducing naturally occurring organic material from the source water, reducing the amount of chlorine added, or using an alternative disinfectant, while not compromising disinfection.

In the USA, the Environmental Protection Agency (1998) regulates a combination of five haloacetic acids, which includes dichloroacetic acid together with monochloracetic acid, trichloroacetic acid and mono- and dibromoacetic acids. The maximum contaminant level for the sum of these five haloacetic acids is 60 μg/L.

The European Union (European Commission, 1998) and Canada (Health Canada, 2001) have not set guideline values, but encourage the reduction of total disinfection by-product concentrations.

2.   Studies of Cancer in Humans

See Introduction to the monographs on chloramine, chloral and chloral hydrate, dichloroacetic acid, trichloroacetic acid and 3-chloro-4-(dichloromethyl)-5-hydroxy-2(5*H*)-furanone.

3.   Studies of Cancer in Experimental Animals

Previous evaluation

In four studies evaluated previously, dichloroacetic acid induced hepatocellular adenomas and carcinomas in male $B6C3F_1$ mice, and studies in experimental animals have since focused on its activity in the liver. The previous evaluation of dichloroacetic acid concluded that there was limited evidence in experimental animals for its carcinogenicity (IARC, 1995).

New studies

3.1   Oral administration

3.1.1   *Mouse*

Groups of 90 female B6C3F₁ mice, 7–8 weeks of age, were administered dichloroacetic acid continuously in the drinking-water at concentrations of 2.0, 6.67 and 20.0 mmol/L [236, 854 or 2350 mg/L], adjusted to pH 6.5–7.5 with sodium hydroxide; a control group of 134 animals received 20.0 mmol/L sodium chloride. Mice were killed after 360 or 576 days of exposure. The livers were weighed and evaluated for foci of altered hepatocytes (eosino-philic and basophilic), adenomas and carcinomas. After 360 or 576 days of exposure, di-chloroacetic acid increased the ratio of liver-to-body weight dose-dependently. Data from mice administered dichloroacetic acid were compared with those from control mice using Fisher's exact test with a p-value < 0.05. At 360 days, only the high concentration of di-chloroacetic acid (20.0 mmol/L) induced a significant increase in the incidence of altered hepatocyte foci (8/20; 40%) and hepatocellular adenomas (7/20; 35%) compared with control values of 0/40 and 1/40 (2.5%), respectively. The multiplicity of hepatocyte foci and adenomas was increased from 0 and 0.03 ± 0.03 to 0.60 ± 0.22 and 0.45 ± 0.17, respectively. After 576 days of exposure, 6.67 mmol/L dichloroacetic acid increased the incidence of foci (11/28; 39.3%) and adenomas (7/28; 25%), while 20.0 mmol/L increased the incidence of foci (17/19; 89.5%), adenomas (16/19; 84.2%) and carcinomas (5/19; 26.3%). In control mice, the incidence of lesions was 1/40 adenomas at 360 days and 10/90 (11.1%) foci, 2/90 adenomas (2.2%) and 2/20 carcinomas (2.2%) at 576 days. An intermittent exposure group [number unspecified] was treated during 72-day cycles of 24 days of exposure to 20.0 mmol/L dichloroacetic acid followed by 48 days without exposure so that the time-weighted average concentration was equivalent to that of the 6.67-mmol/L group. Inter-mittent exposure to 20.0 mmol/L dichloroacetic acid resulted in a much lower incidence and multiplicity of hepatocellular adenomas than those in mice that received the same time-weighted average concentration of 6.67 mmol/L continuously (3/34 [8.8%] versus 7/28 [25%] and 0.09 ± 0.05 versus 0.32 ± 0.13). The dose–response relationship for total liver lesions per mouse did not appear to be linear for dichloroacetic acid (Pereira, 1996).

Groups of 25 female B6C3F₁ mice, 28 days of age, were given 0.5 or 3.5 g/L dichloro-acetic acid in the drinking-water adjusted to 6.9–7.1 with sodium hydroxide for 104 weeks; a group of 39 controls received 1.5% acetic acid. A complete necropsy was performed. The high dose of dichloroacetic acid decreased body weights of the mice and increased the liver-to-body weight ratio. The high dose, but not the low dose, of dichloroacetic acid increased the incidence and multiplicity of liver carcinomas. The incidence of hepatocellular carci-nomas was 2.6 (1/39), 4.0 (1/25) and 92% (23/25) and the multiplicity was 0.05 ± 0.32, 0.04 ± 0.20 and 2.96 ± 1.67 in control, low-dose and high-dose mice, respectively (Schroeder *et al.*, 1997).

Groups of male B6C3F₁ mice, 28–30 days of age, were given 0 (50 animals), 0.5 (50 animals), 1 (71 animals), 2 (55 animals) or 3.5 (46 animals) g/L dichloroacetic acid in the drinking-water adjusted to pH 6.9–7.1 with sodium hydroxide. Starting at a different time, other mice were exposed to 0 (33 animals) or 0.05 (50 animals) g/L dichloroacetic acid. The mice were killed between 90 and 100 weeks. Body, liver, kidney, testis and spleen weights were measured and livers, kidneys, spleens and testes were examined microscopically. No statistical differences with respect to tumour incidence or multiplicity were found between the control groups placed on study at different times or the two groups combined. Five animals from the high-dose group had a complete pathological examination. Animals that received the two higher doses of dichloroacetic acid (2 and 3.5 g/L) had significantly decreased body weights (18% versus control animals). At 1, 2 and 3.5 g/L, dichloroacetic acid increased the incidence and multiplicity of hepatocellular adenomas and carcinomas, while 0.5 g/L increased the multiplicity, but not the incidence of both lesions. After 90–100 weeks of treatment, the percentage of animals with hepatocellular adenomas was 10, 20, 51.4, 42.9 and 45% [p-value ≤ 0.05 for the three highest percentages] and that of animals with carcinomas was 26, 48, 71, 95 and 100% [p-value ≤ 0.05 for the three highest percentages] following administration of 0, 0.5, 1, 2 and 3.5 g/L dichloroacetic acid, respectively. [The numbers of animals that survived to 79–100 weeks and underwent histopathological examination were 50 (controls), 33 (0.05 g/L), 35 (0.5 g/L), 35 (1 g/L), 21 (2 g/L) and 11 (3 g/L) (personal communication from the author)] (DeAngelo et al., 1999).

The ability of mixtures of dichloroacetic acid and trichloroacetic acid to induce liver tumours was studied in 6-week-old male B6C3F$_1$ mice. Treatments administered included 0.1, 0.5 and 2.0 g/L dichloroacetic acid and 0.5 and 2.0 g/L trichloroacetic acid, and selected combinations of these treatments. Twenty animals were assigned to each of 10 groups that received the above concentrations in the drinking-water for 52 weeks. Dose-related increases in liver tumour incidence (adenomas and carcinomas combined) were observed with the individual compounds and, when the animals were exposed to mixtures of dichloroacetic acid and trichloroacetic acid, it appeared that there was an additive effect in terms of tumour incidence (see Table 2) (Bull et al., 2002).

3.1.2   Rat

Groups of 60 male Fischer 344/N rats, 28–30 days of age, received 0.05, 0.5 or 5 g/L dichloroacetic acid daily in the drinking-water for 100 weeks; a control group of 50 mice received 2 g/L sodium chloride for 104 weeks. In another experiment, a group of 78 males received a dose of 2.5 g/L dichloroacetic acid (neutralized) daily, which, due to toxicity, was lowered to 1.5 g/L at 8 weeks and 1.0 g/L at 26 weeks (time-weighted average daily concentration, 1.6 g/L) and a control group of 78 animals received deionized water. Rats were killed after 103 weeks of exposure. The highest concentration of dichloroacetic acid (5 g/L) caused severe, irreversible peripheral neuropathy and rats were killed at 60 weeks and excluded from the analysis. The neurotoxicity observed at 2.5 g/L abated when the dose was lowered to 1.0 g/L. The liver, kidneys, testes, thyroid, stomach, rectum, duodenum, ileum, jejunum, colon, urinary bladder and spleen were examined for gross lesions. Body, liver, kidney, testes and spleen weights were measured and the organs were exa-

Table 2. Effect on liver tumour incidence of administration of dichloroacetic acid (DCA) and trichloroacetic acid (TCA) to B6C3F₁ male mice in the drinking-water for 52 weeks

Treatment	Tumour incidence (adenomas and carcinomas)	Tumour multiplicity[a]
Control (drinking-water)	1/20	0.05 ± 0.0
0.1 g/L DCA	2/20	0.10 ± 0.07
0.5 g/L DCA	5/20[b]	0.35 ± 0.15[b]
2 g/L DCA	12/19[b]	1.7 ± 0.5[b]
0.5 g/L TCA	11/20[b]	0.70 ± 0.16[b]
2 g/L TCA	9/20[b]	0.60 ± 0.18[b]
0.1 DCA + 0.5 TCA g/L	9/20[b]	0.65 ± 0.22[b]
0.1 DCA + 2 TCA g/L	15/20[b]	1.3 ± 0.2[b]
0.5 DCA + 0.5 TCA g/L	13/19[b]	1.4 ± 0.3[b]
0.5 DCA + 2 TCA g/L	13/20[b]	1.5 ± 0.3[b]

From Bull *et al.* (2002)
[a] Total number of tumours divided by total number of animals
[b] Significantly different from control at $p < 0.05$

mined microscopically. Exposure to 1.6 g/L dichloroacetic acid reduced the terminal body weight to 73% that of the control value (308 ± 9 g versus 424 ± 4 g; $p < 0.05$) and reduced the relative liver and kidney weights. No other effect on body and organ weight was reported in this or the other treatment groups. A significant increase in the incidence of hepatocellular carcinomas was noted in animals administered 1.6 g/L dichloroacetic acid (21.4% [6/28] versus 3.0% [1.33]; p-value $\leq$ 0.05). Animals treated with 0.5 g/L dichloroacetic acid did not show a significant increase in the incidences of adenomas (5/29; 17.2%) or carcinomas (3/29; 10.3%), although the combined incidence of adenomas and carcinomas (7/19; 24.1%) was greater than that in controls (1/23; 4.4%) [p-value $\leq$ 0.05]. No liver lesions were found in rats exposed to 0.05 g/L dichloroacetic acid (DeAngelo *et al.*, 1996). [The data for the treatments with 2 g/L sodium chloride and 0.05 and 0.5 g/L dichloroacetic acid were published previously, but the incidences of adenoma and proliferative lesions were reported to be slightly higher (Richmond *et al.*, 1995).]

3.2 Administration with known carcinogens or modifying factors

Tumour-promotion studies

Groups of 10–40 female B6C3F₁ mice, 15 days of age, were initiated with an intraperitoneal injection of 25 mg/kg bw *N*-methyl-*N*-nitrosourea (MNU). At 49 days of age, the animals received 2.0, 6.67 or 20.0 mmol/L [256, 854 or 2560 mg/L] dichloroacetic acid adjusted to pH 6.5–7.5 with sodium hydroxide or 20.0 mmol/L sodium chloride as a control for the sodium salt in the drinking-water. At 31 weeks, administration of 20.0 mmol/L dichloroacetic acid in the drinking-water was stopped for 12 mice that were held untreated until 52 weeks. Some mice were killed after 31 weeks of exposure and the remainder after 52 weeks. Dichloroacetic acid did not significantly increase the incidence or multiplicity of hepatocellular adenocarcinomas. The high dose of dichloroacetic acid increased the incidence of hepatocellular adenomas in MNU-initiated mice from 0/10 to 5/10 (50%) at 31 weeks and from 7/40 (17.5%) to 19/26 (73.1%) at 52 weeks. The multiplicity of hepato-cellular adenomas was increased ($p < 0.01$) from 0.00 to 1.80 ± 0.83 and from 0.28 ± 0.11 to 3.62 ± 0.70 at 31 and 52 weeks, respectively. The two lower doses of dichloroacetic acid did not significantly increase tumour incidence or multiplicity. The high dose of dichloro-acetic acid also promoted MNU-initiated altered hepatocyte foci. The foci and tumours pro-moted by dichloroacetic acid were eosinophilic and contained glutathione S-transferase-π. When exposure to dichloroacetic acid was stopped after 31 weeks, the tumours that it had promoted appeared to regress (1.80 ± 0.83 at week 31 and 0.69 ± 0.26 at week 52) (Pereira & Phelps, 1996).

Combinations of dichloroacetic acid and trichloroacetic acid have been evaluated for tumour-promoting activity (Pereira *et al.*, 1997). Groups of 20–45 female B6C3F$_1$ mice, 15 days of age, were initiated with an intraperitoneal injection of 25 mg/kg bw MNU on day 15 of age. At 4 weeks of age, they received 0, 7.8, 15.6 or 25 mmol/L dichloroacetic acid in the drinking-water with or without 6.0 mmol/L trichloroacetic acid or 25 mmol/L trichloroacetic acid with or without 15.6 mmol/L dichloroacetic acid. The pH of the dose solutions was adjusted to 6.5–7.5 with sodium hydroxide. Treatment was continued up to 48 weeks of age, at which time the mice were killed. The high dose of dichloroacetic acid (25 mmol/L) or trichloroacetic acid (25 mmol/L) significantly ($p < 0.05$) increased the multiplicity of hepatocellular adenomas from 0.07 ± 0.05 (no dichloroacetic acid or trichloroacetic acid) to 1.79 ± 0.29 and 0.52 ± 0.11, respectively. The lower doses of dichloroacetic acid and trichloroacetic acid did not significantly increase the incidence or multiplicity of adenomas. Although the combination containing 25 mmol/L dichloroacetic acid and 6.0 mmol/L trichloroacetic acid produced a less than synergistic increase in the liver-to-body weight ratio, it produced an additive increase in the multiplicity of hepato-cellular adenomas. The multiplicity of adenomas in mice treated with the combination was 2.33 ± 0.42 compared with 1.79 ± 0.29 and 0.15 ± 0.08 for dichloroacetic acid and trichloroacetic acid alone. The results following the combination containing 15.6 mmol/L dichloroacetic acid and 25 mmol/L trichloroacetic acid were also consistent with an addi-tive increase in the multiplicity of adenomas: 0.32 ± 0.11 (15.6 mmol/L dichloroacetic acid), 0.53 ± 0.20 (25.0 mmol/L trichloroacetic acid) and 0.52 ± 0.11 (combination of di-chloroacetic acid and trichloroacetic acid).

The effect of chloroform on the promotion of liver and kidney tumours by dichloro-acetic acid and trichloroacetic acid was evaluated in groups of 6–29 male and female B6C3F$_1$ mice initiated at 15 days of age with an intraperitoneal injection of 30 mg/kg bw MNU. At 5 weeks of age, the mice received 0 or 3.2 g/L dichloroacetic acid with 0, 800 or 1600 mg/L chloroform in the drinking-water. The pH of the dose solutions was neutralized with sodium hydroxide. The concentrations of chloroform were chosen because they prevented dichloroacetic acid-induced DNA hypomethylation and increased mRNA expression of the c-*myc* gene. The mice were killed at 36 weeks of age. The results were analysed for statistical significance by a one-way ANOVA followed by the Tukey test with a *p*-value < 0.05. In MNU-initiated mice that did not receive dichloroacetic acid, liver adenomas were found in 2/29 (6.9%) females and 2/8 males (25%). No hepatocellular adenocarcinomas were found. Dichloroacetic acid increased the incidence of liver adenomas (17/24 [70.8%] females, 21/25 [84.0%] males). The multiplicity of liver tumours (adenomas plus adenocarcinomas) was increased by dichloroacetic acid from 0.25 ± 0.16 to 3.17 ± 0.76 tumours per mouse (females) and from 0.07 ± 0.04 to 3.92 ± 0.54 (males). In mice administered dichloroacetic acid, co-administration of chloroform decreased the incidence of adenomas from 17/24 (70.8%) to 2/8 (25%; 800 mg/L) and 0/6 (0%; 1600 mg/L) in females and from 21/25 (84.0%) to 7/12 (58.3%; 1600 mg/L) in males. The multiplicity of tumours was also decreased by chloroform to 0.8 and 0 for doses of 800 and 1600 mg/L in females and to 1.1 for the dose of 1600 mg/L in males. No altered hepatocyte foci, adenomas or adenocarcinomas were found in six MNU-initiated male mice that were administered 1600 mg/L chloroform. Renal tumours of tubular origin were found in male mice. The majority (> 70%) were papillary cystic adenomas and the rest were cystic adenomas and, to a lesser extent, adenocarcinomas (~5%). No kidney tumours were found in the eight MNU-initiated mice. In MNU-initiated mice that received dichloroacetic acid, a significant increase in the incidence or multiplicity of kidney tumours was observed: 24% (6/25) with tumours and 0.28 ± 0.11 tumours per mouse. However, co-administration of 1600 mg/L chloroform with dichloroacetic acid significantly increased the incidence of kidney tumours in male mice to 100% (12/12) and the multiplicity to 1.75 ± 0.39 tumours per mouse (*p*-value < 0.01). No kidney tumours were found in the six MNU-initiated male mice administered 1600 mg/L chloroform alone. In female mice, the incidence of kidney tumours in all treatment groups was not significantly increased and ranged from 0 to 28.6% and the multiplicity ranged from 0 to 0.29 tumours per mouse. Hence, dichloroacetic acid promoted kidney tumours only in mice that were also administered chloroform (Pereira *et al.*, 2001).

4. Other Data Relevant to an Evaluation of Carcinogenicity and its Mechanisms

4.1 Absorption, distribution, metabolism and excretion

4.1.1 *Humans*

The disposition of dichloroacetic acid (10 or 20 mg/kg bw given intravenously) was studied in four human volunteers (26, 38, 42 and 52 years old [sex not stated]). In subjects given 10 mg/kg bw dichloroacetic acid, the average plasma half-life was 0.34 h (range,

0.33–0.36 h), the average volume-of-distribution was 337 mL/kg (range, 308–366 mL/kg) and the average plasma clearance was 11.31 mL/min/kg (range, 10.86–11.76 mL/min/kg); in subjects given 20 mg/kg bw dichloroacetic acid, the average plasma half-life was 0.51 h (range, 0.41–0.61 h), the average volume-of-distribution was 190 mL/kg (range, 186–195 mL/kg) and the average plasma clearance was 4.55 mL/min/kg (range, 3.53–5.58 mL/min/kg) (Lukas et al., 1980).

Dichloroacetic acid was given intravenously (50 mg/kg bw) or orally (50 mg/kg bw or 50 mg/kg bw plus 50 mg thiamine [vitamin B_1]) to healthy human volunteers (eight men and four women, aged 18–45 years), and a range of pharmacokinetic parameters were measured (Curry et al., 1991). On average, there was no evidence of an effect of vitamin B_1 on the kinetics of dichloroacetic acetic. The average plasma half-life was 2.7 ± 0.4 h, the average volume of distribution was 19.9 ± 1.7 L, the average area-under-the-curve (AUC) was 608.9 ± 61.0 µg/h/mL and the average renal clearance was 53.0 ± 15.9 mL/h. There was no difference in the AUC or elimination half-life between men and women. Only 0.7 ± 0.5% dichloroacetic acid was excreted unchanged in the urine. Urinary excretion of oxalic acid was similar after oral or intravenous administration of dichloroacetic acid (2.1 ± 0.8 mg versus 2.3 ± 0.5 mg). However, the elimination half-life was markedly prolonged after a second dose of dichloroacetic acid (Curry et al., 1991).

The pharmacokinetics of dichloroacetic acid has been studied in humans with a range of diseases. In children (four boys and four girls, aged 1.5–10 years) with lactic acidosis due to severe malaria, dichloroacetic acid given intravenously at a dose of 50 mg/kg bw showed an average plasma half-life of 1.8 ± 0.4 h, a volume of distribution of 0.32 ± 0.09 L/kg and an average AUC of 378 ± 65 mg/L/h (Krishna et al., 1995).

Two studies were conducted on the pharmacokinetics of dichloroacetic acid in patients with severe falciparum malaria. In one study that included 13 adults ([sex not stated]; average age, 27 ± 8 years) who were given 46 mg/kg bw dichloroacetic acid intravenously over 30 min, the elimination half-life was 2.3 ± 1.8 h, the clearance was 0.32 ± 0.16 L/h/kg and the volume of distribution was 0.75 ± 0.35 L/kg (Krishna et al., 1994). In a second study, 11 adults (eight men and three women; average age, 32 ± 10 years) were given 46 mg/kg bw dichloroacetic acid intravenously and a second dose (46 mg/kg bw) was given 12 h later. The mean plasma half-life was 3.4 ± 2 h after the first dose and 4.4 ± 2 h after the second dose, the volume of distribution was 0.44 ± 0.2 L/kg and the plasma clearance was 0.13 ± 0.03 L/h/kg (Krishna et al., 1996).

The effect of end-stage liver disease and liver transplantation on the pharmacokinetics of dichloroacetic acid was studied in 33 subjects [sex and age not stated] who were given 40 mg/kg bw dichloroacetic acid by a 60-min intravenous perfusion, then a second dose (40 mg/kg bw) by intravenous perfusion 4 h later, before and during the anhepatic stage. The clearance of dichloroacetic acid during the paleohepatic, anhepatic and neohepatic stages was 1.0, 0.0 and 1.7 mL/kg/min, respectively, indicating a major role of the liver in the metabolism of dichloroacetic acid (Shangraw & Fisher, 1996). The effect of cirrhosis on the pharmacokinetics of dichloroacetic acid was reported in six healthy volunteers (five men and one woman, 30 ± 3 years old) and seven subjects with end-stage cirrhosis (five men and two women, 47 ± 3 years old) who were given 35 mg/kg bw dichloroacetic acid by intravenous perfusion over 30 min. The clearance of dichloroacetic acid was 2.14 mL/kg/min in control subjects and 0.78 mL/kg/min in patients with cirrhosis (Shangraw & Fisher, 1999).

The pharmacokinetics of dichloroacetic acid was studied in 111 patients (66 men, 56.0 ± 18.4 years old) with lactic acidosis, who received dichloroacetic acid (50 mg/kg bw) by intravenous perfusion over 30 min, then a second perfusion of 50 mg/kg bw 2 h after the beginning of the first. The pharmacokinetics was complex in the acutely ill patients studied and differed markedly from those observed in healthy volunteers. In healthy volunteers, the pharmacokinetics fitted a one-compartment model, whereas in the patients the data fitted one-, two- and three-compartment models. In the two-compartment model, the plasma half-life and plasma clearance were 18.15 ± 3.12 h (mean $\pm$ standard error [SE]) and 0.041 L/kg/h, respectively, after the first treatment, whereas the two values were 68.30 ± 14.50 h (mean $\pm$ SE) and 0.017 L/kg/h, respectively, after the second treatment. Plasma clearance of dichloroacetic acid tended to decrease as either the number of compartments or the number of treatments increased. The prolonged half-life and decreased plasma clearance indicate that repeated administration of dichloroacetic acid impairs its metabolism (Henderson *et al.*, 1997).

The pharmacokinetics of dichloroacetic acid was compared in healthy volunteers (27 subjects) and in patients with traumatic brain injury (25 subjects; average age, 52.8 ± 18.1 years). The healthy volunteers were given cumulative intravenous doses (two doses 8 h apart) of 45, 90 or 150 mg/kg bw dichloroacetic acid; 16 patients with acute traumatic brain injury were given 60, 100 or 200 mg/kg bw dichloroacetic acid as a single intravenous dose; six other patients were given three intravenous doses [dose not stated] of dichloroacetic acid at 24-h intervals; and three patients were given six intravenous doses [dose not stated] at 12-h intervals. The initial clearance of dichloroacetic acid (4.82 L/h) declined (1.07 L/h) after repeated doses in patients with traumatic brain injury. Although the authors suggested several mechanisms by which the clearance of dichloroacetic acid might be decreased after repeated doses, they proposed that the enzyme responsible for the metabolism of dichloroacetic acid might be destroyed after repeated treatment with the compound (Williams *et al.*, 2001).

4.1.2 *Experimental systems*

The pharmacokinetics of dichloroacetic acid was studied in rats and dogs. Three male Sprague-Dawley rats and two male beagle dogs were given 100 mg/kg bw dichloroacetic acid intravenously; the average plasma elimination half-life was 2.97 h (range, 2.1–4.4 h) in rats and 20.8 h (range, 17.1–24.6 h) in dogs; the average volume of distribution was 932 mL/kg (range, 701–1080 mL/kg) in rats and 256 mL/kg (range, 249–262 mL/kg) in dogs; and the average plasma clearance was 4.22 mL/min/kg (range, 1.84–5.94 mL/min/kg) in rats and 0.146 mL/min/kg (range, 0.123–0.168 mL/min/kg) in dogs (Lukas *et al.*, 1980).

The disposition and elimination kinetics of [^{14}C]dichloroacetic acid were studied in male Fischer 344 rats and B6C3F$_1$ mice. In rats given 5, 20 or 100 mg/kg bw [^{14}C]dichloroacetic acid by gavage, 23.9–29.3% of the dose was eliminated as carbon dioxide and 19.6–24.4% was excreted in the urine. Only 1.0–2.2% of the dose was excreted unchanged in the urine. The major urinary metabolites in rats were glyoxylic, oxalic and glycolic acids, which amounted to 10.5–15.0% of the dose, and thiodiacetic acid amounted to 6.3–6.8% of the dose. In mice given 20 and 100 mg/kg bw dichloroacetic acid by gavage, 2.2 and 2.4% of the dose was eliminated as carbon dioxide and 2.2–2.3% was excreted as unchanged dichloroacetic acid. The major urinary metabolites of dichloroacetic acid in mice were glyoxylic, oxalic and glycolic acids and thiodiacetic acid, which amounted to 13.4–18.4% and 7.9–12.3% of the dose, respectively (Larson & Bull, 1992).

The adduction of haemoglobin and albumin by metabolites of dichloroacetic acid was investigated in male Fischer 344 rats and male B6C3F$_1$ mice given 5 mg/kg bw [1,2-^{14}C]-dichloroacetic acid orally. Adduct levels were determined by measuring the difference between total level incorporated and the fraction of the label present in serine and glycine. In rats, haemoglobin and albumin adducts amounted to 3–9 and 177–228 pmol equivalents/mg protein, respectively, whereas in mice, these adducts ranged from undetectable to < 1 and 65–152 pmol equivalents/mg protein, respectively (Stevens et al., 1992).

The effect of prior treatment with dichloroacetic acid on its pharmacokinetics was studied in male Fischer 344 rats. Animals were given unlabelled dichloroacetic acid (0.2 or 2.0 g/L) in the drinking-water for 14 days before receiving [1,2-^{14}C]dichloroacetic acid intravenously (bolus dose of 5, 20 or 100 mg/kg bw) or orally by gavage (100 mg/kg bw). In pretreated rats, the half-life of dichloroacetic acid after intravenous administration was significantly increased (10.8 versus 2.4 h) and the total body clearance was decreased (42.7 versus 267.4 mL/h/kg) compared with controls. The fraction of [1,2-^{14}C]dichloroacetic acid (100 mg/kg bw given orally or intravenously) excreted as carbon dioxide appeared to be decreased in pretreated rats. The urinary excretion of dichloroacetic acid, glycolic acid, glyoxylic acid, chloroacetic acid and acetic acid was significantly increased in rats pre-treated with 2.0 g/L dichloroacetic acid for 14 days before receiving the labelled compound (the excretion of oxalic acid remained unchanged). The in-vitro biotransformation of [1,2-^{14}C]dichloroacetic acid to glyoxylic, oxalic and glycolic acids required glutathione for maximal activity and was decreased in rat liver cytosol isolated from rats pretreated with dichloroacetic acid (Gonzalez-Leon et al., 1997).

These data were reanalysed and extended in another study. Rats were administered dichloroacetic acid (65 mg/kg bw) via jugular vein cannula and whole blood and plasma concentrations of the parent compound were determined. The unbound fraction of dichloro-acetic acid in plasma amounted to 0.94% and its renal clearance (corrected for plasma protein binding) was 3.1 mL/h/kg (Schultz et al., 1999).

The pharmacokinetics and oral bioavailability of dichloroacetic acid were studied in control naive male Fischer 344 rats administered 1–20 mg/kg bw and in glutathione S-trans-ferase (GST)-zeta (GSTZ1-1)-depleted male Fischer 344 rats administered 0.05–20 mg/kg bw intravenously or by gavage (Saghir & Schultz, 2002). GSTZ1-1 activity was depleted by exposing rats to 0.2 g/L dichloroacetic acid in drinking-water for 7 days before pharmaco-kinetic studies (dichloroacetic acid being an inhibitor of the GSTZ1-1 enzyme). The authors also compared the in-vitro metabolism of dichloroacetic acid in human liver cytosol with that in cytosol obtained from naive and GSTZ1-1-depleted rats. The half-life for the elimination of dichloroacetic acid from the plasma was dose-dependent and increased with increasing dose of dichloroacetic acid; it was also increased in GSTZ1-1-depleted rats given dichloroacetic acid in the drinking-water. The oral bioavailability of dichloroacetic acid was 0–13% in control naive rats and 14–75% in GSTZ1-1-depleted rats. The authors predicted that the human oral bioavailability of dichloroacetic acid in drinking-water would be low (< 1%). The intrinsic metabolic clearance (Cl_{int}) from control naive rat liver cytosol was 3.86 mL/h/mg protein compared with 0.25 mL/h/mg protein in GSTZ1-1-depleted rats; the Cl_{int} from human liver cytosol was not significantly different from that observed in the GSTZ1-1-depleted rats. The authors concluded that exposure of rodents to a high level of dichloroacetic acid depletes GSTZ1-1 activity, causing the pharmacokinetics in rats to become comparable to that in humans (Saghir & Schultz, 2002).

The metabolic fate of dichloroacetic acid was investigated in male Fischer 344 rats given either 282 mg/kg bw [1-^{14}C]dichloroacetic acid or 282 mg/kg bw [2-^{14}C]dichloroacetic acid or 28.2 mg/kg [2-^{14}C]dichloroacetic acid by gavage. The disposition of [1-^{14}C]- or [2-^{14}C]dichloroacetic acid at 282 mg/kg bw was similar, except that the fraction of the dose eliminated through expiration as carbon dioxide decreased (25.0 versus 34.4%) whereas the fraction of the administered radioactivity excreted in the urine increased (35.2 versus 12.7%) as the dose of [2-^{14}C]dichloroacetic acid was increased from 28.2 to 282 mg/kg bw. Most of the increase in excreted radioactivity was attributable to unchanged dichloroacetic acid. The major urinary metabolites identified were glycolic acid, glyoxylic acid and oxalic acid. The fraction of administered radioactivity measured in tissues 48 h after treatment ranged from 20.8% to 36.4%; liver and muscle contained the most radioactivity (Lin *et al.*, 1993).

The pharmacokinetics and metabolic fate of dichloroacetic acid were studied in male Sprague-Dawley rats given 50 mg/kg bw dichloroacetic acid containing 280–400 μCi/kg bw [1-^{14}C]dichloroacetic acid, 50 mg/kg bw [1,2-^{13}C]dichloroacetic acid or a mixture of [1-^{14}C]- and [1,2-^{13}C]dichloroacetic acid (1:99) by gavage as sodium dichloroacetate in water. The plasma elimination half-life of dichloroacetic acid was 0.11 ± 0.02 or 5.38 ± 0.76 h in rats given one or two doses of dichloroacetic acid, respectively. The fraction of the dose eliminated as carbon dioxide ranged from 17 to 46% of the dose. Oxalic acid, glyoxylic acid and the glycine conjugates hippuric acid and phenylacetylglycine were identified as metabolites; the glycine conjugates apparently arise from the transamination of glyoxylic acid to give glycine followed by conjugation with benzoic acid or phenylacetic acid. The fraction of the dose excreted in urine as unchanged dichloroacetic acid ranged from 0.36 to 20.2%, depending on the size of the rats and on whether they were fed or fasted (James *et al.*, 1998). The metabolic fate of dichloroacetic acid in rats is shown in Figure 1.

The kinetics of the biotransformation of dichloroacetic acid was studied in livers isolated from male Fischer 344 rats and perfused with Krebs-Ringer buffer containing bovine serum albumin, glucose and taurocholate and an initial concentration of 25 or 250 μM (3.2

Figure 1. Metabolic fate of dichloroacetic acid in rats

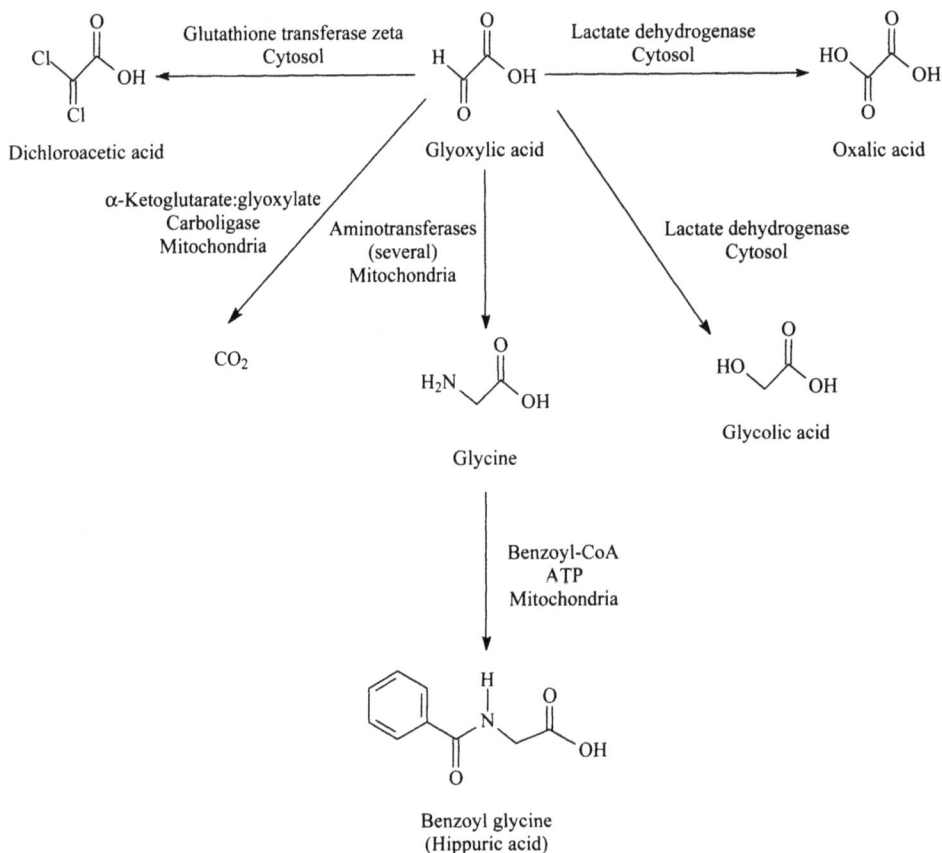

Modified from James *et al.* (1998)

or 32 µg/mL). In livers perfused with 250 µM (32 µg/mL), the free dichloroacetic acid concentration after 5 min of perfusion was 112 µM (14.5 µg/mL) or 47% of the total concentration, indicating binding to bovine serum albumin, whereas at the end of the 120-min perfusion, the concentration of free dichloroacetic acid was 7 µM (0.9 µg/mL) or 41% of the total concentration present at the end of the perfusion. A total of 0.2% of the total concentration of dichloroacetic acid was excreted in the bile during the 120-min perfusion. The half-life and the elimination rate constant for dichloroacetic acid were 32 min and 0.022/min, respectively (Toxopeus & Frazier, 1998).

The in-vitro degradation of dichloroacetic acid followed the order: liver >> lung > kidney > intestine ≈ muscle in mouse $700 \times g$ supernatant fractions (cytosol, microsomes and mitochondria). The highest rate of degradation of dichloroacetic acid was found in liver cytosol, and reduced nicotinamide adenine dinucleotide phosphate (NADPH) and oxygen were not required for maximal activity in microsomes. The degradation of dichloroacetic acid was decreased when diethyl maleate or chlorodinitrobenzene, which deplete gluta-thione (GSH) concentrations, was included in the reaction mixture and increased when GSH was added. Because the enzyme that catalysed the degradation of dichloroacetic acid was not retained on a GSH–sepharose column, it was concluded that a GST was not involved; carbon monoxide and SKF525-A, inhibitors of cytochrome P450 activity, did not inhibit the degradation of dichloroacetic acid, indicating that cytochromes P450 played no role in this degradation (Lipscomb *et al.*, 1995).

The proposed formation of glyoxylic acid as an intermediate in the metabolism of di-chloroacetic acid to oxalic acid, carbon dioxide and glycine was confirmed by the identi-fication of [1-^{14}C]- and [1,2-^{13}C]glyoxylic acid as metabolites of [1-^{14}C]- and [1,2-^{13}C]di-chloroacetic acid. The enzymes that catalysed this biotransformation were located in hepatic cytosolic fractions isolated from male Sprague-Dawley rats and from humans and required GSH, but not NADPH or NADH, for maximal activity. The rate of biotransfor-mation to glyoxylic acid was decreased in hepatic cytosolic fractions from rats pretreated with 50 mg/kg bw dichloroacetic acid for 2 days compared with rats given water (James *et al.*, 1997).

A GSH-dependent enzyme that catalysed the oxygenation of dichloroacetic acid to glyoxylic acid as the sole product was purified to homogeneity from male Fischer 344 rat liver (Tong *et al.*, 1998a). Antibodies to human GSTZ cross-reacted with the rat liver enzyme, thereby identifying it as the rat orthologue of human GSTZ1-1 (hGSTZ1-1), which was identified by interrogating expressed sequence tag databases (Board *et al.*, 1997). hGSTZ1-1 is identical to maleylacetoacetate isomerase, which catalyses the penultimate step in the degradation of phenylalanine and tyrosine (Fernández-Cañón & Peñalva, 1998). Rat liver GSTZ1-1 catalysed the biotransformation of [2-^{13}C]dichloroacetic acid to [2-^{13}C]glyoxylic acid, which was identified by ^{13}C nuclear magnetic resonance spectroscopy. GSH was required for the GSTZ-catalysed oxygenation of dichloroacetic acid metabolism, but it was neither consumed nor oxidized; it was completely recovered from the incubation mixtures, and no formation of GSH disulfide was associated with oxygenation of dichloro-acetic acid, as compared with controls. Immunoblotting with anti-hGSTZ1-1 antibodies demonstrated the presence of immunoreactive GSTZ in rat, mouse and human hepatic cyto-solic fractions. Kinetic studies with human, rat (male Fischer 344) and mouse (male B6C3F$_1$) liver cytosolic fractions with dichloroacetic acid as the variable substrate showed that the V_{max}/K_m followed the order human < rat < mouse. These data identified GSTZ1-1 as the cytosolic enzyme that catalyses the biotransformation of dichloroacetic acid to glyo-xylic acid (Tong *et al.*, 1998b).

Immunohistochemical studies on the localization of GSTZ1-1 in male Fischer 344 rat tissues reported intense staining in the liver, testis and prostate, and moderate-to-sparse staining in a range of tissues. Hepatic GSTZ1-1 activities with maleylacetone or chloro-fluoroacetic acid as substrates were markedly decreased after administration of 1.2 mmol/kg (154.8 mg/kg bw) dichloroacetic acid intraperitoneally for 5 days, but no change in residual GSTZ1-1 activities in the testis and other tissues was observed. The tissue-dependent differences in activities with both substrates reflected the pattern of expression of GSTZ1-1 observed by immunohistochemistry (Lantum *et al.* 2002).

The biotransformation of dichloroacetic acid to glyoxylic acid in rat liver cytosol (male Fischer 344 rats) was not linear with time and reached a plateau after 20–30 min of incubation; the reaction proceeded at the initial rate when a second dose of rat liver cytosol was added, whereas the addition of a second dose of dichloroacetic acid did not result in further glyoxylic acid formation, indicating that dichloroacetic acid inactivated GSTZ1-1. In rats given 0.3 mmol/kg (38.7 mg/kg) bw dichloroacetic acid for 12 days, GSTZ1-1 activity with dichloroacetic acid as the substrate reached a nadir 12 h after the first dose and did not return to control values until 10–12 days after treatment. The loss of GSTZ1-1 activity was paralleled by the loss of immunoreactive GSTZ1-1 protein. Data on dichloroacetic acid-induced inactivation of GSTZ1-1 were used to determine the rate of turnover of GSTZ *in vivo*. GSTZ1-1 was degraded in rat liver at the rate of –0.21/day, which corresponded to a half-life of 3.3 days (Anderson *et al.*, 1999).

The kinetics of in-vitro inactivation of mouse (male B6C3F$_1$), rat (male Fischer 344) and human (male) GSTZ1-1 and of recombinant human GSTZ1-1 (hGSTZ1-1) by dichloroacetic acid was studied. The half-life for dichloroacetic acid-induced inactivation of GSTZ1-1 in mouse, rat and human liver cytosol was 6.61, 5.44 and 22 min, respectively (Tzeng *et al.*, 2000). Four polymorphic variants of hGSTZ1-1 have been identified (Blackburn *et al.*, 2000, 2001). The half-lives for dichloroacetic acid-induced inactivation of hGSTZ1a-1a, -1b-1b, -1c-1c and -1d-1d were 23, 9.6, 10.1 and 9.5 min, respectively (Tzeng *et al.*, 2000). The inactivation of hGSTZ1c-1c was accompanied by a covalent modification of the enzyme: when [1-^{14}C]dichloroacetic acid was incubated with hGSTZ1c-1c in the presence of GSH, [1-^{14}C]dichloroacetic acid-derived radioactivity was irreversibly bound to the protein; similarly, when [^{35}S]GSH was incubated with hGSTZ1c-1c in the presence of dichloroacetic acid, [^{35}S]GSH-derived radioactivity was irreversibly bound to the protein. These data show that dichloroacetic acid-induced inactivation of hGSTZ1c-1c is accompanied by the covalent modification of the protein by both dichloroacetic acid and GSH (Tzeng *et al.*, 2000). Additional studies showed that the partition ratio (turnover number) for dichloroacetic acid-induced, mechanism-based inactivation of hGSTZ1c-1c was $5.7 \pm 0.5 \times 10^2$ (Anderson *et al.*, 2002). hGSTZ1c-1c, which is modified covalently by both dichloroacetic acid and GSH, has been characterized by matrix-assisted laser-desorption-ionization time-of-flight mass spectrometry (MALDI–TOF–MS) and by liquid chromatography–MS (LC–MS) (Anderson *et al.*, 2002). The stoichiometry of dichloroacetic acid binding to hGSTZ1c-1c was ~0.5 mol dichloroacetic acid/mol enzyme monomer. A single dichloroacetic acid-derived adduct was observed and was assigned to cysteine-16 by a combination of MALDI–TOF–MS and electrospray ionization quadrupole ion-trap LC–MS analyses, and by analysis of [1-^{14}C]dichloroacetic acid binding to hGSTZS1c-1c. The dichloroacetic acid-derived adduct with hGSTZ1c-1c contained both GSH and the carbon skeleton of dichloroacetic acid, presumably in a dithioacetal linkage. Hence, the mechanism of biotransformation of dichloroacetic acid to glyoxylic acid and of inactivation of GSTZ1-1 involves the displacement of chloride from dichloroacetic acid to give *S*-(α-chlorocarboxymethyl)GSH, which may react with cysteine-16 of hGSTZ1c-1c to give the covalently modified protein, or with water to give glyoxylic acid and GSH (Figure 2) (Anderson *et al.*, 2002).

Figure 2. Mechanisms of the hGSTZ1-1-catalysed biotransformation of dichloro-acetic acid to glyoxylic acid and of the mechanism-based inactivation of hGSTZ1-1 by dichloroacetic acid

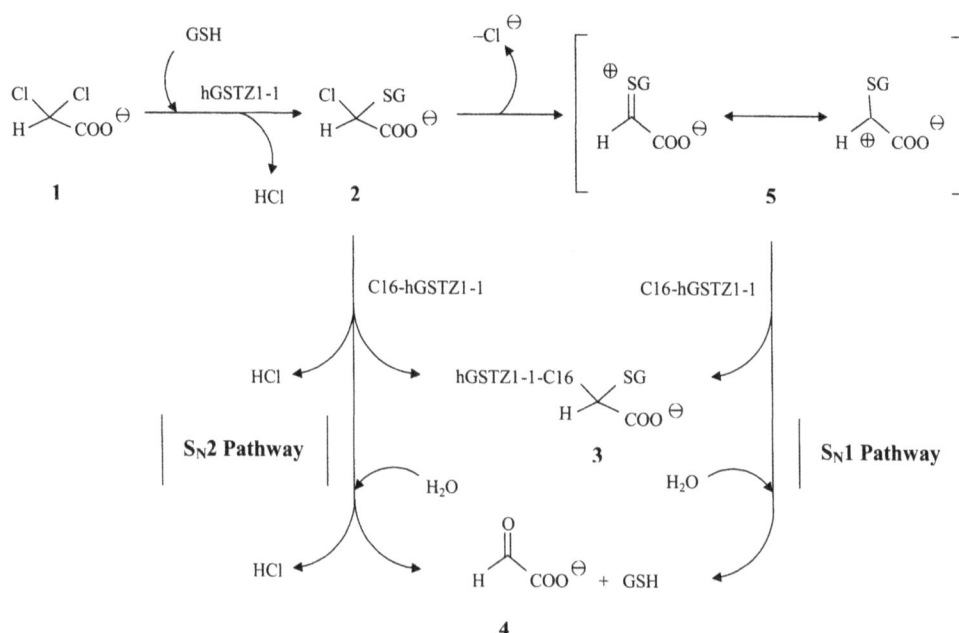

Modified from Anderson *et al.* (2002)
1, dichloroacetic acid; 2, *S*-(α-chlorocarboxymethyl)glutathione; 3, hGSTZ1-1 covalently modified at cysteine-16; 4, glyoxylic acid; 5, sulfonium-carbocation intermediate

4.1.3   *Comparison of humans and animals*

The kinetics of elimination of dichloroacetic acid in humans and rats is quantitatively similar: the half-lives range from 0.3 to 3.5 h. The half-life of dichloroacetic acid in dogs is 20.8 h (Lukas *et al.*, 1980). In both humans and rats, the half-life of dichloroacetic acid is prolonged by prior treatment with the compound. Studies with expressed recombinant hGSTZ1-1 showed that dichloroacetic acid is a mechanism-based inactivator of the enzyme, which explains the observed change in half-life in rodents and humans caused by prior treatment with the compound (Tzeng *et al.*, 2000; Anderson *et al.*, 2002).

4.2   Toxic effects

4.2.1   *Humans*

The pharmacological and toxic effects of dichloroacetic acid in humans have been studied extensively because of its potential use in the treatment of various disorders (see Section 1.2.2). Dichloroacetic acid lowers blood sugar levels in animals and humans with diabetes mellitus by stimulating peripheral use of glucose and inhibiting gluconeogenesis. In addition, long-term administration of dichloroacetic acid reduces plasma triglyceride and cholesterol levels (a particularly important effect in patients with congenital hyper-cholesterolaemia, who have no cholesterol receptors), and it facilitates oxidation of lactate by activating pyruvate dehydrogenase in patients with acquired and congenital forms of lactic acidosis (Stacpoole, 1989).

Drowsiness is a fairly frequent side-effect of dichloroacetic acid and has been observed in healthy volunteers, adults with type II diabetes and patients with lactic acidosis (Stacpoole *et al.*, 1978). A patient with homozygous familial hypercholesterolaemia who received single oral doses of 3 g [concentration not specified] dichloroacetic acid daily for 4 months developed reversible peripheral neuropathy characterized by loss of reflexes and muscle weakness; the effect subsided several weeks after cessation of treatment (Moore *et al.*, 1979). A second case of peripheral neuropathy in the lower extremities was reported in a 13-year-old girl treated with dichloroacetate for 1 year (Saitoh *et al.*, 1998). A more systematic study of nerve conduction velocities was carried out in 27 patients with congenital lactic acidaemia who were treated with dichloroacetate for 1 year (Spruijt *et al.*, 2001). For 25 cases, electrophysiological results were available before administration of dichloroacetate. Ten male and four female patients developed abnormal nerve conduction velocities and amplitudes within 3–6 months of the start of treatment. Motor neurons were more greatly than sensory neurons. The patients were all co-medicated with thiamine to reduce the incidence of polyneuropathy due to oxalate, a metabolite of dichloroacetic acid (Bilbao *et al.*, 1976; Stacpoole *et al.*, 1984).

4.2.2   *Experimental systems*

Exposure of male and female Sprague-Dawley rats to dichloroacetic acid at target doses of 10–600 mg/kg bw per day in the drinking-water for 14 days resulted in reduced weight gain only in the group given the highest dose. Treatment also increased urinary excretion of ammonia and changed the activities of ammoniagenesis enzymes (phosphate-dependent and -independent glutaminase), indicating renal compensation for an acid load (Davis, 1986).

Male Sprague-Dawley rats administered dichloroacetic acid in the drinking-water for 90 days at concentrations providing daily doses of about 3.9, 25.5 or 345.0 mg/kg bw had decreased body weights. Animals given the high dose also showed histological and biochemical signs of liver and kidney damage and increased hepatic peroxisomal β-oxidation activity (Mather *et al.*, 1990).

Ocular toxicity was observed in beagle dogs (which are susceptible to drug-induced cataract formation) that were treated for 13 weeks with approximate doses of 50, 75 and 100 mg/kg bw dichloroacetic acid in the drinking-water. No similar organ-specific effect has been seen in other studies or in other species (rats or humans) (Katz *et al.*, 1981).

Administration of the sodium salt of dichloroacetic acid at a target dose of 50 or 1100 mg/kg bw to male Sprague-Dawley rats in the drinking-water for 7 weeks resulted in severe hindlimb weakness, vacuolation and demyelinization of cerebral and cerebellar parenchyma and thiamine depletion in the high-dose group. The neurotoxic effects were partially prevented by providing thiamine supplementation during the treatment period (Stacpoole *et al.*, 1990). These results confirmed the observed association between neurotoxicity induced by dichloroacetic acid and the histopathological changes in the brain seen with thiamine deficiency (McCandless *et al.*, 1968). The underlying mechanism may involve stimulation of the thiamine-dependent enzymes pyruvate dehydrogenase and α-ketoacid dehydrogenase by dichloroacetic acid, resulting in increased turnover of this vitamin (Stacpoole *et al.*, 1990). Oxalate, a metabolite of dichloroacetic acid in humans and rodents, has been shown to cause both peripheral neuropathy and cataracts (Bilbao *et al.*, 1976; Fielder *et al.*, 1980); however, the renal and testicular oxalate crystals commonly seen in such cases have not been observed after administration of high doses of dichloroacetic acid (Katz *et al.*, 1981; Yount *et al.*, 1982; Stacpoole *et al.*, 1990).

A behavioural study documented effects of dichloroacetate on gait and grip strength in Fischer 344 and Long-Evans rats. Gait was affected at doses as low as 16 mg/kg bw per day within 9 weeks. Hindlimb grip strength was affected within 8 weeks after doses in excess of 90 mg/kg bw per day. Other stereotypical behaviours were also observed. Effects at high doses were irreversible, but there was partial recovery from the effects of a 13-week intake of 172 mg/kg bw, 14 weeks after treatment ended (Moser *et al.*, 1999).

Exposure of male and female B6C3F$_1$ mice to dichloroacetic acid at 1000 and 2000 mg/L in drinking-water for up to 52 weeks induced severe cytomegaly associated with extensive accumulation of glycogen, the effects progressing to multiple focal areas of necrosis, regenerative cell division and hepatomegaly (Bull *et al.*, 1990; Sanchez & Bull, 1990; Bull *et al.*, 1993).

The occurrence of infrequent and scattered acinar necrosis and a small initial increase in cell division have been reported in normal liver after treatment with dichloroacetic acid, but studies are not consistent. Histological examination of liver in other studies found little or no evidence of such damage or of overt cytotoxicity. In all cases, however, cell replication rates in normal liver decreased with chronic treatment (Stauber & Bull, 1997; Bull, 2000). Decreased cell replication rates were paralleled by decreased rates of spontaneous apoptosis (Snyder *et al.*, 1995). However, dichloroacetate significantly increased cell replication rates in a dose-dependent manner in altered hepatic foci and small tumours when chronic treatment was followed by continued administration of dichloroacetic acid at different doses (Stauber & Bull, 1997). These studies indicate that dichloroacetate has selective effects on cell replication. Another experiment, conducted *in vivo*, demonstrated that the growth of tumours, as measured by magnetic resonance imaging, slowed to a rate not statistically different from zero when treatment with dichloroacetate was suspended (Miller et al., 2000). This effect was also demonstrated as increased growth of colonies when isolated anchorage-independent hepatocytes from $B6C3F_1$ mice were treated with dichloroacetate (Stauber et al., 1998).

Induction of peroxisome proliferation has been repeatedly associated with the chronic toxicity and carcinogenicity of dichloroacetic acid in the liver (DeAngelo et al., 1989). It induced peroxisome proliferation in the livers of both mice and rats, as indicated by increased activities of palmitoyl-coenzyme A oxidase and carnitine acetyl transferase, the appearance of a peroxisome proliferation-associated protein and increased volume density of peroxisomes after exposure to dichloroacetic acid for 14 days. With further treatment, peroxisome markers returned to control levels after 45–60 weeks (DeAngelo et al., 1999). Dichloroacetate concentrations in the range of 0.5–2 mM (64.5–258 μg/mL) increased peroxisome proliferation in cultured hepatocytes derived from $B6C3F_1$ mice and Long-Evans rats (Everhart et al., 1998). However, systemic concentrations of dichloroacetate with minimally carcinogenic effects were found to be in the range of 1g/L (8 mM) in mice (Daniel et al., 1992), while a concentration of 3.5 g/L (27 mM) dichloroacetate produced a transitory increase in liver palmitoyl-coenzyme A oxidase activity, suggesting that peroxisome proliferation does not contribute significantly to the development of tumours induced by lower doses of dichloroacetate (DeAngelo et al., 1999).

As described below, dichloroacetate induced changes in carbohydrate metabolism at all doses that gave rise to tumours. The most notable change was an increase in hepatic glycogen (Bull et al., 1990; Sanchez & Bull, 1990; Carter et al., 1995; Kato-Weinstein et al., 1998). In-vitro studies in isolated mouse hepatocytes demonstrated that this effect was mediated through phosphatidylinositol 3-kinase, but was independent of insulin (Lingohr et al., 2002). In intact $B6C3F_1$ mice, treatment with dichloroacetic acid resulted in a substantial and dose-related decrease in serum insulin concentrations and in the levels of insulin receptor expression in the liver (Kato-Weinstein et al., 2001a; Lingohr et al., 2001). Downregulation of the insulin receptor was not observed in hepatic tumours induced by dichloroacetate, since the amount of insulin receptor protein was higher in liver tumours than in normal liver tissue taken from the same animal and in the livers of mice that received no dichloroacetate treatment (Lingohr et al., 2001).

Short-term treatment (11 days) of mice with dichloroacetate produced hypomethylation of DNA in liver, but this effect disappeared with longer-term treatment (44 weeks) (Tao et al., 1998). However, the extent of methylation at 5-methylcytosine sites in DNA of liver tumours was reduced by chronic treatment (44 weeks) with dichloroacetic acid, but returned to normal if treatment was suspended 1 week prior to sacrifice. Hypomethylation of DNA and of the specific proto-oncogenes c-jun and c-myc was associated with an increase in their expression in dichloroacetate-promoted tumours when compared with surrounding non-tumorous liver (Tao et al., 2000a). The hypomethylation was reversed by prior administration of methionine, suggesting that dichloroacetate acts by depleting the availability of S-adenosyl-methionine for methylation (Tao et al., 2000b). It appears that hypomethylation of promoter region for the c-myc gene occurs in several tissues (liver, kidney and bladder) in mice and precedes cell replication within these tissues; dichloroacetate prevents the methylation of hemimethylated sites in newly synthesized strands of DNA (Ge *et al.*, 2001).

As discussed in Section 4.1.1, dichloroacetate inhibits its own metabolism. The enzyme involved in its metabolism, GSTZ1-1, is also known as maleylacetoacetate iso-merase (MAAI) and converts maleylacetoacetate to fumarylacetoacetate and maleyl-acetone to fumarylacetone. The enzyme is part of the tyrosine degradation pathway and interference with tyrosine is associated with human disease, including the development of hepatocellular carcinomas (Tanguay *et al.*, 1996). Cornett *et al.* (1999) demonstrated that as little as 4 mg/kg bw per day dichloroacetate given to rats for 5 days significantly inhi-bited this enzyme. A dose rate of 200 mg/kg bw per day for the same period inhibited it by > 90%. This latter dose was shown to increase significantly excretion of maleylacetone in the urine. The authors found that the human enzyme was insensitive to inhibition by di-chloroacetate *in vitro* while the rat enzyme was inhibited; they suggested that accumulation of toxic metabolites of tyrosine may be responsible for some toxicities seen in animals. While this hypothesis remains viable for some of the toxic effects of dichloroacetate, it does not appear to account for the carcinogenic effects of dichloroacetate in mouse liver. Schultz *et al.* (2002) found that activity of MAAI was only affected by dichloroacetate in young mice (10 weeks old) and not in 60-week-old animals. These observations suggest that, if MAAI plays a role, it would only be in the early stages of treatment and that animals may be able to adapt to inhibition of this enzyme. This is consistent with the observation that *MAAI/GSTZ1-1* knockout mice do not develop hepatocellular carcinomas sponta-neously (Fernández-Cañón *et al.*, 2002).

4.3   Reproductive and prenatal effects

4.3.1   *Humans*

No data were available to the Working Group.

4.3.2   *Experimental systems*

The developmental toxicity of dichloroacetic acid has been reviewed previously (IARC, 1995).

To identify the most sensitive period of organogenesis, single doses of dichloroacetic acid were administered to pregnant Long-Evans rats. Anomalies (particularly defects in the audiovascular system) were produced with treatment on gestational days 9–12 (Epstein *et al.*, 1992). When 300 mg/kg bw dichloroacetate was administered to Sprague-Dawley rats on days 6–15 of gestation, a decrease of 5% in fetal body weight was noted, but no increase in fetal heart malformations (Fisher *et al.*, 2001).

The developmental toxicity of dichloroacetate has also been evaluated in whole-embryo cultures. In rat embryos explanted on day 10 of gestation and cultured for 46 h, malforma-tions were not observed until concentrations of dichloroacetate of 2.5 mM [322.5 mg/L] and above were used in the incubation medium (Saillenfait *et al.*, 1995). In a mouse embryo explanted on day 9 of gestation and cultured for 24–26 h, dichloroacetate produced an increased incidence of neural tube defects only when concentrations in the medium exceeded 5.8 mM [748.2 mg/L]. In comparison, acetic acid induced significant increases in neural tube defects at a lower concentration (4 mM) [516 mg/L] (Hunter *et al.*, 1996).

A subsequent evaluation in Sprague-Dawley rats showed delayed spermiation and formation of atypical residual bodies at doses as low as 54 mg/kg bw per day for 14 days, and sperm fusion and other abnormalities at doses of 160 mg/kg bw or more (Linder *et al.*, 1997).

4.4   Genetic and related effects

4.4.1   *Humans*

No data were available to the Working Group.

4.4.2   *Experimental systems*

(a)   *DNA adduct formation*

Dichloroacetic acid given by gavage caused a slight increase in 8-hydroxydeoxy-guanosine DNA adduct formation in B6C3F$_1$ mouse hepatocytes *in vivo* (Austin *et al.*, 1996), but it had no effect in this assay when given in drinking-water (Parrish *et al.*, 1996).

(b)   *Mutagenic and allied effects* (see Table 3 for details and reference)

Table 3 is an attempt to provide a comprehensive review of the literature on the genotoxic effects of dichloroacetic acid to date, and the text below reviews primarily the data not previously reviewed in IARC (1995).

In a single study, dichloroacetic acid caused a weak induction of SOS repair in *E. coli* strain PQ37. Dichloroacetic acid did not induce differential toxicity in DNA repair-deficient strains of *Salmonella typhimurium* but did induce prophage in *Escherichia coli* in one study. It was mutagenic in three strains of *S. typhimurium*: strain TA100 in three of five studies, strain RSJ100 in a single study, and strain TA98 in two of three studies. Most of the mutations in 400 revertants of dichloroacetic acid-treated *S. typhimurium* TA100 cultures were GC→AT transitions (DeMarini *et al.*, 1994). Dichloroacetic acid failed to induce point mutations in other strains of *S. typhimurium* (TA104, TA1535, TA1537, TA1538) or in *E. coli* strain WP2*uvrA*.

DNA strand breaks were not induced in mammalian cells *in vitro* in the absence of an exogenous metabolic activation system. Dichloroacetic acid failed to induce DNA damage as detected by the single-cell gel electrophoresis assay in Chinese hamster ovary cells *in vitro*.

In the gene mutation assay in mouse lymphoma cells *in vitro*, dichloroacetic acid did not show an effect in one study, but gave a positive response at 10-fold lower doses in a test in sealed tubes in another. The compound failed to induce micronuclei in a single study in the

Table 3. Genetic and related effects of dichloroacetic acid

Test system	Result[a]		Dose[b] (LED or HID)	Reference
	Without exogenous metabolic activation	With exogenous metabolic activation		
λ Prophage induction, *Escherichia coli* WP2s	–	+	2500	DeMarini *et al.* (1994)
SOS chromotest, *Escherichia coli* PQ37	(+)	–	500	Giller *et al.* (1997)
Salmonella typhimurium, DNA repair-deficient strains TS24, TA2322, TA1950	–	–	31 000	Waskell (1978)
Salmonella typhimurium TA100, TA1535, TA1537, TA1538, reverse mutation	–	–	NR	Herbert *et al.* (1980)
Salmonella typhimurium TA100, reverse mutation	+	+	50	DeMarini *et al.* (1994)
Salmonella typhimurium TA100,TA1535, TA1537, TA98, reverse mutation	–	–	5000	Fox *et al.* (1996)
Salmonella typhimurium TA100, reverse mutation, liquid medium	+	+	100	Giller *et al.* (1997)
Salmonella typhimurium RSJ100, reverse mutation	+	–	1935	Kargalioglu *et al.* (2002)
Salmonella typhimurium TA104, reverse mutation, microsuspension	–	–	150 µg/plate	Nelson *et al.* (2001)
Salmonella typhimurium TA98, reverse mutation	–	(+)	10 µg/plate	Herbert *et al.* (1980)
Salmonella typhimurium TA98, reverse mutation	+	–	5160	Kargalioglu *et al.* (2002)
Salmonella typhimurium TA100, reverse mutation	+	+	1935	Kargalioglu *et al.* (2002)
Escherichia coli WP2*uvr*A, reverse mutation	–	–	5000	Fox *et al.* (1996)
DNA strand breaks and alkali-labile damage, Chinese hamster ovary cells *in vitro* (single-cell gel electrophoresis assay)	–	NT	3225 µg/mL	Plewa *et al.* (2002)
DNA strand breaks, B6C3F$_1$ mouse hepatocytes *in vitro*	–	NT	2580	Chang *et al.* (1992)
DNA strand breaks, Fischer 344 rat hepatocytes *in vitro*	–	NT	1290	Chang *et al.* (1992)
Gene mutation, mouse lymphoma cell line L5178Y/TK$^{+/-}$ *in vitro*	–	–	5000	Fox *et al.* (1996)
Gene mutation, mouse lymphoma cell line L5178Y/TK$^{+/-}$-3.7.2C *in vitro*	+	NT	400	Harrington-Brock *et al.* (1998)
Micronucleus formation, mouse lymphoma L5178Y/TK$^{+/-}$-3.7.2C cell line *in vitro*	–	NT	800	Harrington-Brock *et al.* (1998)
Chromosomal aberrations, Chinese hamster ovary *in vitro*	–	–	5000	Fox *et al.* (1996)

Table 3 (contd)

Test system	Result[a] Without exogenous metabolic activation	Result[a] With exogenous metabolic activation	Dose[b] (LED or HID)	Reference
Chromosomal aberrations, mouse lymphoma L5178Y/Tk$^{+/-}$ -3.7.2C cell line in vitro	+	NT	600	Harrington-Brock et al. (1998)
Aneuploidy, mouse lymphoma L5178Y/Tk$^{+/-}$-3.7.2C cell line in vitro	-	NT	800	Harrington-Brock et al. (1998)
DNA strand breaks, human CCRF-CEM lymphoblastoid cells in vitro	-	NT	1290	Chang et al. (1992)
DNA strand breaks, male B6C3F$_1$ mouse liver in vivo	+		13 po × 1	Nelson & Bull (1988)
DNA strand breaks, male B6C3F$_1$ mouse liver in vivo	+		10 po × 1	Nelson et al. (1989)
DNA strand breaks, male B6C3F$_1$ mouse liver in vivo	-		1290 po × 1	Chang et al. (1992)
DNA strand breaks, male B6C3F$_1$ mouse splenocytes in vivo	-		1290 po × 1	Chang et al. (1992)
DNA strand breaks, male B6C3F$_1$ mouse epithelial cells from stomach and duodenum in vivo	-		1290 po × 1	Chang et al. (1992)
DNA strand breaks, male B6C3F$_1$ mouse liver in vivo	-		5000 dw × 7–14 d	Chang et al. (1992)
DNA strand breaks, alkali-labile sites, cross linking, male B6C3F$_1$ mouse blood leukocytes in vivo (single-cell gel electrophoresis assay)	+		3500 dw × 28 d	Fuscoe et al. (1996)
DNA strand breaks, male Sprague-Dawley rat liver in vivo	+		30 po × 1	Nelson & Bull (1988)
DNA strand breaks, male Fischer 344 rat liver in vivo	-		645 po × 1	Chang et al. (1992)
DNA strand breaks, male Fischer 344 rat liver in vivo	-		2000 dw × 30 w	Chang et al. (1992)
Gene mutation, lacI transgenic male B6C3F$_1$ mouse liver assay in vivo	+		1000 dw × 60 w	Leavitt et al. (1997)
Micronucleus formation, male B6C3F$_1$ mouse peripheral erythrocytes in vivo	+		3500 dw × 9 d	Fuscoe et al. (1996)
Micronucleus formation, male B6C3F$_1$ mouse peripheral erythrocytes in vivo	-		3500 dw × 28 d	Fuscoe et al. (1996)
Micronucleus formation, male B6C3F$_1$ mouse peripheral erythrocytes in vivo	+		3500 dw × 10 w +21[c]	Fuscoe et al. (1996)

Table 3 (contd)

Test system	Result[a] Without exogenous metabolic activation	With exogenous metabolic activation	Dose[b] (LED or HID)	Reference
Micronucleus formation, male and female Crl:CD (SD) BR rat bone-marrow erythrocytes in vivo	−		1100 iv × 3	Fox et al. (1996)
Micronucleus formation, Pleurodeles waltl newt larvae peripheral erythrocytes in vivo	−		80[d]	Giller et al. (1997)
Inhibition of intercellular communication, male Sprague-Dawley rat liver clone 9 cells in vitro	+	NT	1290	Benane et al. (1996)

[a] +, positive; (+), weakly positive; −, negative; NT, not tested
[b] LED, lowest effective dose; HID, highest ineffective dose; in-vitro tests, µg/mL; in-vivo tests, mg/kg bw; NR, dose not reported; po, orally; dw, drinking-water (in mg/L); d, day; w, week; iv, intravenous
[c] 10 weeks of dichloroacetic acid-containing drinking-water, followed by 21 weeks on dichloroacetic acid-free drinking-water
[d] Larvae reared in dichloroacetic acid-containing water

same system. Dichloroacetic acid increased the frequency of chromosomal aberrations in mouse lymphoma cell cultures but not in Chinese hamster ovary cells. In a single study, dichloroacetic acid did not induce aneuploidy in mouse lymphoma cells *in vitro*.

Contradictory results were obtained for the induction of DNA strand breaks in mammals *in vivo*. No effects were seen in either mouse or rat hepatic cells after single or repeated dosing, and no effects were observed in epithelial cells from spleen, stomach or duodenum after a single dose. In one study *in vivo*, dichloroacetic acid induced a significant decrease in DNA migration, consistent with the presence of DNA cross-linking in leukocytes *in vivo*, as detected by the single cell gel electrophoresis assay. Dichloroacetic acid caused mutations in male transgenic B6C3F$_1$ mice harbouring the bacterial *lacI* gene. In the latter study, in which mice received dichloroacetic acid in the drinking-water for 60 weeks, the mutation spectrum recovered from treated mice showed a significant decrease in GC→AT transitions (32.8% versus 53.2% for controls) and an increase in mutations at TA sites (32.79% versus 19.15% for controls) (Leavitt *et al.*, 1997). [The Working Group noted that the doses used (3.5 or 1.0 g/L) were high and tumorigenic.]

In one study in male B6C3F$_1$ mice polychromatic erythrocytes *in vivo*, dichloroacetic acid induced the formation of micronuclei in two of three treatments: a dose-related increase after a 9-day exposure; no significant increase after a 28-day exposure [the value for control in that experiment was rather high (higher than after 9 days of exposure)]; and a small but significant increase in the frequency of micronucleated normochromatic erythrocytes following exposure for more than 10 weeks. Coadministration of the antioxidant vitamin E did not affect the ability of dichloroacetic acid to induce this damage, indicating that the small induction of micronuclei by dichloroacetic acid was probably not caused by oxidative damage. Based on the lack of any difference observed in the proportion of kinetochore-positive micronuclei between the treated and control animals, micronuclei were assumed to arise from clastogenic events (Fuscoe *et al.*, 1996). After intravenous administration to male and female Crl:CD (SD) BR rats *in vivo*, dichloroacetic acid failed to induce micronuclei in bone-marrow erythrocytes in a single study. Dichloroacetic acid did not induce micronuclei in erythrocytes of new larvae *in vivo*.

Dichloroacetic acid given in drinking-water (0.5 and 5 g/L) to male B6C3F$_1$ mice significantly reduced apoptosis in hepatocytes in a dose-dependent manner compared to untreated controls (Snyder *et al.*, 1995).

In the livers of female B6C3F$_1$ mice given dichloroacetic acid in drinking-water for 11 days, the level of 5-methylcytosine in DNA was decreased, it was also decreased in liver tumours. In the livers of female B6C3F$_1$ mice initiated with *N*-methyl-*N*-nitrosourea before receiving dichloroacetic acid in drinking-water for 44 weeks, termination of exposure to dichloroacetic acid 1 week prior to sacrifice resulted in an increase in the level of 5-methylcytosine in adenomas to the level found in non-involved livers (Tao *et al.*, 1998). These authors noted that the restauration of DNA methylation in adenomas upon removal of dichloroacetic acid corresponded to the regression of adenomas upon removal of dichloroacetic acid (Bull *et al.*, 1990; Pereira & Phelps, 1996). Gap-junctional intercellular communication was inhibited in rat hepatocytes *in vitro* (Benane *et al.*, 1996).

Mutations of proto-oncogenes in tumours induced by dichloroacetic acid

In a tumour assay by Anna *et al.* (1994), numerous foci of cellular alteration (presumed preneoplastic lesions) were noted in the livers of mice treated with dichloroacetic acid for 76 weeks, but only rare foci were found in the livers of controls. No neoplasms related to treatment were found at other sites. The frequency of mutations in codon 61 of H-*ras* (CAA) was not significantly different in the hepatocellular tumours from 64 treated mice (62%) and in those from 74 combined historical and concurrent controls (69%); however, the spectrum of these mutations showed a significant decrease in AAA and an increase in CTA in the treated mice in comparison with the controls. No other H-*ras* mutations were found, and only one K-*ras* mutation was detected in tumours from the treated and concurrent control groups. The authors interpreted these findings as suggesting that exposure to dichloroacetic acid provides the environment for a selective growth advantage for spontaneous CTA mutations in codon 61 of H-*ras* (Anna *et al.*, 1994).

Point mutations in exons 1, 2 and 3 of K- and H-*ras* proto-oncogenes were studied in dichloroacetic acid-induced liver tumours of male B6C3F$_1$ mice (104-week treatment). Dichloroacetic acid did not modify the incidence of mutations for exon 2 of *H-ras* in carcinomas (50% versus 58% for control). Only three liver carcinomas from dichloroacetate-treated mice showed mutations in the other exons of *H-ras* or in *K-ras*. In tumours with mutation in exon 2 of *H-ras*, treatment with dichloroacetic acid induced a decrease in the frequency of mutations CAA→AAA (21% versus 80% for control), an increase in the frequency of mutations CAA→CGA (50% versus 20% for control) and the appearance of a new mutation CAA→CTA (29% versus 0% for control, about one third of the mutations) (Ferreira-Gonzalez *et al.*, 1995).

H-*ras* codon 61 mutations were studied in female B6C3F$_1$ mouse liver tumours given 3.5 g/L dichloroacetic acid in the drinking-water over a period of 104 weeks. Only one mutation in exon 2 was found among the 22 tumours analysed (4.5%), and it was a CAA→CTA transversion in H-*ras* codon 61 (Schroeder *et al.*, 1997).

In *N*-methyl-*N*-nitrosourea-initiated and dichloroacetic acid-promoted female B6C3F$_1$ mice, no loss of heterozygosity on chromosome 6 was observed in the 24 liver tumours promoted by dichloroacetic acid (Tao *et al.*, 1996).

4.5   Mechanistic considerations

Section 4.4 provides a consistent data set to indicate that dichloroacetic acid induces mutations and chromosomal effects. However, the concentrations of dichloroacetic acid that are required to produce these effects raise serious questions as to whether damage to DNA is involved in carcinogenic responses at low doses (Moore & Harrington-Brock, 2000). This conclusion is reinforced by observations suggesting that alternative mechanisms may adequately account for the carcinogenic responses that have been observed in rodents. These data are discussed in some detail in the following paragraphs.

A series of studies examined the extent to which the mutation frequency and spectra in *ras* genes might provide evidence of a genotoxic mechanism. The first of these studies examined liver tumours that were induced by administering 5 g/L dichloroacetic acid in drinking-water to male B6C3F$_1$ mice for 76 weeks (Anna *et al.*, 1994). Tumours induced by dichloroacetic acid did not have mutations in codons 12 or 13 of the H-*ras* oncogene, whereas genotoxic carcinogens generally induce mutations at these two codons. The incidence of mutations at codon 61 of the H-*ras* oncogene in tumours from dichloroacetic acid-treated mice was not different from that of spontaneous tumours in control mice: 62 and 69% of the animals, respectively. However, the types of mutation were differed: dichloroacetic acid-induced tumours had 27% CAA→AAA and 37% CAA→CTA transversions of codon 61. Of the spontaneous tumours, 58% had a CAA→AAA transversion at this locus and only 13% had CAA→CTA transversions. These differences in the mutation spectrum were statistically significant. Thus, the mutation analysis of the H-*ras* oncogene in dichloroacetic acid-induced tumours is consistent with a non-genotoxic mechanism for carcinogenesis, with dichloroacetic acid promoting the development of a subclass of spontaneous tumours with CAA→CTA transversions.

In a subsequent study in male B6C3F$_1$ mice, dichloroacetic adid did not induce mutations in codons 12 or 13 of H-*ras* but induced liver tumours with mutations at codon 61 (Ferreira-Gonzalez *et al.*, 1995). Similar to the results of Anna *et al.* (1994), the incidence of tumours with mutations in codon 61 was not different from that of spontaneous liver tumours. In addition, dichloroacetic acid-induced tumours had a decreased incidence of CAA→AAA transversions.

In a more recent study, tumours induced in male B6C3F$_1$ mice by dichloroacetic acid at a concentration of 2 g/L were examined at 52 weeks of treatment, whereas those induced by 0.5 g/L were examined at 87 weeks of treatment to ensure a sufficient number of tumours. Concomitant control animals were used in these studies, but the spontaneous rate of tumour production was too small to provide adequate measurements of mutation frequency and spectra. A total of 64 dichloroacetic acid-induced tumours were evaluated: mutation frequencies in these tumours were significantly lower than those in historical controls (33% versus 56%). There was an excess of CTA mutations relative to control, consistent with earlier studies. However, most tumours contained less than 50% H-*ras* mutant sequences and only one was completely without a wild-type sequence, and the frequency of mutations appeared to increase with duration of treatment (or age) (Bull *et al.*, 2002). These data indicated that H-*ras* mutations were most probably a late event rather than an initiating event in dichloroacetic acid-induced liver carcinogenesis. Despite the lack of mutations, however, it was found that expression of H-*ras* was substantially and uniformly elevated in both spontaneous and dichloroacetic acid-induced tumours (Bull *et al.*, 2002). In addition, the lower mutation frequency observed in dichloroacetic acid-induced tumours was found to be consistent with tumours induced by other non-genotoxic carcinogens, which also have lower H-*ras* mutation frequency than spontaneous tumours (Fox *et al.*, 1990; Hegi *et al.*, 1993; Stanley *et al.*, 1994).

The mutation spectrum in the H-*ras* oncogene in dichloroacetic acid-induced liver tumours in female B6C3F₁ mice has also been reported (Schroeder *et al.*, 1997). Female B6C3F₁ mice were exposed to 1.5% acetic acid (control) or 3.5 g/L dichloroacetic acid in the drinking-water for 104 weeks. Only 1/22 (45%) tumours in the dichloroacetic acid-treated female mice had a mutation at codon 61 (Schroeder *et al.*, 1997), in contrast to the 62% in dichloroacetic acid-treated male mice (Anna *et al.*, 1994). The incidence of H-*ras* mutations in spontaneous tumours in control female mice was not determined (only one spontaneous tumour was available). Nevertheless, these data are consistent with the finding that female mice have a lower incidence of spontaneous liver tumours than male mice. Moreover, dichloroacetic acid-treated male mice have a higher percentage of carcinomas relative to adenomas and higher frequency of H-*ras* mutations in carcinomas compared with adenomas, suggesting that codon 61 mutations in H-*ras* may be late events in male mice (Anna *et al.*, 1994; Bull *et al.*, 2002).

Studies vary in their ability to detect a small initial increase in cell division within the liver after the beginning of treatment with dichloroacetic acid. In all cases, however, cell replication rates within the non-neoplastic portions of the liver were seen to decrease with chronic treatment (Carter *et al.*, 1995; Stauber & Bull, 1997; Bull, 2000). Decreased cell replication rates were paralleled by decreased rates of spontaneous apoptosis (Snyder *et al.*, 1995). However, dichloroacetic acid significantly increased cell replication rates in a dose-dependent manner within foci and small tumours when the treatment used to induce the tumours (2 g/L) was followed by continued administration of doses ranging from 0.1 to 2 g/L for 2 weeks (Stauber & Bull, 1997). A parallel experiment demonstrated that the growth of tumours induced by a dose of 2 g/L, as measured *in situ* by magnetic resonance imaging, slowed to a rate not statistically different from zero when treatment with dichloroacetic acid was suspended (Miller *et al.*, 2000). Dichloroacetic acid duplicated these effects by stimulating the growth of anchorage-independent colonies with the same phenotype from hepatocytes isolated from B6C3F₁ mice (Stauber *et al.*, 1998; Kato-Weinstein *et al.*, 2001a,b).

Several studies have documented changes in the expression of different genes following treatment with dichloroacetic acid. Lingohr *et al.* (2001) found that expression of insulin receptor was decreased in the liver of male B6C3F₁ mice following administration of dichloroacetic acid in the drinking-water at concentrations of 0.5 g/L and above for 2 and 10 weeks of treatment. Depressed expression of insulin receptor was also observed in hepatocytes derived from B6C3F₁ mice treated with dichloroacetic acid in culture (Lingohr *et al.*, 2002). These changes were accompanied by substantial decreases in concentrations of serum insulin in the animal. Because the latent period for the development of the changes in serum insulin coincided with accumulation of glycogen in the liver, it was suggested that they may be indirectly mediated by changes in the regulation of carbohydrate metabolism. As insulin receptor concentrations in tumours induced by dichloroacetic acid remained elevated, this could provide these cells with a selective advantage over normal hepatocytes (Bull *et al.*, 2002; Lingohr *et al.*, 2002).

Increased expression of stearoyl-coenzyme A desaturase and depressed expression of α-1 protease inhibitor, cytochrome b5 and carboxylesterase were observed by an RNA differential display technique in the liver of mice treated with dichloroacetic acid at 2 g/L in drinking-water for 4 weeks (Thai *et al.*, 2001). Increased expression of c-*jun* and c-*myc* in liver has also been observed in female mice treated with 500 mg/kg bw dichloroacetic acid by gavage (Tao *et al.*, 2000b). The increases in c-*jun* and c-*myc* expression were associated with changes in the methylation status of the promoter regions of the respective genes in liver DNA (Tao *et al.*, 2000a). Treatment with dichloroacetic acid at a concentration of 25 mmol/L (3.2 g/L) in drinking-water for 11 days was shown to produce hypomethylation of DNA in liver, but this effect disappeared with longer-term treatment (44 weeks) (Tao *et al.*, 1998). However, the extent of methylation at 5-methylcytosine sites in DNA of liver tumours was reduced after chronic treatment with dichloroacetic acid, but returned to normal when treatment was suspended. Hypomethylation of DNA was associated with increased expression of c-*jun* and c-*myc* proto-oncogenes in dichloroacetate-promoted tumours when compared with surrounding non-tumorous liver (Tao *et al.*, 2000a). The hypomethylation could be reversed by prior administration of methionine, suggesting that depletion of *S*-adenosyl-methionine may be responsible for the effect (Tao *et al.*, 2000b). It appeared that hypomethylation of promotor region for the c-*myc* gene occurred in several tissues (liver, kidney and bladder) of mice and preceded cell replication within these tissues (Ge *et al.*, 2001).

5.   Summary of Data Reported and Evaluation

5.1   Exposure data

Dichloroacetic acid is used as an intermediate in the production of glyoxylic acid, dialkoxy and diaroxy acids, sulfonamides and iron chelates. It is used to a lesser extent as a cauterizing agent in dermatology. Wider exposure to dichloroacetic acid occurs at microgram-per-litre levels in drinking-water and swimming pools as a result of chlorination and chloramination.

5.2   Human carcinogenicity data

Several studies were identified that analysed risk with respect to one or more measures of exposure to complex mixtures of disinfection by-products that are found in most chlorinated and chloraminated drinking-water. No data specifically on dichloroacetic acid were available to the Working Group.

5.3   Animal carcinogenicity data

In eight studies, neutralized dichloroacetic acid administered in the drinking-water to male and/or female mice increased the incidences of hepatocellular adenomas and/or carcinomas. Following oral administration of dichloroacetic acid in the drinking-water to male rats, an increased incidence of hepatocellular carcinomas was found at a dose that decreased body weight and an increase in the combined incidence of adenomas and carcinomas was found at a lower dose. When administered in the drinking-water, dichloroacetic acid promoted hepatocellular carcinomas in carcinogen-initiated male and female mice in three studies.

5.4   Other relevant data

Dichloroacetic acid is metabolized to glyoxylic acid, which may be oxidized to oxalic acid, reduced to glycolic acid and transaminated to glycine. The metabolism of dichloroacetic acid to glyoxylic acid is catalysed by glutathione S-transferase zeta-1. Dichloroacetic acid is a mechanism-based inactivator of this enzyme, which decreases its own metabolism in humans or rats treated with the compound.

Clinically, administration of dichloroacetic acid for the treatment of congenital lactic acidosis has been associated with nervous system toxicity, which has also been observed in experimental animals. Animal studies have demonstrated toxic effects in the liver and testis. Treatment of rats with dichloroacetic acid has also given rise to developmental effects, primarily in the cardiovascular system. However, these have not been observed consistently.

Dichloroacetic acid produces a variety of effects on intermediary metabolism, including increases in hepatic glycogen at low concentrations (≤ 1 µM) in the blood and inhibition of pyruvate dehydrogenase kinase at higher concentrations (≥ 100 µM). Dichloroacetic acid also affects gene expression, including various proto-oncogenes and enzymes involved in lipid metabolism. In some cases, changes in gene expression have been associated with decreased methylation of DNA in the promoter region of the gene.

Studies on proto-oncogenes in mice have compared mutation induction in codon 61 of H-ras proto-oncogenes in hepatic tumours from dichloroacetic acid-treated mice and untreated mice. The spectrum of mutations showed a decrease in the frequency of CAA→AAA mutations and an increase in the frequency of mutations from CAA→CGA and from CAA→CTA in treated mice compared with controls. No loss of heterozygosity on chromosome 6 was observed in liver tumours promoted by dichloroacetic acid in female mice.

The evidence for induction of DNA strand breaks in liver cells of rodents exposed to dichloroacetic acid *in vivo* was inconclusive, as were the results of measurements of dichloroacetic acid-induced 8-hydroxydeoxyguanosine DNA adducts in mouse liver. Dichloroacetic acid caused a decrease in the level of 5-methylcytosine in DNA of liver cells and liver tumours of female mice. In peripheral blood cells of mice *in vivo*, dichloroacetic acid induced DNA damage in the single-cell gel electrophoresis assay. It caused mutations (decrease in G:C→A:T and increase in mutations at T:A sites) in male transgenic mice harbouring the bacterial *lacI* gene. Dichloroacetic acid induced the formation of micronuclei *in vivo* in mouse polychromatic erythrocytes but not in rat bone-marrow cells.

DNA strand breaks were not induced in human or rodent cells *in vitro*. The results of assays for mutagenesis in bacteria and in mouse lymphoma cells were inconsistent. Dichloroacetic acid did not induce micronuclei in a mouse lymphoma cell line *in vitro* or in erythrocytes of newt larvae *in vivo*. It caused chromosomal aberrations in one of two studies *in vitro*. Dichloroacetic acid induced no aneuploidy in mouse lymphoma cells *in vitro*.

Dichloroacetic acid affects cell proliferation and cell death both in normal livers and tumours throughout the dose range that induces liver tumours in mice. These changes are associated with differential effects on intermediary metabolism in preneoplastic lesions versus normal liver and by changes in gene expression and DNA hypomethylation.

Dichloroacetic acid is genotoxic *in vivo* and *in vitro*. It also causes DNA hypomethylation *in vivo*. Thus, a genotoxic effect, possibly involving an indirect, epigenetic mechanism, may contribute to the carcinogenic mode of action of dichloroacetic acid.

5.5   Evaluation

There is *inadequate evidence* in humans for the carcinogenicity of dichloroacetic acid.
There is *sufficient evidence* in experimental animals for the carcinogenicity of dichloroacetic acid.

Overall evaluation

Dichloroacetic acid is *possibly carcinogenic to humans (Group 2B)*.

6.   References

American Public Health Association/American Water Works Association/Water Environment Federation (1999) *Standard Methods for the Examination of Water and Wastewater*, 20th Ed., Washington, DC [CD-ROM]

Anderson, W.B., Board, P.G., Gargano, B. & Anders, M.W. (1999) Inactivation of glutathione transferase zeta by dichloroacetic acid and other fluorine-lacking α-haloalkanoic acids. *Chem. Res. Toxicol.*, **12**, 1144–1149

Anderson, W.B., Liebler, D.C., Board, P.G. & Anders, M.W. (2002) Mass spectral characterization of dichloroacetic acid-modified human glutathione transferase zeta. *Chem. Res. Toxicol.*, **15**, 1387–1397

Anna, C.H., Maronpot, R.R., Pereira, M.A., Foley, J.F., Malarkey, D.E. & Anderson, M.W. (1994) *ras* Proto-oncogene activation in dichloroacetic acid-, trichloroethylene- and tetrachloroethylene-induced liver tumors in B6C3F1 mice. *Carcinogenesis*, **15**, 2255–2261

Austin, E.W., Parrish, J.M., Kinder, D.H. & Bull, R.J. (1996) Lipid peroxidation and formation of 8-hydroxydeoxyguanosine from acute doses of halogenated acetic acids. *Fundam. appl. Toxicol.*, **31**, 77–82

Beilstein Online (2002) *Dialog Corporation, File 390* (Beilstein Chemidaten und Software GmbH), Cary, NC

Benane, S.G., Blackman, C.F. & House, D.E. (1996) Effects of perchloroethylene and its metabolites on intercellular communication in clone 9 rat liver cells. *J. Toxicol. environ. Health*, **48**, 427–437

Bilbao, J.M., Berry, H., Marotta, J. & Ross, R.C. (1976) Peripheral neuropathy in oxalosis. A case report with electron microscopic observations. *Can. J. neurol. Sci.*, **3**, 63–67

Blackburn, A.C., Tzeng, H.-F., Anders, M.W. & Board, P.G. (2000) Discovery of a functional polymorphism in human glutathione transferase zeta by expressed sequence tag database analysis. *Pharmacogenetics*, **10**, 49–57

Blackburn, A.C., Coggan, M., Tzeng, H.-F., Lantum, H., Polekhina, G., Parker, M.W., Anders, M.W. & Board, P.G. (2001) GSTZ1d: A new allele of glutathione transferase zeta and maleylacetoacetate isomerase. *Pharmacogenetics*, **11**, 671–678

Board, P.G., Baker, R.T., Chelvanayagam, G. & Jermiin, L.S. (1997) Zeta, a novel class of glutathione transferases in a range of species from plants to humans. *Biochem. J.*, **328**, 929–935

Bull, R.J. (2000) Mode of action of liver tumor induction by trichloroethylene and its metabolites, trichloroacetate and dichloroacetate. *Environ. Health Perspect.*, **108**, (Suppl. 2), 241–259

Bull, R.J., Sanchez, I.M., Nelson, M.A., Larson, J.L. & Lansing, A.J. (1990) Liver tumor induction in B6C3F1 mice by dichloroacetate and trichloroacetate. *Toxicology*, **63**, 341–359

Bull, R.J., Templin, M., Larson, J.L. & Stevens, D.K. (1993) The role of dichloroacetate in the hepatocarcinogenicity of trichloroethylene. *Toxicol. Lett.*, **68**, 203–211

Bull, R.J., Orner, G.A., Cheng, R.S., Stillwell, L., Stauber, A.J., Sasser, L.B., Lingohr, M.K. & Thrall, B.D. (2002) Contribution of dichloroacetate and trichloroacetate to liver tumor induction in mice by trichloroethylene. *Toxicol. appl. Pharmacol.*, **182**, 55–65

Cancho, B., Ventura, F. & Galceran, M.T. (1999) Behavior of halogenated disinfection by-products in the water treatment plant of Barcelona, Spain. *Bull. environ. Contam. Toxicol.*, **63**, 610–617

Carter, J.H., Carter, H.W. & DeAngelo, A.B. (1995) Biochemical, pathologic and morphometric alterations induced in male B6C3F1 mouse liver by short-term exposure to dichloroacetic acid. *Toxicol. Lett.*, **81**, 55–71

Chang, L.W., Daniel, F.B. & DeAngelo, A.B. (1992) Analysis of DNA strand breaks induced in rodent liver in vivo, hepatocytes in primary culture, and a human cell line by chlorinated acetic acids and chlorinated acetaldehydes. *Environ. mol. Mutag.*, **20**, 277–288

Chemical Information Services (2002) *Directory of World Chemical Producers (Version 2002.1)*, Dallas, TX

Chen, W.J. & Weisel, C.P. (1998) Halogenated DBP concentrations in a distribution system. *J. Am. Water Works Assoc.*, **90**, 151–163

Christman, R.F., Norwood, D.L., Millington, D.S. & Johnson, J.D. (1983) Identity and yields of major halogenated products of aquatic fulvic acid chlorination. *Environ. Sci. Technol.*, **17**, 625–628

Clariant Corp. (2001) *Specification Sheet: Dichloroacetic Acid (DCAA) 98%*, Charlotte, NC, USA

Clariant GmbH (2002) *Specification Sheet: Dichloroacetic Acid (DCAA) 98%*, Sulzbach, Germany

Clemens, M. & Schöler, H.F. (1992a) Determination of halogenated acetic acids and 2,2-dichloro-propionic acid in water samples. *Fresenius J. anal. Chem.*, **344**, 47–49

Clemens, M. & Schöler, H.-F. (1992b) [Halogenated organic compounds in swimming pool waters.] *Zbl. Hyg.*, **193**, 91–98 (in German)

Cornett, R., James, M.O., Henderson, G.N., Cheung, J., Shroads, A.L. & Stacpoole, P.W. (1999) Inhibition of glutathione *S*-transferase ζ and tyrosine metabolism by dichloroacetate: A potential unifying mechanism for its altered biotransformation and toxicity. *Biochem. biophys. Res. Commun.*, **262**, 752–756

Curry, S.H., Lorenz, A., Chu, P.-I., Limacher, M. & Stacpoole, P.W. (1991) Disposition and pharmacodynamics of dichloroacetate (DCA) and oxalate following oral DCA doses. *Biopharm. Drug Dispos.*, **12**, 375–390

Daniel, F.B., DeAngelo, A.B., Stober, J.A., Olson, G.R. & Page, N.P. (1992) Hepatocarcinogenicity of chloral hydrate, 2-chloroacetaldehyde, and dichloroacetic acid in the male B6C3F1 mouse. *Fundam. appl. Toxicol.*, **19**, 159–168

Davis, M.E. (1986) Effect of chloroacetic acids on the kidneys. *Environ. Health Perspect.*, **69**, 209–214

DeAngelo, A.B., Daniel, F.B., McMillan, L., Wernsing, P. & Savage, R.E., Jr (1989) Species and strain sensitivity to the induction of peroxisome proliferation by chloroacetic acids. *Toxicol. appl. Pharmacol.*, **101**, 285–298

DeAngelo, A.B., Daniel, F.B., Most, B.M. & Olson, G.R. (1996) The carcinogenicity of dichloro-acetic acid in the male fischer 344 rat. *Toxicology*, **114**, 207–221

DeAngelo, A.B., George, M.H. & House, D.E. (1999) Hepatocarcinogenicity in the male B6C3F1 mouse following a lifetime exposure to dichloroacetic acid in the drinking water: Dose–response determination and modes of action. *J. Toxicol. environ. Health*, **A58**, 485–507

DeMarini, D.M., Perry, E. & Shelton, M.L. (1994) Dichloroacetic acid and related compounds: Induction of prophage in *E. coli* and mutagenicity and mutation spectra in Salmonella TA100. *Mutagenesis*, **9**, 429–437

Dojlido, J., Zbiec, E. & Swietlik, R. (1999) Formation of the haloacetic acids during ozonation and chlorination of water in Warsaw waterworks (Poland). *Water Res.*, **33**, 3111–3118

Ellis, D.A., Hunson, M.L., Sibley, P.K., Shahid, T., Fineberg, N.A., Solomon, K.R., Muir, D.C.G. & Mabury, S.A. (2001) The fate and persistence of trifluoroacetic and chloroacetic acids in pond waters. Chemosphere, *42, 309–318*

Environmental Protection Agency (1995) *Method 552.2. Determination of Haloacetic Acids and Dalapon in Drinking Water by Liquid–liquid Extraction, Derivatization and Gas Chromatography with Electron Capture Detection* (Revision 1.0), Cincinnati, OH, National Exposure Research Laboratory, Office of Research and Development

Environmental Protection Agency (1998) National primary drinking water regulations; disinfectants and disinfection byproducts; final rule. *Fed. Regist.*, December 16

Epstein, D.L., Nolen, G.A., Randall, J.L., Christ, S.A., Read, E.J., Stober, J.A. & Smith, M.K. (1992) Cardiopathic effects of dichloroacetate in the fetal Long-Evans rat. *Teratology*, **46**, 225–235

European Commission (1998) *Council Directive 98/83/EC of 3 November 1998 on the Quality of Water Intended for Human Consumption*, Office for Official Publications of the European Communities, Luxemburg, 27 pp.

Everhart, J.L., Kurtz, D.T. & McMillan, J.M. (1998) Dichloroacetic acid induction of peroxisome proliferation in cultured hepatocytes. *J. Biochem. mol. Toxicol.*, **12**, 351–359

Fernández-Cañón, J.M. & Peñalva, M.A. (1998) Characterization of a fungal maleylacetoacetate isomerase gene and identification of its human homologue. *J. biol. Chem.*, **273**, 329–337

Fernández-Cañón, J.M., Baetscher, M.W., Finegold, M., Burlingame, T., Gibson, K.M. & Grompe, M. (2002) Maleylacetoacetate isomerase (*MAAI/GSTZ*)-deficient mice reveal a glutathione-dependent nonenzymatic bypass in tyrosine catabolism. *Mol. cell. Biol.*, **22**, 4943–4951

Ferreira-Gonzalez, A., DeAngelo, A.B., Nasim, S. & Garrett, C.T. (1995) *Ras* oncogene activation during hepatocarcinogenesis in B6C3F1 male mice by dichloroacetic and trichloroacetic acids. *Carcinogenesis*, **16**, 495–500

Fielder, A.R., Garner, A. & Chambers, T.L. (1980) Ophthalmic manifestations of primary oxalosis. *Br. J. Ophthalmol.*, **64**, 782–788

Fisher, J.W., Channel, S.R., Eggers, J.S., Johnson, P.D., MacMahon, K.L., Goodyear, C.D., Sudberry, G.L., Warren, D.A., Latendresse, J.R. & Graeter, L.J. (2001) Trichloroethylene, trichloroacetic acid, and dichloroacetic acid: Do they affect fetal rat heart development? *Int. J. Toxicol.*, **20**, 257–267

Fox, T.R., Schumann, A.M., Watanabe, P.G., Yano, B.L., Maher, V.M. & McCormick, J.J. (1990) Mutational analysis of the H-*ras* oncogene in spontaneous C57BL/6 × C3H/He mouse liver tumors and tumors induced with genotoxic and nongenotoxic hepatocarcinogens. *Cancer Res.*, **50**, 4014–4019

Fox, A.W., Yang, X., Murli, H., Lawlor, T.E., Cifone, M.A. & Reno, F.E. (1996) Absence of mutagenic effects of sodium dichloroacetate. *Fundam. appl. Toxicol.*, **32**, 87–95

Frank, H., Vital, J. & Frank, W. (1989) Oxidation of airborne C_2-chlorocarbons to trichloroacetic and dichloroacetic acid. *Fresenius Z. anal. Chem.*, **333**, 713

Fuscoe, J.C., Afshari, A.J., George, M.H., DeAngelo, A.B., Tice, R.R., Salman, T. & Allen, J.W. (1996) In vivo genotoxicity of dichloroacetic acid: Evaluation with the mouse peripheral blood micronucleus assay and the single cell gel assay. *Environ. mol. Mutag.*, **27**, 1–9

Ge, R., Yang, S., Kramer, P.M., Tao, L. & Pereira, M.A. (2001) The effect of dichloroacetic acid and trichloroacetic acid on DNA methylation and cell proliferation in B6C3F1 mice. *J. Biochem. mol. Toxicol.*, **15**, 100–106

Geist, S., Lesemann, C., Schütz, C., Seif, P. & Frank, E. (1991) Determination of chloroacetic acids in surface water. In: Angeletti, G. & Bjørseth, A., eds, *Organic Micropollutants in the Aquatic Environment*, Dordrecht, Kluwer Academic, pp. 393–397

Gennaro, A.R. (2000) *Remington: The Science and Practice of Pharmacy*, 20th Ed., Baltimore, MD, Lippincott Williams & Wilkins, p. 1210

Giller, S., Le Curieux, F., Erb, F. & Marzin, D. (1997) Comparative genotoxicity of halogenated acetic acids found in drinking water. *Mutagenesis*, **12**, 321–328

Glaze, W.H., Kenneke, J.F. & Ferry, J.L. (1993) Chlorinated byproducts from the TiO_2-mediated photodegradation of trichloroethylene and tetrachloroethylene in water. *Environ. Sci. Technol.*, **27**, 177–184

Gonzalez-Leon, A., Schultz, I.R., Xu, G. & Bull, R.J. (1997) Pharmacokinetics and metabolism of dichloroacetate in the F344 rat after prior administration in drinking water. *Toxicol. appl. Pharmacol.*, **146**, 189–195

Hansch, C., Leo, A. & Hoekman, D. (1995) *Exploring QSAR: Hydrophobic, Electronic, and Steric Constants*, Washington, DC, American Chemical Society, p. 4

Hargesheimer, E.E. & Satchwill, T. (1989) Gas-chromatographic analysis of chlorinated acids in drinking water. *J. Water Sewage Res. Treat.-Aqua*, **38**, 345–351

Harrington-Brock, K., Doerr, C.L. & Moore, M.M. (1998) Mutagenicity of three disinfection by-products: Di- and trichloroacetic acid and chloral hydrate in L5178Y/TK$^{+/-}$ -3.7.2C mouse lymphoma cells. *Mutat. Res.*, **413**, 265–276

Health Canada (1995) *A National Survey of Chlorinated Disinfection By-products in Canadian Drinking Water* (95-EHD-197), Ottawa, Ontario, Minister of Supply and Services Canada

Health Canada (2001) *Summary of Guidelines for Canadian Drinking Water Quality*, Ottawa, Ontario

Hegi, M.E., Fox, T.R., Belinsky, S.A., Devereux, T.R. & Anderson, M.W. (1993) Analysis of activated protooncogenes in B6C3F1 mouse liver tumors induced by ciprofibrate, a potent peroxisome proliferator. *Carcinogenesis*, **14**, 145–149

Henderson, G.N., Curry, S.H., Derendorf, H., Wright, E.C. & Stacpoole, P.W. (1997) Pharmacokinetics of dichloroacetate in adult patients with lactic acidosis. *J. clin. Pharmacol.*, **37**, 416–425

Herbert, V., Gardner, A. & Colman, N. (1980) Mutagenicity of dichloroacetate, an ingredient of some formulations of pangamic acid (trade-named 'vitamin B$_{15}$'). *Am. J. clin. Nutr.*, **33**, 1179–1182

Hunter, E.S., III, Rogers, E.H., Schmid, J.E. & Richard, A. (1996) Comparative effects of haloacetic acids in whole embryo culture. *Teratology*, **54**, 57–64

IARC (1991a) *IARC Monographs on the Evaluation of Carcinogenic Risks to Humans*, Vol. 52, *Chlorinated Drinking-water; Chlorination By-products; Some Other Halogenated Compounds; Cobalt and Cobalt Compounds*, Lyon, IARCPress, pp. 55–141

IARC (1991b) *IARC Monographs on the Evaluation of the Carcinogenic Risks to Humans*, Vol. 53, *Occupational Exposures in Insecticide Application, and Some Pesticides*, Lyon, IARCPress, pp. 267–307

IARC (1995) *IARC Monographs on the Evaluation of Carcinogenic Risks to Humans*, Vol. 63, *Dry Cleaning, Some Chlorinated Solvents and Other Industrial Chemicals*, Lyon, IARCPress, pp. 271–290

Jacangelo, J.G., Patania, N.L., Reagan, K.M., Aieta, E.M., Krasner, S.W. & McGuire, M.J. (1989) Ozonation: Assessing its role in the formation and control of disinfection by-products. *J. Am. Water Works Assoc.*, **81**, 74–84

James, M.O., Cornett, R., Yan, Z., Henderson, G.N. & Stacpoole, P.W. (1997) Glutathione-dependent conversion to glyoxylate, a major pathway of dichloroacetate biotransformation in hepatic cytosol from humans and rats, is reduced in dichloroacetate-treated rats. *Drug Metab. Dispos.*, **25**, 1223–1227

James, M.O., Yan, Z., Cornett, R., Jayanti, V.M.K.M., Henderson, G.N., Davydova, N., Katovich, M.J., Pollock, B. & Stacpoole, P.W. (1998) Pharmacokinetics and metabolism of [^{14}C]dichloroacetate in male Sprague-Dawley rats. Identification of glycine conjugates, including hippurate, as urinary metabolites of dichloroacetate. *Drug Metab. Dispos.*, **26**, 1134–1143

Kargalioglu, Y., McMillan, B.J., Minear, R.A. & Plewa, M.J. (2002) Analysis of the cytotoxicity and mutagenicity of drinking water disinfection by-products in *Salmonella typhimurium*. *Teratog. Carcinog. Mutagen.*, **22**, 113–128

Kato-Weinstein, J., Lingohr, M.K., Orner, G.A., Thrall, B.D. & Bull, R.J. (1998) Effects of dichloroacetate on glycogen metabolism in B6C3F1 mice. *Toxicology*, **130**, 141–154

Kato-Weinstein, J., Stauber, A.J., Orner, G.A., Thrall, B.D. & Bull, R.J. (2001a) Differential effects of dihalogenated and trihalogenated acetates in the liver of B6C3F1 mice. *J. appl. Toxicol.*, **21**, 81–89

Kato-Weinstein, J., Orner, G.A., Thrall, B.D. & Bull, R.J. (2001b) Differences in the detection of c-Jun/ubiquitin immunoreactive proteins by different c-Jun antibodies. *Toxicol. Meth.*, **11**, 189–207

Katz, R., Tai, C.N., Diener, R.M., McConnell, R.F. & Semonick, D.E. (1981) Dichloroacetate, sodium: 3-Month oral toxicity studies in rats and dogs. *Toxicol. appl. Pharmacol.*, **57**, 273–287

Kim, H. & Weisel, C.P. (1998) Dermal absorption of dichloro- and trichloroacetic acids from chlorinated water. *J. Expo. Anal. environ. Epidemiol.*, **8**, 555–575

Koenig, G., Lohmar, E. & Rupprich, N. (1986) Chloroacetic acids. In: Gerhartz, W., Yamamoto, Y.S., Campbell, F.T., Pfefferkorn, R. & Rounsaville, J.F., eds., *Ullmann's Encyclopedia of Industrial Chemistry*, 5th rev. Ed., Vol. A6, New York, VCH Publishers, pp. 537–552

Krasner, S.W., McGuire, M.J., Jacangelo, J.G., Patania, N.L., Reagan, K.M. & Aieta, E.M. (1989) The occurrence of disinfection by-products in US drinking water. *J. Am. Water Works Assoc.*, **81**, 41–53

Krishna, S., Supanaranond, W., Pukrittayakamee, S., Karter, D., Supputamongkol, Y., Davis, T.M.E., Holloway, P.A. & White, N.J. (1994) Dichloroacetate for lactic acidosis in severe malaria: A pharmacokinetic and pharmacodynamic assessment. *Metabolism*, **43**, 974–981

Krishna, S., Agbenyega, T., Angus, B.J., Bedu-Addo, G., Ofori-Amanfo, G., Henderson, G., Szwandt, I.S.F., O'Brien, R. & Stacpoole, P.W. (1995) Pharmacokinetics and pharmacodynamics of dichloroacetate in children with lactic acidosis due to severe malaria. *Q. J. Med.*, **88**, 341–349

Krishna, S., Supanaranond, W., Pukrittayakamee, S., ter Kuile, F., Ruprah, M. & White, N.J. (1996) The disposition and effects of two doses of dichloroacetate in adults with severe falciparum malaria. *Br. J. clin. Pharmacol.*, **41**, 29–34

Lantum, H.B.M., Baggs, R.B., Krenitsky, D.M., Board, P.G. & Anders, M.W. (2002) Immunohistochemical localization and activity of glutathione transferase zeta (GSTZ1-1) in rat tissues. *Drug Metab. Dispos.*, **30**, 616–625

Larson, J.L. & Bull, R.J. (1992) Metabolism and lipoperoxidative activity of trichloroacetate and dichloroacetate in rats and mice. *Toxicol. appl. Pharmacol.*, **115**, 268–277

Leavitt, S.A., DeAngelo, A.B., George, M.H. & Ross, J.A. (1997) Assessment of the mutagenicity of dichloroacetic acid in *lacI* transgenic B6C3F1 mouse liver. *Carcinogenesis*, **18**, 2101–2106

LeBel, G.L., Benoit, F.M. & Williams, D.T. (1997) A one-year survey of halogenated disinfection by-products in the distribution system of treatment plants using three different disinfection processes. *Chemosphere*, **34**, 2301–2317

Legube, B., Croue, J.P. & Dore, M. (1985) Chlorination of humic substances in aqueous solution: Yields of volatile and major non-volatile organic halides. *Sci. total Environ.*, **47**, 217–222

Lin, E.L.C., Mattox, J.K. & Daniel, F.B. (1993) Tissue distribution excretion and urinary metabolites of dichloroacetic acid in the male Fisher 344 rat. *J. Toxicol. Environ. Health*, **38**, 19–32

Linder, R.E., Klinefelter, G.R., Strader, L.F., Suarez, J.D. & Roberts, N.L. (1997) Spermatotoxicity of dichloroacetic acid. *Reprod. Toxicol.*, **11**, 681–688

Lingohr, M.K., Thrall, B.D. & Bull, R.J. (2001) Effects of dichloroacetate (DCA) on serum insulin levels and insulin-controlled signaling proteins in livers of male B6C3F1 mice. *Toxicol. Sci.*, **59**, 178–184

Lingohr, M.K., Bull, R.J., Kato-Weinstein, J. & Thrall, B.D. (2002) Dichloroacetate stimulates glycogen accumulation in primary hepatocytes through an insulin independent mechanism. *Toxicol. Sci.*, **68**, 508–515

Lipscomb, J.C., Mahle, D.A., Brashear, W.T. & Barton, H.A. (1995) Dichloroacetic acid: Metabolism in cytosol. *Drug Metab. Dispos.*, **23**, 1202–1205

Lopez-Avila, V., Liu, Y. & Charan, C. (1999) Determination of haloacetic acids in water by ion chromatography — Method development. *J. Assoc. off. anal. Chem. int.*, **82**, 689–704

Lukas, G., Vyas, K.H., Brindle, S.D., Le Sher, A.R. & Wagner, W.E., Jr (1980) Biological disposition of sodium dichloroacetate in animals and humans after intravenous administration. *J. pharm. Sci.*, **69**, 419–421

Mather, G.G., Exon, J.H. & Koller, L.D. (1990) Subchronic 90 day toxicity of dichloroacetic and trichloroacetic acid in rats. *Toxicology*, **64**, 71–80

Mazze, R.I. & Cousins, M.J. (1974) Biotransformation of methoxyflurane. *Int. Anesthesiol. Clin.*, **12**, 93–105

McCandless, D.W., Schenker, S. & Cook, M. (1968) Encephalopathy of thiamine deficiency: Studies of intracerebral mechanisms. *J. clin. Invest.*, **47**, 2268–2280

Miller, J.W. & Uden, P.C. (1983) Characterization of nonvolatile aqueous chlorination products of humic substances. *Environ. Sci. Technol.*, **17**, 150–157

Miller, J.H., Minard, K., Wind, R.A., Orner, G.A., Sasser, L.B. & Bull, R.J. (2000) In vivo MRI measurements of tumor growth induced by dichloroacetate: Implications for mode of action. *Toxicology*, **145**, 115–125

Mohamed, M., Matayun, M. & Lim, T.S. (1989) Chlorinated organics in tropical hardwood kraft pulp and paper mill effluents and their elimination in an activated sludge treatment system. *Pertanika*, **12**, 387–394

Moore, M.M. & Harrington-Brock, K. (2000) Mutagenicity of trichloroethylene and its metabolites: Implications for the risk assessment of trichloroethylene. *Environ. Health Perspect.*, **108** (Suppl. 2), 215–223

Moore, G.W., Swift, L.L., Rabinowitz, D., Crofford, O.B., Oates, J.A. & Stacpoole, P.W. (1979) Reduction of serum cholesterol in two patients with homozygous familial hypercholesterolemia by dichloroacetate. *Atherosclerosis*, **33**, 285–293

Morris, E.D. & Bost, J.C. (1991) Acetic acid and derivatives (halogenated). In: Kroschwitz, J.I. & Howe-Grant, M., eds., *Kirk-Othmer Encyclopedia of Chemical Technology*, 4th Ed., Vol. 1, New York, John Wiley & Sons, pp. 165–175

Moser, V.C., Phillips, P.M., McDaniel, K.L. & MacPhail, R.C. (1999) Behavioral evaluation of the neurotoxicity produced by dichloroacetic acid in rats. *Neurotoxicol. Teratol.*, **21**, 719–731

National Health and Medical Research Council and Agriculture and Resource Management Council of Australia and New Zealand (1996) *Australian Drinking Water Guidelines Summary*, Canberra

National Institute for Occupational Safety and Health (1994) National *Occupational Exposure Survey (1981–1983)*, Cincinnati, OH

Nelson, M.A. & Bull, R.J. (1988) Induction of strand breaks in DNA by trichloroethylene and metabolites in rat and mouse liver *in vivo*. *Toxicol. appl. Pharmacol.*, **94**, 45–54

Nelson, M.A., Lansing, A.J., Sanchez, I.M., Bull, R.J. & Springer, D.L. (1989) Dichloroacetic acid and trichloroacetic acid-induced DNA strand breaks are independent of peroxisome proliferation. *Toxicology*, **58**, 239–248

Nelson, G.M., Swank, A.E., Brooks, L.R., Bailey, K.C. & George, S.E. (2001) Metabolism, microflora effects, and genotoxicity in haloacetic acid-treated cultures of rat cecal microbiota. *Toxicol. Sci.*, **60**, 232–241

Nissinen, T.K., Miettinen, I.T., Martikainen, P.J. & Vartiainen, T. (2002) Disinfection by-products in Finnish drinking waters. *Chemosphere*, **48**, 9–20

Obolensky, A., Shukairy, H. & Blank, V. (2003) Occurrence of haloacetic acids in ICR finished water and distribution systems. In: McGuire, M.J., McLain, J.L. & Obolensky, A., eds, *Information Collection Rule Data Analysis*, Denver, CO, AWWA Research Foundation, pp. 1–36

Parrish, J.M., Austin, E.W., Stevens, D.K., Kinder, D.H. & Bull, R.J. (1996) Haloacetate-induced oxidative damage to DNA in the liver of male B6C3F1 mice. *Toxicology*, **110**, 103–111

Pereira, M.A. (1996) Carcinogenic activity of dichloroacetic acid and trichloroacetic acid in the liver of female B6C3F1 mice. *Fundam. appl. Toxicol.*, **31**, 192–199

Pereira, M.A. & Phelps, J.B. (1996) Promotion by dichloroacetic acid and trichloroacetic acid of N-methyl-N-nitrosourea-initiated cancer in the liver of female B6C3F1 mice. *Cancer Lett.*, **102**, 133–141

Pereira, M.A., Li, K. & Kramer, P.M. (1997) Promotion by mixtures of dichloroacetic acid and trichloroacetic acid of N-methyl-N-nitrosourea-initiated cancer in the liver of female B6C3F1 mice. *Cancer Lett.*, **115**, 15–23

Pereira, M.A., Kramer, P.M., Conran, P.B. & Tao, L. (2001) Effect of chloroform on dichloroacetic acid and trichloroacetic acid-induced hypomethylation and expression of the c-*myc* gene and on their promotion of liver and kidney tumors in mice. *Carcinogenesis*, **22**, 1511–1519

Plewa, M.J., Kargalioglu, Y., Vankerk, D., Minear, R.A. & Wagner, E.D. (2002) Mammalian cell cytotoxicity and genotoxicity analysis of drinking water disinfection by-products. *Environ. mol. Mutag.*, **40**, 134–142

Reckhow, D.A. & Singer, P.C. (1990) Chlorination by-products in drinking waters: From formation potentials to finished water concentrations. *J. Am. Water Works Assoc.*, **92**, 173–180

Reckhow, D.A., Singer, P.C. & Malcom, R.L. (1990) Chlorination of humic materials: Byproduct formation and chemical interpretations. *Environ. Sci. Technol.*, **24**, 1655–1664

Richmond, R.E., Carter, J.H., Carter, H.W., Daniel, F.B. & DeAngelo, A.B. (1995) Immunohistochemical analysis of dichloroacetic acid (DCA)-induced hepatocarcinogenesis in male Fischer (F344) rats. *Cancer Lett.*, **92**, 67–76

Römpp, A., Klemm, O., Fricke, W. & Frank, H. (2001) Haloacetates in fog and rain. *Environ. Sci. Technol.*, **35**, 1294–1298

Saghir, S.A. & Schultz, I.R. (2002) Low-dose pharmacokinetics and oral bioavailability of dichloroacetate in naive and GST-ζ-depleted rats. *Environ. Health Perspect.*, **110**, 757–763

Saillenfait, A.M., Langonné, I. & Sabaté, J.P. (1995) Developmental toxicity of trichloroethylene, tetrachloroethylene and four of their metabolites in rat whole embryo culture. *Arch. Toxicol.*, **70**, 71–82

Saitoh, S., Momoi, M.Y., Yamagata, T., Mori, Y. & Imai, M. (1998) Effects of dichloroacetate in three patients with MELAS. *Neurology*, **50**, 531–534

Sanchez, I.M. & Bull, R.J. (1990) Early induction of reparative hyperplasia in the liver of B6C3F1 mice treated with dichloroacetate and trichloroacetate. *Toxicology*, **64**, 33–46

Schroeder, M., DeAngelo, A.B. & Mass, M.J. (1997) Dichloroacetic acid reduces Ha-*ras*-codon 61 mutations in liver tumors from female B6C3F1 mice. *Carcinogenesis*, **18**, 1675–1678

Schultz, D.R., Marxmiller, R.L. & Koos, B.A. (1971) Residue determination of dichlorvos and related metabolites in animal tissues and fluids. *J. agric. Food Chem.*, **19**, 1238–1243

Schultz, I.R., Merdink, J.L., Gonzalez-Leon, A. & Bull, R.J. (1999) Comparative toxicokinetics of chlorinated and brominated haloacetates in F344 rats. *Toxicol. appl. Pharmacol.*, **158**, 103–114

Schultz, I.R., Merdink, J.L., Gonzalez-Leon, A. & Bull, R.J. (2002) Dichloroacetate toxicokinetics and disruption of tyrosine catabolism in B6C3F1 mice: Dose–response relationships and age as a modifying factor. *Toxicology*, **173**, 229–247

Scott, B.F., Mactavisch, D., Spencer, C., Strachan, W.M.J. & Muir, D.C.G. (2000) Haloacetic acids in Canadian lake waters and precipitation. *Environ. Sci. Technol.*, **34**, 4266–4272

Scott, B.F., Spencer, C., Marvin, C.H., Mactavish, D.C. & Muir, D.G.G. (2002) Distribution of halo-acetic acids in the water columns of the Laurentian Great Lakes and Lake Malawi. *Environ. Sci. Technol.*, **36**, 1893–1898

Shangraw, R.E. & Fisher, D.M. (1996) Pharmacokinetics of dichloroacetate in patients undergoing liver transplantation. *Anesthesiology*, **84**, 851–858

Shangraw, R.E. & Fisher, D.M. (1999) Pharmacokinetics and pharmacodynamics of dichloroacetate in patients with cirrhosis. *Clin. Pharmacol. Ther.*, **66**, 380–390

Shin, D., Chung, Y., Choi, Y., Kim, J., Park, Y. & Kum, H. (1999) Assessment of disinfection by-products in drinking water in Korea. *J. Expo. Anal. environ. Epidemiol.*, **9**, 192–199

Simpson, K.L. & Hayes, K.P. (1998) Drinking water disinfection by-products: An Australian perspective. *Water Res.*, **32**, 1522–1528

Snyder, R.D., Pullman, J., Carter, J.H., Carter, H.W. & DeAngelo, A.B. (1995) *In vivo* administration of dichloroacetic acid suppresses spontaneous apoptosis in murine hepatocytes. *Cancer Res.*, **55**, 3702–3705

Spruijt, L., Naviaux, R.K., McGowan, K.A., Nyhan, W.L., Sheean, G., Haas, R.H. & Barshop, B.A. (2001) Nerve conduction changes in patients with mitochondrial diseases treated with dichloro-acetate. *Muscle Nerve*, **24**, 916–924

Stacpoole, P.W. (1989) The pharmacology of dichloroacetate. *Metabolism*, **38**, 1124–1144

Stacpoole, P.W., Moore, G.W. & Kornhauser, D.M. (1978) Metabolic effects of dichloroacetate in patients with diabetes mellitus and hyperlipoproteinemia. *New Engl. J. Med.*, **298**, 526–530

Stacpoole, P.W., Harwood, H.J., Jr, Curry, S.H., Schneider, M., Cockrill, A. & Sauberlich, H.E. (1984) Induction of thiamine deficiency by dichloroacetate (abstract). *Clin. Res.*, **32**, 236A

Stacpoole, P.W., Harwood, H.J., Jr, Cameron, D.F., Curry, S.H., Samuelson, D.A., Cornwell, P.E. & Sauberlich, H.E. (1990) Chronic toxicity of dichloroacetate: Possible relation to thiamine deficiency in rats. *Fundam. appl. Toxicol.*, **14**, 327–337

Stanley, L.A., Blackburn, D.R., Devereaux, S., Foley, J., Lord, P.G., Maronpot, R.R., Orton, T.C. & Anderson, M.W. (1994) *Ras* mutations in methylclofenapate-induced B6C3F1 and C57BL/10J mouse liver tumours. *Carcinogenesis*, **15**, 1125–1131

Stauber, A.J. & Bull, R.J. (1997) Differences in phenotype and cell replicative behavior of hepatic tumors induced by dichloroacetate (DCA) and trichloroacetate (TCA). *Toxicol. appl. Pharmacol.*, **144**, 235–246

Stauber, A.J., Bull, R.J. & Thrall, B.D. (1998) Dichloroacetate and trichloroacetate promote clonal expansion of anchorage-independent hepatocytes *in vivo* and *in vitro*. *Toxicol. appl. Pharmacol.*, **150**, 287–294

Stevens, D.K., Eyre, R.J. & Bull, R.J. (1992) Adduction of hemoglobin and albumin in vivo by meta-bolites of trichloroethylene, trichloroacetate, and dichloroacetate in rats and mice. *Fundam. appl. Toxicol.*, **19**, 336–342

Stottmeister, E. & Naglitsch, F. (1996) [*Human Exposure to Other Disinfection By-products than Trihalomethanes in Swimming Pools*] Berlin, Federal Environmental Agency (in German)

Tanguay, R.M., Jorquera, R., Poudrier, J. & St-Louis, M. (1996) Tyrosine and its catabolites: From disease to cancer. *Acta biochim. pol.*, **43**, 209–216

Tao, L., Li, K., Kramer, P.M. & Pereira, M.A. (1996) Loss of heterozygosity on chromosome 6 in dichloroacetic acid and trichloroacetic acid-induced liver tumors in female B6C3F1 mice. *Cancer Lett.*, **108**, 257–261

Tao, L.., Kramer, P.M., Ge, R. & Pereira, M.A. (1998) Effect of dichloroacetic acid and trichloro-acetic acid on DNA methylation in liver and tumors of female B6C3F1 mice. *Toxicol. Sci.*, **43**, 139–144

Tao, L., Yang, S., Xie, M., Kramer, P.M. & Pereira, M.A. (2000a) Hypomethylation and over-expression of c-*jun* and c-*myc* protooncogenes and increased DNA methyltransferase activity in dichloroacetic and trichloroacetic acid-promoted mouse liver tumors. *Cancer Lett.*, **158**, 185–193

Tao, L., Yang, S., Xie, M., Kramer, P.M. & Pereira, M.A. (2000b) Effect of trichloroethylene and its metabolites, dichloroacetic acid and trichloroacetic acid, on the methylation and expression of c-*jun* and c-*myc* protooncogenes in mouse liver: Prevention by methionine. *Toxicol. Sci.*, **54**, 399–407

Thai, S.-F., Allen, J.W., DeAngelo, A.B., George, M.H. & Fuscoe, J.C. (2001) Detection of early gene expression changes by differential display in the livers of mice exposed to dichloroacetic acid. *Carcinogenesis*, **22**, 1317–1322

Tong, Z., Board, P.G. & Anders, M.W. (1998a) Glutathione transferase Zeta catalyses the oxy-genation of the carcinogen dichloroacetic acid to glyoxylic acid. *Biochem. J.*, **331**, 371–374

Tong, Z., Board, P.G. & Anders, M.W. (1998b) Glutathione transferase Zeta-catalyzed biotrans-formation of dichloroacetic acid and other α-haloacids. *Chem. Res. Toxicol.*, **11**, 1332–1338

Toxopeus, C. & Frazier, J.M. (1998) Kinetics of trichloroacetic acid and dichloroacetic acid in the isolated perfused rat liver. *Toxicol. appl. Pharmacol.*, **152**, 90–98

Tzeng, H.-F., Blackburn, A.C., Board, P.G. & Anders, M.W. (2000) Polymorphism- and species-dependent inactivation of glutathione transferase zeta by dichloroacetate. *Chem. Res. Toxicol.*, **13**, 231–236

Uden, P.C. & Miller, J.W. (1983) Chlorinated acids and chloral in drinking water. *J. Am. Water Works Assoc.*, **75**, 524–527

Villanueva, C.M., Kogevinas, M. & Grimalt, J.O. (2003) Haloacetic acids and trihalomethanes in finished drinking waters from heterogeneous sources. *Water Res.*, **37**, 953–958

Waskell, L. (1978) A study of the mutagenicity of anesthetics and their metabolites. *Mutat. Res.*, **57**, 141–153

Weast, R.C. & Astle, M.J. (1985) *CRC Handbook of Data on Organic Compounds*, Vol. II, Boca Raton, FL, CRC Press, p. 575

Weisel, C.P., Kim, H., Haltmeier, P. & Klotz, J.B. (1999) Exposure estimates to disinfection by-products of chlorinated drinking water. *Environ. Health Perspect.*, **107**, 103–110

WHO (1989) *Dichlorvos* (Environmental Health Criteria 79), Geneva

WHO (1998) *Guidelines for Drinking-water Quality*, 2nd Ed., Vol. 2, *Health Criteria and Other Supporting Information and Addendum to Vol. 2*, Geneva

Williams, D.T., LeBel, G.L. & Benoit, F.M. (1997) Disinfection by-products in Canadian drinking water. *Chemosphere*, **34**, 299–316

Williams, P.J., Lane, J.R., Turkel, C.C., Capparelli, E.V., Dziewanowska, Z. & Fox, A.W. (2001) Dichloroacetate: Population pharmacokinetics with a pharmacodynamic sequential link model. *J. clin. Pharmacol.*, **41**, 259–267

Yount, E.A., Felten, S.Y., O'Connor, B.L., Peterson, R.G., Powell, R.S., Yum, M.N. & Harris, R.A. (1982) Comparison of the metabolic and toxic effects of 2-chloropropionate and dichloro-acetate. *J. Pharmacol. exp. Ther.*, **222**, 501–508

TRICHLOROACETIC ACID

This substance was considered by a previous Working Group, in February 1995 (IARC, 1995). Since that time, new data have become available and these have been incorporated into the monograph and taken into consideration in the present evaluation.

1. Exposure Data

1.1 Chemical and physical data

1.1.1 *Nomenclature*

Chem. Abstr. Serv. Reg. No.: 76-03-9
Chem. Abstr. Name: Trichloroacetic acid
IUPAC Systematic Name: Trichloroacetic acid
Synonyms: TCA; TCA (acid); trichloracetic acid; trichloroethanoic acid; trichloromethane carboxylic acid

1.1.2 *Structural and molecular formulae and relative molecular mass*

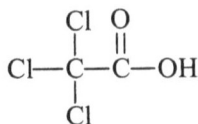

$$\begin{array}{c} Cl\ \ \ O \\ |\ \ \ \ \| \\ Cl-C-C-OH \\ | \\ Cl \end{array}$$

$C_2HCl_3O_2$

Relative molecular mass: 163.39

1.1.3 *Chemical and physical properties of the pure substance*

(a) *Description*: Colourless to white deliquescent crystals with characteristic odour (Morris & Bost, 1991; Gennaro, 2000)
(b) *Boiling-point*: 197.5 °C (Morris & Bost, 1991)
(c) *Melting-point*: 59 °C (Morris & Bost, 1991)
(d) *Density*: 1.6218 at 64 °C/4 °C (Morris & Bost, 1991)

(e) *Spectroscopy data*: Infrared [2376], ultraviolet [1-6], nuclear magnetic resonance [6] and mass [1026] spectral data have been reported (Weast & Astle,
1985; Sadtler Research Laboratories, 1991)

(f) *Solubility*: Very soluble in water (1306 g/100 g at 25 °C) and most organic solvents, including acetone, benzene, ethyl ether, methanol and *ortho*-xylene (Morris
& Bost, 1991)

(g) *Volatility*: Vapour pressure, 1 mm/Hg at 51 °C (Verschueren, 2001)

(h) *Stability*: Dissociation constant (K_a), 0.2159; undergoes decarboxylation when
heated with caustics or amines to yield chloroform (Morris & Bost, 1991)

(i) *Octanol/water partition coefficient (P)*: log P, 1.33 (Hansch et al., 1995)

(j) *Conversion factor*: $mg/m^3 = 6.68 \times ppm$[a]

1.1.4   *Technical products and impurities*

Trade names for trichloroacetic acid include Aceto-Caustin and Amchem Grass Killer.
Trichloroacetic acid is marketed at various degrees of purity. Typical specifications for
commercially available trichloroacetic acid are presented in Table 1. Trichloroacetic acid
is also available as aqueous solutions with concentrations ranging from 3 to 100% (w/v)
(Spectrum Chemical, 2002).

1.1.5   *Analysis*

Trichloroacetic acid has been determined in water using liquid–liquid extraction,
conversion to its methyl ester and gas chromatography with electron capture detection. This
method has been applied to drinking-water, groundwater, water at intermediate stages of
treatment and raw source water, with a limit of detection of 0.08 µg/L (Environmental Protection Agency, 1995; American Public Health Association/American Water Works Asso-
ciation/Water Environment Federation, 1999).

A similar method was used in a 1993 national survey of chlorinated disinfection by-
products in Canadian drinking-water. Methyl esters were analysed by gas chromatography–mass spectrometry with selected ion monitoring. The minimum quantifiable limit
for this method was 0.01 µg/L (Health Canada, 1995; Williams et al., 1997).

Modifications of these methods have been used in an analytical survey of 16 drinking-
water sources in Australia (Simpson & Hayes, 1998) and in a survey of treated water from
35 Finnish waterworks during different seasons (Nissinen et al., 2002).

---

[a] Calculated from: mg/m^3 = (molecular weight/24.45) × ppm, assuming normal temperature (25 °C) and
pressure (760 mm Hg)

Table 1. Typical quality specifications for trichloroacetic acid[a]

Property	Grade					
	Crude[a]		Technical[b]		Ph. Eur.[c]	ACS[b]
Trichloroacetic acid, % min.	96.5	86.0[d]	98	87.6[d]	98	99.0
Dichloroacetic acid, % max.	2.5	2.5	1.2	1.1	–	0.5
Sulfuric acid, % max.	0.5	0.5	0.3	0.3	–	0.02
Sulfated ash, % max.	–	–	–	–	0.1	0.03
Water, % max.	0.5	–	0.2	–	–	0.5
Heavy metals in the form of:						
Lead, ppm max.	10	–	–	–	–	20
Iron, ppm (mg/kg) max.	20	20	10	10	–	10
Halogenides, mg/kg max.	–	–	–	–	–	10
Sulfates, mg/kg max.	–	–	–	–	–	200
Phosphate, ppm (mg/kg) max.	–	–	–	–	–	5
Chloride, ppm (mg/kg) max.	–	–	–	–	100	10
Nitrate, ppm (mg/kg) max.	–	–	–	–	–	20

[a] From Koenig *et al.* (1986); Clariant GmbH (2002a,c)

[b] From Clariant Corp. (2001, 2002); Clariant GmbH (2002a,b,c); cfm Oskar Tropitzsch (2002)

[c] From Council of Europe (2002)

[d] 90% trichloroacetic acid in water (Clariant GmbH, 2002c)

1.2 Production and use

Trichloroacetic acid was reported to have been first synthesized in 1840 by chlorination of acetic acid in sunlight (Beilstein Online, 2002). Haloacetic acids were first detected in 1983 as disinfection by-products in chlorinated drinking-waters, 9 years after the discovery of trihalomethanes in chlorinated waters (Nissinen *et al.*, 2002).

1.2.1 *Production*

Trichloroacetic acid is produced on an industrial scale by chlorination of acetic acid or chloroacetic acid at 140–160 °C. Calcium hypochlorite may be added as a chlorination accelerator, and metal catalysts have been used in some cases. Trichloroacetic acid is isolated from the crude product by crystallization (Koenig *et al.*, 1986; Morris & Bost, 1991).

Available information indicates that trichloroacetic acid is produced by nine companies in India, two companies each in China, Germany and Mexico and one company each in France, Israel, Italy, Japan, Russia and Spain (Chemical Information Services, 2002a).

Available information indicates that trichloroacetic acid is formulated into pharmaceutical products by five companies in Italy, three companies in France, two companies in Poland and one company each in Argentina, Spain and Turkey (Chemical Information Services, 2002b).

1.2.2   *Use*

The main application of trichloroacetic acid, usually as its sodium salt, is as a selective herbicide and, historically, in herbicidal formulations with 2,4-D and 2,4,5-T (IARC, 1977, 1987). Trichloroacetic acid is also used as an etching or pickling agent in the surface treatment of metals, as a swelling agent and solvent in the plastics industry, as an auxiliary in textile finishing, as an additive to improve high-pressure properties in mineral lubricating oils and as an analytical reagent. Trichloroacetic acid, and particularly its esters, are important starting materials in organic syntheses (Koenig *et al.*, 1986; Morris & Bost, 1991; Clariant GmbH, 2002a,b,c).

Trichloroacetic acid is used as a caustic on the skin or mucous membranes to treat local lesions and for the treatment of various dermatological diseases. Its chief medicinal use is in the treatment of ordinary warts and juvenile flat warts (escharotic), although there are reports of its use in removing tattoos, treating genital warts and in dermal peeling. It is also used extensively as a precipitant of protein in the chemical analysis of body fluids and tissue extracts, and as a decalcifier and fixative in microscopy (Gennaro, 2000; Royal Pharmaceutical Society of Great Britain, 2002).

1.3   Occurrence

1.3.1   *Natural occurrence*

Trichloroacetic acid is not known to occur as a natural product.

1.3.2   *Occupational exposure*

The National Occupational Exposure Survey conducted between 1981 and 1983 indicated that 35 124 employees in the USA had potential occupational exposure to trichloroacetic acid in seven industries and 1562 plants (National Institute for Occupational Safety and Health, 1994). The estimate is based on a survey of companies and did not involve measurement of actual exposures.

Because trichloroacetic acid is a major end-metabolite of trichloroethylene (IARC, 1976, 1979, 1987, 1995) and tetrachloroethylene (IARC, 1979, 1987, 1995) in humans, it has been used for many years as a biological marker of exposure to those compounds. It is also a metabolite of 1,1,1-trichloroethane (see IARC, 1979, 1987, 1999), and chloral hydrate (see the monograph in this volume) is rapidly oxidized to trichloroacetic acid in humans. The levels of trichloroacetic acid reported in human blood and urine after occupational and environmental exposure to trichloroacetic acid, trichloroethylene, tetrachloroethylene or 1,1,1-trichloroethane are summarized in Table 2.

The average concentration of trichloroacetic acid in 177 urinary measurements made in various industries in 1986–88 at the Finnish Institute of Occupational Health was 77.1 µmol/L [12.6 mg/L], with a range of < 50–860 µmol/L [< 8.2–140.5 mg/L]. The

Table 2. Concentrations of trichloroacetic acid in human blood and urine following exposure to chlorinated solvents

Job description (country)	Exposure	Concentration of trichloroacetic acid	Reference
Occupational			
Metal degreasing (Switzerland)	Trichloroethylene, 10–300 ppm [54–1611 mg/m^3]	57–980 mg/L (urine)	Boillat (1970)
Metal degreasing (USA)	Trichloroethylene, 170–420 mg/m^3	3–116 mg/g creatinine (urine)	Lowry et al. (1974)
Workshop (Japan)	Trichloroethylene, 3–175 ppm [16.1–940 mg/m^3]	9–297 mg/L (urine)	Ikeda et al. (1972)
Printing factory (Japan)	1,1,1-Trichloroethane, 4.3–53.5 ppm [23–289 mg/m^3]	0.5–5.5 mg/L	Seki et al. (1975)
Workshop (Japan)	Trichloroethylene	Range of means, 108–133 mg/L (urine) (trichloroacetate)	Itoh (1989)
Automobile workshop (Japan)	Trichloroethylene, 1–50 ppm [5–269 mg/m^3]	Average, 136 mg/g creatinine (urine)	Ogata et al. (1987)
Dry cleaning, degreasing (former Yugoslavia)	Trichloroethylene	0.43–154.92 μmol/L (blood) [0.07–25.3 mg/L] 0.58–42.44 mmol/mol creatinine (urine) [0.84–61 mg/g]	Skender et al. (1988)
Solvent exposure (Republic of Korea)	Tetrachloroethylene, 0–61 ppm [0–414 mg/m^3]	0.6–3.5 mg/L (urine)	Jang et al. (1993)
Degreasing (Sweden)	Trichloroethylene, 3–114 mg/m^3	2–260 μmol/L (urine) [0.3–42.5 mg/L]	Ulander et al. (1992)

Table 2 (contd)

Job description (country)	Exposure	Concentration of trichloroacetic acid	Reference
Dry cleaning (former Yugoslavia)	Trichloroethylene, 25–40 ppm [134–215 mg/m³] Tetrachloroethylene, 33–53 ppm [224–359 mg/m³]	13.47–393.56 μmol/L (blood) [2.2–64 mg/L] 1.92–77.35 mmol/mol creatinine (urine) [2.8–112 mg/g] 1.71–20.93 μmol/L (blood) [0.3–3.4 mg/L] 0.81–15.76 mmol/mol creatinine (urine) [1.2–23 mg/g]	Skender et al. (1991)
Printing workshop (Japan)	1,1,1-Trichloroethane, 5–65 ppm [27–351 mg/m³]	2–5 mg/L (urine)	Kawai et al. (1991)
Printing and ceramics workshop (Germany)	Trichloroethylene, 5–70 ppm [26.9–376 mg/m³]	2.0–201.0 mg/g creatinine (urine)	Triebig et al. (1982)
Environment			
Environmental levels (Italy)	Trichloroethylene: air: 1.7–26.9 μg/m³ water: 12–123 μg/L Tetrachloroethylene: air: 2.9–40 μg/m³ water: 2–68 μg/L	8.1–60.0 μg/L (plasma) 6.2–72.0 μg/g creatinine (urine)	Ziglio et al. (1985)
Environmental levels (Australia)	Trichloroacetic acid: drinking-water: 1.1–52 μg/L	0.8–38 μg/day urine	Froese et al. (2002)
Environmental levels (USA)	Trichloroacetic acid: drinking-water: 0.25–120 μg/L	1–~42 μg/day urine	Weisel et al. (1999); Kim et al. (1999)
	Trichloroacetic acid: swimming pools: 57–871 μg/L	Background, 0.15–1.18 μg/one urine void After exposure, 0.29–1.59 μg/one urine void	Kim & Weisel (1998)

samples were usually taken upon request, and seven exceeded the Finnish biological action level of 360 µmol/L [59 mg/L] (Rantala *et al.*, 1992).

Raaschou-Nielsen *et al.* (2001) examined 2397 measurements of trichloroacetic acid in urine collected between 1947 and 1985 from workers in various industries in Denmark. The urine samples were usually taken following a request from the local labour inspection agency or medical officer and the concentration of trichloroacetic acid ranged from 0.5 to 150 mg/L. The data showed that (*a*) a fourfold decrease in concentrations of trichloro-acetic acid occurred from 1947 to 1985; (*b*) the highest concentrations were observed in the iron and metal, chemical and dry cleaning industries; (*c*) levels of trichloroacetic acid were twice as high among men than among women in the iron and metal and dry cleaning industries; (*d*) concentrations of trichloroacetic acid concentrations were higher among younger than among older workers; and (*e*) persons working in an area in which trichloro-ethylene was used, but not working with trichloroethylene themselves, also showed uri-nary levels of trichloroacetic acid indicative of exposure.

1.3.3   *Air*

No data were available to the Working Group.

1.3.4   *Water*

Trichloroacetic acid is produced as a by-product during chlorination of water con-taining humic substances (Christman *et al.*, 1983; Miller & Uden, 1983; Legube *et al.*, 1985; Reckhow *et al.*, 1990). Consequently, it may occur in drinking-water after chlorine-based disinfection of raw waters containing natural organic substances (Hargesheimer & Satchwill, 1989; see IARC, 1991) and in swimming pools (Kim & Weisel, 1998; Stottmeister & Naglitsch, 1996). The concentrations of trichloroacetic acid measured in various water sources are summarized in Table 3.

Geist *et al.* (1991) measured concentrations of trichloroacetic acid ranging from < 3 to 558 µg/L in surface water downstream from a paper mill in Austria, and Mohamed *et al.* (1989) measured concentrations ranging from 838 to 994 µg/L in effluent from a kraft pulp mill in Malaysia.

Clemens and Schöler (1992a) measured concentrations of trichloroacetic acid of 0.9 µg/L in rainwater and 0.05 µg/L in groundwater in Germany. Plümacher and Renner (1993) measured concentrations ranging from 0.1–20 µg/L in rainwater.

Levels of trichloroacetic acid tend to decline with length of residence in the distri-bution system (Chen & Weisel, 1998), and tend to be higher in warmer seasons (LeBel *et al.*, 1997; Chen & Weisel, 1998). Trichloroacetic acid has been identified as a major chlorinated by-product of the photocatalytic degradation of tetrachloroethylene (IARC, 1979, 1987, 1995) in water but a minor by-product of that of trichloroethylene (IARC, 1976, 1979, 1987, 1995) in water (Glaze *et al.*, 1993).

Table 3. Concentrations of trichloroacetic acid in water

Water type (location)	Concentration range (μg/L)	Reference
Treatment plant and distribution system (Canada)	0.2–36.8	LeBel et al. (1997)
Treatment plant and distribution system (Poland)	8.18–15.20	Dojlido et al. (1999)
Treatment plant and distribution system (Republic of Korea)	0.6–4.2[a]	Shin et al. (1999)
Treatment plant (Canada)	0.1–273.2	Williams et al. (1997)
Drinking-water (USA)[b]	33.6–161	Uden & Miller (1983)
Drinking-water (USA)[b]	3.2–67	Norwood et al. (1986)
Drinking-water (USA)[b]	4.0–6.0	Krasner et al. (1989)
Drinking-water (USA)[b]	1.3–22	Jacangelo et al. (1989)
Drinking-water (USA)[b]	15–64	Reckhow & Singer (1990)
Drinking-water (Switzerland)[b]	3.0	Artho et al. (1991)
Drinking-water (Spain)[b]	0.3–2.5	Cancho et al. (1999)
Drinking-water and distribution system (USA)[b]	1–170	Obolensky et al. (2003)
Drinking-water (Germany)	Not detected–3	Lahl et al. (1984)
Distribution system (Canada)	0.1–473.1	Williams et al. (1997)
Distribution system (USA)	5.5–7.5	Chen & Weisel (1998)
Distribution system (Australia)	0.2–14	Simpson & Hayes (1998)
Drinking-water (USA)	6.4–14	Lopez-Avila et al. (1999)
Drinking-water (USA)	0.25–120	Weisel et al. (1999)
Drinking-water (Canada)	3.0–8.4	Scott et al. (2000)
Drinking-water (Australia)	1.1–52	Froese et al. (2002)
Drinking-water (Finland)	< 2.5–210	Nissinen et al. (2002)
Drinking-water (Spain)	0.1–25.5	Villanueva et al. (2003)
Irrigation water (USA)[c]	0–297	Comes et al. (1975)
Swimming pools (Germany)	18–136	Lahl et al. (1984)
Swimming pools (Germany)	Indoor: 3.3–9.1 Open-air: 46.5–100.6	Clemens & Schöler (1992b)
Swimming pools (Germany)	1.5–64.2	Mannschott et al. (1995)
Swimming pools (Germany)		Stottmeister & Naglitsch (1996)
Indoor	3.5–199	
Hydrotherapy	1.1–45	
Outdoor	8.2–887	
Swimming pools (USA)	57–871	Kim & Weisel (1998)

[a] Based on the assumption that trichloroacetic acid makes up approximately 32% of haloacetic acids
[b] Samples taken in water leaving the treatment plant
[c] Trichloroacetic acid was present in the irrigation water from its use as a herbicide.

Trichloroacetic acid has also been detected in the Great Lakes, Canada (Scott *et al.*, 2002) and in fog samples (0.02–2.0 µg/L) at ecological research sites in north-eastern Bavaria, Germany (Römpp *et al.*, 2001). Precipitation samples in Canada contained trichloroacetic acid concentrations ranging from < 0.0006 to 0.87 µg/L and concentrations in the Canadian lakes varied from < 0.0001 to 0.037 µg/L (Scott *et al.*, 2000). Trichloroacetic acid has a residence time of approximately 40 days in pond waters (Ellis *et al.*, 2001).

1.3.5 *Food*

Residues of trichloroacetic acid have been found in the seed of wheat, barley and oats after its use as a postemergence herbicide (Kadis *et al.*, 1972). Trace concentrations (0.01–0.20 ppm [0.01–0.20 mg/kg) have been detected in vegetables and fruits from fields irrigated with water containing trichloroacetic acid; slightly higher levels (0.13–0.43 mg/kg) were detected in field bean pods and seeds (Demint *et al.*, 1975).

1.3.6 *Other*

Concentrations of trichloroacetic acid were determined in the urine of people living in the vicinity of dry cleaning shops in Germany where tetrachloroethylene was used. The mean values were 105 µg/L for 29 neighbours and 682 µg/L for 12 workers (maximum, 1720 µg/L) (Popp *et al.*, 1992). In Zagreb, Croatia, the levels of trichloroacetic acid in fluids from 39 people with no known exposure to solvents were 14–160 µg/L in plasma and 2–292 µg/24 h in urine (Skender *et al.*, 1993). In 66 students in Japan, the levels ranged from not detected to 930 µg/g creatinine (urine) (Ikeda & Ohtsuji, 1969); those in 94 unexposed subjects in Germany were 5–221 µg/L in serum and 0.6–261 µg/24 h in urine (Hajimiragha *et al.*, 1986).

Trichloroacetic acid was detected at levels of 10–150 µg/kg in spruce needles from the Black Forest in Germany and the Montafon region in Austria, both considered to be relatively unpolluted areas (Frank *et al.*, 1989; Frank, 1991). The concentrations of trichloroacetic acid in pine needles from an urban area in Germany were 0.7–175 µg/kg fresh wt (Plümacher & Renner, 1993); those in conifer needles in Finland were 3–126 µg/kg fresh wt (Frank *et al.*, 1992). In Sitka spruce needles from Scotland, concentrations of up to approximately 22 µg/kg fresh needles were measured (Reeves *et al.*, 2000). In the vicinity of a pulp mill in Finland, concentrations of 2–135 µg/kg were found in pine needles (Juuti *et al.*, 1993). Trichloroacetic acid was also detected in earthworms (at 150–400 µg/kg wet wt) from a contaminated forest site (Back & Süsser, 1992).

1.4   Regulations and guidelines

The WHO (1998) has established a provisional guideline of 100 µg/L for trichloroacetic acid in drinking-water. A provisional guideline is established when there is some evidence of a potential health hazard but where available data on health effects are limited, or where an uncertainty factor greater than 1000 has been used in the derivation of the tolerable daily intake.

In Australia and New Zealand, the guideline for trichloroacetic acid in drinking-water is 100 µg/L (National Health and Medical Research Council and Agriculture and Resource Management Council of Australia and New Zealand, 1996). This guideline also notes that the minimization of the concentration of all chlorination by-products is encouraged by reducing the quantity of naturally occurring organic material in the source water, reducing the amount of chlorine added or using an alternative disinfectant, without compromising disinfection.

In the USA, the Environmental Protection Agency (1998) regulates trichloroacetic acid as one of a combination of five haloacetic acids, which also include monochloracetic acid, dichloroacetic acid and mono- and dibromoacetic acids. The maximum contaminant level for the sum of these five haloacetic acids is 60 µg/L.

The European Union (European Commission, 1998) and Canada (Health Canada, 2003) have not set guideline values but encourage the reduction of concentrations of total disinfection by-product.

2.   Studies of Cancer in Humans

See Introduction to the monographs on chloramine, chloral and chloral hydrate, dichloroacetic acid, trichloroacetic acid and 3-chloro-4-(dichloromethyl)-5-hydroxy-2(5H)-furanone.

3.   Studies of Cancer in Experimental Animals

Previous evaluation

Trichloroacetic acid has been evaluated previously (IARC, 1995) and was found to induce hepatocellular adenomas and carcinomas in male B6C3F$_1$ mice and to possess promoter activity. The previous evaluation of trichloroacetic acid indicated that there was limited evidence in experimental animals for its carcinogenicity (IARC, 1995).

New studies

3.1   Oral administration

3.1.1   *Mouse*

Groups of 93, 46 and 38 female B6C3F$_1$ mice, 7–8 weeks of age, were administered trichloroacetic acid in the drinking-water at concentrations of 2.0, 6.67 and 20.0 mmol/L [324, 1080 or 3240 mg/L], adjusted to pH 6.5–7.5 with sodium hydroxide; a control group of 134 animals received 20.0 mmol/L sodium chloride. Mice were killed after 360 or 576 days (when high-dose mice became moribund) of exposure. The livers were weighed and evaluated for foci of altered hepatocytes (basophilic and eosinophilic foci), adenomas and carcinomas. After 360 or 576 days of exposure, the liver-to-body weight ratio was increased dose-dependently following treatment with trichloroacetic acid. Data from mice administered trichloroacetic acid were compared with those from control mice using Fisher's exact test with a p-value < 0.05. The high dose of trichloroacetic acid (20.0 mmol/L) increased, in comparison with controls, the incidence of foci (11/18 versus 10/90 at 576 days), adenomas (7/18 versus 2/90 at 576 days) and carcinomas (5/20 versus 0/40 at 360 days and 5/18 versus 2/90 at 576 days). The mid dose of trichloroacetic acid (6.67 mmol/L) increased the incidence of foci (9/27) and hepatocellular carcinomas (5/27) at 576 days, while the low dose of 2.0 mmol/L did not alter the incidence of any liver lesion. In control mice, the incidence of lesions was 1/40 adenoma (2.5%) at 360 days and 10/90 foci (11.1%), 2/90 adenomas (2.2%) and 2/90 carcinomas (2.2%) at 576 days (Pereira, 1996).

The ability of mixtures of di- and trichloroacetic acid to induce liver tumours was studied in 6-week-old B6C3F$_1$ male mice. Treatments administered included 0.1, 0.5 and 2.0 g/L dichloroacetic acid, and 0.5 and 2.0 g/L trichloroacetic acid, and selected combinations of these treatments. Twenty animals were assigned to each of 10 groups that received the above concentrations in their drinking-water for 52 weeks. Dose-related increases in the incidence of liver tumours (adenomas and carcinomas combined) were observed with the individual compounds and, when the animals were exposed to mixtures of di- and trichloroacetic acid, it appeared that there was an approximately additive effect in terms of tumour incidences (see Table 4) (Bull et al., 2002).

3.1.2 Rat

Groups of 50 male Fischer 344/N rats, 28–30 days of age, received 0.05, 0.5 and 5 g/L neutralized trichloroacetic acid, adjusted to pH 6.9–7.1 with sodium hydroxide, or 2 g/L sodium chloride in the drinking-water for a total 104 weeks. Interim sacrifices were made at 15, 30, 45 and 60 weeks. A complete necropsy of the animals was performed. The liver, kidney, spleen, testes and gross lesions were examined microscopically. A complete pathological examination was carried out on all tissues from all animals in the high-dose group. The high dose of trichloroacetic acid but not the low or mid dose decreased body weight (~11%). Trichloroacetic acid did not affect the absolute or relative (to body weight) weights of the liver, kidneys, spleen or testes except for a decrease in the absolute liver weight in rats administered 5.0 g/L ($p \leq 0.05$). At 104 weeks, the number of animals per treatment group ranged from 20 to 24 including one rat that died after 76 weeks. The number of rats with hepatocellular adenomas varied between one and three among the treatment groups (4.2–15.0%). A single hepatocellular carcinoma (1/22, 4.6%) was found in the high-dose trichloroacetic acid-treated group). None of the treatment groups had a significant increase in the incidence of any tumour in other organs (DeAngelo et al., 1997).

Table 4. Effect on liver tumour incidence of di-chloroacetic acid (DCA) and trichloroacetic acid (TCA) administered in the drinking-water for 52 weeks to male B6C3F$_1$ mice

Treatment	Tumour incidence (adenomas and carcinomas)	Tumour multiplicity[a]
Control (drinking-water)	1/20	0.05 ± 0.0
0.1 g/L DCA	2/20	0.10 ± 0.07
0.5 g/L DCA	5/20[b]	0.35 ± 0.15[b]
2 g/L DCA	12/19[b]	1.7 ± 0.5[b]
0.5 g/L TCA	11/20[b]	0.70 ± 0.16[b]
2 g/L TCA	9/20[b]	0.60 ± 0.18[b]
0.1 DCA + 0.5 TCA g/L	9/20[b]	0.65 ± 0.22[b]
0.1 DCA + 2 TCA g/L	15/20[b]	1.3 ± 0.2[b]
0.5 DCA + 0.5 TCA g/L	13/19[b]	1.4 ± 0.3[b]
0.5 DCA + 2 TCA g/L	13/20[b]	1.5 ± 0.3[b]

From Bull *et al.* (2002)
[a] Total number of tumours divided by total number of animals
[b] Significantly different from control at $p < 0.05$

3.2   Administration with known carcinogens or modifying factors

Tumour-promotion studies

Groups of 6–40 female B6C3F$_1$ mice, 15 days of age, were initiated with an intraperi-toneal injection of 25 mg/kg N-methyl-N-nitrosourea (MNU). At 19 days of age, the animals received 2.0, 6.67 or 20.0 mmol/L [324, 1080 or 3240 mg/L] trichloroacetic acid, adjusted to pH 6.5–7.5 with sodium hydroxide, or 20.0 mmol/L sodium chloride as a control for the sodium salt in the drinking-water. Mice were killed after 31 or 52 weeks of exposure. At 52 weeks, the mid (6.67 mmol/L) and high (20.0 mmol/L) doses of tri-chloroacetic acid significantly ($p < 0.01$) increased the incidence of carcinomas in MNU-initiated mice from 4/40 (10.0%) to 5/6 (83.3%) and 20/24 (83.3%), and multiplicity from 0.10 ± 0.05 to 1.33 ± 0.42 and 2.79 ± 0.48, respectively. At 31 weeks, the high dose of tri-chloroacetic acid increased the incidence of hepatocellular adenomas in MNU-initiated mice from 0/10 to 6/10 (60%) and multiplicity from 0.00 to 1.30 ± 0.045. At 52 weeks, the mid and high doses of trichloroacetic acid increased the incidence of adenomas from 7/40 (17.5%) to 16/24 (66.7%) and 5/6 (83.3%), respectively, and the multiplicity from 0.28 ± 0.11 to 2.00 ± 0.82 and 1.29 ± 0.24. In mice that were not administered MNU, the high dose of trichloroacetic acid significantly increased the incidence of carcinomas from 0/40 to 5/20 (25.0%) (Pereira & Phelps, 1996).

Combinations of dichloroacetic acid and trichloroacetic acid have been evaluated for tumour-promoting activity. Female B6C3F$_1$ mice, 15 days of age, were initiated with MNU (25 mg/kg bw) followed by exposure to 0, 7.8, 15.6 and 25 mmol/L dichloroacetic acid with or without 6.0 mmol/L trichloroacetic acid or 0, 6.0 and 25 mmol/L trichloroacetic acid with or without 15.6 mmol/L dichloroacetic acid. The pH of the dose solutions was adjusted to 6.5–7.5 with sodium hydroxide. Exposure was from week 4 to 48 of age, at which time the mice were killed. The high dose of dichloroacetic acid (25 mmol/L) and trichloroacetic acid (25 mmol/L) significantly increased ($p < 0.05$) the multiplicity of hepatocellular adenomas from 0.07 ± 0.05 (no dichloroacetic acid or trichloroacetic acid) to 1.79 ± 0.29 and 0.52 ± 0.11, respectively. The lower doses of dichloroacetic acid and trichloroacetic acid did not significantly increase the incidence or multiplicity of adenomas (Pereira *et al.*, 1997).

The effect of chloroform on liver and kidney tumour promotion by trichloroacetic acid as well as dichloroacetic acid has been investigated. The concentrations of chloroform were chosen because they prevented dichloroacetic acid-induced DNA hypomethylation and increased mRNA expression of the c-*myc* gene. However, chloroform did not alter trichloroacetic acid-induced DNA hypomethylation and expression of the c-*myc* gene. Groups of male and female B6C3F$_1$ mice, 15 days of age, were initiated with 30 mg/kg MNU. At 5 weeks of age, the mice started to receive in the drinking-water 4.0 g/L trichloroacetic acid or 3.2 g/L dichloroacetic acid neutralized with sodium hydroxide with 0, 800 or 1600 mg/L chloroform and were killed at 36 weeks of age. The results were analysed for statistical significance by a one-way ANOVA followed by the Tukey test with a *p*-value < 0.05. In MNU-initiated mice that did not receive trichloroacetic acid, hepatocellular adenomas were found in 2/29 (6.9%) females and 2/8 (25%) males, while no hepatocellular carcinomas were found. Trichloroacetic acid increased the incidence of liver adenocarcinomas (10/16 [62.5%]) and adenomas (12/16 [75%]) in male mice. In female mice administered trichloroacetic acid, the incidence of mice with hepatocellular adenocarcinomas and adenomas was not significantly altered: 4/14 (28.6%) and 2/14 (14.3%). In male mice administered trichloroacetic acid plus 0, 800 and 1600 mg/L chloroform, the incidence of hepatocellular adenocarcinomas was 10/16 (62.5%), 7/9 (87.5%) and 6/8 (75%) and that of hepatocellular adenomas was 12/16 (75%), 6/9 (75%) and 1/8 (12.5%), respectively. The incidence of mice with hepatocellular adenomas was significantly lower in mice administered trichloroacetic acid plus 1600 mg/L chloroform than in mice administered trichloroacetic acid ($p < 0.05$). No altered hepatocyte foci, adenomas or adenocarcinomas were found in six MNU-initiated male mice that were administered 1600 mg/L chloroform. Multiplicity of tumours (adenomas plus adenocarcinomas) was increased in male mice from 0.25 ± 0.16 to 3.81 ± 0.82 ($p < 0.001$), but not in female mice, with 0.07 ± 0.04 and 0.64 ± 0.22 for control and trichloroacetic acid-exposed mice, respectively. Sixty per cent of the tumours were adenocarcinomas, indicating that the multiplicity of adenocarcinomas was significantly increased in male mice exposed to trichloroacetic acid. Renal tumours of tubular origin were found only in male mice. The majority (more than 70%) were papillary cystic adenomas with the rest were cystic adenomas and to a lesser extent adenocarcinomas (~5%). Trichloroacetic acid increased the incidence and the multiplicity of MNU-initiated kidney tumours from 0 (0/8) to 87.5% (14/16) and 1.68 tumours per mouse. Chloroform did not alter the promotion of kidney tumours by trichloroacetic acid; the incidence of mice with tumours was 71.4–87.5% and multiplicity was 1.00–1.68 kidney tumours per mouse for mice administered trichloroacetic acid with 0, 800 or 1600 mg/L chloroform. No kidney tumours were found in the six MNU-initiated male mice administered 1600 mg/L chloroform only. In female mice, the incidence of kidney tumours among all the treatment groups ranged from 0 to 28.6% and the multiplicity ranged from 0 to 0.29 tumours per mouse (Pereira et al., 2001).

4.   Other Data Relevant to an Evaluation of Carcinogenicity and its Mechanisms

4.1   Absorption, distribution, metabolism and excretion

4.1.1   Humans

A physiologically based pharmacokinetic model for trichloroacetic acid in humans has been developed on the basis of data from studies on the kinetics of trichloroacetic acid in rodents (Allen & Fisher, 1993). The rate of systemic clearance of trichloroacetic acid in humans was slower than that in rodents (0.045–0.1/h/kg in rats and mice). Trichloroacetic acid is eliminated largely (93%) unchanged (Allen & Fisher, 1993).

A second-generation physiologically based pharmacokinetic model for trichloroethylene that includes trichloroacetic acid as a metabolite has been developed (Fisher et al., 1998). Nine male (20–36 years old) and eight female (20–30 years old) volunteers were exposed by inhalation to 50 or 100 ppm [269 or 538 µg/L] trichloroethylene for 4 h. The first-order rate constants for the excretion of trichloroacetic acid in urine averaged 1.54/h (range, 0.7–3.0/h) and 1.1/h (range, 0.13–2.0/h) in women and men, respectively. The pharmacokinetic model successfully simulated the cumulative excretion of trichloroacetic acid in the urine of both men and women (Fisher et al., 1998).

Several studies have examined the elimination half-life of trichloroacetic acid in humans. The plasma half-life of trichloroacetic acid ranged from 4 to 5 days after oral ingestion of 15 mg/kg bw chloral hydrate (Breimer et al., 1974) and was 50.6 h after oral administration of 3 mg/kg bw trichloroacetic acid (Müller et al., 1972, 1974).

Froese et al. (2002) estimated the half-life of trichloroacetic acid in humans after drinking tap-water containing a range of disinfection products, including trichloroacetic acid. The intake of trichloroacetic acid was 20–82 µg/day during the 12-day study period. Although 10 volunteers (eight men, two women) were enrolled in the study, useful elimination data were obtained for only three, in whom the elimination half-lives ranged from 2.3 to 3.67 days.

The fraction of trichloroacetic acid bound to human plasma proteins ranged from 74.8 to 84.3% at concentrations of 6, 61 and 612 nmol/mL trichloroacetic acid; the binding of trichloroacetic acid to human plasma proteins is greater than that to rat (38.3–53.5%) or dog (54.2–64.8%) plasma protein (Templin *et al.*, 1995).

4.1.2   *Experimental systems*

The metabolism of trichloroacetic acid was studied in male Fischer 344 rats and male B6C3F$_1$ mice (Larson & Bull, 1992). The animals were given 5, 20 or 100 mg/kg bw [^{14}C]trichloroacetic acid orally, and the ^{14}C content of the urine, faeces, exhaled air and carcass was measured. Approximately 50% (47.8–64.6%) of any dose of trichloroacetic acid was excreted unchanged in the urine of both rats and mice. The half-life of trichloroacetic acid in rats and mice given 20 or 100 mg/kg bw ranged from 4.2 to 7.0 h, and the clearance ranged from 36 to 66 mL/kg/h. The combined excretion of glyoxylic acid, oxalic acid and glycolic acid in urine amounted to 4.9–10.8% of the administered dose. Although not stated by the authors, glyoxylic acid, oxalic acid and glycolic acid are known metabolites of dichloroacetic acid and may have been formed from trichloroacetic acid-derived dichloroacetic acid. Dichloroacetic acid was detected in the urine of rats and mice, indicating the reduction of trichloroacetic acid. The authors proposed that trichloroacetic acid undergoes reduction to the dichloroacetyl radical ($\bullet$CCl$_2$COOH), which may abstract a hydrogen atom to form dichloroacetic acid or may react with oxygen to form a hydroperoxyl radical ($\bullet$OOCCl$_2$COOH) that may yield oxalic acid (Larson & Bull, 1992).

The kinetics of the elimination of trichloroacetic acid as a metabolite of inhaled trichloroethylene or following intravenous administration has been reported in pregnant rats and in lactating rats and nursing pups (Fisher *et al.*, 1989, 1990). In pregnant rats exposed by inhalation to 618 ppm [3.3 mg/L] trichloroethylene for 4 h on day 12 of gestation or given 4 mg/kg bw trichloroacetic acid intravenously on days 14–15 of pregnancy, the elimination rate constant was 0.045/h. Fetal exposure to trichloroacetic acid was estimated at 63–64% of the maternal dose (Fisher *et al.*, 1989). The elimination rate constants in lactating dams exposed by inhalation to 600.4 ppm [3.2 mg/L] trichloroethylene for 4 h or given 4.4 mg/kg bw trichloroacetic acid intravenously were 0.063/h and 0.086/h, respectively, whereas that in nursing pups whose dams had been exposed to trichloroethylene by inhalation was 0.014/h. It was estimated that 4.2–6.8% of the trichloroacetic acid formed as a metabolite of trichloroethylene was eliminated by lactational transfer (Fisher *et al.*, 1989).

The kinetics of the elimination was studied in male B6C3F$_1$ mice given 0.03, 0.12 and 0.61 mmol/kg bw [5, 20 and 100 mg/kg bw] trichloroacetic acid by gavage (Templin *et al.*, 1993). The half-life ranged from 5.4 to 6.4 h. A comparison of the area-under-the-curve for distribution of trichloroacetic acid to the blood and liver following exposure to trichloroethylene showed that distribution favoured the blood over the liver. Analysis of the binding of trichloroacetic acid to plasma proteins gave a K_D and B_{max} of 248 and 310 μM [40.5 and 50.6 mg/L], respectively.

The half-lives for the elimination of trichloroacetic acid in male Fischer 344 rats given 0.15 and 0.76 mmol/kg bw [24.5 and 124 mg/kg bw] orally were 7.9 and 13 h, respectively, whereas those in male beagle dogs given 0.15, 0.38 and 0.76 mmol/kg bw [24.5, 62 and 124 mg/kg bw] orally were 200, 175 and 238 h, respectively (Templin et al., 1995).

The half-life of trichloroethanol-derived trichloroacetic acid in male Fischer 344 rats that had no enterohepatic circulation because of the placement of a bile shunt and were given 5, 20 or 100 mg/kg bw trichloroethanol intravenously ranged from 7.7 to 11.6 h (Stenner et al., 1997). The areas-under-the-curve for trichloroacetic acid in the blood and bile of rats without enterohepatic circulation that were given 5, 20 or 100 mg/kg bw trichloroethanol were 1.6, 12.1 or 38.3 and 5.4, 12.8 or 28.3 mg/h/mL, respectively. In rats with intact entero-hepatic circulation given 100 mg/kg bw trichloroacetic acid intravenously, the half-life for the elimination of trichloroacetic acid from the blood was 11.6 h. These data were incorpo-rated into a physiologically based pharmacokinetic model for trichloroethylene and its meta-bolites that included enterohepatic circulation of metabolites of trichloroethylene (Stenner et al., 1998). The authors concluded that the elimination of trichloroacetic acid is best described by a multi-exponential decay in which the long half-life of trichloroacetic acid is associated with its renal re-absorption.

Second-generation physiologically based pharmacokinetic models for trichloroethylene that include the kinetics of trichloroacetic acid formation and excretion in $B6C3F_1$ mice have been developed (for review, see Fisher, 2000). These models differ from first-gene-ration models in that a proportionality constant was not used to describe the stoichiometry of the metabolism of trichloroethylene to trichloroacetic acid. A physiologically based phar-macokinetic model for trichloroethylene that includes kinetic constants for the urinary excretion of trichloroacetic acid has been reported (Abbas & Fisher, 1997). In male $B6C3F_1$ mice given 1200 mg/kg bw trichloroethylene orally, the first-order rate constant values for the metabolism of trichloroacetic acid to dichloroacetic acid and for the urinary excretion of trichloroacetic acid were 0.35 and 1.55/h/kg, respectively. The authors noted that 66% of the trichloroethylene-derived trichloroacetic acid was excreted in the urine and the remaining 34% was presumed to be metabolized to dichloroacetic acid. Because dichloroacetic acid was not detected in the urine of mice given trichloroethylene orally, it was speculated that dichloroacetic acid must be metabolized extensively (Abbas & Fisher, 1997). In a study of the rates of elimination of trichloroacetic acid and its metabolism to dichloroacetic acid in male $B6C3F_1$ mice exposed by inhalation to 100 or 600 ppm trichloroethylene [0.538 or 3.2 mg/L], the first-order rate constant values for the conversion of trichloroacetic acid to di-chloroacetic acid and for the urinary excretion of trichloroacetic acid were 0.004 and 2.50/h/kg, respectively (Greenberg et al., 1999).

The metabolic fate of trichloroacetic acid has been investigated in male $B6C3F_1$ mice given 100 mg/kg bw trichloroacetic acid containing [1,2-^{14}C]trichloroacetic acid by gavage; ~5, ~55 and < 10% of the dose was eliminated as carbon dioxide and in the urine and faeces, respectively, and ~25% was found in the carcass. Trichloroacetic acid, dichloroacetic acid, chloroacetic acid, glyoxylic acid, glycolic acid, oxalic acid and unidentified metabolites accounted for 44.5, 0.2, 0.03, 0.06, 0.11, 1.5 and 10.2% of the urinary metabolites (Xu *et al.*, 1995).

The metabolism of trichloroacetic acid to dichloroacetic acid was studied in control and dichloroacetic acid-treated male B6C3F$_1$ mice given 100 mg/kg bw trichloroacetic acid intravenously (Merdink *et al.*, 1998). Prolonged treatment with dichloroacetic acid depletes GSTZ1-1, which catalyses its metabolism which is thereby reduced (Saghir & Schultz, 2002). In contrast with other reports (Larson & Bull, 1992; Xu *et al.*, 1995), quantifiable concentrations of dichloroacetic acid were not detected in the blood of mice given trichloroacetic acid. The authors concluded that, although there is uncertainty about the metabolism of trichloroacetic acid to dichloroacetic acid, pharmacokinetic simulations indicate that dichloroacetic acid is probably formed as a short-lived metabolite of trichloroacetic acid and that its rapid elimination compared with its relatively slow formation prevents its ready detection (Merdink *et al.*, 1998). The artefactual formation of dichloroacetic acid from trichloroacetic acid has also been noted: for example, trichloroacetic acid was converted to dichloroacetic acid in freshly drawn blood samples (Ketcha *et al.*, 1996).

The tissue disposition and elimination of [1-^{14}C]trichloroacetic acid was studied in male Fischer 344 rats given 6.1, 61 or 306 μmol/kg bw [1, 10 and 50 mg/kg bw] trichloroacetic acid intravenously (Yu *et al.*, 2000). The fraction of the initial dose excreted in the urine increased from 67 to 84% as the dose increased and faecal excretion decreased from 7 to 4%. The elimination of trichloroacetic acid as carbon dioxide decreased from 12 to 8% of the total dose. The authors noted that the hepatic intracellular concentrations of trichloroacetic acid were significantly greater than the free plasma concentrations, indicating concentrative uptake by hepatocytes, and that trichloroacetic acid filtered at the glomerulus appears to be reabsorbed from either the renal tubular urine or the bladder (Yu *et al.*, 2000).

Gonthier and Barret (1989) reported that trichloroacetic acid (60 μL/mL neat compound) did not yield a spin adduct when incubated with liver or brain microsomes from male Sprague-Dawley rats in the presence of reduced nicotinamide adenine dinucleotide phosphate (NADPH) and the spin trap α-(4-pyridyl-1-oxide)-*N-tert*-butylnitrone. The absence of a hydroxyl radical signal during microsomal metabolism of trichloroacetic acid remains a controversial point (Gonthier & Barret, 1989). [The Working Group noted that trichloroacetic acid may have inactivated enzymes responsible for its reduction. Moreover, the reaction mixtures were apparently incubated in the presence of air, which may also have prevented its reduction.]

The biotransformation of trichloroacetic acid was studied in hepatic microsomal fractions isolated from control and pyrazole-treated male B6C3F$_1$ mice (Ni *et al.*, 1996). When trichloroacetic acid (5 mM [817 μg/mL]) was incubated with a microsomal fraction and a NADPH-generating system in the presence of the spin trap *N-tert*-butyl-α-phenylnitrone [the concentration of oxygen in the closed reaction flasks was not stated], analysis by electron spin resonance spectroscopy indicated the presence of a carbon-centred radical, which was not characterized (Ni *et al.*, 1996).

The mechanism of the reduction of trichloroacetic acid to dichloroacetic acid has been investigated (Merdink *et al.*, 2000). Microsomal fractions from male B6C3F$_1$ mice or

Fischer 344 rats were incubated with a NADPH-generating system, trichloroacetic acid (1 mM) [163.4 µg/mL] and the spin trap phenyl-*tert*-butyl nitroxide (PBN) in an argon (anaerobic) atmosphere. Gas chromatographic–mass spectrometric analysis of methylated extracts of the reaction mixture revealed the formation of 2-*tert*-butyl-4,4-dichloro-3-phenylisoxazolidin-5-one derived from the dichloroacetate radical, whose mass spectrum was interpreted to be a cyclized form of the expected PBN/dichloroacetyl radical. The same product was formed when trichloroacetic acid was incubated with PBN, ferrous sulfate and hydrogen peroxide (Fenton reaction system) (Merdink *et al.*, 2000).

4.1.3   *Comparison of humans and animals*

The results of several studies indicate that the urinary elimination or plasma clearance of trichloroacetic acid is slower in humans than in rodents.

4.2   Toxic effects

This section reviews the literature on the toxic effects of trichloroacetic acid not cited in the previous monograph (IARC, 1995).

4.2.1   *Humans*

No new data were available to the Working Group.

4.2.2   *Experimental systems*

As reported in the previous monograph on trichloroacetic acid (IARC, 1995), short-term treatment (≤ 14 days) resulted in increases in cell replication rates in the liver of mice. The elevated rates of replication were not sustained and became substantially reduced compared with controls with and without chronic pretreatment (Pereira, 1996; Stauber & Bull, 1997). In an experiment in which treatment of male B6C3F$_1$ mice with 2 g/L trichloroacetic acid was terminated after 1 year (50 weeks), cell replication rates within tumours were not dependent upon continued treatment (for an additional 2 weeks). Trichloroacetic acid did not stimulate replication of initiated cells. As only one time-point was measured, the possibility that trichloroacetic acid affected replication rates of preneoplastic lesions cannot be ruled out (Stauber & Bull, 1997).

The previous review showed that trichloroacetic acid induced peroxisome proliferation in rodents (IARC, 1995). A recent in-vitro study of COS-1 cells transiently co-transfected with a peroxisome proliferator-activated receptor (PPAR) expression plasmid, pCMV-mPPARα, together with a reporter plasmid containing a peroxisome proliferator response element, Pluc4A6-880, clearly demonstrated that trichloroacetate directly activates the PPARα (Zhou & Waxman, 1998). In other studies, trichloroacetic acid induced peroxisome proliferation in primary cultures of hepatocytes from rats and mice but not in those from humans (Elcombe, 1985; Walgren *et al.*, 2000a). In addition, human hepatocytes that expressed endogenous human PPARα did not respond to trichloroacetic acid, whereas human cells co-transfected with mouse PPARα and mouse retinoid X receptor plasmids displayed increased activity of the peroxisome proliferator response element reporter after treatment with trichloroacetic acid and other peroxisome proliferators. Retinoid X receptor that forms a heterodimer with PPAR enhanced PPAR–DNA binding and transcriptional activation (retinoid X receptor is a common partner for many steroid receptors) (Walgren *et al.*, 2000b). The phenotype of trichloroacetic acid-induced liver tumours in mice is also consistent with a mechanism of action that is parallel to or dependent upon activation of peroxisome synthesis (Nakano *et al.*, 1994; Latendresse & Pereira, 1997; Stauber *et al.*, 1998; Bull *et al.*, 2002).

Chronic treatment with trichloroacetic acid produced relatively mild changes in carbohydrate metabolism compared with dichloroacetic acid. In male mice, it reduced liver glycogen content (Kato-Weinstein *et al.*, 2001), whereas dichloroacetate increased it. It had no measurable effects on serum insulin concentrations, but produced an elevation in serum glucose at high doses (3 g/L). These findings also contrast with dichloroacetate, which has been shown to decrease serum insulin levels substantially with no effect on serum glucose concentrations (Kato-Weinstein *et al.* 2001; Lingohr *et al.*, 2001).

Short-term oral treatment (11 days) of mice with trichloroacetate (25 mmol/L) inhibited methylation of DNA in liver, an effect that was not observed with long-term treatment (44 weeks) (Tao *et al.*, 1998). However, methylation of DNA was depressed in trichloroacetate-promoted liver tumours at 44 weeks and suspension of treatment 1 week prior to sacrifice did not reverse this effect. An increased expression of *c-jun* and *c-myc* proto-oncogenes was observed when the 5-methylcytosine levels in their respective promoter regions decreased (Tao *et al.*, 2000a) and administration of methionine 30 min after trichloroacetate inhibited expression of both proto-oncogenes (Tao *et al.*, 2000b). Increased cell replication rates and decreased methylation of the c-*myc* gene were first observed simultaneously in mice 72 h after the start of exposure to trichloroacetic acid. Trichloroacetic acid induced DNA hypomethylation by inducing DNA replication and preventing the methylation of the newly synthesized strands of DNA (Ge *et al.*, 2001). The authors speculated that trichloroacetate depleted *S*-adenosylmethionine levels. Depressed levels of 5-methylcytosine were observed in the kidney and bladder as well as the liver.

4.3   Reproductive and prenatal effects

4.3.1   *Humans*

No data were available to the Working Group.

4.3.2   *Experimental systems*

Sprague-Dawley rats administered 2730 ppm trichloroacetic acid in the drinking-water (mean daily dose, 291 mg/kg) throughout pregnancy had significant increases in the number of fetal resorptions (2.7 per litter compared with 0.70 per litter for controls) (Johnson *et al.*,

1998a,b). These were accompanied by a significant increase in the incidence of heart ano-malies (from 13/605 to 12/114 fetuses) that were described somewhat differently from those reported by Smith *et al.* (1989) because of the more detailed dissection technique used.

A subsequent study in which trichloroacetate was administered by gavage at a daily dose of 300 mg/kg on days 6–15 of gestation during organogenesis did not produce clear evidence of a teratogenic effect (Fisher *et al.*, 2001). The dissections were the same as those used by Johnson *et al.* (1998a,b). However, the study by Fisher *et al.* (2001) was compli-cated by widely divergent incidences of anomalies in the two different vehicle control groups (soya bean oil versus water). Consequently, the in-vivo information on the terato-genic effects of trichloroacetate appears to be inconclusive.

Trichloroacetate was evaluated for effects on the growth and development of Sprague-Dawley rat embryos in whole-embryo culture (Saillenfait *et al.*, 1995). Malformed embryos were observed when the concentration of trichloroacetate reached 2.5 mM (no effect at 1 mM). There appeared to be some effects on growth at 1 mM, as total DNA and protein content per embryo was reduced (but not significantly) at this concentration. Hunter *et al.* (1996) reported a significant dose-dependent increase in the incidence of malformations, including neural tube defects, rotational defects, eye defects, pharyngeal arch defects and heart defects, in mouse whole-embryo cultures treated with 1–5 mM trichloroacetate.

4.4   Genetic and related effects

4.4.1   *Humans*

No data were available to the Working Group.

4.4.2   *Experimental systems*

(a)   *DNA binding*

The level of 8-hydroxydeoxyguanosine–DNA adducts in the liver of B6C3F$_1$ mice was not modified after administration of trichloroacetic acid through drinking-water (Parrish *et al.*, 1996), was slightly increased after administration by gavage (Austin *et al.*, 1996) and was clearly increased after intraperitoneal injection (Von Tungeln *et al.*, 2002). After treatment with trichloroacetic acid, the level of malondialdehyde-derived adducts was increased *in vivo* (Von Tungeln *et al.*, 2002) and *in vitro* (Beland, 1999).

(b)   *Mutagenic and allied effects* (see Table 5 for details and references)

Trichloroacetic acid did not induce λ prophage or SOS repair in *Escherichia coli*. It was not mutagenic to *Salmonella typhimurium* strains in the presence or absence of meta-bolic activation, except in a single study on strain TA100 using a modified protocol in liquid medium.

The frequency of chlorophyll mutations was increased in *Arabidopsis* after treatment of seeds with trichloroacetic acid.

Table 5. Genetic and related effects of trichloroacetic acid

Test system	Result[a] Without exogenous metabolic system	With exogenous metabolic activation	Dose[b] (LED or HID)	Reference
λ Prophage induction, *Escherichia coli* WP2s	–	–	10 000	DeMarini *et al.* (1994)
SOS chromotest, *Escherichia coli* PQ37	–	–	10 000	Giller *et al.* (1997)
Bacillus subtilis H17 *rec*+ and M45 *rec*–	–	NT	20 μg/plate	Shirasu *et al.* (1976)
Escherichia coli, B/r *try* WP2, reverse mutation	–	NT	20 μg/plate	Shirasu *et al.* (1976)
Salmonella typhimurium TA100, TA98, reverse mutation	–	–	450 μg/plate	Waskell (1978)
Salmonella typhimurium TA100, TA1535, reverse mutation	–	–	4000 μg/plate	Nestmann *et al.* (1980)
Salmonella typhimurium TA100, reverse mutation	–	NT	520 μg/plate	Rapson *et al.* (1980)
Salmonella typhimurium TA100, TA98, reverse mutation	–	–	5000 μg/plate	Moriya *et al.* (1983)
Salmonella typhimurium TA100, reverse mutation	–	–	600 ppm [600]	DeMarini *et al.* (1994)
Salmonella typhimurium TA100, reverse mutation, liquid medium	+	+	1750	Giller *et al.* (1997)
Salmonella typhimurium TA100, RSJ100, reverse mutation	–	–	16 300	Kargalioglu *et al.* (2002)
Salmonella typhimurium TA104, reverse mutation, microsuspension	–	–	250 μg/plate	Nelson *et al.* (2001)
Salmonella typhimurium TA1535, TA1536, TA1537, TA1538, reverse mutation	–	NT	20 μg/plate	Shirasu *et al.* (1976)
Salmonella typhimurium TA1537, TA1538, TA98, reverse mutation	–	–	2000 μg/plate	Nestmann *et al.* (1980)
Salmonella typhimurium TA98, reverse mutation	–	–	13 100	Kargalioglu *et al.* (2002)
Arabidopsis species, mutation	+	NT	1000	Plotnikov & Petrov (1976)
Micronucleus formation, *Pleurodeles waltl* newt larvae peripheral erythrocytes *in vivo*	+		80[d]	Giller *et al.* (1997)
DNA strand breaks, B6C3F$_1$ mouse and Fischer 344 rat hepatocytes *in vitro*	–	NT	1630	Chang *et al.* (1992)
DNA damage, Chinese hamster ovary cells *in vitro*, single-cell gel electrophoresis assay	–	NT	3 mM [490]	Plewa *et al.* (2002)

Table 5 (contd)

Test system	Result[a]		Dose[b] (LED or HID)	Reference
	Without exogenous metabolic system	With exogenous metabolic activation		
Gene mutation, mouse lymphoma L5178Y/TK$^{+/-}$-3.7.2.C cells *in vitro*	?	(+)	3000	Harrington-Brock *et al.* (1998)
Formation of anchorage-independent colonies, B6C3F1 mouse hepatocytes *in vitro*	+	NT	82	Stauber *et al.* (1998)
DNA strand breaks, human CCRF-CEM lymphoblastic cells *in vitro*	–	NT	1630	Chang *et al.* (1992)
Chromosomal aberrations, human lymphocytes *in vitro*	–	–[c]	5000	Mackay *et al.* (1995)
DNA strand breaks, B6C3F$_1$ mouse liver *in vivo*	+		1.0 po × 1	Nelson & Bull (1988)
DNA strand breaks, B6C3F$_1$ mouse liver *in vivo*	+		500 po × 1	Nelson *et al.* (1989)
DNA strand breaks, B6C3F$_1$ mouse liver *in vivo*	–		500 po × 10	Nelson *et al.* (1989)
DNA strand breaks, B6C3F$_1$ mouse liver and epithelial cells from stomach and duodenum *in vivo*	–		1630 po × 1	Chang *et al.* (1992)
DNA strand breaks, Sprague-Dawley rat liver *in vivo*	+		100 po × 1	Nelson & Bull (1988)
DNA strand breaks, Fischer 344 rat liver *in vivo*	–		1630 po × 1	Chang *et al.* (1992)
Micronucleus formation, Swiss mice *in vivo*	+		125 ip × 2	Bhunya & Behera (1987)
Micronucleus formation, female C57BL/6JfBL10/Alpk mouse bone-marrow erythrocytes *in vivo*	–		1300 ip × 2	Mackay *et al.* (1995)
Micronucleus formation, male C57BL/6JfBL10/Alpk mouse bone-marrow erythrocytes *in vivo*	–		1080 ip × 2	Mackay *et al.* (1995)
Chromosomal aberrations, Swiss mouse bone-marrow cells *in vivo*	+		125 ip × 1	Bhunya & Behera (1987)
Chromosomal aberrations, Swiss mouse bone-marrow cells *in vivo*	+		100 ip × 5	Bhunya & Behera (1987)
Chromosomal aberrations, Swiss mouse bone-marrow cells *in vivo*	+		500 po × 1	Bhunya & Behera (1987)
Chromosomal aberrations, chicken *Gallus domesticus* bone marrow *in vivo*	+		200 ip × 1	Bhunya & Jena (1996)

Table 5 (contd)

Test system	Result[a]		Dose[b] (LED or HID)	Reference
	Without exogenous metabolic system	With exogenous metabolic activation		
Inhibition of intercellular communication, B6C3F$_1$ mouse hepatocytes in vitro	+	NT	16.3	Klaunig et al. (1990)
Inhibition of intercellular communication, Sprague-Dawley rat liver clone 9 cells in vitro	+	NT	163	Benane et al. (1996)
Sperm morphology, Swiss mice in vivo	+		125 ip × 5	Bhunya & Behera (1987)

[a] +, positive; (+), weakly positive; –, negative; ?, inconclusive; NT, not tested
[b] LED, lowest effective dose; HID, highest ineffective dose; in-vitro tests, μg/mL; in-vivo tests, mg/kg bw; po, orally; ip, intraperitoneally
[c] Neutralized trichloroacetic acid
[d] Larvae reared in water containing trichloroacetic acid

Trichloroacetic acid did not induce DNA strand breaks or DNA damage in mouse, rat or hamster cells *in vitro* but was weakly mutagenic in mouse lymphoma cells. DNA strand breaks were not induced by trichloroacetic acid *in vitro* in human cells nor were chromosomal aberrations in human lymphocytes exposed *in vitro* to trichloroacetic acid neutralized to avoid the effects of low pH seen in cultured mammalian cells.

In one study, trichloroacetic acid induced micronuclei and chromosomal aberrations in bone-marrow cells and abnormal sperm morphology after injection into Swiss mice *in vivo*. In another study, in which a 10-fold higher dose was injected into C57BL/6JfBL10/Alpk mice, no micronucleus formation was observed. Trichloroacetic acid induced the formation of micronuclei in erythrocytes of newt larvae *in vivo*. It also induced chromosomal aberrations *in vivo* in the bone marrow of the chicken *Gallus domesticus*. Gap-junctional intercellular communication was inhibited in mouse and rat hepatocytes *in vitro*.

Mutation of proto-oncogenes in tumours induced by trichloroacetic acid

Point mutations in exons 1, 2 and 3 of K- and H-*ras* proto-oncogenes were studied in trichloroacetic acid-induced liver tumours of male B6C3F$_1$ mice (104-week treatment with 4.5 g/L in drinking-water). Trichloroacetic acid did not modify the incidence of mutations in exon 2 of H-*ras* in carcinomas (45% versus 58% for control). Only four carcinomas showed mutations in the other exons of H-*ras* or in K-*ras*. In tumours with mutation in exon 2 of H-*ras*, treatment with trichloroacetic acid did not modify the mutational spectrum compared with that of spontaneous liver tumours, that is to say 80% of the mutations in codon 61 were CAA→AAA, and 20% were CAA→CGA (Ferreira-Gonzalez *et al.*, 1995).

In *N*-methyl-*N*-nitrosourea (MNU) initiated female B6C3F$_1$ mice, 27% of liver tumours promoted by trichloroacetic acid exhibited loss of heterozygosity for at least two loci on chromosome 6 (Tao *et al.*, 1996).

4.5   Mechanistic considerations

There is little evidence that trichloroacetic acid induces mutation (see Section 4.4). Therefore, it is improbable that it acts through genotoxic mechanisms.

High doses of trichloroacetic acid were required to induce proliferation of peroxisomes in B6C3F$_1$ mice (palmitoyl-coenzyme A oxidase activity increased eightfold), but the compound was less effective in Fischer 344 rats (approximately twofold induction of palmitoyl-coenzyme A oxidase) (DeAngelo *et al.*, 1989). Maloney and Waxman (1999) demonstrated that trichloroacetic acid directly activated hPPAR-α and mPPAR-α, but only weakly activated the mPPAR-γ receptor. The external doses administered as well as the systemic concentrations that resulted in a carcinogenic response in the liver of B6C3F$_1$ mice were generally consistent with those required to activate PPAR receptors and induce peroxisome proliferation [interaction was not examined].

Tumours from trichloroacetic acid-treated male B6C3F$_1$ mice did not have mutations in codons 12 or 13 of the H-*ras* oncogene, which are frequently observed with genotoxic carcinogens. Liver tumours from these mice were found to have mutations at codon 61 of the H-*ras* gene (Ferreira-Gonzalez *et al.*, 1995), the incidence of which was not different from that in spontaneous liver tumours (see also Bull *et al.*, 2002). The frequency of codon 61 mutations in the H-*ras* gene was higher than that expected of other activators of the PPAR-α receptor such as ciprofibrate (Fox *et al.*, 1990; Hegi *et al.* 1993) but was consistent with that of peroxisome proliferators that activate PPAR-γ, such as LY-171883 (Helvering *et al.*, 1994; Kliewer *et al.*, 1994). In contrast to mice, rats did not develop liver tumours when treated with trichloroacetic acid at concentrations of up to 5 g/L in the drinking-water, which is consistent with the weak peroxisome proliferative activity of this compound in rats (DeAngelo *et al.*, 1997).

Treatment for up to 14 days with trichloroacetic acid enhanced cell proliferation in the liver of B6C3F$_1$ mice, but this increase was not apparent after longer exposures when chronic treatment even depressed cell division rates relative to controls (Pereira, 1996; Stauber & Bull, 1997). The growth of anchorage-independent colonies with the same phenotype as that observed *in vivo* were stimulated by appropriate concentrations of trichloroacetic acid (Stauber *et al.*, 1998; Kato-Weinstein *et al.*, 2001). These data suggest that trichloroacetic acid promotes the clonal expansion of a subset of abnormal cells that arise spontaneously *in vivo*.

Expression of the mRNA for the c-*myc* gene was increased by trichloroacetic acid in both tumorous and non-tumorous liver in B6C3F$_1$ mice (Tao *et al.*, 2000a,b). The c-*myc* proto-oncogene is a cellular transcription factor that plays a pivotal role in apoptosis, and cell replication and differentiation (Holden *et al.*, 1998; Christensen *et al.*, 1999). In female B6C3F$_1$ mice administered daily doses of 500 mg/kg trichloroacetic acid by gavage, enhancement of liver cell proliferation and decreased methylation of the *c-myc* gene were observed simultaneously 72 h after the start of exposure (Ge *et al.*, 2001). Methylation of CpG sites in the promoter region of a gene regulates in part the expression of its mRNA (Jones & Buckley, 1990; Wainfan & Poirier, 1992; Razin & Kafri, 1994). Trichloroacetic acid has been shown to decrease the methylation of DNA and of CCGG sites in the promoter region of the c-*myc* and c-*jun* genes and to induce their overexpression (Tao *et al.*, 1998, 2000a). A c-jun phenotype is typical of tumours induced in male B6C3F1 mice by peroxisome proliferators (Nakano *et al.* 1994). In the liver of female B6C3F$_1$ mice given trichloroacetic acid in drinking-water for 11 days, the level of 5-methylcytosine in DNA was decreased. When female B6C3F$_1$ mice were initiated with a single dose of MNU before receiving trichloroacetic acid in the drinking-water for 44 weeks, the amount of 5-methylcytosine in adenomas 1 week after termination of the treatment did not recover to the level observed in non-cancerous liver (Tao *et al.*, 1998), which is consistent with an aggressive rate of cell replication within trichloroacetic acid-induced tumours after cessation of treatment (Stauber & Bull, 1997).

In summary, studies suggest that the carcinogenic activity of trichloroacetic acid in B6C3F$_1$ mice is consistent with its activity as a peroxisome proliferator. This is further supported by a lack of a carcinogenic response in rats, in which the extent of peroxisome proliferation after treatment with trichloroacetic acid is much lower than in mice. In addition, various other effects on cell replication rates, decreased DNA methylation and increased proto-oncogene expression are consistent with an epigenetic mechanism of carcinogenesis in mice.

5. Summary of Data Reported and Evaluation

5.1   Exposure data

Trichloroacetic acid is mainly used as a selective herbicide. It also finds use in the metal, plastics and textile industries and as an analytical reagent. It is used in the topical treatment of warts, cervical lesions and other dermatological conditions. Trichloroacetic acid is a major end metabolite of trichloroethylene and tetrachloroethylene. Wider exposure to trichloroacetic acid occurs at microgram-per-litre levels in drinking-water and swimming pools as a result of chlorination or chloramination.

5.2   Human carcinogenicity data

Several studies analysed risk with respect to one or more measures of exposure to complex mixtures of disinfection by-products that are found in most chlorinated and chloraminated drinking-water. No data specifically on trichloroacetic acid were available to the Working Group.

5.3   Animal carcinogenicity data

In four studies, neutralized trichloroacetic acid, when administered in the drinking-water to female and/or male mice, increased the incidences of hepatocellular adenomas and carcinomas. In a study in male rats, trichloroacetic acid did not increase the incidence of liver tumours or tumours at any other site. When administered in the drinking-water, tri-chloroacetic acid promoted the induction of hepatocellular adenomas and/or carcinomas in carcinogen-initiated male and female mice and of kidney tumours in male mice.

5.4   Other relevant data

The half-life of trichloroacetic acid, given orally or formed as a metabolite of tri-chloroethylene or trichloroethanol, is longer in humans than in rodents. Trichloroacetic acid may be reduced *in vivo* to dichloroacetic acid, but the artefactual conversion of tri-chloroacetic acid to dichloroacetic acid hinders any clear conclusions. A fraction of tri-chloroacetic acid is metabolized to carbon dioxide.

Trichloroacetic acid induces peroxisome proliferation in the livers of mice at doses within the same range as those that induce hepatic tumours. A brief stimulation of cell division is observed in the liver during the first days of treatment, but depressed cell replication results from chronic treatment. The initial increase in cell proliferation was correlated with decreased methylation of the promoter regions of the c-*jun* and c-*myc* proto-oncogenes and increased expression of these genes.

Effects of trichloroacetic acid on reproduction and development in rats have been reported, but were not confirmed in a subsequent study. In-vitro results suggest that trichloroacetic acid can produce teratogenic effects at high doses.

In male mice, trichloroacetic acid modified neither the incidence of mutations in exon 2 of H-*ras* in carcinomas, nor the mutational spectrum observed in tumours that bore a mutation in exon 2. In female mice, 27% of tumours promoted by trichloroacetic acid exhibited loss of heterozygosity at a minimum of two loci on chromosome 6.

In mouse liver *in vivo*, measurements of trichloroacetic acid-induced 8-hydroxydeoxy-guanosine DNA adducts gave different results depending on the route of administration. Trichloroacetic acid induced abnormal sperm in mice *in vivo* in one study and chromosomal aberrations in mouse and chicken bone marrow *in vivo*. The results of in-vivo studies in rodents on the induction of DNA strand breaks and micronuclei were inconsistent. It induced the formation of micronuclei in newt larvae *in vivo*.

In human cells *in vitro*, trichloroacetic acid did not induce chromosomal aberrations or DNA strand breaks in single studies. In single studies on cultured rodent cells, trichloroacetic acid was weakly mutagenic; no effect was observed in a DNA strand-break assay or a single-cell gel assay. It also inhibited intercellular communication in cultured rodent cells. Trichloroacetic acid caused neither mutation in bacteria nor SOS repair.

5.5   Evaluation

There is *inadequate evidence* in humans for the carcinogenicity of trichloroacetic acid.
There is *limited evidence* in experimental animals for the carcinogenicity of trichloroacetic acid.

Overall evaluation

Trichloroacetic acid is *not classifiable as to its carcinogenicity to humans (Group 3)*.

6.   References

Abbas, R. & Fisher, J.W. (1997) A physiologically based pharmacokinetic model for trichloro-ethylene and its metabolites, chloral hydrate, trichloroacetate, dichloroacetate, trichloroethanol, and trichloroethanol glucuronide in B6C3F1 mice. *Toxicol. appl. Pharmacol.*, **147**, 15–30
Allen, B.C. & Fisher, J.W. (1993) Pharmacokinetic modeling of trichloroethylene and trichloro-acetic acid in humans. *Risk Anal.*, **13**, 71–86

American Public Health Association/American Water Works Association/Water Environment Federation (1999) *Standard Methods for the Examination of Water and Wastewater*, 20th Ed., Washington DC [CD-ROM]

Artho, A., Grob, K. & Giger, P. (1991) [Trichloroacetic acid in surface, ground, and drinking waters.] *Mitt. geb. Lebensm. Hyg.*, **82**, 487–491 (in German)

Austin, E.W., Parrish, J.M., Kinder, D.H. & Bull, R.J. (1996) Lipid peroxidation and formation of 8-hydroxydeoxyguanosine from acute doses of halogenated acetic acids. *Fundam. appl. Toxicol.*, **31**, 77–82

Back, H. & Süsser, P. (1992) Concentrations of volatile chlorinated hydrocarbons and trichloroacetic acid in earthworms. *Soil Biol. Biochem.*, **24**, 1745–1748

Beilstein Online (2002) *Dialog Corporation, File 390*, Cary, NC, Beilstein Chemiedaten und Software GmgH

Beland, F.A. (1999) NTP technical report on the toxicity and metabolism studies of chloral hydrate (CAS No. 302-17-0). Administered by gavage to F344/N rats and B6C3F$_1$ mice. *Toxic. Rep. Ser.*, **59**, 1–66, A1–E7

Benane, S.G., Blackman, C.F. & House, D.E. (1996) Effect of perchloroethylene and its metabolites on intercellular communication in clone 9 rat liver cells. *J. Toxicol. environ. Health*, **48**, 427–437

Bhunya, S.P. & Behera, B.C. (1987) Relative genotoxicity of trichloroacetic acid (TCA) as revealed by different cytogenetic assays: Bone marrow chromosome aberration, micronucleus and sperm-head abnormality in the mouse. *Mutat. Res.*, **188**, 215–221

Bhunya, S.P. & Jena, G.B. (1996) The evaluation of clastogenic potential of trichloroacetic acid (TCA) in chick in vivo test system. *Mutat. Res.*, **367**, 253–259

Boillat, M.-A. (1970) [Usefulness of the determination of trichloroacetic acid in urine for detecting poisoning due to trichloroethylene and perchloroethylene. Study of several firms.] *Z. Präventivmed.*, **15**, 447–456 (in French)

Breimer, D.D., Ketelaars, H.C.J. & Van Rossum, J.M. (1974) Gas chromatographic determination of chloral hydrate, trichloroethanol, and trichloroacetic acid in blood and urine employing head-space analysis. *J. Chromatogr.*, **88**, 55–63

Bull, R.J., Orner, G.A., Cheng, R.S., Stillwell, L., Stauber, A.J., Sasser, L.B., Lingohr, M.K. & Thrall, B.D. (2002) Contribution of dichloroacetate and trichloroacetate to liver tumor induction in mice by trichloroethylene. *Toxicol. appl. Pharmacol.*, **182**, 55–65

Cancho, B., Ventura, F. & Galceran, M.T. (1999) Behavior of halogenated disinfection by-products in the water treatment plant of Barcelona, Spain. *Bull. environ. Contam. Toxicol.*, **63**, 610–617

cfm Oskar Tropitzsch (2002) *Inspection Certificate: Trichloroacetic Acid (TCAA) tchn, flakes*, Marktredwitz

Chang, L.W., Daniel, F.B. & DeAngelo, A.B. (1992) Analysis of DNA strand breaks induced in rodent liver in vivo, hepatocytes in primary culture, and a human cell line by chlorinated acetic acids and chlorinated acetaldehydes. *Environ. mol. Mutag.*, **20**, 277–288

Chemical Information Services (2002a) *Directory of World Chemical Producers (Version 2002.1)*, Dallas, TX [CD-ROM]

Chemical Information Services (2002b) *Worldwide Bulk Drug Users Directory (Version 2002)*, Dallas, TX [http://www.chemical.info.com]

Chen, W.J. & Weisel, C.P. (1998) Halogenated DBP concentrations in a distribution system. *J. Am. Water Works Assoc.*, **90**, 151–163

Christensen, J.G., Goldsworthy, T.L. & Cattley, R.C. (1999) Dysregulation of apoptosis by c-*myc* in transgenic hepatocytes and effects of growth factors and nongenotoxic carcinogens. *Mol. Carcinog.*, **25**, 273–284

Christman, R.F., Norwood, D.L., Millington, D.S., Johnson, J.D. & Stevens, A.A. (1983) Identity and yields of major halogenated products of aquatic fulvic acid chlorination. *Environ. Sci. Technol.*, **17**, 625–628

Clariant Corp. (2001) *Specification Sheet: Trichloroacetic Acid (TCAA) Tchn Flakes*, Charlotte, NC

Clariant Corp. (2002) *Specification Sheet: Trichloroacetic Acid (TCAA) Pharma*, Charlotte, NC

Clariant GmbH (2002a) *Specification Sheet: Trichloroacetic Acid*, Sulzbach

Clariant GmbH (2002b) *Specification Sheet: Trichloroacetic Acid Pharma*, Sulzbach, Germany

Clariant GmbH (2002c) *Specification Sheet: Trichloroacetic Acid 90% in Water*, Sulzbach, Germany

Clemens, M. & Schöler, H.F. (1992a) Determination of halogenated acetic acids and 2,2-dichloropropionic acid in water samples. *Fresenius J. anal. Chem.*, **344**, 47–49

Clemens, M. & Schöler, H.F. (1992b) [Halogenated organic compounds in swimming-pool waters.] *Zbl. Hyg.*, **193**, 91–98 (in German)

Comes, R.D., Frank, P.A. & Demint, R.J. (1975) TCA in irrigation water after bank treatments for weed control. *Weed Sci.*, **23**, 207–210

Council of Europe (2002) [*European Pharmacopeia*], 4th Ed., Strasbourg, p. 2260 (in French)

DeAngelo, A.B., Daniel, F.B., McMillan, L., Wernsing, P. & Savage, R.E., Jr (1989) Species and strain sensitivity to the induction of peroxisome proliferation by chloroacetic acids. *Toxicol. appl. Pharmacol.*, **101**, 285–298

DeAngelo, A.B., Daniel, F.B., Most, B.M. & Olson, G.R. (1997) Failure of monochloroacetic acid and trichloroacetic acid administered in the drinking water to produce liver cancer in male F344/N rats. *J. Toxicol. environ. Health*, **52**, 425–445

DeMarini, D.M., Perry, E. & Shelton, M.L. (1994) Dichloroacetic acid and related compounds: Induction of prophage in *E. coli* and mutagenicity and mutation spectra in Salmonella TA100. *Mutagenesis*, **9**, 429–437

Demint, R.J., Pringle, J.C., Jr, Hattrup, A., Bruns, V.F. & Frank, P.A. (1975) Residues in crops irrigated with water containing trichloroacetic acid. *J. agric. Food Chem.*, **23**, 81–84

Dojlido, J., Zbiec, E. & Swietlik, R. (1999) Formation of the haloacetic acids during ozonation and chlorination of water in Warsaw waterworks (Poland). *Water Res.*, **33**, 3111–3118

Elcombe, C.R. (1985) Species differences in carcinogenicity and peroxisome proliferation due to trichloroethylene: A biochemical human hazard assessment. *Arch. Toxicol.*, **Suppl. 8**, 6–17

Ellis, D.A., Hanson, M.L., Sibley, P.K., Shahid, T., Fineberg, N.A., Solomon, K.R., Muir, D.C.G. & Mabury, S.A. (2001) The fate and persistence of trifluoroacetic and chloroacetic acids in pond waters. *Chemosphere*, **42**, 309–318

Environmental Protection Agency (1995) *Method 552.2. Determination of Haloacetic Acids and Dalapon in Drinking Water by Liquid–liquid Extraction, Derivatization and Gas Chromatography with Electron Capture Detection (Revision 1.0)*, Cincinnati, OH, National Exposure Research Laboratory, Office of Research and Development

Environmental Protection Agency (1998) National primary drinking water regulations; disinfectants and disinfection byproducts; final rule. *Fed. Regist.*, December 16

European Commission (1998) *Council Directive 98/83/EC of 3 November 1998 on the Quality of Water Intended for Human Consumption*, Luxemburg, Office for Official Publications of the European Communities

Ferreira-Gonzalez, A., DeAngelo, A.B., Nasim, S. & Garrett, C.T. (1995) *Ras* oncogene activation during hepatocarcinogenesis in B6C3F1 male mice by dichloroacetic and trichloroacetic acids. *Carcinogenesis*, **16**, 495–500

Fisher, J.W. (2000) Physiologically based pharmacokinetic models for trichloroethylene and its oxidative metabolites. *Environ. Health Perspect.*, **108** (Suppl. 2), 265–273

Fisher, J.W., Whittaker, T.A., Taylor, D.H., Clewell, H.J., III & Andersen, M.E. (1989) Physiologically based pharmacokinetic modeling of the pregnant rat: A multiroute exposure model for trichloroethylene and its metabolite, trichloroacetic acid. *Toxicol. appl. Pharmacol.*, **99**, 395–414

Fisher, J.W., Whittaker, T.A., Taylor, D.H., Clewell, H.J., III & Andersen, M.E. (1990) Physiologically based pharmacokinetic modeling of the lactating rat and nursing pup: A multiroute exposure model for trichloroethylene and its metabolite, trichloroacetic acid. *Toxicol. appl. Pharmacol.*, **102**, 497–513

Fisher, J.W., Mahle, D. & Abbas, R. (1998) A human physiologically based pharmacokinetic model for trichloroethylene and its metabolites, trichloroacetic acid and free trichloroethanol. *Toxicol. appl. Pharmacol.*, **152**, 339–359

Fisher, J.W., Channel, S.R., Eggers, J.S., Johnson, P.D., MacMahon, K.L., Goodyear, C.D., Sudberry, G.L., Warren, D.A., Latendresse, J.R. & Graeter, L.J. (2001) Trichloroethylene, trichloroacetic acid, and dichloroacetic acid: Do they affect fetal rat heart development? *Int. J. Toxicol.*, **20**, 257–267

Fox, T.R., Schumann, A.M., Watanabe, P.G., Yano, B.L., Maher, V.M. & McCormick, J.J. (1990) Mutational analysis of the H-*ras* oncogene in spontaneous C57BL/6 × C3H/He mouse liver tumors and tumors induced with genotoxic and nongenotoxic hepatocarcinogens. *Cancer Res.*, **50**, 4014–4019

Frank, H. (1991) Airborne chlorocarbons, photooxidants, and forest decline. *Ambio*, **20**, 13–18

Frank, H., Vital, J. & Frank, W. (1989) Oxidation of airborne C_2-chlorocarbons to trichloroacetic and dichloroacetic acid. *Fresenius Z. anal. Chem.*, **333**, 713

Frank, H., Scholl, H., Sutinen, S. & Norokorpi, Y. (1992) Trichloroacetic acid in conifer needles in Finland. *Ann. bot. Fenici*, **29**, 263–267

Froese, K.L., Sinclair, M.I. & Hrudey, S.E. (2002) Trichloroacetic acid as a biomarker of exposure to disinfection by-products in drinking water: A human exposure trial in Adelaide, Australia. *Environ. Health Perspect.*, **110**, 679–687

Ge, R., Yang, S., Kramer, P.M., Tao, L. & Pereira, M.A. (2001) The effect of dichloroacetic acid and trichloroacetic acid on DNA methylation and cell proliferation in B6C3F1 mice. *J. Biochem. mol. Toxicol.*, **15**, 100–106

Geist, S., Lesemann, C., Schütz, C., Seif, P. & Frank, E. (1991) Determination of chloroacetic acids in surface water. In: Angeletti, G. & Bjørseth, A., eds, *Organic Micropollutants in the Aquatic Environment*, Dordrecht, Kluwer Academic, pp. 393–397

Gennaro, A.R. (2000) *Remington: The Science and Practice of Pharmacy*, 20th Ed., Baltimore, MD, Lippincott Williams & Wilkins, p. 1211

Giller, S., Le Curieux, F., Erb, F. & Marzin, D. (1997) Comparative genotoxicity of halogenated acetic acids found in drinking water. *Mutagenesis*, **12**, 321–328

Glaze, W.H., Kenneke, J.F. & Ferry, J.L. (1993) Chlorinated byproducts from the TiO_2-mediated photodegradation of trichloroethylene and tetrachloroethylene in water. *Environ. Sci. Technol.*, **27**, 177–184

Gonthier, B.P. & Barrett, L.G. (1989) In-vitro spin-trapping of free radicals produced during tri-chloroethylene and diethylether metabolism. *Toxicol. Lett.*, **47**, 225–234

Greenberg, M.S., Burton, G.A. & Fisher, J.W. (1999) Physiologically based pharmacokinetic modeling of inhaled trichloroethylene and its oxidative metabolites in B6C3F$_1$ mice. *Toxicol. appl. Pharmacol.*, **154**, 264–278

Hajimiragha, H., Ewers, U., Jansen-Rosseck, R. & Brockhaus, A. (1986) Human exposure to vola-tile halogenated hydrocarbons from the general environment. *Int. Arch. occup. environ. Health*, **58**, 141–150

Hansch, C., Leo, A. & Hoekman, D. (1995) *Exploring QSAR: Hydrophobic, Electronic, and Steric Constants*, Washington DC, American Chemical Society, p. 4

Hargesheimer, E.E. & Satchwill, T. (1989) Gas-chromatographic analysis of chlorinated acids in drinking water. *J. Water Sewage Res. Treat. Aqua*, **38**, 345–351

Harrington-Brock, K., Doerr, C.L. & Moore, M.M. (1998) Mutagenicity of three disinfection by-products: Di- and trichloroacetic acid and chloral hydrate in L5178Y/TK$^{+/-}$-3.7.2.C mouse lym-phoma cells. *Mutat. Res.*, **413**, 265–276

Health Canada (1995) *A National Survey of Chlorinated Disinfection By-products in Canadian Drinking Water* (95-EHD-197), Ottawa, Minister of National Health and Welfare

Health Canada (2003) *Summary of Guidelines for Canadian Drinking Water Quality*, Ottawa

Hegi, M.E., Fox, T.R., Belinsky, S.A., Devereux, T.R. & Anderson, M.W. (1993) Analysis of acti-vated protooncogenes in B6C3F1 mouse liver tumors induced by ciprofibrate, a potent peroxi-some proliferator. *Carcinogenesis*, **14**, 145–149

Helvering, L.M., Richardson, F.C., Horn, D.M., Rexroat, M.A., Engelhardt, J.A. & Richardson, K.K. (1994) H-*ras* 61st codon activation in archival proliferative hepatic lesions isolated from female B6C3F1 mice exposed to the leukotriene D$_4$-antagonist, LY171883. *Carcinogenesis*, **15**, 331–333

Holden, P.R., Odum, J., Soames, A.R., Foster, J.R., Elcombe, C.R. & Tugwood, J.D. (1998) Imme-diate-early gene expression during regenerative and mitogen-induced liver growth in the rat. *J. Biochem. mol. Toxicol.*, **12**, 79–82

Hunter, E.S., III, Rogers, E.H., Schmid, J.E. & Richard, A. (1996) Comparative effects of halo-acetic acids in whole embryo culture. *Teratology*, **54**, 57–64

IARC (1976) *IARC Monographs on the Evaluation of Carcinogenic Risk of Chemicals to Man*, Vol. 11, *Cadmium, Nickel, Some Epoxides, Miscellaneous Industrial Chemicals and General Considerations on Volatile Anaesthetics*, Lyon, IARC*Press*

IARC (1977) *IARC Monographs on the Evaluation of the Carcinogenic Risk of Chemicals to Man*, Vol. 15, *Some Fumigants, the Herbicides 2,4-D and 2,4,5-T, Chlorinated Dibenzodioxins and Miscellaneous Industrial Chemicals*, Lyon, IARC*Press*

IARC (1979) *IARC Monographs on the Evaluation of the Carcinogenic Risks to Humans*, Vol. 20, *Some Halogenated Hydrocarbons*, Lyon, IARC*Press*, pp. 491, 525, 545

IARC (1987) *IARC Monographs on the Evaluation of Carcinogenic Risks to Humans*, Suppl. 7, *Overall Evaluations of Carcinogenicity: An Updating of* IARC Monographs *Volumes 1 to 42*, Lyon, IARC*Press*

IARC (1991) *IARC Monographs on the Evaluation of the Carcinogenic Risks to Humans*, Vol. 52, *Chlorinated Drinking-water; Chlorination By-products; Some Other Halogenated Com-pounds; Cobalt and Cobalt Compounds*, Lyon, IARC*Press*, pp. 55–141

IARC (1995) *IARC Monographs on the Evaluation of Carcinogenic Risks to Humans*, Vol. 63, *Dry Cleaning, Some Chlorinated Solvents and Other Industrial Chemicals*, Lyon, IARC*Press*, pp. 75, 159, 291–314

IARC (1999) *IARC Monographs on the Evaluation of Carcinogenic Risks to Humans*, Vol. 71, *Re-Evaluation of Some Organic Chemicals, Hydrazine and Hydrogen Peroxide (Part Three)*, Lyon, IARC*Press*, p. 1153

Ikeda, M. & Ohtsuji, H. (1969) Hippuric acid, phenol, and trichloroacetic acid levels in the urine of Japanese subjects with no known exposure to organic solvents. *Br. J. ind. Med.*, **26**, 162–164

Ikeda, M., Ohtsuji, H., Imamura, T. & Komoike, Y. (1972) Urinary excretion of total trichloro-compounds, trichloroethanol, and trichloroacetic acid as a measure of exposure to trichloro-ethylene and tetrachloroethylene. *Br. J. ind. Med.*, **29**, 328–333

Itoh, H. (1989) Determination of trichloroacetate in human serum and urine by ion chromato-graphy. *Analyst*, **114**, 1637–1640

Jacangelo, J.G., Patania, N.L., Reagan, K.M., Aieta, E.M., Krasner, S.W. & McGuire, M.J. (1989) Ozonation: Assessing its role in the formation and control of disinfection by-products. *J. Am. Water Works Assoc.*, **81**, 74–84

Jang, J.-Y., Kang, S.-K. & Chung, H.K. (1993) Biological exposure indices of organic solvents for Korean workers. *Int. Arch. occup. environ. Health*, **65**, S219–S222

Johnson, P.D., Dawson, B.V. & Goldberg, S.J. (1998a) Cardiac teratogenicity of trichloroethylene metabolites. *J. Am. Coll. Cardiol.*, **32**, 540–545

Johnson, P.D., Dawson, B.V. & Goldberg, S.J. (1998b) A review: Trichloroethylene metabolites: Potential cardiac teratogens. *Environ. Health Perspect.*, **106** (Suppl. 4), 995–999

Jones, P.A. & Buckley, J.D. (1990) The role of DNA methylation in cancer. *Adv. Cancer Res.*, **54**, 1–23

Juuti, S., Hirvonen, A., Tarhanen, J., Holopainen, J.K. & Ruuskanen, J. (1993) Trichloroacetic acid in pine needles in the vicinity of a pulp mill. *Chemosphere*, **26**, 1859–1868

Kadis, V.W., Yarish, W., Molberg, E.S. & Smith, A.E. (1972) Trichloroacetic acid residues in cereals and flax. *Can. J. plant Sci.*, **52**, 674–676

Kargalioglu, Y., McMillan, B.J., Minear, R.A. & Plewa, M.J. (2002) Analysis of the cytotoxicity and mutagenicity of drinking water disinfection by-products in *Salmonella typhimurium*. *Tera-tog. Carcinog. Mutag.*, **22**, 113–128

Kato-Weinstein, J., Stauber, A.J., Orner, G.A., Thrall, B.D. & Bull, R.J. (2001) Differential effects of dihalogenated and trihalogenated acetates in the liver of B6C3F1 mice. *J. Appl. Toxicol.*, **21**, 81–89

Kawai, T., Yamaoka, K., Uchida, Y. & Ikeda, M. (1991) Exposure to 1,1,1-trichloroethane and dose-related excretion of metabolites in urine of printing workers. *Toxicol. Lett.*, **55**, 39–45

Ketcha, M.M., Stevens, D.K., Warren, D.A., Bishop, C.T. & Brashear, W.T. (1996) Conversion of trichloroacetic acid to dichloroacetic acid in biological samples. *J. anal. Toxicol.*, **20**, 236–241

Kim, H. & Weisel, C.P. (1998) Dermal absorption of dichloro- and trichloroacetic acids from chlorinated water. *J. Expo. Anal. environ. Epidemiol.*, **8**, 555–575

Kim, H., Haltmeier, P., Klotz, J.B. & Weisel, C.P. (1999) Evaluation of biomarkers of environ-mental exposures: Urinary haloacetic acids associated with ingestion of chlorinated drinking water. *Environ. Res.*, **A80**, 187–195

Klaunig, J.E., Ruch, R.J., Hampton, J.A., Weghorst, C.M. & Hartnett, J.A. (1990) Gap-junctional intercellular communication and murine hepatic carcinogenesis. *Prog. clin. biol. Res.*, **331**, 277–291

Kliewer, S.A., Forman, B.M., Blumberg, B., Ong, E.S., Borgmeyer, U., Mangelsdorf, D.J., Umesono, K. & Evans, R.M. (1994) Differential expression and activation of a family of murine peroxisome proliferator-activated receptors. *Proc. natl Acad. Sci USA,* **91**, 7355–7359

Koenig, G., Lohmar, E. & Rupprich, N. (1986) Chloroacetic acids. In: Gerhartz, W., Yamamoto, Y.S., Campbell, F.T., Pfefferkorn, R. & Rounsaville, J.F., eds., *Ullmann's Encyclopedia of Industrial Chemistry*, 5th rev. Ed., Volume A6, New York, VCH Publishers, pp. 537–552

Krasner, S.W., McGuire, M.J., Jacangelo, J.G., Patania, N.L., Reagan, K.M. & Aieta, E.M. (1989) The occurrence of disinfection by-products in US drinking water. *J. Am. Water Works Assoc.*, **81**, 41–53

Lahl, U., Stachel, B., Schröer, W. & Zeschmar, B. (1984) [Determination of organohalogenic acids in water samples.] *Z. Wasser Abwasser-Forsch.*, **17**, 45–49 (in German)

Larson, J.L. & Bull, R.J. (1992) Metabolism and lipoperoxidative activity of trichloroacetate and dichloroacetate in rats and mice. *Toxicol. appl. Pharmacol.*, **115**, 268–277

Latendresse, J.R. & Pereira, M.A. (1997) Dissimilar characteristics of *N*-methyl-*N*-nitrosourea-initiated foci and tumors promoted by dichloroacetic acid or trichloroacetic acid in the liver of female B6C3F1 mice. *Toxicol. Pathol.*, **25**, 433–440

LeBel, G.L., Benoit, F.M. & Williams, D.T. (1997) A one-year survey of halogenated disinfection by-products in the distribution system of treatment plants using three different disinfection processes. *Chemosphere*, **34**, 2301–2317

Legube, B., Croue, J.P. & Dore, M. (1985) Chlorination of humic substances in aqueous solution: Yields of volatile and major non-volatile organic halides. *Sci. total Environ.*, **47**, 217–222

Lingohr, M.K., Thrall, B.D. & Bull, R.J. (2001) Effects of dichloroacetate (DCA) on serum insulin levels and insulin-controlled signaling proteins in livers of male B6C3F1 mice. *Toxicol. Sci.*, **59**, 178–184

Lopez-Avila, V., Liu, Y. & Charan, C. (1999) Determination of haloacetic acids in water by ion chromatography — Method development. *J. Assoc. off. anal. Chem. int.*, **82**, 689–704

Lowry, L.K., Vandervort, R. & Polakoff, P.L. (1974) Biological indicators of occupational exposure to trichloroethylene. *J. occup. Med.*, **16**, 98–101

Mackay, J.M., Fox, V., Griffiths, K., Fox, D.A., Howard, C.A., Coutts, C., Wyatt, I. & Styles, J.A. (1995) Trichloroacetic acid: Investigation into the mechanism of chromosomal damage in the *in vitro* human lymphocyte cytogenetic assay and the mouse bone marrow micronucleus test. *Carcinogenesis*, **16**, 1127–1133

Maloney, E.K. & Waxman, D.J. (1999) *trans*-Activation of PPARα and PPARγ by structurally diverse environmental chemicals. *Toxicol. appl. Pharmacol.*, **161**, 209–218

Mannschott, P., Erdinger, L. & Sonntag, H.-G. (1995) [Halogenated organic compounds in swimming pool water.] *Zbl. Hyg.*, **197**, 516–533 (in German)

Merdink, J.L., Gonzalez-Leon, A., Bull, R.J. & Schultz, I.R. (1998) The extent of dichloroacetate formation from trichloroethylene, chloral hydrate, trichloroacetate, and trichloroethanol in B6C3F1 mice. *Toxicol. Sci.*, **45**, 33–41

Merdink, J.L., Bull, R.J. & Schultz, I.R. (2000) Trapping and identification of the dichloroacetate radical from the reductive dehalogenation of trichloroacetate by mouse and rat liver microsomes. *Free Radic. Biol. Med.*, **29**, 125–130

Miller, J.W. & Uden, P.C. (1983) Characterization of nonvolatile aqueous chlorination products of humic substances. *Environ. Sci. Technol.*, **17**, 150–157

Mohamed, M., Matayun, M. & Lim, T.S. (1989) Chlorinated organics in tropical hardwood kraft pulp and paper mill effluents and their elimination in an activated sludge treatment system. *Pertanika*, **12**, 387–394

Moriya, M., Ohta, T.,Watanabe, K., Miyazawa, T., Kato, K. & Shirasu, Y. (1983) Further mutagenicity studies on pesticides in bacterial reversion assay systems. *Mutat. Res.*, **116**, 185–216

Morris, E.D. & Bost, J.C. (1991) Acetic acid and derivatives (halogenated). In: Kroschwitz, J.I. & Howe-Grant, M., eds., *Kirk-Othmer Encyclopedia of Chemical Technology*, 4th Ed., Vol. 1, New York, John Wiley & Sons, pp. 165–175

Müller, G., Spassovski, M. & Henschler, D. (1972) Trichloroethylene exposure and trichloroethylene metabolites in urine and blood. *Arch. Toxikol.*, **29**, 335–340

Müller, G., Spassovski, M. & Henschler, D. (1974) Metabolism of trichloroethylene in man. II. Pharmacokinetics of metabolites. *Arch. Toxicol.*, **32**, 283–295

Nakano, H., Hatayama, I., Satoh, K., Suzuki, S., Sato, K. & Tsuchida, S. (1994) c-Jun expression in single cells and preneoplastic foci induced by diethylnitrosamine in B6C3F1 mice: Comparison with the expression of pi-class glutathione *S*-transferase. *Carcinogenesis*, **15**, 1853–1857

National Health and Medical Research Council and Agriculture and Resource Management Council of Australia and New Zealand (1996) *Australian Drinking Water Guidelines Summary*, Canberra

National Institute for Occupational Safety and Health (1994) *National Occupational Exposure Survey (1981–1983)*, Cincinnati, OH, p. 9

Nelson, M.A. & Bull, R.J. (1988) Induction of strand breaks in DNA by trichloroethylene and metabolites in rat and mouse liver *in vivo*. *Toxicol. appl. Pharmacol.*, **94**, 45–54

Nelson, M.A., Lansing, A.J., Sanchez, I.M., Bull, R.J. & Springer, D.L. (1989) Dichloroacetic acid and trichloroacetic acid-induced DNA strand breaks are independent of peroxisome proliferation. *Toxicology*, **58**, 239–248

Nelson, G.M., Swank, A.E., Brooks, L.R., Bailey, K.C. & George, S.E. (2001) Metabolism, microflora effects, and genotoxicity in haloacetic acid-treated cultures of rat cecal microbiota. *Toxicol. Sci.*, **60**, 232–241

Nestmann, E.R., Chu, I., Kowbel, D.J. & Matula, T.I. (1980) Short-lived mutagen in *Salmonella* produced by reaction of trichloroacetic acid and dimethyl sulphoxide. *Can. J. Genet. Cytol.*, **22**, 35–40

Ni, Y.-C., Wong, T.-Y., Lloyd, R.V., Heinze, T.M., Shelton, S., Casciano, D., Kadlubar, F.F. & Fu, P.P. (1996) Mouse liver microsomal metabolism of chloral hydrate, trichloroacetic acid, and trichloroethanol leading to induction of lipid peroxidation *via* a free radical mechanism. *Drug Metab. Dispos.*, **24**, 81–90

Nissinen, T.K., Miettinen, I.T., Martikainen, P.J. & Vartiainen, T. (2002) Disinfection by-products in Finnish drinking waters. *Chemosphere*, **48**, 9–20

Norwood, D.L., Christman, R.F., Johnson, J.D. & Hass, J.R. (1986) Using iostope dilution mass spectrometry to determine aqueous trichloroacetic acid. *J. Am. Water Works Assoc.*, **78**, 175–180

Obolensky, A., Shukairy, H. & Blank, V. (2003) Occurrence of haloacetic acids in ICR finished water and distribution systems. In: McGuire, M.J., McLain, J.L. & Obolensky, A., eds, *Information Collection Rule Data Analysis*, Denver, CO, AWWA Research Foundation, pp. 1–36

Ogata, M., Shimada, Y. & Taguchi, T. (1987) A new microdetermination method used in an analysis of the excretion of trichloro-compounds in the urine of workers exposed to trichloroethylene vapour. *Ind. Health*, **25**, 103–112

Parrish, J.M., Austin, E.W., Stevens, D.K., Kinder, D.H. & Bull, R.J. (1996) Haloacetate-induced oxidative damage to DNA in the liver of male B6C3F1 mice. *Toxicology*, **110**, 103–111

Pereira, M.A. (1996) Carcinogenic activity of dichloroacetic acid and trichloroacetic acid in the liver of female B6C3F$_1$ mice. *Fundam. appl. Toxicol.*, **31**, 192–199

Pereira, M.A. & Phelps, J.B. (1996) Promotion by dichloroacetic acid and trichloroacetic acid of *N*-methyl-*N*-nitrosourea-initiated cancer in the liver of female B6C3F1 mice. *Cancer Lett.*, **102**, 133–141

Pereira, M.A., Li, K. & Kramer, P.M. (1997) Promotion by mixtures of dichloroacetic acid and trichloroacetic acid of *N*-methyl-*N*-nitrosourea-initiated cancer in the liver of female B6C3F1 mice. *Cancer Lett.*, **115**, 15–23

Pereira, M.A., Kramer, P.M., Conran, P.B. & Tao, L. (2001) Effect of chloroform on dichloroacetic acid and trichloroacetic acid-induced hypomethylation and expression of the c-*myc* gene and on their promotion of liver and kidney tumors in mice. *Carcinogenesis*, **22**, 1511–1519

Plewa, M.J., Kargalioglu, Y., Vankerk, D., Minear, R.A. & Wagner, E.D. (2002) Mammalian cell cytotoxicity and genotoxicity analysis of drinking water disinfection by-products. *Environ. mol. Mutag.*, **40**, 134–142

Plotnikov, V.A. & Petrov, A.P. (1976) Combined action of trichloroacetic acid and dimethyl sulfate on *Arabidopsis* in relation to the theory of the role of histones in the mutation process. *Tsitol. Genet.*, **10**, 35–39

Plümacher, J. & Renner, I. (1993) Determination of volatile chlorinated hydrocarbons and trichloroacetic acid in conifer needles by headspace gas-chromatography. *Fresenius J. anal. Chem.*, **347**, 129–135

Popp, W., Müller, G., Baltes-Schmitz, B., Wehner, B., Vahrenholz, C., Schmieding, W., Benninghoff, M. & Norpoth, K. (1992) Concentrations of tetrachloroethene in blood and trichloroacetic acid in urine in workers and neighbours of dry-cleaning shops. *Int. Arch. occup. environ. Health*, **63**, 393–395

Raaschou-Nielsen, O., Hansen, J., Christensen, J.M., Blot, W.J., McLauglin, J.K. & Olsen, J.O. (2001) Urinary concentrations of trichloroacetic acid in Danish workers exposed to trichloroethylene, 1947–1985. *Am. J. ind. Med.*, **39**, 320–327

Rantala, K., Riipinen, H. & Anttila, A. (1992) *Altisteet Työssä, 22. Halogeenihiilivedyt* [Exposure at work: 22. Halogenated hydrocarbons], Helsinki, Finnish Institute of Occupational Health–Finnish Work Environment Fund, pp. 18–19 (in Finnish)

Razin, A. & Kafri, T. (1994) DNA methylation from embryo to adult. *Progr. Nucleic Acid Res. mol. Biol.*, **48**, 53–81

Reckhow, D.A. & Singer, P.C. (1990) Chlorination by-products in drinking waters: From formation potentials to finished water concentrations. *J. Am. Water Works Assoc.*, **92**, 173–180

Reckhow, D.A., Singer, P.C. & Malcom, R.L. (1990) Chlorination of humic materials: Byproduct formation and chemical interpretations. *Environ. Sci. Technol.*, **24**, 1655–1664

Reeves, N.M., Heal, M.R. & Cape, J.N. (2000) A new method for the determination of trichloroacetic acid in spruce foliage and other environmental media. *J. Environ. Monit.*, **2**, 447–450

Römpp, A., Klemm, O., Fricke, W. & Frank, H. (2001) Haloacetates in fog and rain. *Environ. Sci. Technol.*, **35**, 1294–1298

Royal Pharmaceutical Society of Great Britain (2002) *Martindale, The Complete Drug Reference,* 33rd Ed., London, The Pharmaceutical Press [MicroMedex online]

Sadtler Research Laboratories (1991) *Sadtler Standard Spectra, 1981–1991 Supplementary Index,* Philadelphia, PA

Saghir, S.A. & Schultz, I.R. (2002) Low-dose pharmacokinetics and oral bioavailability of di-chloroacetate in naive and GSTζ-depleted rats. *Environ. Health Perspect.,* **110,** 757–763

Saillenfait, A.M., Langonne, I. & Sabaté, J.P. (1995) Developmental toxicity of trichloroethylene, tetrachloroethylene and four of their metabolites in rat whole embryo culture. *Arch. Toxicol.,* **70,** 71–82

Scott, B.F., Mactavisch, D., Spencer, C., Strachan, W.M.J. & Muir, D.C.G. (2000) Haloacetic acids in Canadian lake waters and precipitation. *Environ. Sci. Technol.,* **34,** 4266–4272

Scott, B.F., Spencer, C., Marvin, C.H., Mactavish, D.C. & Muir, D.G.G. (2002) Distribution of haloacetic acids in the water columns of the Laurentian Great Lakes and Lake Malawi. *Environ. Sci. Technol.,* **36,** 1893–1898

Seki, Y., Urashima, Y., Aikawa, H., Matsumura, H., Ichikawa, Y., Hiratsuka, F., Yoshioka, Y., Shimbo, S. & Ikeda, M. (1975) Trichloro-compounds in the urine of humans exposed to methyl chloroform at sub-threshold levels. *Int. Arch. Arbeitsmed.,* **34,** 39–49

Shin, D., Chung, Y., Choi, Y., Kim, J., Park, Y. & Kum, H. (1999) Assessment of disinfection by-products in drinking water in Korea. *J. Expo. Anal. environ. Epidemiol.,* **9,** 192–199

Shirasu, Y., Moriya, M., Kato, K., Furuhashi, A. & Kada, T. (1976) Mutagenicity screening of pesticides in the microbial system. *Mutat. Res.,* **40,** 19–30

Simpson, K.L. & Hayes, K.P. (1998) Drinking water disinfection by-products: An Australian perspective. *Water Res.,* **32,** 1522–1528

Skender, L., Karacic, V. & Prpic-Majic, D. (1988) Metabolic activity of antipyrine in workers occupationally exposed to trichloroethylene. *Int. Arch. occup. environ. Health,* **61,** 189–195

Skender, L.J., Karacic, V. & Prpic-Majic, D. (1991) A comparative study of human levels of tri-chloroethylene and tetrachloroethylene after occupational exposure. *Arch. environ. Health,* **46,** 174–178

Skender, L., Karacic, V., Bosner, B. & Prpic-Majic, D. (1993) Assessment of exposure to trichloro-ethylene and tetrachloroethylene in the population of Zagreb, Croatia. *Int. Arch. occup. environ. Health,* **65,** S163–S165

Smith, M.K., Randall, J.L., Read, E.J. & Stober, J.A. (1989) Teratogenic activity of trichloroacetic acid in the rat. *Teratology,* **40,** 445–451

Spectrum Chemical (2002) *Online Catalog: Trichloroacetic Acids* [www.spectrumchemical.com]

Stauber, A.J. & Bull, R.J. (1997) Differences in phenotype and cell replicative behavior of hepatic tumors induced by dichloroacetate (DCA) and trichloroacetate (TCA). *Toxicol. appl. Pharmacol.,* **144,** 235–246

Stauber, A.J., Bull, R.J. & Thrall, B.D. (1998) Dichloroacetate and trichloroacetate promote clonal expansion of anchorage-independent hepatocytes *in vivo* and *in vitro. Toxicol. appl. Pharmacol.,* **150,** 287–294

Stenner, R.D., Merdink, J.L., Stevens, D.K., Springer, D.L. & Bull, R.J. (1997) Enterohepatic recir-culation of trichloroethanol glucuronide as a significant source of trichloroacetic acid. Meta-bolites of trichloroethylene. *Drug Metab. Dispos.,* **25,** 529–535

Stenner, R.D., Merdink, J.L., Fisher, J.W. & Bull, R.J. (1998) Physiologically-based pharmaco-kinetic model for trichloroethylene considering enterohepatic recirculation of major meta-bolites. *Risk Anal.*, **18**, 261–269

Stottmeister, E. & Naglitsch, F. (1996) [Human exposure to other disinfection by-products than tri-halomethanes in swimming pools.] In: *Annual Report of the Federal Environmental Agency*, Berlin (in German)

Tao, L., Li, K., Kramer, P.M. & Pereira, M.A. (1996) Loss of heterozygosity on chromosome 6 in dichloroacetic acid and trichloroacetic acid-induced liver tumors in female B6C3F1 mice. *Cancer Lett.*, **108**, 257–261

Tao, L., Kramer, P.M., Ge, R. & Pereira, M.A. (1998) Effect of dichloroacetic acid and trichloroacetic acid on DNA methylation in liver and tumors of female B6C3F1 mice. *Toxicol. Sci.*, **43**, 139–144

Tao, L., Yang, S., Xie, M., Kramer, P.M. & Pereira, M.A. (2000a) Hypomethylation and over-expression of c-*jun* and c-*myc* protooncogenes and increased DNA methyltransferase activity in dichloroacetic and trichloroacetic acid-promoted mouse liver tumors. *Cancer Lett.*, **158**, 185–193

Tao, L., Yang, S., Xie, M., Kramer, P.M. & Pereira, M.A. (2000b) Effect of trichloroethylene and its metabolites, dichloroacetic acid and trichloroacetic acid, on the methylation and expression of c-*Jun* and c-*Myc* protooncogenes in mouse liver: Prevention by methionine. *Toxicol. Sci.*, **54**, 399–407

Templin, M.V., Parker, J.C. & Bull, R.J. (1993) Relative formation of dichloroacetate and trichloro-acetate from trichloroethylene in male B6C3F1 mice. *Toxicol. appl. Pharmacol.*, **123**, 1–8

Templin, M.V., Stevens, D.K., Stenner, R.D., Bonate, P.L., Tuman, D. & Bull, R.J. (1995) Factors affecting species differences in the kinetics of metabolites of trichloroethylene. *J. Toxicol. environ. Health*, **44**, 435–447

Triebig, G., Trautner, P., Weltle, D., Saure, E. & Valentin, H. (1982) [Investigations on neurotoxi-city of chemical substances at the workplace. III. Determination of the motor and sensory nerve conduction velocity in persons occupationally exposed to trichloroethylene.] *Int. Arch. occup. environ. Health*, **51**, 25–34 (in German)

Uden, P.C. & Miller, J.W. (1983) Chlorinated acids and chloral in drinking water. *J. Am. Water Works Assoc.*, **75**, 524–527

Ulander, A., Seldén, A. & Ahlborg, G., Jr (1992) Assessment of intermittent trichloroethylene expo-sure in vapor degreasing. *Am. ind. Hyg. Assoc. J.*, **53**, 742–743

Verschueren, K. (2001) *Handbook of Environmental Data on Organic Chemicals*, 4th Ed., Vol. 2, New York, John Wiley & Sons, pp. 2041–2042

Villanueva, C.M., Kogevinas, M. & Grimalt, J.O. (2003) Haloacetic acids and trihalomethanes in finished drinking waters from heterogeneous sources. *Water Res.*, **37**, 953–958

Von Tungeln, L.S., Yi, P., Bucci, T.J., Samokyszyn, V.M., Chou, M.W., Kadlubar, F.F. & Fu, P.P. (2002) Tumorigenicity of chloral hydrate, trichloroacetic acid, trichloroethanol, malondi-aldehyde, 4-hydroxy-2-nonenal, crotonaldehyde, and acrolein in the B6C3F$_1$ neonatal mouse. *Cancer Lett.*, **185**, 13–19

Wainfan, E. & Poirier, L.A. (1992) Methyl groups in carcinogenesis: Effects on DNA methylation and gene expression. *Cancer Res.*, **52** (Suppl.), 2071s–2077s

Walgren, J.E., Kurtz, D.T. & McMillan, J.M. (2000a) The effect of the trichloroethylene meta-bolites trichloroacetate and dichloroacetate on peroxisome proliferation and DNA synthesis in cultured human hepatocytes. *Cell Biol. Toxicol.*, **16**, 257–273

Walgren, J.E., Kurtz, D.T. & McMillan, J.M. (2000b) Expression of PPARα in human hepatocytes and activation by trichloroacetate and dichloroacetate. *Res. Commun. mol. Pathol. Pharmacol.*, **108**, 116–132

Waskell, L. (1978) A study of the mutagenicity of anesthetics and their metabolites. *Mutat. Res.*, **57**, 141–153

Weast, R.C. & Astle, M.J. (1985) *CRC Handbook of Data on Organic Compounds*, Volumes I and II, Boca Raton, Florida, CRC Press

Weisel, C.P., Kim, H., Haltmeier, P. & Klotz, J.B. (1999) Exposure estimates to disinfection by-products of chlorinated drinking water. *Environ. Health Perspect.*, **107**, 103–110

WHO (1998) *Guidelines for Drinking-water Quality*, 2nd Ed., Vol. 2, *Health Criteria and Other Supporting Information, Addendum to Vol. 2*, Geneva

Williams, D.T., LeBel, G.L. & Benoit, F.M. (1997) Disinfection by-products in Canadian drinking water. *Chemosphere*, **34**, 299–316

Xu, G., Stevens, D.K. & Bull, R.J. (1995) Metabolism of bromodichloroacetate in B6C3F1 mice. *Drug Metab. Dispos.*, **23**, 1412–1416

Yu, K.O., Barton, H.A., Mahle, D.A. & Frazier, J.M. (2000) *In vivo* kinetics of trichloroacetate in male Fischer 344 rats. *Toxicol. Sci.*, **54**, 302–311

Zhou, Y.-C. & Waxman, D.J. (1998) Activation of peroxisome proliferator activated receptors by chlorinated hydrocarbons and endogenous steroids. *Environ. Health Perspect.*, **106** (Suppl. 4), 983–988

Ziglio, G., Beltramelli, G., Pregliasco, F. & Ferrari, G. (1985) Metabolites of chlorinated solvents in blood and urine of subjects exposed at environmental level. *Sci. total Environ.*, **47**, 473–477

3-CHLORO-4-(DICHLOROMETHYL)-5-HYDROXY-2(5*H*)-FURANONE (MX)

1. Exposure Data

1.1 Chemical and physical data

1.1.1 *Nomenclature*

Chem. Abstr. Serv. Reg. No.: 77439-76-0
Deleted CAS Reg. No.: 124054-17-7
Chem. Abstr. Name: 3-Chloro-4-(dichloromethyl)-5-hydroxy-2(5*H*)-furanone
IUPAC Systematic Name: 3-Chloro-4-(dichloromethyl)-5-hydroxy-2(5*H*)-furanone
Synonyms: Chloro(dichloromethyl)-5-hydroxy-2(5*H*)-furanone; MX

1.1.2 *Structural and molecular formulae and relative molecular mass*

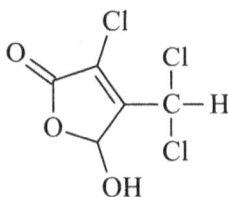

$C_5H_3Cl_3O_3$ Relative molecular mass: 217.43

1.1.3 *Chemical and physical properties of the pure substance*

(*a*) *Description*: Pale yellow to brown viscous oil (Padmapriya *et al.*, 1985; Sigma Chemical Co., 2002)
(*b*) *Boiling-point*: Not reported
(*c*) *Melting-point*: Not reported
(*d*) *Spectroscopy data*: Infrared, ultraviolet, nuclear magnetic resonance (proton, C-13) and mass spectral data have been reported (Padmapriya *et al.*, 1985)

(e) *Solubility*: Soluble in water (50.8 mg/mL at pH 2; 43.7 mg/mL at pH 7); soluble in organic solvents (Vartiainen *et al.*, 1991; Sigma Chemical Co., 2002)

(f) *Stability*: Stable in ethyl acetate and acidic water solutions (Vartiainen *et al.*, 1991); may decompose on exposure to light or heat (Sigma Chemical Co., 2002)

(g) *Octanol/water partition coefficient (P)*: 1.13 at pH 2; –0.44 at pH 7 (Vartiainen *et al.*, 1991)

1.1.4   *Technical products and impurities*

3-Chloro-4-(dichloromethyl)-5-hydroxy-2(5*H*)-furanone (MX) is not known to be produced commercially except for research purposes (Sigma-Aldrich Co., 2002).

1.1.5   *Analysis*

Analysis in water samples is difficult because MX is usually present at trace levels (nanograms per litre) and it is thermolabile. Analytical methods for MX require pre-concentration of several litres of water, a clean-up of the water extract (adsorption on XAD resins, desorption with ethyl acetate, evaporation of the solvent), its derivatization (methylation with an acidic methanol solution) and high-resolution gas chromatography (GC) coupled to low- or high-resolution mass spectrometry (MS). An alternative to high-resolution MS is an ion-trap mass spectrometer with MS capabilities. An ion-trap detector with electron ionization and MS–MS fragmentation was used for the selective determination of MX and its chlorinated and brominated analogues (Charles *et al.*, 1992; Romero *et al.*, 1997; Onstad & Weinberg, 2001; Zwiener & Kronberg, 2001).

In potable water, MX is routinely determined by derivatization with methanol, followed by GC–MS. Electron capture detection is usually only suitable to detect MX present in clean matrices; chlorinated tap-water cannot be analysed for MX by this method due to the presence of many interfering contaminants. Derivatization with other alcohols can improve the detection of MX. Results achieved using propyl alcohols have been described. Results obtained with butyl alcohols as MX derivatization agents have also been reported; derivatization with *sec*-butanol was presented as a method which significantly lowers GC–MS detection levels of MX (Nawrocki *et al.*, 1997, 2001).

1.2   Production and use

1.2.1   *Production*

Padmapriya *et al.* (1985) and Franzén and Kronberg (1995) have reported methods for the synthesis of MX.

MX is produced commercially only in small quantities for research purposes. It is a product of the chlorine-based disinfection of drinking-water (see Section 1.3.4). Like other disinfection by-products, MX is formed at the first stage of the treatment process (prechlorination), then decreases in subsequent treatment stages, is totally removed in the carbon filters and finally may be formed again in the post-chlorination step (Romero *et al.*, 1997).

MX and chlorinated acetic acids such as dichloroacetic acid and trichloroacetic acid have been the focus of a number of laboratory studies of disinfection by-products. To determine the effects of reaction time, total organic carbon, chlorine dose, pH and temperature on the formation of MX, dichloroacetic acid and trichloroacetic acid, fulvic acid was extracted from the sediment of Tai Lake (China), and simulated chlorination of samples rich in fulvic acid was conducted. Results showed a positive correlation between total organic carbon and the yields of MX, dichloroacetic acid and trichloroacetic acid. The influences of pH, chlorine dose, reaction time and temperature were quite complex. The optimal chlorination condition for the formation of MX was pH 2, a temperature of 45 °C, a C/Cl ratio of 1/4 and a reaction time of 12 h. Lower pH, a longer reaction time and a higher chlorine dose resulted in greater yields of both dichloroacetic acid and trichloroacetic acid; there was a strong linear relation between the formation of dichloroacetic acid and trichloroacetic acid, but there was no direct correlation between levels of the haloacetic acids and MX (Chen *et al.*, 2001).

1.2.2   *Use*

MX has no known commercial uses.

1.3   Occurrence

1.3.1   *Natural occurrence*

MX is not known to occur as a natural product.

1.3.2   *Occupational exposure*

No data were available to the Working Group.

1.3.3   *Air*

No data were available to the Working Group.

1.3.4   *Water*

MX has been detected in water in Canada (Andrews *et al.*, 1990), China (Zou *et al.*, 1995), Finland (Kronberg & Vartiainen, 1988; Kronberg & Franzén, 1993; Kronberg, 1999), Japan (Suzuki & Nakanishi, 1990), Spain (Romero *et al.*, 1997), the United Kingdom (Horth, 1990) and the USA (Meier *et al.*, 1987a; Kronberg *et al.*, 1991). Concentrations of MX in water are given in Table 1.

Table 1. Concentrations of MX in water

Water type (location)	Concentration range (ng/L)	Reference
Chlorinated raw water[a] (Finland)	280–510	Hemming et al. (1986)
Chlorinated raw water (Netherlands)	< 15	Backlund et al. (1989)
Treatment plant (United Kingdom)	< 3–41	Horth (1990)
Drinking-water (Australia)	< 0.5–33	Simpson & Hayes (1998)
Drinking-water (China)	3.8–58.4	Zou et al. (1995)
Drinking-water (Finland)	< 4–67	Kronberg & Vartiainen (1988)
Drinking-water (Japan)	< 3–9	Suzuki & Nakanishi (1990)
Drinking-water (USA)	2–33	Meier et al. (1987a)
Drinking-water (USA)	4.0–79.9	Wright et al. (2002)
Various waters including drinking-water (Spain)	0.1–56	Romero et al. (1997)

[a] Chlorinated raw water = water chlorinated in laboratory to estimate formation of MX

MX is produced as a by-product during disinfection of water containing humic substances using chlorine, chlorine dioxide or chloramines (Backlund et al., 1988; Kanniganti et al., 1992; Xu et al., 1997). Consequently, it may occur in drinking-water after chlorine-based disinfection of raw waters containing natural organic substances. Before it was first synthesized in 1985 (Padmapriya et al., 1985), it was referred to simply as MX (mutagen X). MX has been reported to account for between 15 and 60% of the mutagenicity in water based on the results of the *Salmonella* mutagenicity assay (Hemming et al., 1986; Meier et al., 1987a,b; Kronberg & Vartiainen, 1988; Horth et al., 1989; Wright et al., 2002). MX was initially detected in pulp chlorination waters in the early 1980s and subsequently in chlorinated drinking-waters (Holmbom et al., 1984; Hemming et al., 1986; Holmbom, 1990). MX in water gradually undergoes pH-dependent isomerization (to E-MX, 2-chloro-3-(dichloromethyl)-4-oxobutenoic acid) and hydrolytic degradation; at pH 8 and 23 °C, the half-life of MX is 6 days (Kronberg & Christman, 1989). In the USA, concentrations of MX increased with multiple chlorine applications, chlorine dose and total organic carbon, decreased with an increase in pH, were lower in chloraminated systems and showed seasonal variation with higher levels occurring in the spring compared with the autumn (Wright et al., 2002).

1.4   Regulations and guidelines

The WHO (1998) and Australia and New Zealand (National Health and Medical Research Council and Agriculture and Resource Management Council of Australia and New Zealand, 1996) note that insufficient data on health effects are available to establish a guideline for MX. The guideline in Australia and New Zealand also encourages a minimization of the concentration of all chlorination by-products by reducing naturally occurring organic material from the source water, reducing the amount of chlorine added, or using an alternative disinfectant, while not compromising disinfection.

The European Union (European Commission, 1998) and Canada (Health Canada, 2003) have not set a guideline but encourage the reduction of concentrations of total disinfection by-products. The Environmental Protection Agency (1998) in the USA controls the formation of unregulated disinfection by-products, which would include MX, with regulatory requirements for the reduction of their precursors, in this case, total organic carbon. The Stage 1 Disinfectants/Disinfection By-Product Rule mandates a reduction in the percentage of total organic carbon between source water and finished water based on the original quantity in source water and alkalinity. Enhanced coagulation or enhanced softening are mandated as a treatment technique for this reduction, unless the levels of total organic carbon or disinfection by-products in source water are low.

2. Studies of Cancer in Humans

See Introduction to the monographs on chloramine, chloral and chloral hydrate, dichloroacetic acid, trichloroacetic acid and 3-chloro-4-(dichloromethyl)-5-hydroxy-2(5*H*)-furanone.

3. Studies of Cancer in Experimental Animals

3.1 Oral administration

3.1.1 *Mouse*

Groups of 10 male and 10 female C57Bl/6J-*Min*$^{-/-}$ mice [age unspecified] were administered 0.20 mg/mL MX (94% pure dissolved at pH 3.4–4.0) in sterile drinking-water for 6 weeks; control groups received sterile water only. The animals were then given sterile water for an additional 4 weeks after which the animals were killed. Calculated daily doses of MX were 33 mg/kg bw for males and 42 mg/kg bw for females. The small intestine, colon and caecum were removed and processed for the measurement of aberrant crypt foci. For continuous data, the Student *t*-test or Mann-Whitney rank sum test was used for comparisons between the two treatment groups. The one-way ANOVA or Kruskal-Wallis ANOVA on ranks was used for comparisons among more than two groups. When a significance was noted, individual differences were tested by the Student-Newman-Keuls multiple comparison procedure. Differences between incidences of lesions were evaluated by the chi-square test or by Fisher's exact probability test (two-tailed). Male mice treated with MX had depressed water consumption (~8%) compared with the control value (3.5 ± 0.4 mL versus 3.8 ± 0.5 mL; *p* = 0.006). All mice had aberrant crypt foci ranging from 4 to 18 foci per animal, and MX in the drinking-water had no effect on aberrant crypt foci in the colon or caecum [data not presented]. No significant differences among treatment groups were found for tumours in the small intestine (Table 2) or in the colon (Table 3) (Steffensen *et al.*, 1999).

Table 2. Incidence of tumours in the small intestine of C57Bl/6J-*Min*$^{-/-}$ mice after exposure to MX in the drinking-water

Sex	Treatment group	Incidence[a]	Number of tumours[b]	Diameter (mm)
Male	Water	10/10	41.9 ± 6.8	0.94 ± 0.11
	33 mg/kg bw MX	10/10	46.8 ± 7.9	0.98 ± 0.10
Female	Water	9/9	53.6 ± 11.5	0.97 ± 0.14
	42 mg/kg bw MX	10/10	57.3 ± 16.0	0.90 ± 0.16

From Steffensen *et al.* (1999)
[a] Number of mice with tumours/total number in group
[b] Mean ± standard deviation

Table 3. Incidence of tumours in the colon of C57Bl/6J-*Min*$^{-/-}$ mice after exposure to MX in the drinking-water

Sex	Treatment group	Incidence[a]	Number of tumours[b]	Diameter (mm)
Male	Water	7/10	1.0 ± 0.9	3.76 ± 1.49
	33 mg/kg bw MX	6/10	0.9 ± 1.0	3.41 ± 1.37
Female	Water	2/9	0.3 ± 0.7	3.25 ± 0.66
	42 mg/kg bw MX	4/10	0.5 ± 0.7	3.28 ± 1.12

From Steffensen *et al.* (1999)
[a] Number of mice with tumours/total number in group
[b] Mean ± standard deviation

3.1.2   *Rat*

Groups of 50 male and 50 female Wistar rats, 5 weeks of age, were exposed to MX (97% pure) in the drinking-water for 104 weeks. The MX solutions, prepared every 1–2 weeks, were adjusted to a pH range of 3.5–5.0. The concentrations of MX were set to yield doses that were at or below a no-observed-effect level to preclude overt toxicity (5.9 µg/mL, 18.7 µg/mL and 70.0 µg/mL), giving average daily doses of MX of 0.4 (low dose), 1.3 (mid dose) and 5.0 mg/kg bw (high dose) for males and 0.6 (low dose), 1.9 (mid dose) and

6.6 mg/kg bw (high dose) for females. Control animals were exposed to the vehicle as well as filtered and ultraviolet-irradiated tap-water. For all parametric data (e.g. body weight, water and feed consumption), the one-way analysis for variance was used to test for significant differences between groups. When differences were found, Dunnett's or Scheffe's test was used to determine significance between control and treatment groups. Mortality was analysed by the Kaplan-Meier method and survival curves by the log-rank test. Water consumption was depressed in mid- and high-dose males (5–20% decrease; $p < 0.01$) and females (6–24% decrease; $p < 0.05$). The decreased water consumption did not affect the water balance as reflected by no differences in urine volume among all treatment groups. No overt effects of MX on toxicity or mortality were found for either males or females. Body weight depression was observed only for high-dose males ($p < 0.01$) and females ($p < 0.05$). For any particular tumour site, the dose–response was analysed by the one-sided trend test. The findings from the study indicated that MX is a multisite carcinogen that induces both benign and malignant tumours in both male (see Table 4) and female rats (see Table 5). A dose-related statistically significant increase in the incidence of tumours was observed in the liver, thyroid gland, adrenal gland, lung and pancreas of males and in the liver, thyroid gland, adrenal gland and mammary gland and for lymphoma and leukaemia in females (Komulainen *et al.* 1997).

3.2   Administration with known carcinogens

3.2.1   *Mouse*

Groups of male BALB/cABOM mice [initial numbers unspecified], 4 weeks of age, were administered an intraperitoneal injection of 7 mg/kg bw azoxymethane or 0.9% sodium chloride (vehicle control) once a week for 2 weeks. One week later, animals were administered 0 or 20 mg/kg bw MX (94% pure dissolved in sterile water at pH 3.4–4.0) intrarectally three times weekly for 6 weeks to give a total MX dose of 360 mg/kg or 40 mg/kg bw MX three times weekly over 4 weeks, giving a total MX dose of 480 mg/kg. All mice were killed 15 weeks after the start of the experiment, 6 or 8 weeks after the last dose. The small intestine, colon and caecum were removed. The colon and caecum were scored for altered crypt foci and small intestine, colon and caecum were scored for tumours [for a description of statistical methods, see Section 3.1.1]. Neither dose of MX increased the incidence of aberrant crypt foci per colon when compared with sodium chloride or azoxymethane. MX (40 mg/kg) without prior initiation with azoxymethane increased the number of crypts per aberrant crypt focus when compared with the control value (Table 6). Most mice in every treatment group, including the vehicle control, had small tumours (0.3–0.6 mm, 1–16 per small intestine). One mouse receiving azoxymethane and 40 mg/kg bw MX had a colon tumour (Steffensen *et al.*, 1999). [The Working Group noted the inadequacy of the strain and dose of azoxymethane.]

Table 4. Incidence of primary tumours having a statistically significantly positive trend in male Wistar rats

Tissue site	Control	0.4 mg/kg bw per day MX	1.3 mg/kg bw per day MX	5.0 mg/kg bw per day MX	*p*-value trend test
Integumentary system					
Basal-cell tumour skin	1/50 (2%)	0/50	1/50 (2%)	3/50 (6%)	0.0314
Respiratory system					
Alveolar and bronchiolar adenoma	2/50 (4%)	1/50 (2%)	1/50 (2%)	7/50 (14%)	0.0015
Liver					
Cholangioma	0/50	0/50	1/50 (2%)	4/50 (8%)	0.0009
Adenoma	0/50	1/50 (2%)	2/50 (4%)	4/50 (8%)	0.0142
Pancreas					
Langerhans' cell adenoma	5/50 (10%)	8/50 (16%)	8/50 (16%)	12/50 (24%)	0.0116
Adrenal gland					
Cortical adenoma	5/50 (10%)	2/50 (4%)	7/50 (14%)	14/50 (28%)	0.0001
Thyroid gland					
Follicular carcinoma	0/49	1/50 (2%)	9/50 (18%)	27/49 (55%)	0.000
Follicular adenoma	2/49 (4%)	20/50 (40%)	34/50 (68%)	21/49 (43%)	0.0045

From Komulainen *et al.* (1997)

Table 5. Incidence of primary tumours having a statistically significantly positive trend in female Wistar rats

Tissue site	Control	0.6 mg/kg bw per day MX	1.9 mg/kg bw per day MX	6.6 mg/kg bw per day MX	p-value
Integumentary system					
Mammary gland					
Adenocarcinoma	3/50 (6%)	2/50 (4%)	5/50 (10%)	11/50 (22%)	0.0012
Fibroadenoma	23/50 (46%)	25/50 (50%)	32/50 (64%)	34/50 (68%)	0.0090
Haematopoietic system					
Lymphoma and leukaemia	1/50 (2%)	1/50 (2%)	2/50 (4%)	4/50 (8%)	0.0474
Liver					
Cholangioma	0/50	4/50 (8%)	10/50 (20%)	33/50 (66%)	0.0000
Adenoma	1/50 (2%)	1/50 (2%)	1/50 (2%)	10/50 (20%)	0.0000
Adrenal gland					
Cortical adenoma	5/50 (10%)	10/50 (20%)	12/50 (24%)	16/50 (32%)	0.0098
Thyroid gland					
Follicular carcinoma	1/50 (2%)	3/49 (6%)	6/50 (12%)	22/50 (44%)	0.0000
Follicular adenoma	4/50 (8%)	16/49 (33%)	36/50 (72%)	36/50 (72%)	0.0000

From Komulainen *et al.* (1997)

Table 6. Aberrant crypt foci (ACF) in the colon of BALB/cA mice after intrarectal administration of MX and intraperitoneal injections of azoxymethane

Treatment (mg/kg bw)	Incidence[a]	ACF/Colon[b]	Crypts/ACF[c]
NaCl + water (control)	8/11	4.0 ± 3.7	1.7 ± 1.5
NaCl + MX (20)	7/11	3.6 ± 4.3	2.6 ± 3.6
NaCl + MX (40)	9/9	3.7 ± 1.9	3.4 ± 1.6[d]
AOM + water	10/11	5.6 ± 4.6	1.9 ± 1.0
AOM + MX (20)	9/11	4.8 ± 4.6	2.3 ± 1.7
AOM + MX (40)	6/6	4.2 ± 3.2	2.2 ± 1.2

From Steffensen *et al.* (1999)

AOM, azoxymethane

[a] Number of animals with ACF/number of animals in the group

[b] Number of ACF/animal (± standard deviation)

[c] Number of aberrant crypts/ACF (± standard deviation)

[d] Statistically significant compared with control value ($p = 0.029$, Student t-test)

3.2.2   Rat

Groups of 30 male Wistar rats, 6 weeks of age, were given 100 ppm [μg/mL] *N*-methyl-*N'*-nitro-*N*-nitrosoguanidine (MNNG) in the drinking-water for 8 weeks (initiation phase), after which they were given drinking-water containing 0, 10 or 30 ppm [μg/mL] MX (97% pure) for 57 weeks (promotion phase). Three groups of 10 animals received the vehicle control only during the initiation phase and, after 8 weeks, were given 0, 10 or 30 ppm MX in the drinking-water for 57 weeks. Tumour incidence was analysed using the Fisher's or chi-square test. Tumour multiplicities and organ weights were analysed using the Student's t-test. No differences in survival were measured for any of the treatment groups when compared with the controls. Four rats in the group that received no initiation before 30 ppm MX died before the termination of the experiment and were not included in the study. Gross and histopathological analyses were determined only for those animals that survived the 57 weeks. Water consumption did not differ among the treatment and control groups (mean water consumption, 21.2–22.7 g per animal per day). Total MX intake for the groups receiving 10 ppm and 30 ppm MX was 0.08 and 0.24 g per animal, respectively. With the exception of the left lung in the MNNG-initiated rats given 30 ppm MX, no significant differences in the final body or selected organ weights were found among the treatment and control groups. MX without prior initiation with MNNG was ineffective in increasing the incidence of cancerous (adenocarcinoma) or precancerous (atypical hyperplasia) lesions in the rat glandular stomach (fundus or pylorus) (Table 7). Other than atypical hyperplasia in 2/9 (30%) rats receiving 30 ppm MX alone, no other lesions were noted. The precancerous and cancerous lesions in the groups treated with

Table 7. Incidence of proliferative lesions in the glandular stomach of rats treated with MNNG and/or MX

Treatment	Adenocarcinoma[a]	Atypical hyperplasia[a]
MNNG/30 μg/mL MX	8/27 (29.6%)[b,c]	25/27 (92.5%)[c]
MNNG/10 μg/mL MX	7/27 (25.9%)	26/27 (96.2%)[d]
MNNG	1/26 (3.8%)	16/26 (61.5%)
30 μg/mL MX	0/9	2/9 (22.2%)
10 μg/mL MX	0/10	0/10
Vehicle control	0/10	0/10

From Nishikawa *et al.* (1999)

MNNG, *N*-methyl-*N'*-nitro-*N*-nitrosoguanidine

[a] Total lesions in the glandular stomach (fundus and pylorus)

[b] Number of animals with a lesion/number of animals examined

[c] Statistically significant when compared with animals treated with MNNG alone ($p < 0.05$)

[d] Statistically significant when compared with animals treated with MNNG alone ($p < 0.01$)

MNNG were found mainly in the pyloric portion of the glandular stomach. MNNG alone induced atypical hyperplasia in the glandular stomach (16/26, 61.5%). Treating the initiated animals with 10 and 30 ppm MX increased atypical hyperplasia in the glandular stomach to 26/27 (96.2%) and 25/27 (92.5%), respectively. The incidence of adenocarcinomas in the glandular stomach was increased by 30 ppm MX (8/27, 29.6%) compared with MNNG alone (1/26, 3.8%). Cholangiocarcinoma, cholangioma and bile duct hyperplasia were found in the livers of all groups except the drinking-water controls. Treatment with MX enhanced the combined incidence of cholangiocarcinoma and cholangioma: 13% (30 ppm) and 17% (10 ppm MX) when compared with the MNNG control (4%), but the increases were not statistically significant. Thyroid follicular-cell hyperplasia was found only in the groups treated with MNNG and 30 ppm MX (8%) and 10 ppm MX (5%) (Nishikawa *et al.*, 1999).

4.   Other Data Relevant to an Evaluation of Carcinogenicity and its Mechanisms

4.1   Absorption, distribution, metabolism and excretion

4.1.1   *Humans*

No data were available for the Working Group.

4.1.2   *Experimental systems*

Pharmacokinetics were evaluated in male Wistar rats after administration of 20 mg/kg of [^{14}C]MX (97% purity) by gavage in deionized water (Komulainen *et al.*, 1992). Blood samples, urine and faeces were collected over 72 h and radioactivity was determined in tissues. Overall, 94% of the administered radioactivity was recovered from the animals after 72 h: 56% in faeces, 35% in urine, 2% in tissues, 1% in blood and 0.1% in the contents of the intestines. Of the radioactivity excreted in urine, 77% appeared within 12 h and 90% within 24 h. At 72 h, the highest concentration of radioactivity in tissues was detected in kidneys, followed by the liver. Also, the level of radioactivity in whole blood was threefold higher than that in serum, suggesting binding of radioactivity to blood cells.

In another experiment (Komulainen *et al.*, 1992), a dose of 2 mg/kg [^{14}C]MX was administered to Wistar rats and radioactivity was determined in tissues after 2 and 6 h. After 2 h, the highest amount of radioactivity was detected in kidneys, followed by the stomach, ileum, urinary bladder and liver. At 6 h, the concentration of radioactivity was markedly decreased in kidneys and urinary bladder and to a lesser extent in the gastrointestinal tract.

After administration of 1 mg/kg [^{14}C]MX intravenously or by gavage to male Wistar rats (Komulainen *et al.*, 1992), elimination of the radioactivity from blood was much slower after intravenous than after oral administration (elimination half-life, 22.9 h versus 3.8 h). The elimination followed bi-phasic kinetics. Approximately 8% of the intravenous dose was detected in faeces, suggesting biliary excretion of MX or its metabolites. In a further study with metabolic cages, no radioactivity was detected in exhaled air ($n = 2$).

Radiolabelled MX (96% pure) was given to male Fischer rats in water by gavage at approximately 40 mg/kg bw (Ringhand *et al.*, 1989). Urine and faecal samples were collected up to 48 h after exposure. Within 48 h, 33.5% of the administered dose was recovered in the urine, mostly within 24 h, 47% in faeces and less than 1% in exhaled air. Some tissues still retained traces of radioactivity after 48 h: 1.6% in muscles, 1.2% in blood, 1.2% in the liver, 0.5% in kidneys and 2.3% in the gastrointestinal contents. None of the radioactivity in urine or faeces was recovered as MX.

Clark and Chipman (1995) studied the intestinal absorption of MX *in vitro* in the everted gut sac system (ileum) of rats. Transport of MX-derived mutagenicity was measured from the mucosal to the serosal side using the standard reverse mutation asssay in *Salmonella typhimurium* TA100. A low, dose-dependent absorption of MX-derived mutagenicity was observed, which became significant at and above 50 µg/mL [230 µM] MX. Preincubation of MX with 1 mM glutathione reduced the number of revertants to below detection levels, and depletion of endogenous glutathione potentiated the mutagenicity, suggesting that mucosal glutathione modulates the absorption and/or mutagenicity of the mutagenic compounds. [The Working Group noted that it is not known whether the mutagenic components are MX or metabolites.]

MX (97% pure) was found to bind bovine serum albumin in phosphate-buffered saline (pH 7.2) *in vitro*. The largest fraction of MX (90%) bound reversibly and was recovered as intact MX, but a minor part bound tightly to protein. Human plasma bound MX in the same manner as bovine albumin (Haataja *et al.*, 1991).

Upon incubation of rat blood with [^{14}C]MX (2 μg/mL) *in vitro*, 42% of the radio-activity was found in plasma, 26% in erythrocyte cell membranes and 32% bound to haemoglobin (Risto *et al.*, 1993). The binding to erythrocytes may partially explain the long elimination half-life of radioactivity from blood observed after systemic adminis-tration (Komulainen *et al.*, 1992).

After a single oral administration of 200 or 300 mg/kg bw MX (98% pure) to male Wistar rats, the fraction of intact MX excreted in the urine within 72 h was 0.03–0.07% of the administered dose, and 90% of that appeared within 24 h (*n* = 3). The metabolites were not identified (Komulainen *et al.*, 1994).

Meier *et al.* (1996) measured MX-derived mutagenicity (Ames test with the TA100 strain) in the urine of male and female Fisher 344 rats administered MX at doses of 0, 8, 16, 32 and 64 mg/kg bw by gavage daily for 14 days. Only urine from the highest-dose group showed significant mutagenic activity, which corresponded to approximately 0.3% of the dose administered daily. Treatment of the urine samples with β-glucuronidase had no effect on the mutagenicity of any samples. The result was interpreted to indicate nearly complete metabolism of MX. [The Working Group noted that it is not known whether the mutagenic components are MX or metabolites.]

4.2   Toxic effects

4.2.1   *Humans*

No data were available to the Working Group.

4.2.2   *Experimental systems*

MX (> 99% pure) in water was administered by gavage to male and female Swiss-Webster mice at doses of 10, 20, 42, 88 or 184 mg/kg bw on 2 consecutive days and the animals were observed for 2 weeks (Meier *et al.*, 1987a). All animals given the highest dose died, and 9/10 died within 24 h. At necropsy, enlarged stomach and moderate haemorrhaging in the forestomach were observed. Doses of 88 mg/kg bw or less did not cause death or any clinical signs of toxicity. The acute 50% lethal dose (LD$_{50}$) was esti-mated to be 128 mg/kg bw.

MX (98% pure) in deionized water was administered orally to groups of 10 male Wistar rats at doses of 200, 300, 400 or 600 mg/kg bw to evaluate acute toxicity (Komulainen *et al.*, 1994). The LD$_{50}$ in 48 h was 230 mg/kg bw. Clinical signs included laborious breathing, wheezing, gasping and dyspnoea and decreased motor activity leading to catalepsia and cyanosis before death. At autopsy, the lungs appeared oedematous and spongy. Histological examination showed a strong irritation and necrosis of the mucosa of the entire gastrointes-tinal tract, and expansion of the stomach. Tubular damage was observed in both kidneys in one animal, characterized by a thin epithelium and dilated tubules.

To evaluate subchronic toxicity in rats, MX (98% pure) in deionized water was administered to groups of 15 male and 15 female Wistar rats on 5 days per week by gavage (Vaittinen *et al.*, 1995). For the low-dose regimen, the dose of 30 mg/kg bw was administered for 18 weeks. For the high-dose regimen, the dose was gradually increased from 45 mg/kg bw for 7 weeks to 60 mg/kg bw for the next 2 weeks to 75 mg/kg bw for the last 5 weeks. Urine and blood samples were collected and full histopathology was performed. The low-dose regimen did not result in any signs of toxicity. At the high-dose regimen, two males and one female died. Food consumption and body weights were significantly decreased in males. Sodium concentration in serum decreased, urine excretion increased and its specific gravity decreased in both sexes, while water consumption increased only in males. Serum cholesterol and triglycerides, as well as the relative weight of the liver and kidneys, increased in a dose-dependent manner in both sexes. Mucosal hyperplasia of the duodenum occurred at both dose levels in both sexes (57% of animals in the high-dose regimen group), accompanied by focal epithelial hyperplasia of the forestomach in males (controls, 1/15; low-dose regimen, 2/15; and high-dose regimen, 3/14) and superficial haemorrhagic necrosis in gastric glandular epithelium in 2/14 females at the high-dose regimen. However, these hyperplasias were thought to result from the local irritating effect of the gavage dosing of MX. Epithelial atypia in the bladder was observed in 1/14 male and 1/14 female in the high-dose group. The target organs of toxicity were the kidneys and the liver.

In the same animals, activities of the xenobiotic metabolizing enzymes 7-ethoxyresorufin-*O*-deethylase (EROD), pentoxyresorufin-*O*-dealkylase (PROD), NADPH-cytochrome-*c*-reductase, UDP-glucuronosyltransferase (UDPGT) and glutathione *S*-transferase (GST) were measured in microsomal and cytosolic fractions of the liver, kidneys, duodenum and lung (Heiskanen *et al.*, 1995). Most changes were observed in the kidneys. MX decreased the activity of the phase I metabolism enzymes EROD and PROD and induced the activity of the phase II conjugation enzymes UDPGT and GST in both sexes. In addition, EROD and GST activities in the liver were similarly affected in both sexes.

In 14-day range-finding studies of a subchronic toxicity study (Vaittinen *et al.*, 1995), male Han-Wistar rats were given MX (> 98% pure) in water by gavage at doses of 12.5, 25, 50, 100 or 200 mg/kg bw daily or at doses of 5, 10 or 20 mg/kg bw on 5 days per week. Daily doses of 100 and 200 mg/kg were lethal to all animals within 5 days and 3/5 animals given 50 mg/kg died within 9 days. Breathing difficulties, nostril discharge and a foamy liquid in the respiratory tract were observed with doses of 25 mg/kg and above. Doses of 5, 10 and 20 mg/kg bw did not cause general toxicity detectable clinically or by macroscopic pathology, but several biochemical changes in serum and urine were observed, particularly in females. Urea and creatinine in serum were increased at all doses and bilirubin at the two highest doses. Inorganic phosphate and potassium were decreased at the two highest doses and chloride concentration at the highest dose only. The pH of urine decreased and the specific gravity increased at the two highest doses in both sexes. The changes in electrolyte balance and urine composition, and the increase in blood urea and creatinine suggested an acute or subacute effect of MX on kidney function in females. However, no notable changes were observed by histopathology.

Male and female Fischer 344 rats received MX (> 99% pure) by gavage in distilled water at doses of 8, 16, 32 or 64 mg/kg bw for 14 consecutive days (Daniel *et al.*, 1994). At the highest dose in both sexes, body weight gain and food and water consumption decreased, and the relative liver and kidney weight increased significantly. In addition, the relative weight of the testes and thymus in males and of the spleen in females was increased, all at the highest dose only. Serum cholesterol and calcium concentrations increased in a dose-dependent manner in both sexes. In females, blood platelets increased in a dose-dependent fashion and the activities of aspartate aminotransferase (AST), alanine aminotransferase (ALT) and alkaline phosphatase (ALP) were significantly decreased at most doses. In males, the number of erythrocytes and platelets, the concentration of haemoglobin, haematocrit and creatinine, and ALP activity were decreased at the highest dose. Lesions in the forestomach, characterized by epithelial hyperplasia, chronic-active inflammation, hyperkeratosis or ulceration, were the most notable histopathological changes, and were observed in both sexes in 40% of the animals in the highest-dose group. Extramedullary haematopoiesis was slightly more common among MX-treated animals. The liver was considered to be the target organ of toxicity. [The Working Group noted that kidney function was not evaluated thoroughly in this study.]

Administration of MX to male Fischer 344 rats by gavage at a dose of 64 mg/kg bw per day for 4 days caused acute toxicity characterized by pyloric blockage, dilatation of the intestine and nose bleeding. No consistent effect on the peroxisome proliferation enzymes was observed in the liver (Meier *et al.*, 1996). The activity of catalase decreased, that of urate oxidase increased and that of fatty acyl coenzyme A oxidase remained unchanged. The enzymes of glutathione metabolism (γ-glutamylcysteine synthetase [glutamate-cysteine ligase], GST and glutathione peroxidase) were not affected. The phase I metabolism enzymes NADPH-cytochrome-P450 reductase, aminopyrine *N*-demethylase and aryl hydrocarbon hydroxylase were all significantly inhibited. Concomitantly, urinary excretion of thioethers and D-glucaric acid were largely reduced, indicating toxic metabolic effects in the liver.

Daniel *et al.* (1994) administered MX (> 99% pure) to female and male B6C3F$_1$ mice by gavage at 8, 16, 32 or 64 mg/kg bw for 14 consecutive days with the same study design as for rats (see above). Because the animals were group-housed, a large number of males in all groups displayed extensive skin lesions and chronic inflammation due to fighting. No changes in food or water consumption were observed in any group. MX caused epithelial hyperplasia of the forestomach at all treatment levels in both sexes (20–60% of animals), and extramedullary haematopoiesis in the spleen of females. Extramedullary haematopoiesis was common in control males. In males, body weight gain, spleen weight, total white blood cell count, the distribution of neutrophils, ALT activity and creatinine concentrations were significantly decreased; distribution of lymphocytes and eosinophils, ALP activity in serum and the relative weight of the thymus, testes and adrenal glands were increased in all treatment groups, generally in a dose-dependent manner. In the highest-dose group only, AST, lactate dehydrogenase and blood urea nitrogen were also decreased in serum. In females, the changes showing a clear dose relation were an increase in neutrophils (except at the highest dose) and a decrease in lymphocytes and haematocrit, ALP activity and thymus weight. At the highest dose only, the number of erythrocytes, the concentration of haemoglobin and blood urea nitrogen, and the weight of adrenal glands had decreased in females. Because of inconsistency in changes, no clear target organ of toxicity could be defined in mice in this study. [The Working Group noted that chronic inflammation and stress were likely to have contributed to the results observed in males.]

After a single dose of 0, 10, 30 or 60 mg/kg bw by gavage in 0.02 M HCl, MX [purity not specified] stimulated DNA single-strand scissions, replicative DNA synthesis and ornithine decarboxylase activity in the pyloric mucosa of the stomach in male Fischer 344 rats. The effect was dose-dependent and indicated an increase in cell proliferation (Furihata et al., 1992).

Male Wistar rats were administered MX (97% pure) at concentrations of 6.25, 12.5, 25 and 50 µg/mL in the drinking-water for 5 weeks. Doses of 12.5 µg/mL and above stimulated cell proliferation (a twofold increase) in the mucosal epithelium of the glandular stomach but not in the pyloric mucosa. This was accompanied by focal gastric erosion at the two highest doses. Lipid peroxidation in the gastric mucosa was marginally increased in a loose dose-dependent manner, and urine volume and urinary lipid peroxidation products were significantly increased at 12.5 and 25 µg/mL (Nishikawa et al., 1994).

In male and female Wistar rats given 0, 1, 10 or 60 mg/kg bw (40 mg/kg for females) for 7 or 21 days, MX did not cause any morphological changes in thyroid glands, adrenal glands or liver after either time-point. No notable changes were seen in staining for proliferating cell nuclear antigen between control and treated animals in these organs (Komulainen et al., 2000a).

MX (> 99% pure) slightly decreased glutathione levels at 100 µM and above in human white blood cells in vitro, but no change was observed in the level of free intracellular calcium (Nunn & Chipman, 1994).

4.3   Reproductive, developmental and hormonal effects

4.3.1   Humans

No data were available to the Working Group.

4.3.2   Experimental systems

MX showed a dose-dependent decrease in differentiation of foci in the micromass test system in vitro in rat embryo midbrain cells and limb bud cells exposed to concentrations of 1, 2, 5 or 10 µg/mL [4.6, 9.2, 23 and 46 µM, respectively], in the absence of metabolic activation. Concomitantly, MX inhibited differentiation of both cell types with a 50% inhibition concentration (IC_{50}) of about 3 µg/mL [13 µM]. Metabolic activation eliminated the effect (Teramoto et al., 1998). Preincubation of MX in tissue culture medium, which is known to degrade MX, also restored differentiation of foci, supporting the concept that degradation products of MX were not the responsible components (Teramoto *et al.*, 1999).

To evaluate the effects of MX on blood levels of thyroid-stimulating hormone (TSH), thyroxine (T_4), triiodothyronine (T_3), prolactin and growth hormone, groups of male and females Wistar rats were administered MX (> 97% pure) by gavage in deionized water at levels of 0, 1, 10 or 60 mg/kg bw (or 40 mg/kg for females) daily for 7 or 21 days (Komulainen *et al.*, 2000a). MX did not affect blood levels of TSH, T_4 or prolactin. In males, levels of T_3 increased transiently at the highest dose, whereas those of growth hormone decreased transiently at all doses. MX did not affect the levels of TSH, prolactin or growth hormone in blood 2 h after a single dose. It was concluded that MX does not cause dysregulation of the thyroid–pituitary axis and that the tumorigenicity of MX to the thyroid follicular epithelium in rats does not result from hormonal imbalance.

Blood concentrations of TSH, T_4 and T_3 were also measured in male and female Wistar rats in a 104-week carcinogenicity study (Komulainen *et al.*, 1997; see Section 3.1.2) in all animals surviving to the end of the study. No significant differences in hormone levels were observed at any dose compared with control groups.

4.4 Genetic and related effects

4.4.1 *Humans*

No data were available to the Working Group.

4.4.2 *Experimental systems* (see Table 8 for details and references)

MX was highly mutagenic to all but one *Salmonella typhimurium* strain tested, and one of the most potent mutagens in the TA100 strain ever tested. MX was also mutagenic in most strains of *Eschericia coli* tested. MX acts as a direct-acting mutagen, the activity of which was largely decreased or eliminated by metabolic activation. MX caused DNA damage (including unscheduled DNA synthesis) in purified DNA and in all animal and human cells studied *in vitro* except in rat hepatocytes in one study. Inhibitors of DNA repair enzymes potentiated the effect of MX in rapidly dividing cells by a factor of 10–100. MX caused gene mutations, sister chromatid exchange and chromosomal aberrations in all animal and human cells studied *in vitro*. Also, it caused apoptosis in human HL-60 cells (Marsteinstredet *et al.*, 1997b). MX was positive in a two-stage cell transformation assay in CH3 10T1/2 cells as either an initiator or promoter. A metabolic cooperation assay in Chinese hamster V79 cells showed a tentative inhibition of intercellular communication *in vitro*. Glutathione, L-cysteine, several other nucleophiles, tissue culture medium and serum strongly decreased or eliminated the mutagenic and genotoxic effects of MX *in vitro* (Ishiguro *et al.*, 1987; Mäki-Paakkanen *et al.*, 1994; Matsumura *et al.*, 1994; Watanabe *et al.*, 1994). MX reacted directly with glutathione, and GST catalysed the reaction (Meier *et al.*, 1990), leading to a further inhibition of MX-induced mutagenicity. Partial depletion

Table 8. Genetic and related effects of MX

Test system	Result[a] Without exogencus metabolic system	Result[a] With exogenous metabolic system	Dose[b] (LED or HID)	Reference
DNA strand breaks and apurinic/apyrimidinic sites, PM2 DNA, in vitro	+	NT	[217 µg/mL]	Hyttinen & Jansson (1995)
DNA damage, cleavage of ΦX174, in vitro	+	NT	0.92 µg/mL[c]	LaLonde & Ramdayal (1997)
DNA damage, Escherichia coli PQ37, SOS chromotest	+	(−)	0.4–3.3 ng/mL	Tikkanen & Kronberg (1990)
Differential DNA repair, Escherichia coli K-12 343/113	+	(+)	4 ng/mL	Fekadu et al. (1994)
Prophage-induction assay, Escherichia coli WP2s (λ)	+	+	10.9 ng/mL	DeMarini et al. (1995)
Salmonella typhimurium TM677, forward mutation for 8-azaguanine resistance	+	NT	10.9 ng/mL	DeMarini et al. (1995)
Salmonella typhimurium TA100, reverse mutation	+	(+)	29 rev/ng	Ishiguro et al. (1987)
Salmonella typhimurium TA100, reverse mutation	+	+	64 rev/ng	Meier et al. (1987b)
Salmonella typhimurium TA100, reverse mutation	+	−	29 rev/ng	Tikkanen & Kronberg (1990)
Salmonella typhimurium TA100, reverse mutation	+	NT	18 rev/ng	LaLonde et al. (1991a)
Salmonella typhimurium TA100, reverse mutation	+	NT	14 rev/ng	LaLonde et al. (1991b)
Salmonella typhimurium TA100, reverse mutation	+	NT	22 rev/ng	DeMarini et al. (1995)
Salmonella typhimurium TA100, reverse mutation	+	NT	20 rev/ng	Hyttinen et al. (1995)
Salmonella typhimurium TA100, reverse mutation	+	NT	18 rev/ng	Jansson et al. (1995)
Salmonella typhimurium TA100, reverse mutation	+	NT	19 rev/ng	Knasmüller et al. (1996)
Salmonella typhimurium TA100, reverse mutation	+	NT	18 rev/ng	LaLonde et al. (1997)
Salmonella typhimurium TA100, reverse mutation	+	NT	38 rev/ng	Yamada et al. (1997)
Salmonella typhimurium TA100, reverse mutation	+	NT	54 rev/ng	Franzén et al. (1998a)

Table 8 (contd)

Test system	Result[a]		Dose[b] (LED or HID)	Reference
	Without exogenous metabolic system	With exogenous metabolic system		
Salmonella typhimurium TA100, reverse mutation	+	–	4.5 rev/ng	Kargalioglu *et al.* (2002)
Salmonella typhimurium TA102, reverse mutation	+	NT	6.35 rev/ng	Meier *et al.* (1987b)
Salmonella typhimurium TA104, reverse mutation	+	NT	39 rev/ng	Franzén *et al.* (1998a)
Salmonella typhimurium TA1535, reverse mutation	+	NT	0.48 rev/ng	Meier *et al.* (1987b)
Salmonella typhimurium TA1535, reverse mutation	(+)	NT	0.04 rev/ng	DeMarini *et al.* (1995)
Salmonella typhimurium TA1535, reverse mutation	+	NT	0.1 rev/ng	Jansson *et al.* (1995)
Salmonella typhimurium TA1538, reverse mutation	+	NT	0.13 rev/ng	Meier *et al.* (1987b)
Salmonella typhimurium TA98, reverse mutation	+	NT	9.18 rev/ng	Meier *et al.* (1987b)
Salmonella typhimurium TA98, reverse mutation	+	–	0.89 rev/ng	Tikkanen & Kronberg (1990)
Salmonella typhimurium TA98, reverse mutation	+	NT	2.60 rev/ng	DeMarini *et al.* (1995)
Salmonella typhimurium TA98, reverse mutation	+	NT	6.1 rev/ng	Franzén *et al.* (1998a)
Salmonella typhimurium TA98, reverse mutation	+	+	0.53 rev/ng	Kargalioglu *et al.* (2002)
Salmonella typhimurium TA92, reverse mutation	+	NT	0.32 rev/ng	Meier *et al.* (1987b)
Salmonella typhimurium TA97, reverse mutation	+	NT	7.26 rev/ng	Meier *et al.* (1987b)
Salmonella typhimurium TA97, reverse mutation	+	–	4.8 rev/ng	Tikkanen & Kronberg (1990)
Salmonella typhimurium UTH8414, reverse mutation	+	NT	0.20 rev/ng	DeMarini *et al.* (1995)
Salmonella typhimurium UTH8413, reverse mutation	+	NT	0.08 rev/ng	DeMarini *et al.* (1995)
Salmonella typhimurium TP2428, reverse mutation	+	NT	0.4 rev/ng	Knasmüller *et al.* (1996)
Salmonella typhimurium TA1950, reverse mutation	+	NT	0.04 rev/ng	Knasmüller *et al.* (1996)

Table 8 (contd)

Test system	Result[a] Without exogenous metabolic system	Result[a] With exogenous metabolic system	Dose[b] (LED or HID)	Reference
Salmonella typhimurium YG7112, reverse mutation	+	NT	26.4 rev/ng	Yamada et al. (1997)
Salmonella typhimurium YG7113, reverse mutation	+	NT	26.2 rev/ng	Yamada et al. (1997)
Salmonella typhimurium YG7119, reverse mutation	+	NT	33 rev/ng	Yamada et al. (1997)
Salmonella typhimurium RSJ100, reverse mutation	–	–	1.63 µg/mL	Kargalioglu et al. (2002)
Escherichia coli CC102, CC104, reverse mutation	–	NT	20 µg/plate	Watanabe-Akanuma & Ohta (1994)
Escherichia coli CC107, CC109, CC111, reverse mutation	+	NT	5 µg/plate	Watanabe-Akanuma & Ohta (1994)
Escherichia coli CC108, CC110, reverse mutation	+	NT	2 µg/plate	Watanabe-Akanuma & Ohta (1994)
Escherichia coli ZA2102, ZA2104, reverse mutation	+	NT	0.05 µg/plate	Watanabe-Akanuma & Ohta (1994)
Escherichia coli ZA2108, ZA2109, ZA2110, reverse mutation	+	NT	0.1 µg/plate	Watanabe-Akanuma & Ohta (1994)
Escherichia coli ZA2107, ZA2111, reverse mutation	+	NT	0.2 µg/plate	Watanabe-Akanuma & Ohta (1994)
Escherichia coli ZA4102, ZA4104, reverse mutation	–	NT	20 µg/plate	Watanabe-Akanuma & Ohta (1994)
Escherichia coli ZA4107, ZA4108, ZA4109, ZA4110, ZA4111, reverse mutation	+	NT	5 µg/plate	Watanabe-Akanuma & Ohta (1994)
Escherichia coli ZA5102, reversion mutation	+	NT	0.1 µg/plate	Watanabe-Akanuma & Ohta (1994)
Eschericia coli ZA5104, ZA5108, ZA5109, reversion mutation	+	NT	0.2 µg/plate	Watanabe-Akanuma & Ohta (1994)

Table 8 (contd)

Test system	Result[a]		Dose[b] (LED or HID)	Reference
	Without exogenous metabolic system	With exogenous metabolic system		
Eschericia coli ZA5107, ZA5110, ZA5111, reverse mutation	–	NT	1 μg/plate	Watanabe-Akanuma & Ohta (1994)
DNA damage, rat hepatocytes, rat testicular cells, Chinese hamster V79 cells *in vitro* (alkaline elution)	+	NT	6.5 μg/mL	Brunborg *et al.* (1991)
DNA strand breaks, rat hepatocytes *in vitro* (DNA alkaline unwinding assay)	–	NT	11 μg/mL	Chang *et al.* (1991)
DNA single-strand breaks or alkali-labile sites, pig LLC-PK₁ cells *in vitro* (without pretreatment with DNA repair enzyme inhibitors, alkaline elution)	+	NT	65 μg/mL	Holme *et al.* (1999)
DNA single-strand breaks or alkali-labile sites, pig LLC-PK₁ cells *in vitro* (with pretreatment with DNA repair enzyme inhibitors, alkaline elution)	+	NT	6.5 μg/mL	Holme *et al.* (1999)
DNA single-strand breaks or alkali-labile sites, rat testicular cells *in vitro* (with and without pretreatment with DNA repair enzyme inhibitors, alkaline elution)	+	NT	22 μg/mL	Holme *et al.* (1999)
DNA strand breaks and alkali-labile damage, Chinese hamster ovary cells *in vitro* (single-cell gel electrophoresis assay)	+	NT	8 μg/mL	Mäki-Paakkanen *et al.* (2001)
Unscheduled DNA synthesis, male Wistar rat and BALB/c mouse hepatocytes *in vitro*	+	NT	0.43 μg/mL	Nunn *et al.* (1997)
Unscheduled DNA synthesis, rat hepatocytes *in vitro*	+	NT	3.6 μg/mL	Le Curieux *et al.* (1999)
Gene mutation, Chinese hamster ovary cells, *Hprt* locus *in vitro*	+	NT	2.5 μg/mL	Jansson & Hyttinen (1994)
Gene mutation, Chinese hamster ovary cells, ouabain resistance *in vitro*	+	NT	2 μg/mL	Mäki-Paakkanen *et al.* (1994)
Gene mutation, Chinese hamster V79 cells, *Hprt* locus *in vitro*	–	NT	1.1 μg/mL	Brunborg *et al.* (1991)
Gene mutation, Chinese hamster V79 cells, 6-TG resistance *in vitro*	+	NT	1 μg/mL	Matsumura *et al.* (1994)
Gene mutation, mouse lymphoma L5178Y cells, *Tk* locus *in vitro*	+	NT	0.5 μg/mL	Harrington-Brock *et al.* (1995)

Table 8 (contd)

Test system	Result[a]		Dose[b] (LED or HID)	Reference
	Without exogenous metabolic system	With exogenous metabolic system		
Sister chromatid exchange, Chinese hamster V79 cells in vitro	+	NT	0.43 µg/mL	Brunborg et al. (1991)
Sister chromatid exchange, Chinese hamster cells in vitro	+	NT	0.19 µg/mL	Mäki-Paakkanen et al. (1994)
Sister chromatid exchange, Chinese hamster ovary cells in vitro	+	NT	0.24 µg/mL	Mäki-Paakkanen et al. (2001)
Sister chromatid exchange, rat peripheral lymphocytes in vitro	+	NT	20 µg/mL	Jansson et al. (1993)
Micronucleus formation, mouse lymphoma L5178Y cells in vitro	+	NT	11 µg/mL	Le Curieux et al. (1999)
Chromosomal aberrations, Chinese hamster ovary cells in vitro	+	+	4 µg/mL	Meier et al. (1987b)
Chromosomal aberrations, Chinese hamster ovary cells in vitro	+	NT	4 µg/mL	Mäki-Paakkanen et al. (1994)
Chromosomal aberrations, Chinese hamster ovary cells in vitro	+	NT	0.5 µg/mL	Mäki-Paakkanen et al. (2001)
Chromosomal aberrations, mouse lymphoma L5178Y cells in vitro	+	NT	0.75 µg/mL	Harrington-Brock et al. (1995)
Chromosomal aberrations, rat peripheral lymphocytes in vitro	+	NT	60 µg/mL	Jansson et al. (1993)
Cell transformation, C3H 10T1/2 mouse cells in vitro	+	NT	10 µg/mL	Laaksonen et al. (2001)
DNA strand breaks, human lymphoblastoid cell line CCRF-CEM in vitro (DNA alkaline unwinding assay)	+	+	9.6 µg/mL	Chang et al. (1991)
DNA strand breaks, human white blood cells in vitro (DNA alkaline unwinding assay)	+	NT	0.22 µg/mL	Nunn & Chipman (1994)
DNA strand breaks or alkali-labile sites, HL-60 cells in vitro (without pretreatment with DNA repair enzyme inhibitors, alkaline elution)	+	NT	21.7 µg/mL	Marsteinstredet et al. (1997a)

Table 8 (contd)

Test system	Result[a]		Dose[b] (LED or HID)	Reference
	Without exogenous metabolic system	With exogenous metabolic system		
DNA strand breaks or alkali-labile sites, HL-60 cells in vitro (with pretreatment with DNA repair enzyme inhibitors, alkaline elution)	+	NT	0.22 µg/mL	Marsteinstredet et al. (1997a)
DNA strand breaks or alkali-labile sites, resting human leukocytes in vitro (with and without pretreatment with DNA repair inhibitors, alkaline elution)	+	NT	21.7 µg/mL	Holme et al. (1999)
DNA strand breaks or alkali-labile sites, proliferating human leukocytes in vitro (without pretreatment with DNA repair enzyme inhibitors, alkaline elution)	+	NT	21.7 µg/mL	Holme et al. (1999)
DNA strand breaks or alkali-labile sites, proliferating human leukocytes in vitro (with pretreatment with DNA repair enzyme inhibitors, alkaline elution)	+	NT	0.65 µg/mL	Holme et al. (1999)
DNA strand breaks or alkali-labile sites, HL-60 cells in vitro (without pretreatment with DNA repair enzyme inhibitors, alkaline elution)	+	NT	65 µg/mL	Holme et al. (1999)
DNA strand breaks or alkali-labile sites, HL-60 cells in vitro (with pretreatment with DNA repair enzyme inhibitors, alkaline elution)	+	NT	0.65 µg/mL	Holme et al. (1999)
Unscheduled DNA synthesis, human hepatocytes in vitro	+	NT	0.43 µg/mL	Nunn et al. (1997)
Gene mutation, human B-lymphoblastoid cell lines MCL-5, AHH-1 TK$^{+/-}$, h1A1v2, TK locus in vitro	+	NT	3 µg/mL	Woodruff et al. (2001)
DNA damage, male Wistar rat liver, kidney, lung, testis, stomach, duodenum, colon, urinary bladder, bone marrow cells in vivo (alkaline elution)	–		125 po, 1 h	Brunborg et al. (1991)
DNA single-strand breaks, male Fischer 344 rat pyloric mucosa of stomach in vivo (alkaline elution)	+		48 po, 2 h	Furihata et al. (1992)

Table 8 (contd)

Test system	Result[a]		Dose[b] (LED or HID)	Reference
	Without exogenous metabolic system	With exogenous metabolic system		
DNA strand breaks and alkali-labile damage, male CD-1 mouse liver, kidney, lung, brain, stomach, jejunum, ileum, colon, bladder cells in vivo (single cell gel electrophoresis assay)	+		100 po, 1, 3, 6, 24 h	Sasaki et al. (1997)
DNA strand breaks and alkali-labile damage, male CD-1 mouse spleen, bone-marrow cells in vivo (single-cell gel electrophoresis assay)	–		100 po, 1, 3, 6, 24 h	Sasaki et al. (1997)
DNA damage, male B6C3F₁ mouse liver, kidney, spleen, colon cells in vivo (with pretreatment with DNA repair enzyme inhibitors, alkaline elution)	–		80 ip, 1 h	Holme et al. (1999)
DNA damage, male B6C3F₁ mouse liver, kidney cells in vivo (with pretreatment with DNA repair enzyme inhibitors, alkaline elution)	+		40 ip, 1 h	Holme et al. (1999)
DNA damage, male B6C3F₁ mouse spleen, colon cells in vivo (with pretreatment with DNA repair enzyme inhibitors, alkaline elution)	–		80 ip, 1 h	Holme et al. (1999)
Differential DNA repair, male Swiss albino mouse liver, lung, kidney, spleen, stomach, intestines in vivo, measured in two strains of E. coli K-12 indicator cells recovered from organs	+		4.3 po, 2 h	Fekadu et al. (1994)
Unscheduled DNA synthesis, BALB/c hepatocytes ex vivo	–		100 po, 16 h	Nunn et al. (1997)
Sister chromatid exchange, Wistar rat peripheral lymphocytes in vivo	+		30 po, 18 weeks	Jansson et al. (1993)
Sister chromatid exchange, male Wistar rat kidney in vivo	+		25, po × 3	Mäki-Paakkanen & Jansson (1995)
Sister chromatid exchange, male Wistar rat peripheral lymphocytes in vivo	+		100, po × 3	Mäki-Paakkanen & Jansson (1995)
Micronucleus formation, Swiss–Webster mouse, bone-marrow polychromatic erythrocytes in vivo	–		90 po × 2	Meier et al. (1987b)
Micronucleus formation, NMRI mouse bone-marrow polychromatic erythrocytes in vivo	–		8.8 ip	Tikkanen & Kronberg (1990)

Table 8 (contd)

Test system	Result[a]		Dose[b] (LED or HID)	Reference
	Without exogenous metabolic system	With exogenous metabolic system		
Nuclear anomalies (including micronuclei), male B6C3F$_1$ mouse forestomach, duodenum *in vivo*	+		80 po	Daniel *et al.* (1991)
Nuclear anomalies (including micronuclei), male B6C3F$_1$ mouse colon *in vivo*	−		100 po	Daniel *et al.* (1991)
Micronucleus formation, B6C3F$_1$ mouse blood polychromatic erythrocytes *in vivo*	−		64 po, 14 days	Meier *et al.* (1996)
Micronucleus formation, male Wistar rat peripheral blood lymphocytes *in vivo*	+		100 po × 3	Mäki-Paakkanen & Jansson (1995)
Micronucleus formation, Wistar rat bone-marrow polychromatic erythrocytes *in vivo*	−		approx. 5, 104 weeks	Jansson (1998)
Inhibition of intercellular communication, Chinese hamster V79 cells *in vitro* (metabolic cooperation assay)	(+)	NT	0.8 µg/mL	Matsumura *et al.* (1994)

[a] +, positive; (+), weak positive; −, negative; NT, not tested
[b] LED, lowest effective dose; HID, highest ineffective dose; in-vitro tests, as indicated; in-vivo tests, mg/kg bw per day; po, oral; ip, intraperitoneal; rev, revertant
[c] Only dose tested

of glutathione in cells had little effect on MX-induced DNA damage (Brunborg *et al.*, 1991; Chang *et al.*, 1991).

The *cis* arrangement of the CHCl$_2$ (position 4) and Cl (position 3) substituents in MX is essential for the high mutagenicity in *S. typhimurium* TA100 (Ishiguro *et al.*, 1987, 1988; LaLonde *et al.*, 1991c). The OH group at position 5 and Cl at position 3 and 6 are also important for the mutagenic potency (LaLonde *et al.*, 1991b,c).

Results on DNA damage *in vivo* are less clear. A single oral dose of MX to mice caused DNA damage in a number of tissues, but not in spleen or bone marrow. Following intraperitoneal administration at a lower dose and with a shorter follow-up period, DNA damage was observed only after pretreatment of the animals with DNA-repair enzyme inhibitors, and only in the liver and kidney. MX-induced DNA damage was repaired rapidly (within 3 h) in mouse liver but more slowly in other tissues (Sasaki *et al.*, 1997). In rats, MX did not induce DNA damage in any of the tissues examined, except in the pyloric mucosa of the stomach following oral administration in a single study. Oral administration of MX to mice caused DNA damage in *E. coli* K-12 indicator cells recovered from a number of tissues.

In rats, MX caused sister chromatid exchange in peripheral blood lymphocytes in two studies and in kidney cells in a single study *in vivo* after repeated doses, but had no effect on micronucleus formation in bone marrow or blood polychromatic erythrocytes in any of four studies. Micronuclei were observed in rat peripheral blood lymphocytes in a single study, and nuclear anomalies (including micronuclei) in mouse forestomach and duodenum in a single study, both after oral administration of MX.

The predominant base-pair substitution induced by MX is a GC→TA transversion, observed in several *S. typhimurium* strains at the *hisG* locus (DeMarini *et al.*, 1995; Hyttinen *et al.*, 1995; Knasmüller *et al.*, 1996; Shaughnessy *et al.*, 2000) and in Chinese hamster ovary cells at the *Hprt*-locus (Hyttinen *et al.*, 1996). In *S. typhimurium* TA100, the hot spot was in the second position of the *hisG46* target CCC codon. MX also caused duplications, frameshift mutations of a two-base deletion and frameshifts with adjacent base substitution in *S. typhimurium* TA98 (DeMarini *et al.*, 1995), and frameshift mutations of one- or two-base insertion or deletion in several strains of *E. coli* (Watanabe-Akanuma & Ohta, 1994). AT→TA transversions, deletions of single base pairs, larger deletions and insertions in cDNA were reported in Chinese hamster ovary cells (Hyttinen *et al.*, 1996).

MX produced DNA adducts *in vitro* under various reaction conditions but not in all experiments (Alhonen-Raatesalmi & Hemminki, 1991). Five different adducts were reported to be formed in buffered solutions with MX *in vitro*, three with 2′-deoxyadenosine (Le Curieux *et al.*, 1997; Munter *et al.*, 1998) and two with guanosine (Franzén *et al.*, 1998b; Munter *et al.*, 1999). The guanosine adduct described by Franzén *et al.* (1998b) could not be confirmed in the later study (Munter *et al.*, 1999). Two of the adducts formed with 2′-deoxyadenosine were also observed in calf thymus DNA reacted with MX *in vitro* (Le Curieux *et al.*, 1997; Munter *et al.*, 1998) but not the guanosine adduct (Munter *et al.*, 1999). All adducts were observed after incubation for several days.

Altogether four point mutations were found in *p53* (exon 4–7) in 47 different MX-induced liver tumours in Wistar rats (Komulainen *et al.*, 1997, 2000b). No consistent pattern of point mutations was observed. Exons 1 and 2 of Ki-*ras*, Ha-*ras* and N-*ras* of the same tumours did not contain any mutations. MX did not alter the expression of p53 protein in the normal liver (Komulainen *et al.*, 2000b) or in thyroid gland tumours of these animals (Hakulinen *et al.*, 2002). The expression of p21 Ki-ras protein was also unaffected in MX-induced thyroid gland tumours (Hakulinen *et al.*, 2002).

4.5   Mechanistic considerations

In-vitro data in bacteria and in mammalian cells indicate that MX is genotoxic. In addition, it causes DNA damage *in vivo*. Data on mutagenicity tests *in vitro*, mutation spectra and adduct formation suggest that the guanine moiety may be one target of MX in DNA. In addition, MX was 100-fold more mutagenic in the form of lactone (closed-ring form) than in the open-ring conformation *in vitro* (LaLonde *et al.,* 1991b), suggesting that the closed-ring conformation may be responsible for the mutagenicity at physiological pH. However, examination of the structural and electronic properties of MX have not yet identified the form of interaction of MX with DNA.

Studies on hormonal effects of MX suggest that it does not cause thyroid gland tumours in rats by the TSH-mediated promotion mechanism (Komulainen *et al.*, 1997, 2000a).

5.   Summary of Data Reported and Evaluation

5.1   Exposure data

3-Chloro-4-(dichloromethyl)-5-hydroxy-2(5*H*)-furanone (MX) is a disinfection by-product that has been found at nanogram-per-litre levels in drinking-water as a result of chlorination or chloramination.

5.2   Human carcinogenicity data

Several studies were identified that analysed risk with respect to one or more measures of exposure to complex mixtures of disinfection by-products that are found in most chlorinated and chloraminated drinking-water. No data specifically on MX were available to the Working Group.

5.3   Animal carcinogenicity data

In a single bioassay, MX induced malignant and benign thyroid and mammary tumours in male and female rats. MX failed to increase the incidence of tumours in the small intestine or colon of C57BL/6J-MiN$^{-/-}$ mice. MX promoted preneoplastic and malignant lesions in the glandular stomach but not aberrant crypt foci (preneoplastic lesions) in the colon of rats initiated with N-methyl-N'-nitro-N-nitrosoguanidine.

5.4   Other relevant data

MX is well absorbed in the gastrointestinal tract of experimental animals. About 40% of the administered dose is excreted rapidly in urine almost entirely as uncharacterized metabolites. Pharmacokinetic studies suggest that MX does not accumulate upon continuous exposure.

The target organs of toxicity of MX in rats are the kidneys and the liver. It affects hepatic lipid metabolism and kidney function. At high doses, MX is highly irritating in the gastrointestinal tract. No data were available on the teratogenicity of MX *in vivo*.

MX is genotoxic. It is a direct-acting mutagen and causes DNA and chromosome damage *in vitro* and DNA damage *in vivo*. MX-induced liver tumours of rats contained only a few point mutations in *p53* and none in *ras* genes. Hormonal data suggest that MX does not cause thyroid gland tumours in rats by the thyroid-stimulating hormone-mediated promotion mechanism.

5.5   Evaluation

There is *inadequate evidence* in humans for the carcinogenicity of MX.

There is *limited evidence* in experimental animals for the carcinogenicity of MX.

MX is a potent, direct-acting mutagen that induces primarily GC→TA transversions in both bacterial and mammalian cells. It induces DNA damage in bacterial and mammalian cells as well as in rodents *in vivo*. MX is a chromosomal mutagen in mammalian cells and in rats, and it induces mammalian cell transformation *in vitro*. The MX-associated thyroid gland tumours in rats are caused by mechanisms other than TSH-mediated hormonal promotion.

Overall evaluation

3-Chloro-4-(dichloromethyl)-5-hydroxy-2(5*H*)-furanone (MX) is *possibly carcinogenic to humans (Group 2B)*.

6.   References

Alhonen-Raatesalmi, A. & Hemminki, K. (1991) Assay for nucleoside and nucleotide binding of a potent mutagen, 3-chloro-4-(dichloromethyl)-5-hydroxy-2(5*H*)-furanone. *Toxicol. Lett.*, **56**, 167–172

Andrews, R.C., Daignault, S.A., Laverdure, C., Williams, D.T. & Huck, P.M. (1990) Occurrence of the mutagenic compound 'MX' in drinking water and its removal by activated carbon. *Environ. Technol.*, **11**, 685–694

Backlund, P., Kronberg, L. & Tikkanen, L. (1988) Formation of Ames mutagenicity and of the strong bacterial mutagen 3-chloro-4-(dichloromethyl)-5-hydroxy-2(5H)-furanone and other halogenated compounds during disinfection of humic waters. *Chemosphere*, **17**, 1329–1336

Backlund, P., Wondergem, E., Voogd, K. & de Jong, A. (1989) Mutagenic activity and presence of the strong mutagen 3-chloro-4-(dichloromethyl)-5-hydroxy-2(5H)-furanone (MX) in chlorinated raw and drinking waters in the Netherlands. *Sci. total Environ.*, **84**, 273–282

Brunborg, G., Holme, J.A., Søderlund, E.J., Hongslo, J.K., Vartiainen, T., Lötjönen, S. & Becher, G. (1991) Genotoxic effects of the drinking water mutagen 3-chloro-4-(dichloromethyl)-5-hydroxy-2[5H]-furanone (MX) in mammalian cells in vitro and in rats in vivo. *Mutat. Res.*, **260**, 55–64

Chang, L.W., Daniel, F.B. & DeAngelo, A.B. (1991) DNA strand breaks induced in cultured human and rodent cells by chlorohydroxyfuranones — Mutagens isolated from drinking water. *Teratog. Carcinog. Mutag.*, **11**, 103–114

Charles, M.J., Chen, G., Kanniganti, R. & Marbury, G.D. (1992) High-resolution mass spectrometry method for the analysis of 3-chloro-4-(dichloromethyl)-5-hydroxy-2(5H)-furanone in waters. *Environ. Sci. Technol.*, **26**, 1030–1035

Chen, Z., Yang, C., Lu, J., Zou, H. & Zhang, J. (2001) Factors on the formation of disinfection by-products MX, DCA and TCA by chlorination of fulvic acid from lake sediments. *Chemosphere*, **45**, 379–385

Clark, N.W.E. & Chipman, J.K. (1995) Absorption of 3-chloro-4-(dichloromethyl)-5-hydroxy-2-[5H] furanone (MX) through rat small intestine in vitro. *Toxicol. Lett.*, **81**, 33–38

Daniel, F.B., Olson, G.R. & Stober, J.A. (1991) Induction of gastrointestinal tract nuclear anomalies in B6C3F1 mice by 3-chloro-4-(dichloromethyl)-5-hydroxy-2[5H]-furanone and 3,4-(dichloro)-5-hydroxy-2[5H]-furanone, mutagenic byproducts of chlorine disinfection. *Environ. mol. Mutag.*, **17**, 32–39

Daniel, F.B., Robinson, M., Olson, G.R., Stober, J.A. & Page, N.P. (1994) Toxicological studies on MX, a disinfection by-product. *J. Am. Water Works Assoc.*, **86**, 103–111

DeMarini, D.M., Abu-Shakra, A., Felton, C.F., Patterson, K.S. & Shelton, M.L. (1995) Mutation spectra in Salmonella of chlorinated, chloraminated, or ozonated drinking water extracts: Comparison to MX. *Environ. mol. Mutag.*, **26**, 270–285

Environmental Protection Agency (1998) National primary drinking water regulations; disinfectants and disinfection byproducts; final rule. *Fed. Regist.*, December 16

European Commission (1998) *Council Directive 98/83/EC of 3 November 1998 on the Quality of Water Intended for Human Consumption*, Luxemburg, Office for Official Publications of the European Communities

Fekadu, K., Parzefall, W., Kronberg, L., Franzen, R., Schulte-Hermann, R. & Knasmüller, S. (1994) Induction of genotoxic effects by chlorohydroxyfuranones, byproducts of water disinfection, in *E. coli* K-12 cells recovered from various organs of mice. *Environ. mol. Mutag.*, **24**, 317–324

Franzén, R. & Kronberg, L. (1995) Synthesis of chlorinated 5-hydroxy 4-methyl-2(5H)-furanones and mucochloric acid. *Tetrahedron Lett.*, **36**, 3905–3908

Franzén, R., Goto, S., Tanabe, K. & Morita, M. (1998a) Genotoxic activity of chlorinated butenoic acids in *Salmonella typhimurium* strains TA98, TA100 and TA104. *Mutat. Res.*, **417**, 31–37

Franzén, R., Tanabe, K. & Morita, M. (1998b) Isolation of a MX–guanosine adduct formed at physiological conditions. *Chemosphere*, **36**, 2803–2808

Furihata, C., Yamashita, M., Kinae, N. & Matsushima, T. (1992) Genotoxity and cell proliferative activity of 3-chloro-4-(dichloromethyl)-5-hydroxy-2(5H)-furanone [MX] in rat glandular stomach. *Water Sci. Tech.*, **25**, 341–345

Haataja, L., Vartiainen, T., Lampelo, S., Lötjönen, S. & Tuomisto, J. (1991) Binding of the strong bacterial mutagen, 3-chloro-4-(dichloromethyl)-5-hydroxy-2(5*H*)-furanone (MX) to bovine serum albumin. *Toxicol. Lett.*, **59**, 187–195

Hakulinen, P., Kosma, V.-M. & Komulainen, H. (2002) Expression of p53 and p21 Ki-ras proteins in rat thyroid gland tumors induced by 3-chloro-4-(dichloromethyl)-5-hydroxy-2(5*H*)-furanone (MX). *Anticancer Res.*, **22**, 703–706

Harrington-Brock, K., Doerr, C.L. & Moore, M.M. (1995) Mutagenicity and clastogenicity of 3-chloro-4-(dichloromethyl)-5-hydroxy-2(5H)-furanone (MX) in L5178Y/TK$^{+/-}$-3.7.2C mouse lymphoma cells. *Mutat. Res.*, **348**, 105–110

Health Canada (2003) *Summary of Guidelines for Canadian Drinking Water Quality*, Ottawa

Heiskanen, K., Lindström-Seppä, P., Haataja, L., Vaittinen, S.-L., Vartiainen, T. & Komulainen, H. (1995) Altered enzyme activities of xenobiotic biotransformation in kidneys after subchronic administration of 3-chloro-4-(dichloromethyl)-5-hydroxy-2(5*H*)-furanone (MX) to rats. *Toxicology*, **100**, 121–128

Hemming, J., Holmbom, B., Reunanen, M. & Kronberg, L. (1986) Determination of the strong mutagen 3-chloro-4-(dichloromethyl)-5-hydroxy-2(5H)-furanone in chlorinated drinking and humic waters. *Chemosphere*, **15**, 549–556

Holmbom, B. (1990) Mutagenic compounds in chlorinated pulp bleaching waters and drinking waters. In: Vainio, H., Sorsa M. & McMichael, A.J., eds, *Complex Mixtures and Cancer Risk* (IARC Scientific Publications No. 104), Lyon, IARC*Press*, pp. 333–340

Holmbom, B., Voss, R.H., Mortimer, R.D. & Wong, A. (1984) Fractionation, isolation, and characterization of Ames mutagenic compounds in kraft chlorination effluents. *Environ. Sci. Technol.*, **18**, 333–337

Holme, J.A., Haddeland, U., Haug, K. & Brunborg, G. (1999) DNA damage induced by the drinking water mutagen 3-chloro-4-(dichloromethyl)-5-hydroxy-2[5*H*]-furanone (MX) in mammalian cells in vitro and in mice. *Mutat. Res.*, **441**, 145–153

Horth, H. (1990) Identification of mutagens in drinking water. *J. fr. Hydrol.*, **21**, 135–145

Horth, H., Fielding, M., Gibson, T., James, H.A. & Ross, H. (1989) *Identification of Mutagens in Drinking Water* (Water Research Centre Technical Report PRD 2038–M), Marlow, Water Research Centre Medmenham

Hyttinen, J.M.T. & Jansson, K. (1995) PM2 DNA damage induced by 3-chloro-4-(dichloromethyl)-5-hydroxy-2(5*H*)-furanone (MX). *Mutat. Res.*, **348**, 183–186

Hyttinen, J.M.T., Niittykoski, M. & Jansson, K. (1995) Lack of uniformity in the mutational spectra of chlorohydroxyfuranones in *Salmonella typhimurium* strain TA100. *Mutagenesis*, **10**, 321–323

Hyttinen, J.M.T., Myöhänen, S. & Jansson, K. (1996) Kinds of mutations induced by 3-chloro-4-(dichloromethyl)-5-hydroxy-2(5*H*)-furanone (MX) in the *hprt* gene of Chinese hamster ovary cells. *Carcinogenesis*, **17**, 1179–1181

Ishiguro, Y., LaLonde, R.T., Dence, C.W. & Santodonato, J. (1987) Mutagenicity of chlorinate-substituted furanones and their inactivation by reaction with nucleophiles. *Environ. Toxicol. Chem.*, **6**, 935–946

Ishiguro, Y., Santodonato, J. & Neal, M.W. (1988) Mutagenic potency of chlorofuranones and related compounds in *Salmonella. Environ. mol. Mutag.*, **11**, 225–234

Jansson, K. (1998) Lack of induction of micronuclei in bone marrow erythrocytes of rats exposed to 3-chloro-4-(dichloromethyl)-5-hydroxy-2(5*H*)-furanone (MX) for two years in a carcino-genicity bioassay. *Environ. mol. Mutag.*, **32**, 185–187

Jansson, K. & Hyttinen, J.M.T. (1994) Induction of gene mutation in mammalian cells by 3-chloro-4-(dichloromethyl)-5-hydroxy-2(5*H*)-furanone (MX), a chlorine disinfection by-product in drinking water. *Mutat. Res.*, **322**, 129–132

Jansson, K., Mäki-Paakkanen, J., Vaittinen, S.-L., Vartiainen, T., Komulainen, H. & Tuomisto, J. (1993) Cytogenetic effects of 3-chloro-4-(dichloromethyl)-5-hydroxy-2(5*H*)-furanone (MX) in rat peripheral lymphocytes in vitro and in vivo. *Mutat. Res.*, **229**, 25–28

Jansson, K., Hyttinen, J.M.T., Niittykoski, M. & Mäki-Paakkanen, J. (1995) Mutagenicity in vitro of 3,4-dichloro-5-hydroxy-2(5*H*)-furanone (mucochloric acid), a chlorine disinfection by-product in drinking water. *Environ. mol. Mutag.*, **25**, 284–287

Kanniganti, R., Johnson, J.D., Ball, L.M. & Charles, M.J. (1992). Identification of compounds in mutagenic extracts of aqueous monochloraminated fulvic acid. *Environ. Sci. Technol.*, **26**, 1998–2004

Kargalioglu, Y., McMillan, B., Minear, R.A. & Plewa, M.J. (2002) Analysis of the cytotoxicity and mutagenicity of drinking water disinfection by-products in *Salmonella typhimurium. Teratog. Carcinog. Mutag.*, **22**, 113–128

Knasmüller, S., Zöhrer, E., Kronberg, L., Kundi, M., Franzén, R. & Schulte-Hermann, R. (1996) Mutational spectra of *Salmonella typhimurium* revertants induced by chlorohydroxyfuranones, byproducts of chlorine disinfection of drinking water. *Chem. Res. Toxicol.*, **9**, 374–381

Komulainen, H., Vaittinen, S.-L., Vartiainen, T., Lötjönen, S., Paronen, P. & Tuomisto, J. (1992) Pharmacokinetics in rat of 3-chloro-4-(dichloromethyl)-5-hydroxy-2(5H)-furanone (MX), a drinking water mutagen, after a single dose. *Pharmacol. Toxicol.*, **70**, 424–428

Komulainen, H., Huuskonen, H., Kosma, V.-M., Lötjönen, S. & Vartiainen, T. (1994) Toxic effects and excretion in urine of 3-chloro-4-(dichloromethyl)-5-hydroxy-2(5H)-furanone (MX) in the rat after a single oral dose. *Arch. Toxicol.*, **68**, 398–400

Komulainen, H., Kosma, V.-M., Vaittinen, S.-L., Vartiainen, T., Kaliste-Korhonen, E., Lötjönen, S., Tuominen, R.K. & Tuomisto, J. (1997) Carcinogenicity of the drinking water mutagen 3-chloro-4-(dichloromethyl)-5-hydroxy-2(5*H*)-furanone in the rat. *J. natl Cancer Inst.*, **89**, 848–856

Komulainen, H., Tuominen, R.K., Kosma, V.-M. & Huuskonen, H. (2000a) 3-Cloro-4-(dichloro-methyl)-5-hydroxy-2(5H)-furanone (MX), a rat thyroid gland carcinogen, does not affect serum levels of TSH and thyroid hormones. *Environ. Toxicol. Pharmacol.*, **8**, 267–273

Komulainen, H., Hakulinen, P., Servomaa, K., Makkonen, K., Vasara, R., Mäki-Paakkanen, J. & Kosma, V.-M. (2000b) No consistent pattern of mutations in *p53* and *ras* genes in liver tumors of rat treated with the drinking water mutagen 3-chloro-4-(dichloromethyl)-5-hydroxy-2(5*H*)-furanone (MX). *Environ. mol. Mutag.*, **36**, 292–300

Kronberg, L. (1999) Water treatment practice and the formation of genotoxic chlorohydroxyfura-nones. *Water Sci. Technol.*, **40**, 31–36

Kronberg, L. & Christman, R.F. (1989) Chemistry of mutagenic by-products of water chlorination. *Sci. total Environ.*, **81/82**, 219–230

Kronberg, L. & Franzén, R. (1993) Determination of chlorinated furanones, hydroxyfuranones, and butenedioic acids in chlorine-treated water and in pulp bleaching liquor. *Environ. Sci. Technol.*, **27**, 1811–1818

Kronberg, L. & Vartiainen, T. (1988) Ames mutagenicity and concentration of the strong mutagen 3-chloro-4-(dichloromethyl)-5-hydroxy-2(5*H*)-furanone and of its geometric isomer *E*-2-chloro-3-(dichloromethyl)-4-oxo-butenoic acid in chlorine-treated tap waters. *Mutat. Res.*, **206**, 177–182

Kronberg, L., Christman, R.F., Singh, R. & Ball, L.M. (1991) Identification of oxidized and reduced forms of the strong bacterial mutagen (*Z*)-2-chloro-3-(dichloromethyl)-4-oxobutenoic acid (MX) in extracts of chlorine-treated water. *Environ. Sci. Technol.*, **25**, 99–104

Laaksonen, M., Mäki-Paakkanen, J. & Komulainen, H. (2001) Enhancement of 3-methylcholan-threne-induced neoplastic transformation by 3-chloro-4-(dichloromethyl)-5-hydroxy-2(5*H*)-furanone in the two-stage transformation assay in C3H 10T1/2 cells. *Arch. Toxicol.*, **75**, 613–617

LaLonde, R.T. & Ramdayal, F. (1997) Mucochloric acid action on ΦX174 DNA: A comparison to other chlorine-substituted 2(5*H*)-furanones. *Chem. Res. Toxicol.*, **10**, 205–210

LaLonde, R.T., Cook, G.P., Perakyla, H. & Dence, C.W. (1991a) Effect of mutagenicity of the step-wise removal of hydroxyl group and chlorine atoms from 3-chloro-4-(dichloromethyl)-5-hydroxy-2(5*H*)-furanone: ^{13}C NMR chemical shifts as determinants of mutagenicity. *Chem. Res. Toxicol.*, **4**, 35–40

LaLonde, R.T., Cook, G.P., Perakyla, H., Dence, C.W. & Babish, J.G. (1991b) *Salmonella typhimurium* (TA100) mutagenicity of 3-chloro-4-(dichloromethyl)-5-hydroxy-2(5*H*)-furanone and its open- and closed-ring analogs. *Environ. mol. Mutag.*, **17**, 40–48

LaLonde, R.T., Cook, G.P., Perakyla, H. & Bu, L. (1991c) Structure–activity relationships of bacterial mutagens related to 3-chloro-4-(dichloromethyl)-5-hydroxy-2(5*H*)-furanone: An emphasis on the effect of stepwise removal of chlorine from the dichloromethyl group. *Chem. Res. Toxicol.*, **4**, 540–545

LaLonde, R.T., Bu, L., Henwood, A., Fiumano, J. & Zhang, L. (1997) Bromine-, chlorine-, and mixed halogen-substituted 4-methyl-2(5*H*)-furanones: Synthesis and mutagenic effects of halogen and hydroxyl group replacements. *Chem. Res. Toxicol.*, **10**, 1427–1436

Le Curieux, F., Munter, T. & Kronberg, L. (1997) Identification of adenine adducts formed in reaction of calf thymus DNA with mutagenic chlorohydroxyfuranones found in drinking water. *Chem. Res. Toxicol.*, **10**, 1180–1185

Le Curieux, F., Nesslany, F., Munter, T., Kronberg, L. & Marzin, D. (1999) Genotoxic activity of chlorohydroxyfuranones in the microscale micronucleus test on mouse lymphoma cells and the unscheduled DNA synthesis assay in rat hepatocytes. *Mutagenesis*, **14**, 457–462

Mäki-Paakkanen, J. & Jansson, K. (1995) Cytogenetic effects in the peripheral lymphocytes and kidney cells of rats exposed to 3-chloro-4-(dichloromethyl)-5-hydroxy-2(5*H*)-furanone (MX) orally on three consecutive days. *Mutat. Res.*, **343**, 151–156

Mäki-Paakkanen, J., Jansson, K. & Vartiainen, T. (1994) Induction of mutation, sister-chromatid exchanges, and chromosome aberrations by 3-chloro-4-(dichloromethyl)-5-hydroxy-2(5*H*)-furanone in Chinese hamster ovary cells. *Mutat. Res.*, **310**, 117–123

Mäki-Paakkanen, J., Laaksonen, M., Munter, T., Kronberg, L. & Komulainen, H. (2001) Comparable DNA and chromosome damage in Chinese hamster ovary cells by chlorohydroxyfuranones. *Environ. mol. Mutag.*, **38**, 297–305

Marsteinstredet, U., Brunborg, G., Bjørås, M., Søderlund, E., Seeberg, E., Kronberg, L. & Holme, J.A. (1997a) DNA damage induced by 3-chloro-4-(dichloromethyl)-5-hydroxy-2[5*H*]-furanone (MX) in HL-60 cells and purified DNA in vitro. *Mutat. Res.*, **390**, 171–178

Marsteinstredet, U., Wiger, R., Brunborg, G., Hongslo, J.K. & Holme, J.A. (1997b) Apoptosis in HL-60 cells induced by 3-chloro-4-(dichloromethyl)-5-hydroxy-2(5*H*)-furanone (MX). *Chem.-biol. Interact.*, **106**, 89–107

Matsumura, H., Watanabe, M., Matsumoto, K. & Ohta, T. (1994) 3-Chloro-4-(dichloromethyl)-5-hydroxy-2(5*H*)-furanone (MX) induces gene mutations and inhibits metabolic cooperation in cultured Chinese hamster cells. *J. Toxicol. environ. Health*, **43**, 65–72

Meier, J.R., Knohl, R.B., Coleman, W.E., Ringhand, H.P., Munch, J.W., Kaylor, W.H., Streicher, R.P. & Kopfler, F.C. (1987a) Studies on the potent bacterial mutagen, 3-chloro-4-(dichloromethyl)-5-hydroxy-2(5*H*)-furanone: Aqueous stability, XAD recovery and analytical determination in drinking water and in chlorinated humic acid solutions. *Mutat. Res.*, **189**, 363–373

Meier, J.R., Blazak, W.F. & Knohl, R.B. (1987b) Mutagenic and clastogenic properties of 3-chloro-4-(dichloromethyl)-5-hydroxy-2 (5H)-furanone: A potent bacterial mutagen in drinking water. *Environ. mol. Mutag.*, **10**, 411–424

Meier, J.R., Knohl, R.B., Merrick, B.A. & Smallwood, C.L. (1990) Importance of glutathione in the in vitro detoxification of 3-chloro-4-(dichloromethyl)-5-hydroxy-2(5H)-furanone, an important mutagenic by-product of water chlorination. In: Jolley, R.L., Condie, L.W., Johnson, J.D., Katz, S., Minear, R.A., Mattice, J.S. & Jacobs, V.A., eds, *Water Chlorination: Chemistry, Environmental Impact and Health Effects*, Vol. 6, Chelsea, Lewis Publishers, pp. 159–170

Meier, J.R., Monarca, S., Patterson, K.S., Villarini, M., Daniel, F.B., Moretti, M. & Pasquini, R. (1996) Urine mutagenicity and biochemical effects of the drinking water mutagen, 3-chloro-4-(dichloromethyl)-5-hydroxy-2[5*H*]-furanone (MX), following repeated oral administration to mice and rats. *Toxicology*, **110**, 59–70

Munter, T., Le Curieux, F., Sjöholm, R. & Kronberg, L. (1998) Reaction of the potent bacterial mutagen 3-chloro-4-(dichloromethyl)-5-hydroxy-2(5*H*)-furanone (MX) with 2′-deoxyadenosine and calf thymus DNA: Identification of fluorescent propenoformyl derivatives. *Chem. Res. Toxicol.*, **11**, 226–233

Munter, T., Le Curieux, F., Sjöholm, R. & Kronberg, L. (1999) Identification of an ethenoformyl adduct formed in the reaction of the potent bacterial mutagen 3-chloro-4-(dichloromethyl)-5-hydroxy-2(5*H*)-furanone with guanosine. *Chem. Res. Toxicol.*, **12**, 46–52

National Health and Medical Research Council and Agriculture and Resource Management Council of Australia and New Zealand (1996) *Australian Drinking Water Guidelines Summary*, Canberra

Nawrocki, J., Andrzejewski, P., Kronberg, L. & Jelén, H. (1997) New derivatization method for the determination of 3-chloro-4-(dichloromethyl)-5-hydroxy-2(5H)-furanone in water. *J. Chromatogr.*, **A790**, 242–247

Nawrocki, J., Andrzejewski, P., Jelén, H. & Wasowicz, E. (2001) Derivatization of the mutagen MX (3-chloro-4-(dichloromethyl)-5-hydroxy-2(5H)-furanone) with butyl alcohols prior to GC–MS analysis. *Water Res.*, **35**, 1891–1896

Nishikawa, A., Kinae, N., Furukawa, F., Mitsui, M., Enami, T., Hasegawa, T. & Takahashi, M. (1994) Enhancing effects of 3-chloro-4-(dichloromethyl)-5-hydroxy-2(5*H*)-furanone (MX) on cell proliferation and lipid peroxidation in the rat gastric mucosa. *Cancer Lett.*, **85**, 151–157

Nishikawa, A., Furukawa, F., Lee, I.-S., Kasahara, K.-I., Tanakamaru, Z.-Y.,Nakamura, H., Miyauchi, M., Kinae, N. & Hirose, M. (1999) Promoting effects of 3-chloro-4-(dichloromethyl)-5-hydroxy-2(5*H*)-furanone on rat glandular stomach carcinogenesis initiated with *N*-methyl-*N'*-nitro-*N*-nitrosoguanidine. *Cancer Res.*, **59**, 2045–2049

Nunn, J.W. & Chipman, J.K. (1994) Induction of DNA strand breaks by 3-chloro-4-(dichloromethyl)-5-hydroxy-2(5H)-furanone and humic substances in relation to glutathione and calcium status in human white blood cells. *Mutat. Res.*, **341**, 133–140

Nunn, J.W., Davies, J.E. & Chipman, J.K. (1997) Production of unscheduled DNA synthesis in rodent hepatocytes in vitro, but not in vivo, by 3-chloro-4-(dichloromethyl)-5-hydroxy-2[5*H*]-furanone (MX). *Mutat. Res.*, **373**, 67–73

Onstad, G.D. & Weinberg, H.S. (2001) Improvements in extraction of MX-analogues from drinking water. In: *Proceedings of the Water Quality Technology Conference*, Denver, CO, American Water Works Association, pp. 988–998

Padmapriya, A.A., Just, G. & Lewis, N.G. (1985) Synthesis of 3-chloro-4-(dichloromethyl)-5-hydroxy-2(5*H*)-furanone, a potent mutagen. *Can. J. Chem.*, **63**, 828–832

Ringhand, H.P., Kaylor, W.H., Miller, R.G. & Kopfler, F.C. (1989) Synthesis of 3-^{14}C-3-chloro-4-(dichloromethyl)-5-hydroxy-2(5H)-furanone and its use in a tissue distribution study in the rat. *Chemosphere*, **18**, 2229–2236

Risto, L., Vartiainen, T. & Komulainen, H. (1993) Formation of methemoglobin by 3-chloro-4-(dichloromethyl)-5-hydroxy-2(5H)-furanone, MX, in rat erythrocytes in vitro. *Toxicol. Lett.*, **68**, 325–332

Romero, J., Ventura, F., Caixach, J., Rivera, J. & Guerrero, R. (1997) Identification and quantification of the mutagenic compound 3-chloro-4-(dichloromethyl)-5-hydroxy-2(5H)-furanone (MX) in chlorine-treated water. *Bull. environ. Contam. Toxicol.*, **59**, 715–722

Sasaki, Y.F., Nishidate, E., Izumiyama, F., Watanabe-Akanuma, M., Kinae, N., Matsusaka, N. & Tsuda, S. (1997) Detection of in vivo genotoxicity of 3-chloro-4-(dichloromethyl)-5-hydroxy-2[5*H*]-furanone (MX) by the alkaline single cell gel electrophoresis (Comet) assay in multiple mouse organs. *Mutat. Res.*, **393**, 47–53

Shaughnessy, D.T., Ohe, T., Landi, S., Warren, S.H., Richard, A.M., Munter, T., Franzén, R., Kronberg, L. & DeMarini, D.M. (2000) Mutation spectra of the drinking water mutagen 3-chloro-4-methyl-5-hydroxy-2(5*H*)-furanone (MCF) in Salmonella TA100 and TA104: Comparison to MX. *Environ. mol. Mutag.*, **35**, 106–113

Sigma Chemical Co. (2002) *MSDS: 3-Chloro-4-(dichloromethyl)-5-hydroxy-2(5H)-furanone [Product No. C5710]*, St Louis, MO

Sigma-Aldrich Co. (2002) *Biochemicals and Reagents for Life Science Research 2002–2003*, St Louis, MO, p. 492

Simpson, K.L. & Hayes, K.P. (1998) Drinking water disinfection by-products. An Australian perspective. *Water Res.*, **32**, 1522–1528

Steffensen, I.-L., Paulsen, J.E., Engeset, D., Kronberg, L. & Alexander, J. (1999) The drinking water chlorination by-products 3,4-dichloro-5-hydroxy-2[5*H*]-furanone (mucochloric acid) and 3-chloro-4-(dichloromethyl)-5-hydroxy-2[5*H*]-furanone do not induce preneoplastic or neoplastic intestinal lesions in F344 rats, Balb/cA mice or C57BL/6J-*Min* mice. *Pharmacol. Toxicol.*, **85**, 56–64

Suzuki, N. & Nakanishi, J. (1990) The determination of strong mutagen, 3-chloro-4-(dichloromethyl)-5-hydroxy-2(5H)-furanone in drinking water in Japan. *Chemosphere*, **21**, 387–392

Teramoto, S., Takahashi, K., Kikuta, M. & Kobayashi, H. (1998) Potential teratogenicity of 3-chloro-4-(dichloromethyl)-5-hydroxy-2(5*H*)-furanone (MX) in micromass in vitro test. *J. Toxicol. environ. Health*, **A53**, 607–614

Teramoto, S., Takahashi, K.L., Kikuta, M. & Kobayashi, H. (1999) 3-Chloro-4-(dichloromethyl)-5-hydroxy-2(5*H*)-furanone (MX) as a direct-acting teratogen in micromass *in vitro* tests. *Cong. Anom.*, **39**, 31–35

Tikkanen, L. & Kronberg, L. (1990) Genotoxic effects of various chlorinated butenoic acids identified in chlorinated drinking water. *Mutat. Res.*, **240**, 109–116

Vaittinen, S.-L., Komulainen, H., Kosma, V.-M., Julkunen, A., Mäki-Paakkanen, J., Jansson, K., Vartiainen, T. & Tuomisto, J. (1995) Subchronic toxicity of 3-chloro-4-(dichloromethyl)-5-hydroxy-2(5*H*)-furanone (MX) in Wistar rats. *Food chem. Toxicol.*, **33**, 1027–1037

Vartiainen, T., Heiskanen, K. & Lötjönen, S. (1991) Analysis and some chemical properties of MX (3-chloro-4-(dichloromethyl)-5-hydroxy-2(5H)-furanone), the potent drinking water mutagen. *Fresenius J. anal. Chem.*, **340**, 230–233

Watanabe, M., Kobayashi, H. & Ohta, T. (1994) Rapid inactivation of 3-chloro-4-(dichloromethyl)-5-hydroxy-2(5*H*)-furanone (MX), a potent mutagen in chlorinated drinking water, by sulfhydryl compounds. *Mutat. Res.*, **312**, 131–138

Watanabe-Akanuma, M. & Ohta, T. (1994) Effects of DNA repair deficiency on the mutational specificity in the *lacZ* gene of *Escherichia coli*. *Mutat. Res.*, **311**, 295–304

WHO (1998) *Guidelines for Drinking-water Quality*, 2nd Ed., Vol. 2, *Health Criteria and Other Supporting Information and Addendum to Vol. 2*, Geneva

Woodruff, N.W., Durant, J.L., Donhoffner, L.L., Penman, B.W. & Crespi, C.L. (2001) Human cell mutagenicity of chlorinated and unchlorinated water and the disinfection byproduct 3-chloro-4-(dichloromethyl)-5-hydroxy-2(5*H*)-furanone (MX). *Mutat. Res.*, **495**, 157–168

Wright, J.M., Schwartz, J., Vartiainen, T., Mäki-Paakkanen, J., Altshul, L., Harrington, J.J., & Dockery, D.W. (2002) 3-Chloro-4-(dichloromethyl)-5-hydroxy-2(5*H*)-furanone (MX) and mutagenic activity in Massachusetts drinking water. *Environ. Health Perspect.*, **110**, 157–164

Xu, X., Lin, L., Huixian, Z., Yongbin, L., Liansheng, W. & Jinqi, Z. (1997) Studies on the precursors of strong mutagen [3-chloro-4-(dichloromethyl)-5-hydroxy-2(5H)- ranone] MX by chlorination of fractions from different waters. *Chemosphere*, **35**, 1709–1716

Yamada, M., Matsui, K., Sofuni, T. & Nohmi, T. (1997) New tester strains of *Salmonella typhimurium* lacking O^6-methylguanine DNA methyltransferases and highly sensitive to mutagenic alkylating agents. *Mutat. Res.*, **381**, 15–24

Zou, H., Xu, X., Zhang, J. & Zhu, Z. (1995) The determination of MX [3-chloro-4-(dichloromethyl)-5-hydroxy-2(5H)-furanone] in drinking water in China. *Chemosphere*, **30**, 2219–2225

Zwiener, C. & Kronberg, L. (2001) Determination of the strong mutagen 3-chloro-4-(dichloromethyl)-5-hydroxy-2(5H)-furanone (MX) and its analogues by GC-ITD-MS-MS. *Fresenius J. anal. Chem.*, **371**, 591–597

SUMMARY OF FINAL EVALUATIONS

Agent	Degree of evidence of carcinogenicity		Overall evaluation of carcinogenicity to humans
	Human	Animal	
Arsenic in drinking-water	S (bladder, lung and skin cancer)		1
Dimethylarsinic acid		S	
Sodium arsenite		L	
Calcium arsenate		L	
Arsenic trioxide		L	
Sodium arsenate		I	
Arsenic trisulfide		I	
Chloral and chloral hydrate	I		
Chloral hydrate		L	3
Dichloroacetic acid	I	S	2B
Trichloroacetic acid	I	L	3
3-Chloro-4-(dichloromethyl)-5-hydroxy-2(5*H*)-furanone (MX)	I	L	2B*
Chloramine	I		3
Monochloramine		I	

S, sufficient evidence of carcinogenicity; L, limited evidence of carcinogenicity; I, inadequate evidence of carcinogenicity; Group 1, carcinogenic to humans; Group 2B, possibly carcinogenic to humans; Group 3, cannot be classified as to carcinogenicity to humans. For definitions of criteria for degrees of evidence, see Preamble.

* Other relevant data were used to upgrade the evaluation.

CUMULATIVE CROSS INDEX TO *IARC MONOGRAPHS ON THE EVALUATION OF CARCINOGENIC RISKS TO HUMANS*

The volume, page and year of publication are given. References to corrigenda are given in parentheses.

Benz[c]acridine	3, 241 (1973); 32, 129 (1983); Suppl. 7, 58 (1987)
Benzal chloride (see also α-Chlorinated toluenes and benzoyl chloride)	29, 65 (1982); Suppl. 7, 148 (1987); 71, 453 (1999)
Benz[a]anthracene	3, 45 (1973); 32, 135 (1983); Suppl. 7, 58 (1987)
Benzene	7, 203 (1974) (corr. 42, 254); 29, 93, 391 (1982); Suppl. 7, 120 (1987)
Benzidine	1, 80 (1972); 29, 149, 391 (1982); Suppl. 7, 123 (1987)
Benzidine-based dyes	Suppl. 7, 125 (1987)
Benzo[b]fluoranthene	3, 69 (1973); 32, 147 (1983); Suppl. 7, 58 (1987)
Benzo[j]fluoranthene	3, 82 (1973); 32, 155 (1983); Suppl. 7, 58 (1987)
Benzo[k]fluoranthene	32, 163 (1983); Suppl. 7, 58 (1987)
Benzo[ghi]fluoranthene	32, 171 (1983); Suppl. 7, 58 (1987)
Benzo[a]fluorene	32, 177 (1983); Suppl. 7, 58 (1987)
Benzo[b]fluorene	32, 183 (1983); Suppl. 7, 58 (1987)
Benzo[c]fluorene	32, 189 (1983); Suppl. 7, 58 (1987)
Benzofuran	63, 431 (1995)
Benzo[ghi]perylene	32, 195 (1983); Suppl. 7, 58 (1987)
Benzo[c]phenanthrene	32, 205 (1983); Suppl. 7, 58 (1987)
Benzo[a]pyrene	3, 91 (1973); 32, 211 (1983) (corr. 68, 477); Suppl. 7, 58 (1987)
Benzo[e]pyrene	3, 137 (1973); 32, 225 (1983); Suppl. 7, 58 (1987)
1,4-Benzoquinone (see para-Quinone)	
1,4-Benzoquinone dioxime	29, 185 (1982); Suppl. 7, 58 (1987); 71, 1251 (1999)
Benzotrichloride (see also α-Chlorinated toluenes and benzoyl chloride)	29, 73 (1982); Suppl. 7, 148 (1987); 71, 453 (1999)
Benzoyl chloride (see also α-Chlorinated toluenes and benzoyl chloride)	29, 83 (1982) (corr. 42, 261); Suppl. 7, 126 (1987); 71, 453 (1999)
Benzoyl peroxide	36, 267 (1985); Suppl. 7, 58 (1987); 71, 345 (1999)
Benzyl acetate	40, 109 (1986); Suppl. 7, 58 (1987); 71, 1255 (1999)
Benzyl chloride (see also α-Chlorinated toluenes and benzoyl chloride)	11, 217 (1976) (corr. 42, 256); 29, 49 (1982); Suppl. 7, 148 (1987); 71, 453 (1999)
Benzyl violet 4B	16, 153 (1978); Suppl. 7, 58 (1987)
Bertrandite (see Beryllium and beryllium compounds)	
Beryllium and beryllium compounds	1, 17 (1972); 23, 143 (1980) (corr. 42, 260); Suppl. 7, 127 (1987); 58, 41 (1993)
Beryllium acetate (see Beryllium and beryllium compounds)	
Beryllium acetate, basic (see Beryllium and beryllium compounds)	
Beryllium-aluminium alloy (see Beryllium and beryllium compounds)	
Beryllium carbonate (see Beryllium and beryllium compounds)	
Beryllium chloride (see Beryllium and beryllium compounds)	
Beryllium-copper alloy (see Beryllium and beryllium compounds)	
Beryllium-copper-cobalt alloy (see Beryllium and beryllium compounds)	

β-Butyrolactone   *11*, 225 (1976); *Suppl. 7*, 59
                                                   (1987); *71*, 1317 (1999)
γ-Butyrolactone   *11*, 231 (1976); *Suppl. 7*, 59
                                                   (1987); *71*, 367 (1999)

C

Cabinet-making (*see* Furniture and cabinet-making)
Cadmium acetate (*see* Cadmium and cadmium compounds)
Cadmium and cadmium compounds   *2*, 74 (1973); *11*, 39 (1976)
                                                   (*corr. 42*, 255); *Suppl. 7*, 139
                                                   (1987); *58*, 119 (1993)

Cadmium chloride (*see* Cadmium and cadmium compounds)
Cadmium oxide (*see* Cadmium and cadmium compounds)
Cadmium sulfate (*see* Cadmium and cadmium compounds)
Cadmium sulfide (*see* Cadmium and cadmium compounds)
Caffeic acid   *56*, 115 (1993)
Caffeine   *51*, 291 (1991)
Calcium arsenate (*see* Arsenic in drinking-water)
Calcium chromate (*see* Chromium and chromium compounds)
Calcium cyclamate (*see* Cyclamates)
Calcium saccharin (*see* Saccharin)
Cantharidin   *10*, 79 (1976); *Suppl. 7*, 59 (1987)
Caprolactam   *19*, 115 (1979) (*corr. 42*, 258);
                                                   39, 247 (1986) (*corr. 42*, 264);
                                                   Suppl. 7, 59, 390 (1987); *71*, 383
                                                   (1999)
Captafol   *53*, 353 (1991)
Captan   *30*, 295 (1983); *Suppl. 7*, 59 (1987)
Carbaryl   *12*, 37 (1976); *Suppl. 7*, 59 (1987)
Carbazole   *32*, 239 (1983); *Suppl. 7*, 59
                                                   (1987); *71*, 1319 (1999)
3-Carbethoxypsoralen   *40*, 317 (1986); *Suppl. 7*, 59 (1987)
Carbon black   *3*, 22 (1973); *33*, 35 (1984);
                                                   Suppl. 7, 142 (1987); *65*, 149
                                                   (1996)
Carbon tetrachloride   *1*, 53 (1972); *20*, 371 (1979);
                                                   Suppl. 7, 143 (1987); *71*, 401
                                                   (1999)
Carmoisine   *8*, 83 (1975); *Suppl. 7*, 59 (1987)
Carpentry and joinery   *25*, 139 (1981); *Suppl. 7*, 378
                                                   (1987)
Carrageenan   *10*, 181 (1976) (*corr. 42*, 255); *31*,
                                                   79 (1983); *Suppl. 7*, 59 (1987)
Cassia occidentalis (*see* Traditional herbal medicines)
Catechol   *15*, 155 (1977); *Suppl. 7*, 59
                                                   (1987); *71*, 433 (1999)
CCNU (*see* 1-(2-Chloroethyl)-3-cyclohexyl-1-nitrosourea)
Ceramic fibres (*see* Man-made vitreous fibres)
Chemotherapy, combined, including alkylating agents (*see* MOPP and
      other combined chemotherapy including alkylating agents)
Chloral (*see also* Chloral hydrate)   *63*, 245 (1995); *84*, 317 (2004)
Chloral hydrate   *63*, 245 (1995); *84*, 317 (2004)

4-Chloro-*meta*-phenylenediamine	*27*, 82 (1982); *Suppl. 7*, 60 (1987)
Chloroprene	*19*, 131 (1979); *Suppl. 7*, 160 (1987); *71*, 227 (1999)
Chloropropham	*12*, 55 (1976); *Suppl. 7*, 60 (1987)
Chloroquine	*13*, 47 (1977); *Suppl. 7*, 60 (1987)
Chlorothalonil	*30*, 319 (1983); *Suppl. 7*, 60 (1987); *73*, 183 (1999)
para-Chloro-*ortho*-toluidine and its strong acid salts (*see also* Chlordimeform)	*16*, 277 (1978); *30*, 65 (1983); *Suppl. 7*, 60 (1987); *48*, 123 (1990); *77*, 323 (2000)
4-Chloro-*ortho*-toluidine (see *para*-chloro-*ortho*-toluidine)	
5-Chloro-*ortho*-toluidine	*77*, 341 (2000)
Chlorotrianisene (*see also* Nonsteroidal oestrogens)	*21*, 139 (1979); *Suppl. 7*, 280 (1987)
2-Chloro-1,1,1-trifluoroethane	*41*, 253 (1986); *Suppl. 7*, 60 (1987); *71*, 1355 (1999)
Chlorozotocin	*50*, 65 (1990)
Cholesterol	*10*, 99 (1976); *31*, 95 (1983); *Suppl. 7*, 161 (1987)
Chromic acetate (*see* Chromium and chromium compounds)	
Chromic chloride (*see* Chromium and chromium compounds)	
Chromic oxide (*see* Chromium and chromium compounds)	
Chromic phosphate (*see* Chromium and chromium compounds)	
Chromite ore (*see* Chromium and chromium compounds)	
Chromium and chromium compounds (*see also* Implants, surgical)	*2*, 100 (1973); *23*, 205 (1980); *Suppl. 7*, 165 (1987); *49*, 49 (1990) (*corr. 51*, 483)
Chromium carbonyl (*see* Chromium and chromium compounds)	
Chromium potassium sulfate (*see* Chromium and chromium compounds)	
Chromium sulfate (*see* Chromium and chromium compounds)	
Chromium trioxide (*see* Chromium and chromium compounds)	
Chrysazin (*see* Dantron)	
Chrysene	*3*, 159 (1973); *32*, 247 (1983); *Suppl. 7*, 60 (1987)
Chrysoidine	*8*, 91 (1975); *Suppl. 7*, 169 (1987)
Chrysotile (*see* Asbestos)	
CI Acid Orange 3	*57*, 121 (1993)
CI Acid Red 114	*57*, 247 (1993)
CI Basic Red 9 (*see also* Magenta)	*57*, 215 (1993)
Ciclosporin	*50*, 77 (1990)
CI Direct Blue 15	*57*, 235 (1993)
CI Disperse Yellow 3 (see Disperse Yellow 3)	
Cimetidine	*50*, 235 (1990)
Cinnamyl anthranilate	*16*, 287 (1978); *31*, 133 (1983); *Suppl. 7*, 60 (1987); *77*, 177 (2000)
CI Pigment Red 3	*57*, 259 (1993)
CI Pigment Red 53:1 (*see* D&C Red No. 9)	
Cisplatin (*see also* Etoposide)	*26*, 151 (1981); *Suppl. 7*, 170 (1987)
Citrinin	*40*, 67 (1986); *Suppl. 7*, 60 (1987)
Citrus Red No. 2	*8*, 101 (1975) (*corr. 42*, 254); *Suppl. 7*, 60 (1987)
Clinoptilolite (*see* Zeolites)	

Estazolam	66, 105 (1996)
Ethinyloestradiol	6, 77 (1974); 21, 233 (1979); Suppl. 7, 286 (1987); 72, 49 (1999)
Ethionamide	13, 83 (1977); Suppl. 7, 63 (1987)
Ethyl acrylate	19, 57 (1979); 39, 81 (1986); Suppl. 7, 63 (1987); 71, 1447 (1999)
Ethylbenzene	77, 227 (2000)
Ethylene	19, 157 (1979); Suppl. 7, 63 (1987); 60, 45 (1994); 71, 1447 (1999)
Ethylene dibromide	15, 195 (1977); Suppl. 7, 204 (1987); 71, 641 (1999)
Ethylene oxide	11, 157 (1976); 36, 189 (1985) (corr. 42, 263); Suppl. 7, 205 (1987); 60, 73 (1994)
Ethylene sulfide	11, 257 (1976); Suppl. 7, 63 (1987)
Ethylenethiourea	7, 45 (1974); Suppl. 7, 207 (1987); 79, 659 (2001)
2-Ethylhexyl acrylate	60, 475 (1994)
Ethyl methanesulfonate	7, 245 (1974); Suppl. 7, 63 (1987)
N-Ethyl-N-nitrosourea	1, 135 (1972); 17, 191 (1978); Suppl. 7, 63 (1987)
Ethyl selenac (see also Selenium and selenium compounds)	12, 107 (1976); Suppl. 7, 63 (1987)
Ethyl tellurac	12, 115 (1976); Suppl. 7, 63 (1987)
Ethynodiol diacetate	6, 173 (1974); 21, 387 (1979); Suppl. 7, 292 (1987); 72, 49 (1999)
Etoposide	76, 177 (2000)
Eugenol	36, 75 (1985); Suppl. 7, 63 (1987)
Evans blue	8, 151 (1975); Suppl. 7, 63 (1987)
Extremely low-frequency electric fields	80 (2002)
Extremely low-frequency magnetic fields	80 (2002)

F

Fast Green FCF	16, 187 (1978); Suppl. 7, 63 (1987)
Fenvalerate	53, 309 (1991)
Ferbam	12, 121 (1976) (corr. 42, 256); Suppl. 7, 63 (1987)
Ferric oxide	1, 29 (1972); Suppl. 7, 216 (1987)
Ferrochromium (see Chromium and chromium compounds)	
Fluometuron	30, 245 (1983); Suppl. 7, 63 (1987)
Fluoranthene	32, 355 (1983); Suppl. 7, 63 (1987)
Fluorene	32, 365 (1983); Suppl. 7, 63 (1987)
Fluorescent lighting (exposure to) (see Ultraviolet radiation)	
Fluorides (inorganic, used in drinking-water)	27, 237 (1982); Suppl. 7, 208 (1987)
5-Fluorouracil	26, 217 (1981); Suppl. 7, 210 (1987)
Fluorspar (see Fluorides)	
Fluosilicic acid (see Fluorides)	
Fluroxene (see Anaesthetics, volatile)	

H

Haematite	*1*, 29 (1972); *Suppl. 7*, 216 (1987)
Haematite and ferric oxide	*Suppl. 7*, 216 (1987)
Haematite mining, underground, with exposure to radon	*1*, 29 (1972); *Suppl. 7*, 216 (1987)
Hairdressers and barbers (occupational exposure as)	*57*, 43 (1993)
Hair dyes, epidemiology of	*16*, 29 (1978); *27*, 307 (1982);
Halogenated acetonitriles	*52*, 269 (1991); *71*, 1325, 1369, 1375, 1533 (1999)
Halothane (*see* Anaesthetics, volatile)	
HC Blue No. 1	*57*, 129 (1993)
HC Blue No. 2	*57*, 143 (1993)
α-HCH (*see* Hexachlorocyclohexanes)	
β-HCH (*see* Hexachlorocyclohexanes)	
γ-HCH (*see* Hexachlorocyclohexanes)	
HC Red No. 3	*57*, 153 (1993)
HC Yellow No. 4	*57*, 159 (1993)
Heating oils (*see* Fuel oils)	
Helicobacter pylori (infection with)	*61*, 177 (1994)
Hepatitis B virus	*59*, 45 (1994)
Hepatitis C virus	*59*, 165 (1994)
Hepatitis D virus	*59*, 223 (1994)
Heptachlor (*see also* Chlordane/Heptachlor)	*5*, 173 (1974); *20*, 129 (1979)
Hexachlorobenzene	*20*, 155 (1979); *Suppl. 7*, 219 (1987); *79*, 493 (2001)
Hexachlorobutadiene	*20*, 179 (1979); *Suppl. 7*, 64 (1987); *73*, 277 (1999)
Hexachlorocyclohexanes	*5*, 47 (1974); *20*, 195 (1979) (*corr. 42*, 258); *Suppl. 7*, 220 (1987)
Hexachlorocyclohexane, technical-grade (*see* Hexachlorocyclohexanes)	
Hexachloroethane	*20*, 467 (1979); *Suppl. 7*, 64 (1987); *73*, 295 (1999)
Hexachlorophene	*20*, 241 (1979); *Suppl. 7*, 64 (1987)
Hexamethylphosphoramide	*15*, 211 (1977); *Suppl. 7*, 64 (1987); *71*, 1465 (1999)
Hexoestrol (*see also* Nonsteroidal oestrogens)	*Suppl. 7*, 279 (1987)
Hormonal contraceptives, progestogens only	*72*, 339 (1999)
Human herpesvirus 8	*70*, 375 (1997)
Human immunodeficiency viruses	*67*, 31 (1996)
Human papillomaviruses	*64* (1995) (*corr. 66*, 485)
Human T-cell lymphotropic viruses	*67*, 261 (1996)
Hycanthone mesylate	*13*, 91 (1977); *Suppl. 7*, 64 (1987)
Hydralazine	*24*, 85 (1980); *Suppl. 7*, 222 (1987)
Hydrazine	*4*, 127 (1974); *Suppl. 7*, 223 (1987); *71*, 991 (1999)
Hydrochloric acid	*54*, 189 (1992)
Hydrochlorothiazide	*50*, 293 (1990)
Hydrogen peroxide	*36*, 285 (1985); *Suppl. 7*, 64 (1987); *71*, 671 (1999)
Hydroquinone	*15*, 155 (1977); *Suppl. 7*, 64 (1987); *71*, 691 (1999)
1-Hydroxyanthraquinone	*82*, 129 (2002)
4-Hydroxyazobenzene	*8*, 157 (1975); *Suppl. 7*, 64 (1987)

K

Kaempferol	*31*, 171 (1983); *Suppl. 7*, 65 (1987)
Kaposi's sarcoma herpesvirus	*70*, 375 (1997)
Kepone (*see* Chlordecone)	
Kojic acid	*79*, 605 (2001)

L

Lasiocarpine	*10*, 281 (1976); *Suppl. 7*, 65 (1987)
Lauroyl peroxide	*36*, 315 (1985); *Suppl. 7*, 65 (1987); *71*, 1485 (1999)
Lead acetate (*see* Lead and lead compounds)	
Lead and lead compounds (*see also* Foreign bodies)	*1*, 40 (1972) (*corr. 42*, 251); *2*, 52, 150 (1973); *12*, 131 (1976); *23*, 40, 208, 209, 325 (1980); *Suppl. 7*, 230 (1987)
Lead arsenate (*see* Arsenic and arsenic compounds)	
Lead carbonate (*see* Lead and lead compounds)	
Lead chloride (*see* Lead and lead compounds)	
Lead chromate (*see* Chromium and chromium compounds)	
Lead chromate oxide (*see* Chromium and chromium compounds)	
Lead naphthenate (*see* Lead and lead compounds)	
Lead nitrate (*see* Lead and lead compounds)	
Lead oxide (*see* Lead and lead compounds)	
Lead phosphate (*see* Lead and lead compounds)	
Lead subacetate (*see* Lead and lead compounds)	
Lead tetroxide (*see* Lead and lead compounds)	
Leather goods manufacture	*25*, 279 (1981); *Suppl. 7*, 235 (1987)
Leather industries	*25*, 199 (1981); *Suppl. 7*, 232 (1987)
Leather tanning and processing	*25*, 201 (1981); *Suppl. 7*, 236 (1987)
Ledate (*see also* Lead and lead compounds)	*12*, 131 (1976)
Levonorgestrel	*72*, 49 (1999)
Light Green SF	*16*, 209 (1978); *Suppl. 7*, 65 (1987)
d-Limonene	*56*, 135 (1993); *73*, 307 (1999)
Lindane (*see* Hexachlorocyclohexanes)	
Liver flukes (*see Clonorchis sinensis, Opisthorchis felineus* and *Opisthorchis viverrini*)	
Lucidin (*see* 1,3-Dihydro-2-hydroxymethylanthraquinone)	
Lumber and sawmill industries (including logging)	*25*, 49 (1981); *Suppl. 7*, 383 (1987)
Luteoskyrin	*10*, 163 (1976); *Suppl. 7*, 65 (1987)
Lynoestrenol	*21*, 407 (1979); *Suppl. 7*, 293 (1987); *72*, 49 (1999)

M

Madder root (*see also Rubia tinctorum*)	*82*, 129 (2002)

8-Methoxypsoralen (*see also* 8-Methoxypsoralen plus ultraviolet radiation)	*24*, 101 (1980)
8-Methoxypsoralen plus ultraviolet radiation	*Suppl. 7*, 243 (1987)
Methyl acrylate	*19*, 52 (1979); *39*, 99 (1986); *Suppl. 7*, 66 (1987); *71*, 1489 (1999)
5-Methylangelicin plus ultraviolet radiation (*see also* Angelicin and some synthetic derivatives)	*Suppl. 7*, 57 (1987)
2-Methylaziridine	*9*, 61 (1975); *Suppl. 7*, 66 (1987); *71*, 1497 (1999)
Methylazoxymethanol acetate (*see also* Cycasin)	*1*, 164 (1972); *10*, 131 (1976); *Suppl. 7*, 66 (1987)
Methyl bromide	*41*, 187 (1986) (*corr. 45*, 283); *Suppl. 7*, 245 (1987); *71*, 721 (1999)
Methyl *tert*-butyl ether	*73*, 339 (1999)
Methyl carbamate	*12*, 151 (1976); *Suppl. 7*, 66 (1987)
Methyl-CCNU (*see* 1-(2-Chloroethyl)-3-(4-methylcyclohexyl)-1-nitrosourea)	
Methyl chloride	*41*, 161 (1986); *Suppl. 7*, 246 (1987); *71*, 737 (1999)
1-, 2-, 3-, 4-, 5- and 6-Methylchrysenes	*32*, 379 (1983); *Suppl. 7*, 66 (1987)
N-Methyl-*N*,4-dinitrosoaniline	*1*, 141 (1972); *Suppl. 7*, 66 (1987)
4,4'-Methylene bis(2-chloroaniline)	*4*, 65 (1974) (*corr. 42*, 252); *Suppl. 7*, 246 (1987); *57*, 271 (1993)
4,4'-Methylene bis(*N,N*-dimethyl)benzenamine	*27*, 119 (1982); *Suppl. 7*, 66 (1987)
4,4'-Methylene bis(2-methylaniline)	*4*, 73 (1974); *Suppl. 7*, 248 (1987)
4,4'-Methylenedianiline	*4*, 79 (1974) (*corr. 42*, 252); *39*, 347 (1986); *Suppl. 7*, 66 (1987)
4,4'-Methylenediphenyl diisocyanate	*19*, 314 (1979); *Suppl. 7*, 66 (1987); *71*, 1049 (1999)
2-Methylfluoranthene	*32*, 399 (1983); *Suppl. 7*, 66 (1987)
3-Methylfluoranthene	*32*, 399 (1983); *Suppl. 7*, 66 (1987)
Methylglyoxal	*51*, 443 (1991)
Methyl iodide	*15*, 245 (1977); *41*, 213 (1986); *Suppl. 7*, 66 (1987); *71*, 1503 (1999)
Methylmercury chloride (*see* Mercury and mercury compounds)	
Methylmercury compounds (*see* Mercury and mercury compounds)	
Methyl methacrylate	*19*, 187 (1979); *Suppl. 7*, 66 (1987); *60*, 445 (1994)
Methyl methanesulfonate	*7*, 253 (1974); *Suppl. 7*, 66 (1987); *71*, 1059 (1999)
2-Methyl-1-nitroanthraquinone	*27*, 205 (1982); *Suppl. 7*, 66 (1987)
N-Methyl-*N'*-nitro-*N*-nitrosoguanidine	*4*, 183 (1974); *Suppl. 7*, 248 (1987)
3-Methylnitrosaminopropionaldehyde [*see* 3-(*N*-Nitrosomethylamino)-propionaldehyde]	
3-Methylnitrosaminopropionitrile [*see* 3-(*N*-Nitrosomethylamino)-propionitrile]	
4-(Methylnitrosamino)-4-(3-pyridyl)-1-butanal [*see* 4-(*N*-Nitrosomethyl-amino)-4-(3-pyridyl)-1-butanal]	
4-(Methylnitrosamino)-1-(3-pyridyl)-1-butanone [*see* 4-(-Nitrosomethyl-amino)-1-(3-pyridyl)-1-butanone]	

1-Naphthylthiourea   *30*, 347 (1983); *Suppl. 7*, 263
                                                            (1987)
Neutrons   *75*, 361 (2000)
Nickel acetate (*see* Nickel and nickel compounds)
Nickel ammonium sulfate (*see* Nickel and nickel compounds)
Nickel and nickel compounds (*see also* Implants, surgical)   *2*, 126 (1973) (*corr. 42*, 252); *11*,
                                                            75 (1976); *Suppl. 7*, 264 (1987)
                                                            (*corr. 45*, 283); *49*, 257 (1990)
                                                            (*corr. 67*, 395)

Nickel carbonate (*see* Nickel and nickel compounds)
Nickel carbonyl (*see* Nickel and nickel compounds)
Nickel chloride (*see* Nickel and nickel compounds)
Nickel-gallium alloy (*see* Nickel and nickel compounds)
Nickel hydroxide (*see* Nickel and nickel compounds)
Nickelocene (*see* Nickel and nickel compounds)
Nickel oxide (*see* Nickel and nickel compounds)
Nickel subsulfide (*see* Nickel and nickel compounds)
Nickel sulfate (*see* Nickel and nickel compounds)
Niridazole   *13*, 123 (1977); *Suppl. 7*, 67 (1987)
Nithiazide   *31*, 179 (1983); *Suppl. 7*, 67 (1987)
Nitrilotriacetic acid and its salts   *48*, 181 (1990); *73*, 385 (1999)
5-Nitroacenaphthene   *16*, 319 (1978); *Suppl. 7*, 67 (1987)
5-Nitro-*ortho*-anisidine   *27*, 133 (1982); *Suppl. 7*, 67 (1987)
2-Nitroanisole   *65*, 369 (1996)
9-Nitroanthracene   *33*, 179 (1984); *Suppl. 7*, 67 (1987)
7-Nitrobenz[*a*]anthracene   *46*, 247 (1989)
Nitrobenzene   *65*, 381 (1996)
6-Nitrobenzo[*a*]pyrene   *33*, 187 (1984); *Suppl. 7*, 67
                                                            (1987); *46*, 255 (1989)
4-Nitrobiphenyl   *4*, 113 (1974); *Suppl. 7*, 67 (1987)
6-Nitrochrysene   *33*, 195 (1984); *Suppl. 7*, 67
                                                            (1987); *46*, 267 (1989)
Nitrofen (technical-grade)   *30*, 271 (1983); *Suppl. 7*, 67 (1987)
3-Nitrofluoranthene   *33*, 201 (1984); *Suppl. 7*, 67 (1987)
2-Nitrofluorene   *46*, 277 (1989)
Nitrofural   *7*, 171 (1974); *Suppl. 7*, 67 (1987);
                                                            50, 195 (1990)
5-Nitro-2-furaldehyde semicarbazone (*see* Nitrofural)
Nitrofurantoin   *50*, 211 (1990)
Nitrofurazone (*see* Nitrofural)
1-[(5-Nitrofurfurylidene)amino]-2-imidazolidinone   *7*, 181 (1974); *Suppl. 7*, 67 (1987)
N-[4-(5-Nitro-2-furyl)-2-thiazolyl]acetamide   *1*, 181 (1972); *7*, 185 (1974);
                                                            Suppl. 7, 67 (1987)
Nitrogen mustard   *9*, 193 (1975); *Suppl. 7*, 269 (1987)
Nitrogen mustard *N*-oxide   *9*, 209 (1975); *Suppl. 7*, 67 (1987)
Nitromethane   *77*, 487 (2000)
1-Nitronaphthalene   *46*, 291 (1989)
2-Nitronaphthalene   *46*, 303 (1989)
3-Nitroperylene   *46*, 313 (1989)
2-Nitro-*para*-phenylenediamine (*see* 1,4-Diamino-2-nitrobenzene)
2-Nitropropane   *29*, 331 (1982); *Suppl. 7*, 67
                                                            (1987); *71*, 1079 (1999)
1-Nitropyrene   *33*, 209 (1984); *Suppl. 7*, 67
                                                            (1987); *46*, 321 (1989)

Norethisterone acetate   *72*, 49 (1999)
Norethynodrel   *6*, 191 (1974); *21*, 461 (1979)
                                                            (*corr. 42*, 259); *Suppl. 7*, 295
                                                            (1987); *72*, 49 (1999)
Norgestrel   *6*, 201 (1974); *21*, 479 (1979);
                                                            Suppl. 7, 295 (1987); *72*, 49 (1999)
Nylon 6   *19*, 120 (1979); *Suppl. 7*, 68 (1987)

O

Ochratoxin A   *10*, 191 (1976); *31*, 191 (1983)
                                                            (*corr. 42*, 262); *Suppl. 7*, 271
                                                            (1987); *56*, 489 (1993)
Oestradiol   *6*, 99 (1974); *21*, 279 (1979);
                                                            Suppl. 7, 284 (1987); *72*, 399
                                                            (1999)

Oestradiol-17β (*see* Oestradiol)
Oestradiol 3-benzoate (*see* Oestradiol)
Oestradiol dipropionate (*see* Oestradiol)
Oestradiol mustard   *9*, 217 (1975); *Suppl. 7*, 68 (1987)
Oestradiol valerate (*see* Oestradiol)
Oestriol   *6*, 117 (1974); *21*, 327 (1979);
                                                            Suppl. 7, 285 (1987); *72*, 399
                                                            (1999)

Oestrogen-progestin combinations (*see* Oestrogens,
    progestins (progestogens) and combinations)
Oestrogen-progestin replacement therapy (*see* Post-menopausal
    oestrogen-progestogen therapy)
Oestrogen replacement therapy (*see* Post-menopausal oestrogen
    therapy)
Oestrogens (*see* Oestrogens, progestins and combinations)
Oestrogens, conjugated (*see* Conjugated oestrogens)
Oestrogens, nonsteroidal (*see* Nonsteroidal oestrogens)
Oestrogens, progestins (progestogens) and combinations   *6* (1974); *21* (1979); *Suppl. 7*, 272
                                                            (1987); *72*, 49, 339, 399, 531
                                                            (1999)

Oestrogens, steroidal (*see* Steroidal oestrogens)
Oestrone   *6*, 123 (1974); *21*, 343 (1979)
                                                            (*corr. 42*, 259); *Suppl. 7*, 286
                                                            (1987); *72*, 399 (1999)

Oestrone benzoate (*see* Oestrone)
Oil Orange SS   *8*, 165 (1975); *Suppl. 7*, 69 (1987)
Opisthorchis felineus (infection with)   *61*, 121 (1994)
Opisthorchis viverrini (infection with)   *61*, 121 (1994)
Oral contraceptives, combined   *Suppl. 7*, 297 (1987); *72*, 49 (1999)
Oral contraceptives, sequential (*see* Sequential oral contraceptives)
Orange I   *8*, 173 (1975); *Suppl. 7*, 69 (1987)
Orange G   *8*, 181 (1975); *Suppl. 7*, 69 (1987)
Organolead compounds (*see also* Lead and lead compounds) *Suppl. 7*, 230 (1987)
Oxazepam   *13*, 58 (1977); *Suppl. 7*, 69 (1987);
                                                            66, 115 (1996)
Oxymetholone (*see also* Androgenic (anabolic) steroids)   *13*, 131 (1977)
Oxyphenbutazone   *13*, 185 (1977); *Suppl. 7*, 69 (1987)

Phillipsite (*see* Zeolites)

PhIP	*56*, 229 (1993)
Pickled vegetables	*56*, 83 (1993)
Picloram	*53*, 481 (1991)

Piperazine oestrone sulfate (*see* Conjugated oestrogens)

Piperonyl butoxide	*30*, 183 (1983); *Suppl. 7*, 70 (1987)

Pitches, coal-tar (*see* Coal-tar pitches)

Polyacrylic acid	*19*, 62 (1979); *Suppl. 7*, 70 (1987)
Polybrominated biphenyls	*18*, 107 (1978); *41*, 261 (1986); *Suppl. 7*, 321 (1987)
Polychlorinated biphenyls	*7*, 261 (1974); *18*, 43 (1978) (*corr. 42*, 258); *Suppl. 7*, 322 (1987)

Polychlorinated camphenes (*see* Toxaphene)

Polychlorinated dibenzo-*para*-dioxins (other than 2,3,7,8-tetrachlorodibenzodioxin)	*69*, 33 (1997)
Polychlorinated dibenzofurans	*69*, 345 (1997)
Polychlorophenols and their sodium salts	*71*, 769 (1999)
Polychloroprene	*19*, 141 (1979); *Suppl. 7*, 70 (1987)
Polyethylene (*see also* Implants, surgical)	*19*, 164 (1979); *Suppl. 7*, 70 (1987)

Poly(glycolic acid) (*see* Implants, surgical)

Polymethylene polyphenyl isocyanate (*see also* 4,4′-Methylenediphenyl diisocyanate)	*19*, 314 (1979); *Suppl. 7*, 70 (1987)
Polymethyl methacrylate (*see also* Implants, surgical)	*19*, 195 (1979); *Suppl. 7*, 70 (1987)

Polyoestradiol phosphate (*see* Oestradiol-17β)

Polypropylene (*see also* Implants, surgical)	*19*, 218 (1979); *Suppl. 7*, 70 (1987)
Polystyrene (*see also* Implants, surgical)	*19*, 245 (1979); *Suppl. 7*, 70 (1987)
Polytetrafluoroethylene (*see also* Implants, surgical)	*19*, 288 (1979); *Suppl. 7*, 70 (1987)
Polyurethane foams (*see also* Implants, surgical)	*19*, 320 (1979); *Suppl. 7*, 70 (1987)
Polyvinyl acetate (*see also* Implants, surgical)	*19*, 346 (1979); *Suppl. 7*, 70 (1987)
Polyvinyl alcohol (*see also* Implants, surgical)	*19*, 351 (1979); *Suppl. 7*, 70 (1987)
Polyvinyl chloride (*see also* Implants, surgical)	*7*, 306 (1974); *19*, 402 (1979); *Suppl. 7*, 70 (1987)
Polyvinyl pyrrolidone	*19*, 463 (1979); *Suppl. 7*, 70 (1987); *71*, 1181 (1999)
Ponceau MX	*8*, 189 (1975); *Suppl. 7*, 70 (1987)
Ponceau 3R	*8*, 199 (1975); *Suppl. 7*, 70 (1987)
Ponceau SX	*8*, 207 (1975); *Suppl. 7*, 70 (1987)
Post-menopausal oestrogen therapy	*Suppl. 7*, 280 (1987); *72*, 399 (1999)
Post-menopausal oestrogen-progestogen therapy	*Suppl. 7*, 308 (1987); *72*, 531 (1999)

Potassium arsenate (*see* Arsenic and arsenic compounds)
Potassium arsenite (*see* Arsenic and arsenic compounds)

Potassium bis(2-hydroxyethyl)dithiocarbamate	*12*, 183 (1976); *Suppl. 7*, 70 (1987)
Potassium bromate	*40*, 207 (1986); *Suppl. 7*, 70 (1987); *73*, 481 (1999)

Potassium chromate (*see* Chromium and chromium compounds)
Potassium dichromate (*see* Chromium and chromium compounds)

Prazepam	*66*, 143 (1996)
Prednimustine	*50*, 115 (1990)

R

Radiation (*see* gamma-radiation, neutrons, ultraviolet radiation, X-radiation)

Radionuclides, internally deposited	*78* (2001)
Radon	*43*, 173 (1988) (*corr. 45*, 283)

Refractory ceramic fibres (*see* Man-made vitreous fibres)

Reserpine	*10*, 217 (1976); *24*, 211 (1980) (*corr. 42*, 260); *Suppl. 7*, 330 (1987)
Resorcinol	*15*, 155 (1977); *Suppl. 7*, 71 (1987); *71*, 1119 (1990)
Retrorsine	*10*, 303 (1976); *Suppl. 7*, 71 (1987)
Rhodamine B	*16*, 221 (1978); *Suppl. 7*, 71 (1987)
Rhodamine 6G	*16*, 233 (1978); *Suppl. 7*, 71 (1987)
Riddelliine	*10*, 313 (1976); *Suppl. 7*, 71 (1987); *82*, 153 (2002)
Rifampicin	*24*, 243 (1980); *Suppl. 7*, 71 (1987)
Ripazepam	*66*, 157 (1996)

Rock (stone) wool (*see* Man-made vitreous fibres)

Rubber industry	*28* (1982) (*corr. 42*, 261); *Suppl. 7*, 332 (1987)
Rubia tinctorum (*see also* Madder root, Traditional herbal medicines)	*82*, 129 (2002)
Rugulosin	*40*, 99 (1986); *Suppl. 7*, 71 (1987)

S

Saccharated iron oxide	*2*, 161 (1973); *Suppl. 7*, 71 (1987)
Saccharin and its salts	*22*, 111 (1980) (*corr. 42*, 259); *Suppl. 7*, 334 (1987); *73*, 517 (1999)
Safrole	*1*, 169 (1972); *10*, 231 (1976); *Suppl. 7*, 71 (1987)
Salted fish	*56*, 41 (1993)

Sawmill industry (including logging) (*see* Lumber and sawmill industry (including logging))

Scarlet Red	*8*, 217 (1975); *Suppl. 7*, 71 (1987)
Schistosoma haematobium (infection with)	*61*, 45 (1994)
Schistosoma japonicum (infection with)	*61*, 45 (1994)
Schistosoma mansoni (infection with)	*61*, 45 (1994)
Selenium and selenium compounds	*9*, 245 (1975) (*corr. 42*, 255); *Suppl. 7*, 71 (1987)

Selenium dioxide (*see* Selenium and selenium compounds)
Selenium oxide (*see* Selenium and selenium compounds)

Semicarbazide hydrochloride	*12*, 209 (1976) (*corr. 42*, 256); *Suppl. 7*, 71 (1987)
Senecio jacobaea L. (*see also* Pyrrolizidine alkaloids)	*10*, 333 (1976)
Senecio longilobus (*see also* Pyrrolizidine alkaloids, Traditional herbal medicines)	*10*, 334 (1976); *82*, 153 (2002)
Senecio riddellii (*see also* Traditional herbal medicines)	*82*, 153 (1982)
Seneciphylline	*10*, 319, 335 (1976); *Suppl. 7*, 71 (1987)
Senkirkine	*10*, 327 (1976); *31*, 231 (1983); *Suppl. 7*, 71 (1987)

Styrene	*19*, 231 (1979) (*corr. 42*, 258); *Suppl. 7*, 345 (1987); *60*, 233 (1994) (*corr. 65*, 549); *82*, 437 (2002)
Styrene–acrylonitrile copolymers	*19*, 97 (1979); *Suppl. 7*, 72 (1987)
Styrene–butadiene copolymers	*19*, 252 (1979); *Suppl. 7*, 72 (1987)
Styrene-7,8-oxide	*11*, 201 (1976); *19*, 275 (1979); *36*, 245 (1985); *Suppl. 7*, 72 (1987); *60*, 321 (1994)
Succinic anhydride	*15*, 265 (1977); *Suppl. 7*, 72 (1987)
Sudan 1	*8*, 225 (1975); *Suppl. 7*, 72 (1987)
Sudan II	*8*, 233 (1975); *Suppl. 7*, 72 (1987)
Sudan III	*8*, 241 (1975); *Suppl. 7*, 72 (1987)
Sudan Brown RR	*8*, 249 (1975); *Suppl. 7*, 72 (1987)
Sudan Red 7B	*8*, 253 (1975); *Suppl. 7*, 72 (1987)
Sulfadimidine (*see* Sulfamethazine)	
Sulfafurazole	*24*, 275 (1980); *Suppl. 7*, 347 (1987)
Sulfallate	*30*, 283 (1983); *Suppl. 7*, 72 (1987)
Sulfamethazine and its sodium salt	*79*, 341 (2001)
Sulfamethoxazole	*24*, 285 (1980); *Suppl. 7*, 348 (1987); *79*, 361 (2001)
Sulfites (*see* Sulfur dioxide and some sulfites, bisulfites and metabisulfites)	
Sulfur dioxide and some sulfites, bisulfites and metabisulfites	*54*, 131 (1992)
Sulfur mustard (*see* Mustard gas)	
Sulfuric acid and other strong inorganic acids, occupational exposures to mists and vapours from	*54*, 41 (1992)
Sulfur trioxide	*54*, 121 (1992)
Sulphisoxazole (*see* Sulfafurazole)	
Sunset Yellow FCF	*8*, 257 (1975); *Suppl. 7*, 72 (1987)
Symphytine	*31*, 239 (1983); *Suppl. 7*, 72 (1987)

T

2,4,5-T (*see also* Chlorophenoxy herbicides; Chlorophenoxy herbicides, occupational exposures to)	*15*, 273 (1977)
Talc	*42*, 185 (1987); *Suppl. 7*, 349 (1987)
Tamoxifen	*66*, 253 (1996)
Tannic acid	*10*, 253 (1976) (*corr. 42*, 255); *Suppl. 7*, 72 (1987)
Tannins (*see* also Tannic acid)	*10*, 254 (1976); *Suppl. 7*, 72 (1987)
TCDD (*see* 2,3,7,8-Tetrachlorodibenzo-*para*-dioxin)	
TDE (*see* DDT)	
Tea	*51*, 207 (1991)
Temazepam	*66*, 161 (1996)
Teniposide	*76*, 259 (2000)
Terpene polychlorinates	*5*, 219 (1974); *Suppl. 7*, 72 (1987)
Testosterone (*see also* Androgenic (anabolic) steroids)	*6*, 209 (1974); *21*, 519 (1979)
Testosterone oenanthate (*see* Testosterone)	
Testosterone propionate (*see* Testosterone)	
2,2′,5,5′-Tetrachlorobenzidine	*27*, 141 (1982); *Suppl. 7*, 72 (1987)

T-2 Toxin (*see* Toxins derived from *Fusarium sporotrichioides*)

Toxins derived from *Fusarium graminearum*, *F. culmorum* and *F. crookwellense*	*11*, 169 (1976); *31*, 153, 279 (1983); *Suppl. 7*, 64, 74 (1987); *56*, 397 (1993)
Toxins derived from *Fusarium moniliforme*	*56*, 445 (1993)
Toxins derived from *Fusarium sporotrichioides*	*31*, 265 (1983); *Suppl. 7*, 73 (1987); *56*, 467 (1993)
Traditional herbal medicines	*82*, 41 (2002)
Tremolite (*see* Asbestos)	
Treosulfan	*26*, 341 (1981); *Suppl. 7*, 363 (1987)
Triaziquone (*see* Tris(aziridinyl)-*para*-benzoquinone)	
Trichlorfon	*30*, 207 (1983); *Suppl. 7*, 73 (1987)
Trichlormethine	*9*, 229 (1975); *Suppl. 7*, 73 (1987); *50*, 143 (1990)
Trichloroacetic acid	*63*, 291 (1995) (*corr. 65*, 549); *84*, 403 (2004)
Trichloroacetonitrile (*see also* Halogenated acetonitriles)	*71*, 1533 (1999)
1,1,1-Trichloroethane	*20*, 515 (1979); *Suppl. 7*, 73 (1987); *71*, 881 (1999)
1,1,2-Trichloroethane	*20*, 533 (1979); *Suppl. 7*, 73 (1987); *52*, 337 (1991); *71*, 1153 (1999)
Trichloroethylene	*11*, 263 (1976); *20*, 545 (1979); *Suppl. 7*, 364 (1987); *63*, 75 (1995) (*corr. 65*, 549)
2,4,5-Trichlorophenol (*see also* Chlorophenols; Chlorophenols, occupational exposures to; Polychlorophenols and their sodium salts)	*20*, 349 (1979)
2,4,6-Trichlorophenol (*see also* Chlorophenols; Chlorophenols, occupational exposures to; Polychlorophenols and their sodium salts)	*20*, 349 (1979)
(2,4,5-Trichlorophenoxy)acetic acid (*see* 2,4,5-T)	
1,2,3-Trichloropropane	*63*, 223 (1995)
Trichlorotriethylamine-hydrochloride (*see* Trichlormethine)	
T₂ Trichothecene (*see* Toxins derived from *Fusarium sporotrichioides*)	
Tridymite (*see* Crystalline silica)	
Triethanolamine	*77*, 381 (2000)
Triethylene glycol diglycidyl ether	*11*, 209 (1976); *Suppl. 7*, 73 (1987); *71*, 1539 (1999)
Trifluralin	*53*, 515 (1991)
4,4′,6-Trimethylangelicin plus ultraviolet radiation (*see also* Angelicin and some synthetic derivatives)	*Suppl. 7*, 57 (1987)
2,4,5-Trimethylaniline	*27*, 177 (1982); *Suppl. 7*, 73 (1987)
2,4,6-Trimethylaniline	*27*, 178 (1982); *Suppl. 7*, 73 (1987)
4,5′,8-Trimethylpsoralen	*40*, 357 (1986); *Suppl. 7*, 366 (1987)
Trimustine hydrochloride (*see* Trichlormethine)	
2,4,6-Trinitrotoluene	*65*, 449 (1996)
Triphenylene	*32*, 447 (1983); *Suppl. 7*, 73 (1987)
Tris(aziridinyl)-*para*-benzoquinone	*9*, 67 (1975); *Suppl. 7*, 367 (1987)
Tris(1-aziridinyl)phosphine-oxide	*9*, 75 (1975); *Suppl. 7*, 73 (1987)
Tris(1-aziridinyl)phosphine-sulphide (*see* Thiotepa)	
2,4,6-Tris(1-aziridinyl)-*s*-triazine	*9*, 95 (1975); *Suppl. 7*, 73 (1987)
Tris(2-chloroethyl) phosphate	*48*, 109 (1990); *71*, 1543 (1999)

W

X

Y

Z

List of IARC Monographs on the Evaluation of Carcinogenic Risks to Humans*

---

*Certain older volumes, marked out-of-print, are still available directly from IARCPress. Further, high-quality photo-copies of all out-of-print volumes may be purchased from University Microfilms International, 300 North Zeeb Road, Ann Arbor, MI 48106-1346, USA (Tel.: +1 313-761-4700, +1 800-521-0600).

Supplement No. 8
**Cross Index of Synonyms and
Trade Names in Volumes 1 to 46
of the *IARC Monographs***
1990; 346 pages (out-of-print)

**All IARC publications are available directly from
IARCPress, 150 Cours Albert Thomas, 69372 Lyon cedex 08, France
(Fax: +33 4 72 73 83 02; E-mail: press@iarc.fr).**

**IARC Monographs and Technical Reports are also available from the
World Health Organization Marketing and Dissemination, 1211 Geneva 27, Switzerland
(Fax: +41 22 791 4857; E-mail: publications@who.int)
and from WHO Sales Agents worldwide.**

**IARC Scientific Publications, IARC Handbooks and IARC CancerBases are also available from
Oxford University Press, Walton Street, Oxford, UK OX2 6DP (Fax: +44 1865 267782).**

**IARC Monographs are also available in an electronic edition,
both on-line by internet and on CD-ROM, from GMA Industries, Inc.,
20 Ridgely Avenue, Suite 301, Annapolis, Maryland, USA
(Fax: +01 410 267 6602; internet: https//www.gmai.com/Order_Form.htm)**

Achevé d'imprimer sur rotative par l'Imprimerie Darantiere
à Dijon-Quetigny en octobre 2004

Dépôt légal : octobre 2004 - N° d'impression : 24-1008

Imprimé en France